VOLUME ONE HUNDRED AND FIFTY ONE

# CURRENT TOPICS IN DEVELOPMENTAL BIOLOGY

## Meiosis in Development and Disease

# CURRENT TOPICS IN DEVELOPMENTAL BIOLOGY

*"A meeting-ground for critical review and discussion of developmental processes"*
**A.A. Moscona and Alberto Monroy (Volume 1, 1966)**

VOLUME ONE HUNDRED AND FIFTY ONE

# Current Topics in DEVELOPMENTAL BIOLOGY

## Meiosis in Development and Disease

Edited by

**FRANCESCA COLE**
*Department of Epigenetics and Molecular Carcinogenesis*
*The University of Texas MD Anderson Cancer Center*
*Houston, TX, United States*

ELSEVIER

ACADEMIC PRESS
An imprint of Elsevier

Academic Press is an imprint of Elsevier
50 Hampshire Street, 5th Floor, Cambridge, MA 02139, United States
525 B Street, Suite 1650, San Diego, CA 92101, United States
The Boulevard, Langford Lane, Kidlington, Oxford OX5 1GB, United Kingdom
125 London Wall, London, EC2Y 5AS, United Kingdom

First edition 2023



ISBN: 978-0-12-820156-5
ISSN: 0070-2153

For information on all Academic Press publications
visit our website at https://www.elsevier.com/books-and-journals

*Publisher:* Zoe Kruze
*Developmental Editor:* Naiza Mendoza
*Production Project Manager:* Vijayaraj Purushothaman
*Cover Designer:* Miles Hitchen

Typeset by STRAIVE, India

# Contents

# Contributors

**Needhi Bhalla**
Department of Molecular, Cell and Developmental Biology, University of California, Santa Cruz, Santa Cruz, CA, United States

**Katherine Kretovich Billmyre**
Stowers Institute for Medical Research, Kansas City, MO, United States

**María de las Mercedes Carro**
Department of Biomedical Sciences; Cornell Reproductive Sciences Center (CoRe), Cornell University, Ithaca, NY, United States

**Yu-Chien Chuang**
Fred Hutchinson Cancer Center, Seattle, WA, United States

**Paula E. Cohen**
Department of Biomedical Sciences; Cornell Reproductive Sciences Center (CoRe), Cornell University, Ithaca, NY, United States

**Elena Damm**
Department Evolutionary Genetics, Research Group Meiotic Recombination and Genome Instability, Max Planck Institute for Evolutionary Biology, Plön, Germany

**Ananya Dodamani**
Department of Biology, Indian Institute of Science Education and Research, Pune, India

**Magdalena Firlej**
Structural Biochemistry of Meiosis Group, Friedrich Miescher Laboratory of the Max Planck Society, Tuebingen, Germany

**Jinmin Gao**
Institute of Biomedical Sciences, College of Life Sciences, Key Laboratory of Animal Resistance Biology of Shandong Province, Shandong Normal University, Jinan, China

**Andrew Grimson**
Cornell Reproductive Sciences Center (CoRe); Department of Molecular Biology & Genetics, Cornell University, Ithaca, NY, United States

**Geert Hamer**
Center for Reproductive Medicine, Reproductive Biology Laboratory, Amsterdam Reproduction and Development Research Institute, Amsterdam UMC, University of Amsterdam, Amsterdam, The Netherlands

**Noriko Hosoya**
Laboratory of Molecular Radiology, Center for Disease Biology and Integrative Medicine, Graduate School of Medicine, The University of Tokyo, Tokyo, Japan

**Kei-ichiro Ishiguro**
Department of Chromosome Biology, Institute of Molecular Embryology and Genetics (IMEG), Kumamoto University, Kumamoto, Japan

**Christophe Lambing**
Plant Science Department, Rothamsted Research, Harpenden, United Kingdom

**Qijing Lei**
Center for Reproductive Medicine, Reproductive Biology Laboratory, Amsterdam Reproduction and Development Research Institute, Amsterdam UMC, University of Amsterdam, Amsterdam, The Netherlands

**Yuanyuan Liu**
Institute of Biomedical Sciences, College of Life Sciences, Key Laboratory of Animal Resistance Biology of Shandong Province, Shandong Normal University, Jinan, China

**Urszula Lucja McClurg**
Institute of Systems, Molecular and Integrative Biology, University of Liverpool, Liverpool, United Kingdom

**Chris Morgan**
John Innes Centre, Norwich, United Kingdom

**Mridula Nambiar**
Department of Biology, Indian Institute of Science Education and Research, Pune, India

**Aditya Nayak**
Department of Biology, Institute of Molecular Plant Biology, Swiss Federal Institute of Technology (ETH) Zurich, Zürich, Switzerland

**Linda Odenthal-Hesse**
Department Evolutionary Genetics, Research Group Meiotic Recombination and Genome Instability, Max Planck Institute for Evolutionary Biology, Plön, Germany

**Vivek B. Raina**
Department of Biochemistry and Molecular Biophysics, Columbia University Medical Center, New York City, NY, United States

**Maud Schoot Uiterkamp**
Center for Reproductive Medicine, Reproductive Biology Laboratory, Amsterdam Reproduction and Development Research Institute, Amsterdam UMC, University of Amsterdam; Cancer Center Amsterdam, Cancer Biology and Immunology; Section of Oncogenetics, Department of Human Genetics, Amsterdam UMC, Vrije Universiteit, Amsterdam, The Netherlands

**Sucharita Sen**
Department of Biology, Indian Institute of Science Education and Research, Pune, India

**Gerald R. Smith**
Fred Hutchinson Cancer Center; Division of Basic Sciences, Fred Hutchinson Cancer Research Center, Seattle, WA, United States

**Ieng Fong Sou**
Institute of Systems, Molecular and Integrative Biology, University of Liverpool, Liverpool, United Kingdom; Chromatin Dynamics and Disease Epigenetics Group, Institute of Molecular and Cell Biology, Agency for Science, Technology and Research (A*STAR), Singapore, Singapore

**Wee-Wei Tee**
Chromatin Dynamics and Disease Epigenetics Group, Institute of Molecular and Cell Biology, Agency for Science, Technology and Research (A*STAR); Department of Physiology, Yong Loo Lin School of Medicine, National University of Singapore, Singapore, Singapore

**Gerben Vader**
Center for Reproductive Medicine, Reproductive Biology Laboratory, Amsterdam Reproduction and Development Research Institute, Amsterdam UMC, University of Amsterdam; Cancer Center Amsterdam, Cancer Biology and Immunology; Section of Oncogenetics, Department of Human Genetics, Amsterdam UMC, Vrije Universiteit, Amsterdam, The Netherlands

**Ans M.M. van Pelt**
Center for Reproductive Medicine, Reproductive Biology Laboratory, Amsterdam Reproduction and Development Research Institute, Amsterdam UMC, University of Amsterdam, Amsterdam, The Netherlands

**John R. Weir**
Structural Biochemistry of Meiosis Group, Friedrich Miescher Laboratory of the Max Planck Society, Tuebingen, Germany

**Ruirui Zhang**
Institute of Biomedical Sciences, College of Life Sciences, Key Laboratory of Animal Resistance Biology of Shandong Province, Shandong Normal University, Jinan, China

# Preface

Meiosis is the specialized cell division program that produces haploid gametes such as sperm and eggs. From an evolutionary and parental perspective, nothing could be more important than transmitting a functional genome to the next generation. It is estimated that 20% of oocytes from young women and 5% of sperm have the wrong chromosome complement (aneuploidy). This alarming frequency of aneuploid gametes increases with age—dramatically in oocytes. As such, errors in meiotic chromosome segregation are the leading cause of infertility, miscarriage, and birth defects in humans.

Faithful chromosome segregation during meiosis requires parental chromosomes (homologs) to become tethered by crossover recombination, DNA repair that exchanges homolog arms. Programmed DNA double-strand breaks (DSBs) provoke recombination during meiosis. Meiotic recombination is initiated and repaired in association with development of meiosis-specific higher-order chromosome structure, a unique transcriptional program, and regulated by extensive checkpoints. The chapters in this book explore the latest research in the field of meiosis with special emphasis on the natural development and variation between organisms and diseases associated with the dysregulation of meiotic processes.

The developmental switch between mitotic and meiotic division cycles requires extensive transcriptional regulation, reviewed here by Kei-ichiro Ishiguro. His review focuses on the collaborative roles of MEIOSIN and STRA8 in boosting the transcription of meiotic genes and describes the phenotypes of mice disrupted in these pathways and their extensive target genes. The review clarifies what is known and still not well understood in the hierarchical regulation of meiotic cell cycle induction. Sexual dimorphism in meiotic induction is also explored, focusing on the differences between male and female germ cell development and the roles of retinoic acid signaling. Finally, the review discusses evolutionary conservation of the mitotic to meiotic transition and highlights both transcriptional and posttranscriptional regulation of the process.

Meiotic recombination is initiated by programmed DNA DSBs. In most mammals, the distribution of meiotic DSBs is largely regulated by the methyltransferase protein PR domain zinc finger protein 9 (PRDM9). The review by Elena Damm and Linda Odenthal-Hesse synthesizes much of the recent work to understand how cellular processes such as DNA replication and chromosome organization influence DSB patterning and formation and subsequent DSB-dependent homolog pairing. The authors particularly focus on the roles of protein complexes that (1) reorganize chromatin accessibility for DSB formation, (2) coordinate PRDM9-targeted sites with chromosome axis-associated DSB factors, and (3) facilitate inter-homolog pairing to provide a comprehensive overview of our current state of knowledge.

Meiosis involves specialized chromatin organization and transcriptional programs. Frequently, meiotic transcripts are inappropriately expressed in cancer cells and are highly correlated with cancer initiation, progression, and metastasis. Meiotic transcripts are a potentially powerful therapeutic target for cancers, and the review by Sou et al. discusses what is known about the similarity between transcriptional regulation in meiotic and cancer cells. This comprehensive review compares the status of histone modifications, DNA demethylation, genome organization, higher-order chromatin structure, enhancer element associations, and transcription dependent R-loop distribution between meiotic and cancer cells.

Phase separation or biomolecular condensation has emerged as a critical driver of cellular processes and in particular chromosome dynamics. This timely review by Zhang et al. describes the unique chromosome behavior in meiosis and how phase separation generates biomolecular condensates *in vivo*. The authors then integrate these concepts by discussing what is known about how phase separation contributes to meiotic DSB formation, homolog pairing, synaptonemal complex formation and dynamics, and the formation of the mammalian sex body. Finally, they pose some questions about how phase separation may be driving additional meiotic events, such as the assurance of crossovers between homologs, that are necessary for accurate chromosome segregation in meiosis I.

The chromosome axis is the proteinaceous structure that forms the linear organization and compaction of chromosomal DNA. In meiosis, unique axis-associated proteins recruit and regulate key processes such as meiotic DSB formation, interhomolog recombination, synaptonemal complex assembly, and crossing over. In the review by Morgan et al., the authors describe how the chromosome axis regulates meiotic chromosome organization and compare axis, cohesion, and synaptonemal complex proteins between species to examine conserved and distinct roles. They further discuss the consequences of axis protein dysregulation under heat stress and in certain cancers.

How chromosomes behave in meiosis differs based upon size, composition, locale, and structure. The review by Katherine Kretovich Billmyre examines chromosome features across multiple model organisms and how differences between chromosomes are associated with altered behavior during meiosis. This comprehensive analysis includes a discussion of the differences between DSBs and synaptonemal complex formation, unique features attributed to ribosomal DNA and the nucleolus, specialized homolog pairing, and the impact of altered chromosomal structures such as heterogametic sex chromosomes.

Crossovers that form at centromeres are highly deleterious for meiotic chromosome segregation, and suppression of crossing over at centromeres and pericentromeres is widely conserved. The review by Sen et al. provides a comprehensive analysis of the current state of knowledge on the molecular mechanisms of the centromere effect on crossover suppression. The review focuses on the regulation of DSB distribution, recombination pathway choice, the proteins that determine the unique structure of the centromeric heterochromatin and how they interface with meiotic recombination, cohesion, and axis proteins. The review concludes with a discussion of what is known about the clinical impact of crossovers near centromeres.

Meiotic recombination relies upon many DNA-modifying enzymes, among them DNA helicases that translocate and/or unwind DNA. The review by Magdalena Firlej and John R. Weir discusses the pro- and antirecombination DNA helicases that ensure successful meiotic recombination and crossing over. The authors discuss the structure and modifications, interacting factors, and regulatory roles of eight helicases that play specific roles in meiotic recombination, such as DSB formation and resection, interhomolog engagement and bias for recombination, and crossing over. They include discussions of what is understood about how helicases impact human health and disease.

In many organisms, crossovers do not form in close juxtaposition, a phenomenon called crossover interference. Crossover interference was first observed more than 100 years ago, and the molecular mechanisms of and reasons for interference are still hotly debated in the meiosis field. The review by Yu-Chien Chuang and Gerald R. Smith describes the methods for measuring and interpreting crossover interference, the differences between organisms, the proteins involved, and potential models for interference.

There are a series of meiosis-specific gene regulatory events such as the global shutdown of transcription in early meiotic prophase I and meiotic sex chromosome inactivation. The review by de las Mercedes Carro et al. brings clarity to the canonical and noncanonical roles of small RNAs and RNA-binding proteins in the regulation of meiosis. The authors discuss what is known about the biogenesis and processing of these small RNAs, their roles in gene silencing, pericentric heterochromatin remodeling, and regulation of meiotic recombination.

In meiotic prophase, unique and specialized cell cycle checkpoints monitor meiotic DSB formation, synaptonemal complex formation, and recombinational repair, which are the topics of the three final reviews in this book. The review by Raina et al. describes how the meiotic checkpoint ensures proper chromosome segregation. The authors also highlight striking mechanistic similarities between the meiotic checkpoint and the spindle assembly checkpoint that ensures proper chromosome orientation and spindle attachment. The review provides an interesting framework to better understand the underlying logic of cell cycle surveillance, despite fundamental differences in the signals involved. Finally, the authors describe what is known about how misregulation of meiotic checkpoints can contribute to tumorigenesis. Another review by Needhi Bhalla further dives into the role of a critical checkpoint protein PCH-2. She describes the evolutionary conservation of the PCH-2/HORMAD module and suggests its fundamental mandate is to coordinate meiotic recombination with synapsis. The review synthesizes the disparate phenotypes observed between model organisms with the variation in how recombination and synapsis are differentially regulated between species. The review by Lei et al. describes the latest advances in *in vitro* spermatogenesis protocols that could potentially provide a method to preserve male fertility. The authors survey meiotic events and checkpoint regulation *in vivo* to provide additional criteria to assess the quality of checkpoint control *in vitro*.

This book provides an in-depth overview of the state of the field of meiosis research, gives context for age-old observations, and poses intriguing questions for future study.

Francesca Cole
Department of Epigenetics and Molecular Carcinogenesis
The University of Texas MD Anderson Cancer Center
Houston, TX, United States

CHAPTER ONE

# Mechanism of initiation of meiosis in mouse germ cells

**Kei-ichiro Ishiguro***

Department of Chromosome Biology, Institute of Molecular Embryology and Genetics (IMEG), Kumamoto University, Kumamoto, Japan

*Corresponding author: e-mail address: ishiguro@kumamoto-u.ac.jp

## Contents



## Abstract

Meiosis is critical for germ cell development in multicellular organisms. Initiation of meiosis coincides with pre-meiotic S phase, which is followed by meiotic prophase, a prolonged G2 phase that ensures numerous meiosis-specific chromosome events. Meiotic prophase is accompanied by robust alterations of gene expression. In mouse germ cells, MEIOSIN and STRA8 direct cell cycle switch from mitosis to meiosis. MEIOSIN and STRA8 coordinate meiotic initiation with cell cycle, by activating the meiotic genes to have meiotic prophase program installed at S phase. This review mainly focuses on the mechanism of meiotic initiation in mouse germ cells from the viewpoint of the transcription of meiotic genes. Furthermore, signaling pathways that regulate meiotic initiation will be discussed in the context of germ cell development, pointing out the sexual differences in the mode of meiotic initiation.

*Current Topics in Developmental Biology*, Volume 151
ISSN 0070-2153
https://doi.org/10.1016/bs.ctdb.2022.04.005

## Abbreviations

**MEIOSIN** meiosis initiator
**STRA8** stimulated by retinoic acid 8

## 1. Introduction

Meiosis accompanies germ cell development in multicellular organisms. Initiation of meiosis is the most noteworthy point of sexual reproduction as mitosis-to-meiosis transition consequently leads to germ cell differentiation into sperm or oocyte. Meiotic entry coincides with pre-meiotic S phase followed by meiotic prophase. Meiotic prophase is equivalent to G2 phase, which is prolonged to ensure numerous meiosis-specific chromosome events such as chromosome axis-loop formation, homolog synapsis, meiotic recombination and telomere-driven chromosome movement along the nuclear envelope (Baudat, Imai, & de Massy, 2013; Cahoon & Hawley, 2016; Ishiguro, 2019; Keeney, Lange, & Mohibullah, 2014; Shibuya & Watanabe, 2014; Zickler & Kleckner, 2015). To accomplish these events, gene expression program for subsequent meiotic prophase must be installed at pre-meiotic S phase. Although the mechanism of meiotic initiation is a fundamental issue in biology, it has remained elusive how germ cells regulate mitosis-to-meiosis transition in any organisms. This review mainly focuses on the mechanism of meiotic initiation and activation of downstream meiotic prophase program in mammalian germ cells, pointing out the definitive roles of transcriptional regulation for cell cycle switching from mitosis to meiosis. This review also discusses the differences in the mode of meiotic initiation in sexual context and in diverse organisms.

## 2. STRA8 plays a key role in the initiation of meiosis

In male, meiotic initiation coincides with the differentiation from spermatogonia into spermatocyte in postnatal testes (Fig. 1). In female, meiotic entry coincides with the time when primordial germ cells complete sex determination and exit from pluripotent state in embryonic day 13–14 (E13–E14) ovaries. In mice, retinoic acid (RA) and BMP signaling, at least in oogonia, synergistically induce meiotic transcription (Bowles et al., 2006; Koubova et al., 2006; Miyauchi et al., 2017; Nagaoka et al., 2020). Upon the

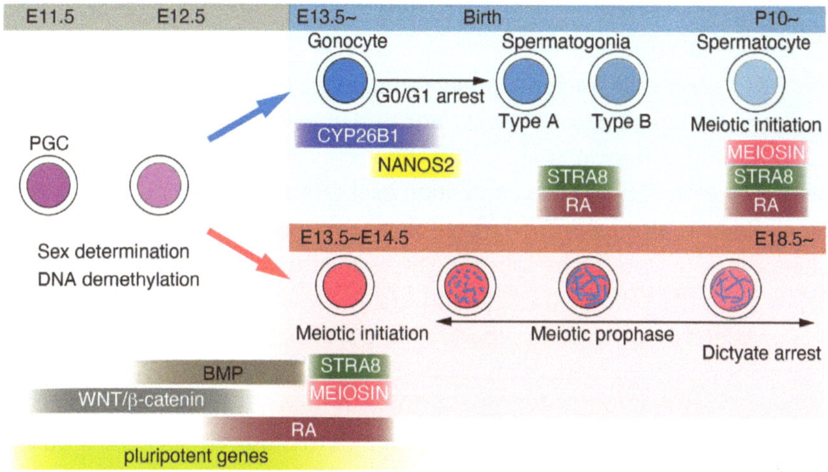


**Fig. 1** Meiotic initiation is associated with germ cell development. Schematic illustration of the key events associated with germ cell development in testis (upper right) and fetal ovary (lower right). The periods of key gene expressions (STRA8, MEIOSIN. CYP26B1, NANOS2, pluripotent genes) and extrinsic signal executions (RA, BMP, WNT/β-catenin) are shown along the developmental stages. Initiation of meiosis is regulated by sexually dimorphic mechanisms in mouse germ cell development. In female, meiotic entry occurs around E13.5–E14.5, and is followed by meiotic prophase and subsequently by dictyate (diplotene) arrest. In male, meiotic entry is suppressed by CYP26B1 and NANOS2 in fetal testis. After birth, meiotic entry accompanies differentiation of spermatocytes at puberty.

stimulation by RA, STRA8 (stimulated by retinoic acid gene 8) is transiently expressed in the germline of embryonic ovaries and postnatal testes prior to meiotic entry (Bowles et al., 2006; Koubova et al., 2006; Oulad-Abdelghani et al., 1996; Zhou, Li, et al., 2008; Zhou, Nie, et al., 2008). Since temporal expression of STRA8 coincides with the start of meiosis, STRA8 has been recognized as a hallmark of meiotic entry.

Meiotic entry is assumed to occur in the middle of preleptotene stage that corresponds to the timing of STRA8 expression (Fig. 1). Since *Stra8* KO germ cells fail to undergo normal meiosis in male and female (Anderson et al., 2008; Baltus et al., 2006; Dokshin, Baltus, Eppig, & Page, 2013; Mark et al., 2008), it has been assumed that STRA8 plays an key role in the entry into meiosis. In female, STRA8 is expressed within a limited period around E13.5–E14.5, which coincides with meiotic entry in the ovary (Bullejos & Koopman, 2004; Feng et al., 2021; Menke, Koubova, & Page, 2003; Yao, DiNapoli, & Capel, 2003). Further, in *Stra8* KO female, germ cells fail to undergo meiosis, exhibiting no signs of pre-meiotic DNA replication (Baltus et al., 2006). Thus, expression of STRA8 well correlates with meiotic entry in female germ cells. In male, *Stra8* KO germ cells reach the preleptotene stage of spermatocyte, exhibiting signs of DNA replication, but do not undergo meiotic entry in C57BL/6 background (Anderson et al., 2008). However, another report showed that *Stra8* KO germ cells in mixed genetic background initiate but fail to complete meiotic prophase I (Mark et al., 2008). Thus, the slightly different phenotypes with tight or loose block of meiotic entry in *Stra8* KO male depends on the genetic background of the mice. Furthermore, STRA8 is also transiently expressed in differentiating spermatogonia far before the differentiation into spermatocytes (Endo et al., 2015; Mark et al., 2008; Zhou, Li, et al., 2008; Zhou, Nie, et al., 2008) (Fig. 1). This implies that *Stra8* expression alone is not sufficient to induce meiosis in spermatogonia. It has been a longstanding enigma why meiosis is induced only in preleptotene spermatocytes, the phase of the second expression of STRA8, but not in spermatogonia with the first expression of STRA8, which makes it elusive how STRA8 is involved in meiotic initiation.

## 3. MEIOSIN interacts with STRA8 during meiotic initiation

To understand the molecular mechanism of meiotic initiation and how STRA8 is involved in this process, STRA8 interacting factors were screened from mouse testis chromatin fraction by mass spectrometry (MS) analysis of STRA8 immunoprecipitation (IP). To facilitate STRA8 IP from testes, where it was transiently expressed in a limited cell population, *Stra8-3xFLAG-HA-p2A-GFP* knock-in mice were generated. This allowed tandem IPs using anti-FLAG and anti-HA antibodies as well as isolation of *Stra8*-expressing germ cells by GFP labeling (Fig. 2A). This proteomic approach identified a protein encoded by a hypothetical gene *Gm4969* as one of the STRA8 interacting factors, which was named Meiosis initiator (MEIOSIN) (Ishiguro et al., 2020). As the gene annotation for *Gm4969* was yet to be correctly assigned in the genome database, even extensive studies in the field with transcriptome approaches might have missed recognizing *Gm4969*.

MEIOSIN protein possesses basic helix–loop–helix (HLH) and high mobility group (HMG) box domains, suggesting its role as a DNA binding protein. The *Meiosin* mRNA shows a specific expression pattern in adult testis and embryonic ovary like *Stra8*. In male, MEIOSIN is expressed in

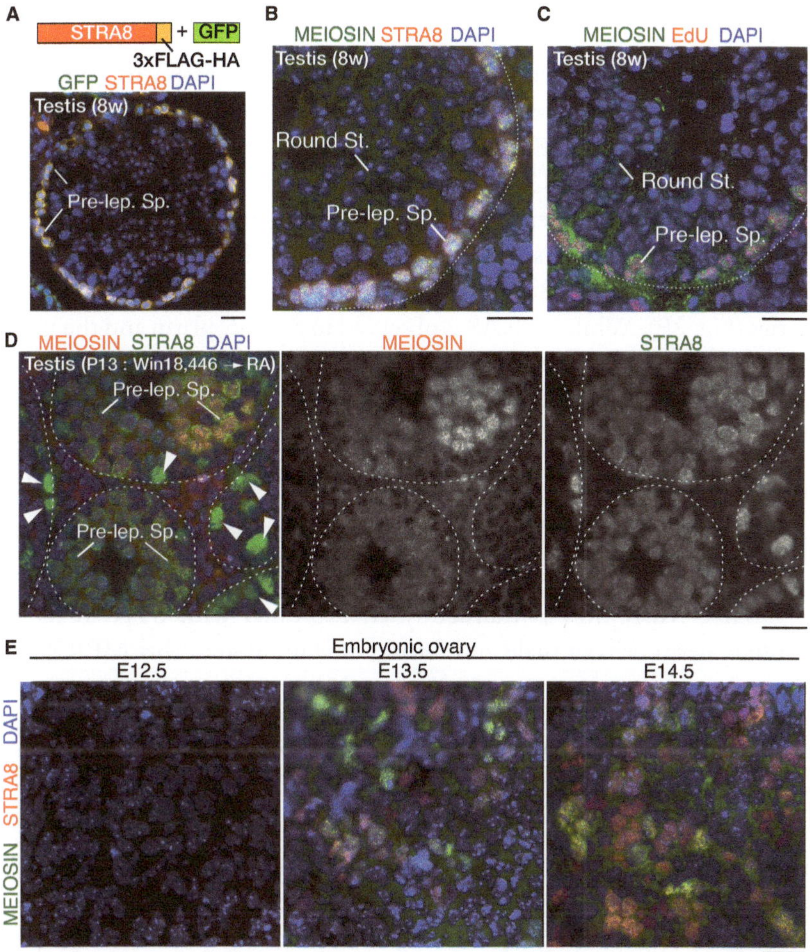


**Fig. 2** MEIOSIN and STRA8 are co-expressed at meiotic entry in male and female germ cells. (A) Seminiferous tubule section from *Stra8-3xFLAG-HA-p2A-GFP* KI mice testes was stained for STRA8, GFP and DAPI. Scale bar: 25 μm. STRA8-3xFLAG-HA fusion protein is expressed in pre-leptotene spermatocyte in a similar manner to normal WT testis. MEIOSIN was identified as one of the STRA8 interacting proteins that were purified from the *Stra8-3xFLAG-HA-p2A-GFP* KI mice testes. (B) Seminiferous tubule sections from adult WT testis (8-week-old) were stained as indicated. STRA8 and MEIOSIN proteins are co-expressed in the pre-leptotene spermatocyte in the seminiferous stages VII–VIII. Pre-lep. Sp.: Pre-leptotene spermatocyte. Round Sp.: Round spermatid. Scale bar: 25 μm. (C) Seminiferous tubule sections from adult WT testis (8-week-old) were stained as in (B). MEIOSIN expression coincides with EdU incorporation. Scale bar: 25 μm. (D) WT neonatal male mice were subjected to consecutive injection of a RA synthesis inhibitor WIN 18,446 from P5 to P11, followed by RA injection at P12 to enrich pre-leptotene spermatocytes. Seminiferous tubule sections at P13 were stained as indicated. Pre-lep: Pre-leptotene spermatocyte. White arrow heads: STRA8-positive/MEIOSIN-negative differentiating spermatogonia. Pre-lep Sp.: Pre-leptotene spermatocyte. Scale bar: 25 μm. (E) Embryonic (E12.5–E14.5) ovary section was stained as in (B). STRA8 and MEIOSIN proteins are co-expressed in the pre-leptotene germ cells. Scale bar: 25 μm. The samples were prepared and the images were acquired as described previously (Ishiguro et al., 2020).

preleptotene spermatocytes in seminiferous stages VII-VIII (Fig. 2B). Since MEIOSIN positive cells overlapped with EdU incorporation, the expression timing of MEIOSIN coincides with the pre-meiotic S phase (Fig. 2C). While STRA8 is expressed both in differentiating spermatogonia and in pre-leptotene spermatocytes, the expression of MEIOSIN is restricted to pre-leptotene spermatocytes (Fig. 2D). In female, MEIOSIN is coexpressed with STRA8 in preleptotene oocytes of E13.5–14.5 embryonic ovaries (Fig. 2E). While STRA8 is observed in the cytoplasm and the nuclei, MEIOSIN is localized mostly in the nuclei of the spermatocytes and oocytes. Thus, MEIOSIN and STRA8 are transiently co-expressed in pre-leptotene germ cells in male and female.

## 4. MEIOSIN and STRA8 direct the switch from mitosis to meiosis

Since MEIOSIN is transiently co-expressed with STRA8 in pre-leptotene germ cells in male and female, it is implicated that MEIOSIN is involved in meiotic initiation. Indeed, in *Meiosin* KO testes, spermatocytes later than pre-leptotene stage do not appear, and consequently postmeiotic spermatids or sperm are absent (Fig. 3A) (Ishiguro et al., 2020). Presumably, commitment to pre-leptotene spermatocytes seems to be independent of the expressions of *Meiosin* and *Stra8*. In *Meiosin* KO testes, STRA8 is still expressed and *vice versa*, implying the expressions of *Meiosin* and *Stra8* are regulated independently of each other. Despite the expression of STRA8, *Meiosin* KO spermatocytes fail to enter meiotic prophase (Fig. 3B). Notably, STRA8 largely remained in the cytosol rather than the nuclei in *Meiosin* KO (Fig. 3B), and MEIOSIN localized in the nuclei of *Stra8* KO (Ishiguro et al., 2020). This suggests that MEIOSIN is required for nuclear localization of STRA8 in pre-leptotene spermatocytes. Consequently, *Meiosin* KO cells do not show any signs of homolog synapsis or meiotic recombination (Fig. 3C). The observed phenotypes in *Meiosin* KO are accompanied by the downregulation of genes associated with meiotic prophase-events such as the processes of meiotic chromosome organization, meiotic recombination and meiotic cell cycle. Thus, MEIOSIN and STRA8 are required for meiotic initiation and the subsequent meiotic program in male.

Overall, the cytological phenotypes of *Meiosin* KO mice were similar to those observed in *Stra8* KO. However, it is worth noting that a phenotypic difference is observed between *Meiosin* KO and *Stra8* KO testes. In *Stra8*

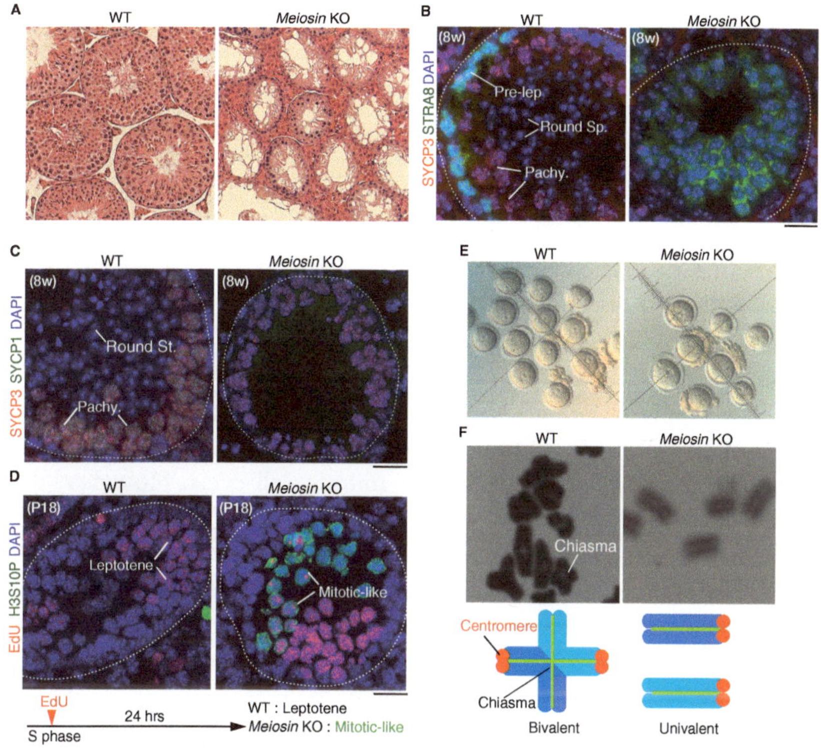


**Fig. 3** MEIOSIN plays an essential role in meiotic entry in testes. (A) Hematoxylin and eosin staining of the sections from WT and *Meiosin* KO testes (8-week-old). Spermatocyte in the meiotic prophase or postmeiotic spermatid are absent in *Meiosin* KO testis. (B) Seminiferous tubule sections from WT and *Meiosin* KO testes (8-week-old) were stained for SYCP3, STRA8 and DAPI. Despite the STRA8 expression, spermatocytes that enter meiotic prophase are absent in *Meiosin* KO testes. STRA8 localized in the nuclei in WT testes, and mostly remained in the cytoplasm in *Meiosin* KO testis. Pre-lep: Pre-leptotene spermatocyte. Pachy.: Pachytene spermatocyte. Round Sp.: Round spermatid. (C) Seminiferous tubule sections from WT and *Meiosin* KO testes (8-week-old) were stained as indicated. Homolog synapsis was not observed in *Meiosin* KO testes. (D) Seminiferous tubule sections from WT and *Meiosin* KO testes (P18) were stained as indicated 24 h after EdU injection. The fate of *Meiosin* KO spermatocyte-like cells was analyzed by EdU labeling and chase. EdU-labeled *Meiosin* KO spermatocyte-like cells showed H3S10P-positive mitotic-like state 24 h after DNA replication, in contrast to WT spermatocytes that reached the leptotene stage. Thus, *Meiosin* KO spermatocyte-like cells underwent mitotic chromosome condensation, but did not progress through canonical meiotic prophase I. Scale bars: 25 μm. (E) GV oocytes isolated from WT (left) and oocyte-like cells isolated from *Meiosin* KO (right). (F) Giemsa staining of metaphase I bivalents in WT (left), and metaphase-like univalent (right) in *Meiosin* KO oocyte-like cells. Schematic illustrations of bivalent and univalent chromosomes are shown on the bottom. The samples were prepared and the images were acquired as described previously (Ishiguro et al., 2020).

KO, some, if not all, spermatocytes still exhibit the ability to enter meiosis but fail to complete meiotic prophase, similar to the previous study (Mark et al., 2008). This could be in part due to leaky expressions of a subset of meiotic genes in *Stra8* KO, as shown in the transcriptome analysis of *Meiosin* KO *versus Stra8* KO (Ishiguro et al., 2020). Thus, although the precise reason is yet to be clarified, more severe pre-leptotene block is observed in *Meiosin* KO compared to *Stra8* KO.

It is curious what happens to *Meiosin* KO testes without entering meiosis. In *Meiosin* KO testes, mitotic metaphase cells marked by histone H3Ser10 phosphorylation are highly accumulated (Fig. 3D). Moreover, mitotic Cyclin A2 is ectopically expressed in *Meiosin* KO spermatocytes, which should be suppressed in WT spermatocytes (Abby et al., 2016; Jain et al., 2018; Soh et al., 2017). Thus, *Meiosin* KO spermatocyte-like cells that fail to enter meiosis are still placed under mitotic cell cycle. However, those mitotic-like *Meiosin* KO cells are consequently eliminated by apoptosis rather than continue growing. This phenotype could be at least in part due to the downregulation of *Meioc* and *Ythdc2*. MEIOC and YTHDC forms a complex that binds to RNA and involved in the destabilization of mitotic cell cycle-associated transcripts. In *Meioc* KO and *Ythdc2* KO mice, germ cells initiate but fail to maintain the meiotic prophase, showing misexpression of mitotic cyclin and metaphase chromosome condensation (Abby et al., 2016; Jain et al., 2018; Soh et al., 2017). Since similar mitotic status is also observed in *Meiosin* KO and *Stra8* KO mice (Ishiguro et al., 2020; Mark et al., 2008), the program for cell cycle switching from mitosis to meiosis might be executed at least in part through the activation of *Meioc* and *Ythdc2*. This may explain why *Meiosin* KO spermatocyte-like cells undergo mitotic status after reaching the preleptotene-like stage. Altogether, MEIOSIN is required for cell cycle switching from mitosis to meiosis in male.

Also in *Meiosin* KO female, ovarian germ cells fail to enter meiosis despite the expression of STRA8. Furthermore, a subpopulation of *Meiosin* KO ovarian germ cells undergo mitotic status, as has been observed in *Meiosin* KO testes. As a result, ovaries are degenerated with apparently fewer mature follicles in *Meiosin* KO female. This suggests that ovarian germ cell loss is accelerated in *Meiosin* KO. Surprisingly, although *Meiosin* KO ovarian germ cells fail to undergo meiotic prophase, a small population of *Meiosin* KO ovarian germ cells survive and differentiate into oocyte-like cells (Fig. 3E), which is similar to the observation in *Stra8* KO female (Dokshin et al., 2013). *Meiosin* KO oocyte-like cells exhibit germinal vesicles (GV) and zona pellucida, whose morphology is apparently indistinguishable from WT oocytes. However, whereas 20 pairs of bivalents with chiasma are observed in WT oocytes, 40 univalents (sister chromatids) are observed in *Meiosin* KO oocytes-like cells (Fig. 3F). Lack of chiasma indicates that *Meiosin* KO oocytes-like cells have skipped the meiotic prophase and not experienced meiotic recombination. Thus, intriguingly, oocytes are developed even without entering meiosis in *Meiosin* KO and *Stra8* KO females. This observation is consistent with the study showing that fully-grown oocyte-like cells can be produced from pluripotent stem cells *in vitro* without entering meiosis (Hamazaki et al., 2021). Therefore, one important implication is the gene regulation in "meiotic cell cycle program" is separable from that in "oocyte developmental program". Collectively, MEIOSIN plays a crucial role in cell cycle switching from mitosis to meiosis in female.

## 5. MEIOSIN together with STRA8 determines meiotic entry

Spermatogonia and pre-leptotene spermatocytes are assumed to respond to RA in seminiferous stages VII–VIII (Endo et al., 2015). STRA8 is expressed both in differentiating spermatogonia and in pre-leptotene spermatocytes of seminiferous stages VII–VIII (Fig. 2D). However, the expression of MEIOSIN is restricted only to pre-leptotene spermatocytes despite the existence of RA responsive element (RARE) in the 5′ region upstream of *Meiosin* locus (Fig. 2D). Since the expression of MEIOSIN is suppressed by a RA synthesis inhibitor WIN 18,446, it may be upregulated at least in part in response to RA in testes. Indeed, RA receptors (RARs) bind to the promoter regions of the *Stra8* and *Meiosin* loci in germline stem (GS) cells that are established from spermatogonial stem cells (Kanatsu-Shinohara et al., 2003). However, the response to RA for *Meiosin* expression seems to be slower compared to that for *Stra8* expression, as shown in GS cells entering into meiotic prophase by the treatment with RA *in vitro* (Ishiguro et al., 2020). The difference in the sensitivity to RA between *Stra8* and *Meiosin* loci may be accounted for at least in part by DMRT1-mediated repressive mechanism. DMRT1 prevents premature meiotic entry in spermatogonia by negatively regulating RA-dependent transcription (Matson et al., 2010). Since reanalysis of previous DMRT1 ChIP-seq data suggests that DMRT1 binds to the 5′ region upstream of *Meiosin* locus (Murphy et al., 2015), DMRT1 may be repressing *Meiosin* expression in spermatogonia. On the other hand, DMRT1 ChIP-seq showed a marginal binding peak at *Stra8* locus suggesting that repression of *Stra8* by DMRT1 in spermatogonia could be less strict compared to that of *Meiosin*. Although the precise mechanism of *Meiosin* expression is yet to be fully investigated, meiotic entry occurs exclusively when MEIOSIN and STRA8 are coexpressed. Therefore, MEIOSIN together with STRA8 determines meiotic entry in response to RA during spermatogenesis.

## 6. MEIOSIN and STRA8 activate meiotic prophase program

Gene expression program for meiotic prophase is executed through STRA8-dependent and STRA8-independent pathways (Koubova et al., 2014; Soh et al., 2015). However, it remained elusive how STRA8 was involved in the gene expression program for meiotic prophase. Previously, it was shown that STRA8 has an ability to transactivate a reporter gene when GAL4-fused STRA8 is expressed in HEK293 cells (Tedesco, La Sala, Barbagallo, De Felici, & Farini, 2009). STRA8 is predicted to possess HLH-like domain and HMG-like domain (Fig. 4A) (Baltus et al., 2006; Kojima, de Rooij, & Page, 2019). Similarly, MEIOSIN possesses HLH and HMG domains (Fig. 4A), and interacts with STRA8 (Ishiguro et al., 2020). Now it becomes critical whether STRA8 and MEIOSIN directly activate the gene expression program for meiotic prophase.

Two studies addressed MEIOSIN and STRA8 binding sites in preleptotene-enriched spermatocytes by chromatin immunoprecipitation followed by sequencing (ChIP-seq) analyses (Ishiguro et al., 2020; Kojima et al., 2019). MEIOSIN and STRA8 share thousands of binding sites with a common DNA-binding motifs: 5′-CNCCTCAG-3′ or its complementary sequence 5′-CTGAGGNG-3′ (Kojima et al., 2019), and 5′-T/CNCCTCA-3′ or its complementary sequence 5′-TGAGGNG/A-3′ (Ishiguro et al., 2020) (Fig. 4B). Importantly, the genes commonly bound by both MEIOSIN and STRA8 are downregulated in *Meiosin* KO and *Stra8* KO spermatocytes. Gene ontology analyses suggest that many of the MEIOSIN and STRA8 target genes associate with meiotic prophase processes such as meiotic chromosome dynamics and meiotic recombination, and reproductive processes (Fig. 4C). Those target genes also include the genes that are involved in other processes. Thus, the MEIOSIN-STRA8 complex directly binds and activates the transcription of a subset of critical

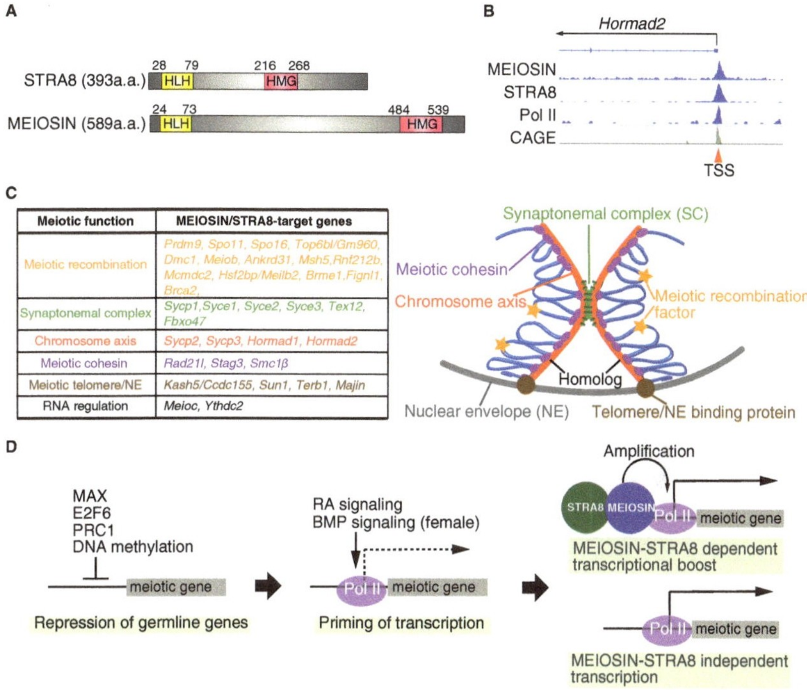


**Fig. 4** MEIOSIN and STRA8 boost the transcription of meiotic genes. (A) A schematic illustration of the domains in mouse STRA8 and MEIOSIN proteins. (B) Genomic view of ChIP-seq data of MEIOSIN, STRA8 and RNA polymerase II over a representative commonly bound meiotic gene (*Hormad2*) locus. CAGE-seq indicates the transcription start site (TSS). RNA polymerase II ChIP-seq and CAGE-seq data in postnatal day 10.5 testis were derived from GEO (GSE44690) (Li et al., 2013), and reanalyzed along with MEIOSIN and STRA8 ChIP-seq data (Ishiguro et al., 2020). (C) Summary of meiotic genes that are activated by both MEIOSIN and STRA8. The gene list is adopted from the data in previous study (Ishiguro et al., 2020). MEIOSIN/STRA8-target genes are defined as MEIOSIN- and STRA8-bound genes that are commonly identified with those downregulated in *Meiosin* KO testes. GO analyses indicated that meiotic genes are enriched in the MEIOSIN/STRA8-target genes. Some of the well-known meiotic genes are not on the list. This is in part due to the strict statistical threshold used in our previous RNA-seq data set of WT *vs Meiosin* KO to define the downregulated genes in *Meiosin* KO. Schematic illustration of chromosome structures during meiotic prophase is shown on the right. Meiotic chromosomal factors and structures are colored corresponding to the gene list. (D) Schematic illustrations of sequential activation of meiotic prophase genes. Germline genes are repressed by multiple mechanisms in somatic cells and mitotically growing germ cells. Meiotic genes are primed prior to meiotic entry in response to RA. BMP signaling also contributes transcriptional priming at least in female. MEIOSIN and STRA8 amplify the transcription of meiotic genes, and the genes that encode themselves. There are also MEIOSIN/STRA8-independent activation for meiotic gene transcription as shown in *Rec8*.

meiotic genes in preleptotene. While the MEIOSIN- and STRA8-bound genes predicted by two ChIP-seq studies are overall similar, E2F responsive genes for G1-S cell cycle regulation are also enriched by STRA8 ChIP-seq (Kojima et al., 2019). Given that meiosis starts from S phase followed by meiotic prophase (Pratto et al., 2021), the MEIOSIN-STRA8 complex coordinates meiotic initiation with S phase by robust activation of target genes to have meiotic gene expression program installed at S phase.

Here are several examples of the MEIOSIN/STRA8-target genes (Fig. 4C). During meiotic prophase, chromosomes are organized into axial element (AE) or chromosome axis (Fig. 4C). The MEIOSIN-STRA8 complex activates the genes encoding AE components such as *Sycp2* (Yang et al., 2006), *Sycp3* (Yuan et al., 2000), *Hormad1* (Shin et al., 2010) and *Hormad2* (Wojtasz et al., 2009). The synaptonemal complex (SC) mediates tight association of homologous chromosomes and promotes recombination (Fig. 4C). The genes encoding SC components are also activated by the MEIOSIN-STRA8 complex, such as *Sycp1* (de Vries et al., 2005), *Syce1* (Costa et al., 2005), *Syce2* (Bolcun-Filas et al., 2007), *Syce3* (Schramm et al., 2011) and *Tex12* (Hamer et al., 2006).

Meiotic cohesin complex plays crucial roles, not only in sister chromatid cohesion but also in meiosis-specific chromosomal dynamics such as AE formation, homolog pairing/synapsis and meiotic recombination (Fig. 4C) (Ishiguro, 2019). The MEIOSIN-STRA8 complex activates the expressions of the genes encoding meiosis-specific cohesin subunits, *Rad21L* (Herran et al., 2011; Ishiguro et al., 2014; Ishiguro, Kim, Fujiyama-Nakamura, Kato, & Watanabe, 2011; Lee & Hirano, 2011), *Smc1β* (Biswas, Stevense, & Jessberger, 2018; Revenkova, Eijpe, Heyting, Gross, & Jessberger, 2001) and *Stag3* (Fukuda et al., 2014), except for *Rec8* whose expression seems to be already regulated by RA at the time of meiotic initiation (Koubova et al., 2014).

Many of the genes involved in meiotic recombination are also activated by the MEIOSIN-STRA8 complex, such as *Prdm9* (Baudat et al., 2010), *Dmc1* (Pittman et al., 1998; Yoshida et al., 1998), *Spo11* (Baudat, Manova, Yuen, Jasin, & Keeney, 2000; Romanienko & Camerini-Otero, 2000), *Top6BL/Gm960* (Robert et al., 2016; Vrielynck et al., 2016), *Meiob* (Luo et al., 2013; Souquet et al., 2013; Xu, Greenberg, Schonbrunn, & Wang, 2017), *Ankrd31* (Boekhout et al., 2019; Papanikos et al., 2019), *Hsf2BP/Meilb2* (Zhang, Fujiwara, Yamamoto, & Shibuya, 2019), *Brme1/4930432K21Rik* (Takemoto et al., 2020) and *Msh5* (de Vries et al., 1999) (Fig. 4C).

During meiotic prophase, telomeres are anchored to the nuclear envelope (NE) by meiosis-specific telomere proteins to facilitate homolog synapsis (Fig. 4C) (Shibuya & Watanabe, 2014). The genes involved in this process, such as *Kash5/Ccdc155* (Morimoto et al., 2012), *Majin* (Shibuya et al., 2015), *Terb1* (Shibuya, Ishiguro, & Watanabe, 2014) and *Sun1* (Ding et al., 2007) are also activated by the MEIOSIN-STRA8 complex.

The MEIOSIN-STRA8 complex activates *Meioc* (Abby et al., 2016; Soh et al., 2017) and *Ythdc2* (Bailey et al., 2017; Jain et al., 2018), whose products are involved in the stability of mitotic cell cycle-associated transcripts. Furthermore, MEIOSIN and STRA8 bind to their own promoter regions, which may form autoactivation feedback loops to promote rapid expressions during the short period of meiotic initiation. Indeed, supporting this idea, *Stra8* expression is significantly decreased in *Meiosin* KO compared to WT, and *vice versa* (Ishiguro et al., 2020). It should be noted that not all the well-known meiotic genes are directly activated by the MEIOSIN-STRA8 complex. It is possible that a subset of the meiotic genes are regulated independently of MEIOSIN and STRA8 like the case of *Rec8*.

## 7. MEIOSIN and STRA8 boost the transcription of the meiotic genes

It has been suggested that a subpopulation of genes involved in meiotic prophase-events such as *Sycp3* already start to be expressed at low levels just before meiosis initiates (Chuma & Nakatsuji, 2001; Evans, Hogarth, Mitchell, & Griswold, 2014; Jan et al., 2017; Soh et al., 2015). This is in part due to the STRA8-independent and RA-independent expressions of those meiotic genes. RA signaling induces the expressions of a subset of meiotic genes (Bowles et al., 2006; Koubova et al., 2006, 2014). BMP signaling also synergistically induces the expressions of a broad range of genes through genome wide activation by ZGLP1 at least in oogonia (Miyauchi et al., 2017; Nagaoka et al., 2020). Indeed, expressions of MEIOSIN/STRA8-target genes that are associated with meiotic prophase-events are still detected at residual levels in *Meiosin* KO and *Stra8* KO, as revealed by RNA-seq analyses in these KO germ cells (Ishiguro et al., 2020; Kojima et al., 2019). For example, patchy signals of SYCP3 still can be detected in *Meiosin* KO and *Stra8* KO cells. This suggests the expressions of meiotic genes is primed prior to meiotic initiation. Although the precise molecular mechanism is yet to be investigated, MEIOSIN and STRA8 boost the transcriptions of those primed genes for meiotic entry rather than turning them on in an all-or-nothing manner.

ChIP-seq analyses suggest that MEIOSIN and STRA8 bind to the transcription start sites (TSSs) rather than the upstream promoter or enhancer regions (Fig. 4B) (Ishiguro et al., 2020; Kojima et al., 2019). In accordance with the basal transcription prior to meiosis, meiotic gene promoters in spermatogonia are often poised by the enrichment of H3K4me2 and accompanied by the loading of RNA polymerase II (Sin, Kartashov, Hasegawa, Barski, & Namekawa, 2015) (Fig. 4B). MEIOSIN and STRA8 may promote the release of paused RNA polymerase II or modify the assembly of the basal transcription machineries for a rapid and synchronous burst of transcription. Given that MEIOSIN and STRA8 bind to thousands of TSSs, transcriptions of a broad range of genes may be amplified by the MEIOSIN-STRA8 complex at meiotic initiation (Fig. 4D).

Co-expression of MEIOSIN and STRA8 is necessary for the decision of meiotic entry in germ cells. However, ectopic expression of MEIOSIN and STRA8 is not sufficient to induce meiosis in *in vitro* cultured mitotic cells. In mitotic cells, germ line genes are suppressed by MAX (Suzuki et al., 2016) and E2F-6, an atypical E2F-family protein that plays a distinct role from cell cycle regulation (Dahlet et al., 2021; Kehoe et al., 2008; Leseva, Santostefano, Rosenbluth, Hamazaki, & Terada, 2013; Pohlers et al., 2005) (Fig. 4D). In addition, epigenetic repressions such as DNA methylation (Yamaguchi et al., 2012) and Polycomb repressive complex (PRC)-mediated H3K27me3 (Endoh et al., 2017; Yokobayashi et al., 2013) are imposed to germ line genes before the initiation of meiosis, at least in female germ cells (Fig. 4D). Prior to meiotic initiation, meiotic genes are released from those repressive mechanisms, by which the germ cells become competent for meiotic initiation (Fig. 4D).

In this way, multiple layers of regulation potentiate the competency for meiotic gene activation, and the MEIOSIN-STRA8 complex may be at the upstream of robust transcriptional activation to boost meiotic transcripts for the mitosis-to-meiosis transition.

## 8. Sexual difference of meiotic initiation

Meiotic initiation is accompanied by germ cell development both in male and female (Fig. 1). Initiation of meiosis is regulated by sexually dimorphic mechanisms in mice, so the transcription and chromatin status are altered in male and female coordinating with the developmental program.

In male, meiotic entry is accompanied by the differentiation of B-type spermatogonia to spermatocytes in adulthood, and continuously occurs in the testis throughout life. In female, meiotic entry is restricted within a limited period around E13.5-E14.5, which propagates in a wave from the anterior to posterior gonadal pole (Bullejos & Koopman, 2004; Menke et al., 2003; Yao et al., 2003), and is subsequently followed by diplotene arrest in fetal ovary. Therefore, the life-span reproductive system in female is conferred by the oocyte pool that has been produced during this limited time window in fetal ovaries.

In fetal gonad, primordial germ cells (PGCs) are sexually bipotent and express pluripotency genes such as *Oct4*, *Sox2* and *Nanog* (Saitou & Yamaji, 2012). At E12.5, PGCs undergo sex determination: XX PGC becomes oogonia and XY PGCs becomes pro-spermatogonia (male gonocyte). During this period, promoters of germline genes such as *Mael*, *Sycp1* and *Sycp3* undergo DNA demethylation by TET1 in female PGCs (Yamaguchi et al., 2012). It was also shown that disruption of Polycomb Repressive Complex 1 (PRC1) led to precocious expression of *Stra8* and meiotic genes, and concomitant downregulation of pluripotency genes, as early as E12.5 in female germ cells (Yokobayashi et al., 2013). Thus, PRC1 suppresses the sensitivity of *Stra8* promoter to RA in female germ cells, which prevents *Stra8* from premature expression at E12.5 (Fig. 5A). At E13.5-E14.5 onward, while male germ cells stop proliferation and enter

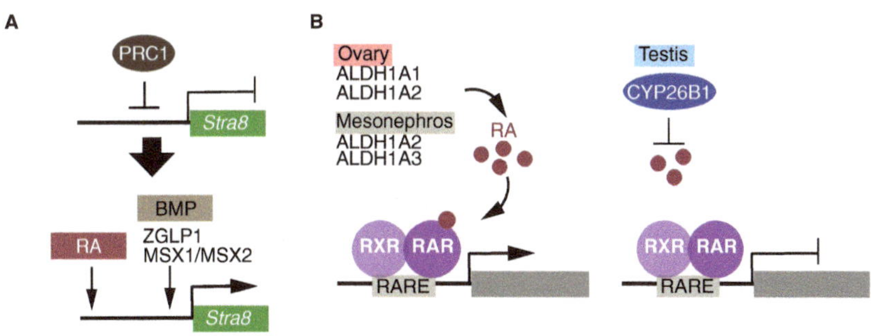


**Fig. 5** Sexually dimorphic mechanisms of meiotic initiation. (A) *Stra8* expression is initially suppressed directly or indirectly by PRC1 pathway (upper), and subsequently activated by BMP and RA signaling pathways (lower). (B) RA is synthesized by ALDH1A1 and ALDH1A2 in the somatic cells of fetal ovary, and by ALDH1A2 and ALDH1A3 in the mesonephros. In female germ cells, RA signaling is mediated by heterodimer of RARs (RARα, RARβ, RARγ) and RXR that binds to retinoic acid responsive elements (RAREs) of the target genes. In the fetal testis, RA is degraded by CYP26B1.

into G0/G1 arrest until they differentiate into spermatogonia and resume cell cycle after birth, female germ cells stop proliferation and subsequently enter into meiosis (Fig. 1).

It has been suggested that sex-specific initiation of meiosis in the fetal ovary requires RA and other signaling pathways (Fig. 1). RA signaling has been assumed as the key pathway that stimulates STRA8 expression (Feng et al., 2021) and in turn meiotic initiation in fetal ovary. Two sources of RA, developing fetal ovary and the mesonephros, a transient organ neighboring the fetal gonad, exist for meiotic initiation (Fig. 5B). Retinaldehyde dehydrogenase ALDH1A synthesizes RA from the precursor retinol (Vitamin A). ALDH1A1 and ALDH1A2 are expressed in the somatic cells of fetal ovary (Bowles et al., 2016; Chassot et al., 2020). ALDH1A2 and ALDH1A3 are expressed in the mesonephros, from which RA is diffused to fetal testis or ovary. It was shown that RA or RAR-agonists upregulate the expressions of *Stra8* and other meiotic genes, and pan-RAR antagonists suppress the expressions of them in embryonic ovaries cultured *ex vivo* (Bowles et al., 2006; Koubova et al., 2006). In the fetal testis, a P450 enzyme CYP26B1 that degrades RA is highly expressed in Sertoli cells, preventing male germ cells from being exposed to RA signaling. In *Cyp26b1* KO fetal testes, male germ cells ectopically express STRA8 and show a sign of meiotic entry (Bowles et al., 2006; Koubova et al., 2006; MacLean, Li, Metzger, Chambon, & Petkovich, 2007; Saba, Wu, & Saga, 2014). Therefore, CYP26B1 suppresses meiotic entry by counteracting RA in the fetal testis, consequently preventing precocious expression of RA-responsive genes that are required for meiotic initiation (Fig. 5B). In addition to CYP26B1, an RNA-binding protein NANOS2 also suppresses *Stra8* expression at posttranscriptional level in fetal testes (Suzuki & Saga, 2008). In this way, RA signaling is suppressed in fetal testis while playing a role in fetal ovary, which confers sex-specific initiation of meiosis.

However, other contradictory evidence arose from genetic mutation studies, showing RA signaling *per se* was dispensable for meiotic initiation (Bellutti et al., 2019; Chassot et al., 2020; Vernet et al., 2020). In embryos lacking *Aldh1a2* and *Aldh1a3* genes, expression of STRA8 is still detectable in some germ cells (Kumar et al., 2011). In *Aldh1a1* KO female, *Stra8* expression is reduced and meiotic initiation is delayed (Bowles et al., 2016). In fetal ovary lacking all the *Aldh1a* genes, *Aldh1a1, Aldh1a2* and *Aldh1a3*, some female germ cells are still able to express *Stra8* and show a sign of meiotic entry despite the lack of RA sources, although *Stra8* expression is delayed (Chassot et al., 2020).

RA signaling is mediated by RARs, which are bound by RA and regulate transcription. RARs and retinoid X receptors (RXRs) form heterodimers and bind to RAREs of the target genes (Fig. 5B). In mice, three paralogs of RARs exist: RARα, RARβ, and RARγ. Notably, although *Stra8* expression is delayed in mice lacking all the paralogs of RARs, germ cells are still capable of entering meiosis in fetal ovary (Vernet et al., 2020). Therefore, RARs are dispensable for meiotic entry, but required for timely expression of *Stra8* in fetal ovary. Altogether, RA signaling is dispensable for triggering *Stra8* expression, but rather contributes to meiotic initiation by ensuring timely expressions of *Stra8* and meiotic genes. These precedent studies imply that RA signaling is not a sole determinant of meiotic initiation, but is one of the key pathways that converge on meiotic initiation in fetal ovary.

In addition to RA signaling, the competency for meiotic initiation is also regulated by several regulatory signals synthesized in the somatic cells of the ovary. BMP, synergistically with RA, primes the competency for meiotic initiation in female germ cells (Miyauchi et al., 2017). Under BMP signaling, ZGLP1 activates the key genes of oogenic program including meiosis (Fig. 5A) (Nagaoka et al., 2020). Msh homeobox transcription factors MSX1 and MSX2 are also upregulated by BMP and RA signaling pathways, and are proposed to be required for full activation of *Stra8* (Le Bouffant et al., 2011) (Fig. 5A). WNT/β-catenin signaling also plays a critical role in oocyte differentiation to have meiotic entry in a timely manner in fetal ovary (Chassot et al., 2011; Le Rolle et al., 2021) (Fig. 5A). In the absence of β-catenin activity, female germ cells lose the expressions of pluripotency genes and precociously enter meiosis (Le Rolle et al., 2021). In addition to the above, an RNA-binding protein DAZL confers competency for interpreting RA as a meiosis-inducing signal both in XX and XY germ cells, suggesting that meiotic entry is also regulated at posttranscriptional level (Lin, Gill, Koubova, & Page, 2008). Thus, multiple layers of regulatory pathways synergistically coordinate female germ cell development and meiotic gene activation, and prime the competency for meiotic initiation in fetal ovary. Although MEIOSIN plays an essential role in meiotic entry in female together with STRA8, how the expression of MEIOSIN is regulated under RA and other signaling pathways is one of the future issues to be analyzed.

## 9. Evolutionary conservation and divergence in the meiotic entry system

Potential MEIOSIN homologs were initially searched in genome database of various organisms using TBLASTN with a query of the first 20-amino-acid sequence of the mouse MEIOSIN HMG box

(Ishiguro et al., 2020). This TBLASTN search successfully identified the genes that encode HMG box in vertebrates, of which the genomic loci showed a partial synteny to that of the mouse genome. It is still unknown whether MEIOSIN homologs act in heterodimers *via* the HMG-box and/or HLH domains in other vertebrate species. Although MEIOSIN is conserved like STRA8 in vertebrates, its potential homolog is not found in invertebrates or plants.

In budding yeast and fission yeast, external environmental signals converge for transcriptional and posttranscriptional regulation of the meiotic program. IME1 and Ste11 are the master transcriptional regulators of meiosis induction in budding and fission yeast, respectively (Sugimoto, Iino, Maeda, Watanabe, & Yamamoto, 1991; van Werven & Amon, 2011). Since significant structural homology is not found between IME1 and Ste11 transcription factors, evolutionary conservation of the meiotic entry system has remained controversial in these yeast (Anandhakumar, Fauquenoy, Materne, Migeot, & Hermand, 2013; van Werven & Amon, 2011). However, it is worth noting that Ste11 possesses an HMG-box which binds to the promoters of the reproduction-related genes (Sugimoto et al., 1991), similar to MEIOSIN and STRA8. Moreover, many of the HMG-box proteins are involved in the control of sexual differentiation in ascomycetes (Nelson, 1996). Thus, ancestral HMG-box protein may have evolutionally diverged to develop a mechanism of cell cycle switching from mitosis to meiosis in eukaryotes.

In *C. elegance* and *Drosophila*, mitosis-to-meiosis transition accompanies the exit from germline stem cell (GSC) (Kimble, 2011). Mitosis-to-meiosis transition is regulated at posttranscriptional level by translational repressor GLD proteins in *C. elegance*, and Bam-Bcgn proteins in *Drosophila*. However, it remains elusive how meiotic genes are regulated at transcriptional level in these multicellular organisms.

## 10. Conclusion

Discovery of MEIOSIN solved several longstanding enigmas in meiotic initiation that could not be simply explained by STRA8 expression. One of the most critical findings was that meiotic initiation occurs only when both MEIOSIN and STRA8 are co-expressed, which solved a longstanding enigma why meiotic initiation occurs only in preleptotene spermatocytes, but not in STRA8-positive spermatogonia.

It should be noted that some of the genes involved in meiotic prophase-events are already expressed at low levels prior to meiotic initiation. Given that MEIOSIN and STRA8 boost the transcriptions of those meiotic genes at preleptotene, it should be discussed carefully what the *bona fide* "meiotic prophase" is. For example, although SYCP3 is used as a marker of germ cells that show a sign of meiotic entry in numerous literatures, its expression should be appropriately interpreted depending on developmental context. RA signaling has been assumed for a long time as the key pathway of meiotic initiation in fetal ovary. Although RA indeed does promote timely expression of STRA8, RA is not the sole pathway to prime meiotic initiation. In addition to RA signaling, multiple pathways including BMP and WNT/β-catenin signalings also positively and negatively converge on meiotic initiation.

Several issues are still yet to be understood in meiotic initiation. How MEIOSIN expression is regulated by the upstream regulatory pathways is a critical question to understand the precise mechanism of meiotic initiation (Sun et al., 2022). Given that MEIOSIN and STRA8 binds to TSSs rather than upstream enhancers, how MEIOSIN and STRA8 transactivate meiotic genes is the next question. There are numerous targets of the MEIOSIN-STRA8 complex that are encoded by uncharacterized genes. It is possible that those uncharacterized MEIOSIN/STRA8 target genes play a crucial role in meiosis. Indeed, among the MEIOSIN/STRA8 target genes, we newly identified several genes that play an essential role in meiotic prophase-events (Horisawa-Takada et al., 2021; Takemoto et al., 2020; Tanno et al., 2022). Further investigation of MEIOSIN/STRA8 target genes will shed light on new players of meiosis. Although MEIOSIN and STRA8 are conserved in vertebrates, it remains largely elusive how meiotic initiation is regulated at transcription level in other species. Further investigation is awaited to know whether similar or distinct system acts for meiotic initiation in other multi-cellular organisms.

## Acknowledgments

This work was supported in part by KAKENHI grants (#19H05743, #20H03265, #20K21504, #JP 16H06276 #22H04922) from MEXT Japan; Grant from AMED PRIME (21gm6310021h0001); Grants from The Sumitomo Foundation; The Naito Foundation, Astellas Foundation for Research on Metabolic Disorders; Daiichi Sankyo Foundation of Life Science; The Uehara Memorial Foundation; The NOVARTIS Foundation (Japan) for the promotion of Science; Takeda Science Foundation.

## Conflict of interest statement

The authors declare no conflicts of interests.

## References

Abby, E., Tourpin, S., Ribeiro, J., Daniel, K., Messiaen, S., Moison, D., et al. (2016). Implementation of meiosis prophase I programme requires a conserved retinoid-independent stabilizer of meiotic transcripts. *Nature Communications*, 7, 10324. https://doi.org/10.1038/ncomms10324.

Anandhakumar, J., Fauquenoy, S., Materne, P., Migeot, V., & Hermand, D. (2013). Regulation of entry into gametogenesis by Ste11: The endless game. *Biochemical Society Transactions*, *41*, 1673–1678. https://doi.org/10.1042/BST20130225.

Anderson, E. L., Baltus, A. E., Roepers-Gajadien, H. L., Hassold, T. J., de Rooij, D. G., van Pelt, A. M., et al. (2008). Stra8 and its inducer, retinoic acid, regulate meiotic initiation in both spermatogenesis and oogenesis in mice. *Proceedings of the National Academy of Sciences of the United States of America*, *105*, 14976–14980. https://doi.org/10.1073/pnas.0807297105.

Bailey, A. S., Batista, P. J., Gold, R. S., Chen, Y. G., de Rooij, D. G., Chang, H. Y., et al. (2017). The conserved RNA helicase YTHDC2 regulates the transition from proliferation to differentiation in the germline. *eLife*, *6*. https://doi.org/10.7554/eLife.26116.

Baltus, A. E., Menke, D. B., Hu, Y. C., Goodheart, M. L., Carpenter, A. E., de Rooij, D. G., et al. (2006). In germ cells of mouse embryonic ovaries, the decision to enter meiosis precedes premeiotic DNA replication. *Nature Genetics*, *38*, 1430–1434. https://doi.org/10.1038/ng1919.

Baudat, F., Buard, J., Grey, C., Fledel-Alon, A., Ober, C., Przeworski, M., et al. (2010). PRDM9 is a major determinant of meiotic recombination hotspots in humans and mice. *Science*, *327*, 836–840. https://doi.org/10.1126/science.1183439.

Baudat, F., Imai, Y., & de Massy, B. (2013). Meiotic recombination in mammals: Localization and regulation. *Nature Reviews. Genetics*, *14*, 794–806. https://doi.org/10.1038/nrg3573.

Baudat, F., Manova, K., Yuen, J. P., Jasin, M., & Keeney, S. (2000). Chromosome synapsis defects and sexually dimorphic meiotic progression in mice lacking Spo11. *Molecular Cell*, *6*, 989–998.

Bellutti, L., Abby, E., Tourpin, S., Messiaen, S., Moison, D., Trautmann, E., et al. (2019). Divergent roles of CYP26B1 and endogenous retinoic acid in mouse fetal gonads. *Biomolecules*, *9*. https://doi.org/10.3390/biom9100536.

Biswas, U., Stevense, M., & Jessberger, R. (2018). SMC1alpha substitutes for many meiotic functions of SMC1beta but cannot protect telomeres from Damage. *Current Biology*, *28*, 249–261 e244. https://doi.org/10.1016/j.cub.2017.12.020.

Boekhout, M., Karasu, M. E., Wang, J., Acquaviva, L., Pratto, F., Brick, K., et al. (2019). REC114 partner ANKRD31 controls number, timing, and location of meiotic DNA breaks. *Molecular Cell*, *74*, 1053–1068 e1058. https://doi.org/10.1016/j.molcel.2019.03.023.

Bolcun-Filas, E., Costa, Y., Speed, R., Taggart, M., Benavente, R., De Rooij, D. G., et al. (2007). SYCE2 is required for synaptonemal complex assembly, double strand break repair, and homologous recombination. *The Journal of Cell Biology*, *176*, 741–747. https://doi.org/10.1083/jcb.200610027.

Bowles, J., Feng, C. W., Miles, K., Ineson, J., Spiller, C., & Koopman, P. (2016). ALDH1A1 provides a source of meiosis-inducing retinoic acid in mouse fetal ovaries. *Nature Communications*, 7, 10845. https://doi.org/10.1038/ncomms10845.

Bowles, J., Knight, D., Smith, C., Wilhelm, D., Richman, J., Mamiya, S., et al. (2006). Retinoid signaling determines germ cell fate in mice. *Science*, *312*, 596–600. https://doi.org/10.1126/science.1125691.

Bullejos, M., & Koopman, P. (2004). Germ cells enter meiosis in a rostro-caudal wave during development of the mouse ovary. *Molecular Reproduction and Development*, *68*, 422–428. https://doi.org/10.1002/mrd.20105.

Cahoon, C. K., & Hawley, R. S. (2016). Regulating the construction and demolition of the synaptonemal complex. *Nature Structural & Molecular Biology*, *23*, 369–377. https://doi.org/10.1038/nsmb.3208.

Chassot, A. A., Gregoire, E. P., Lavery, R., Taketo, M. M., de Rooij, D. G., Adams, I. R., et al. (2011). RSPO1/beta-catenin signaling pathway regulates oogonia differentiation and entry into meiosis in the mouse fetal ovary. *PLoS One*, *6*, e25641. https://doi.org/10.1371/journal.pone.0025641.

Chassot, A. A., Le Rolle, M., Jolivet, G., Stevant, I., Guigonis, J. M., Da Silva, F., et al. (2020). Retinoic acid synthesis by ALDH1A proteins is dispensable for meiosis initiation in the mouse fetal ovary. *Science Advances*, *6*, eaaz1261. https://doi.org/10.1126/sciadv.aaz1261.

Chuma, S., & Nakatsuji, N. (2001). Autonomous transition into meiosis of mouse fetal germ cells in vitro and its inhibition by gp130-mediated signaling. *Developmental Biology*, *229*, 468–479. https://doi.org/10.1006/dbio.2000.9989.

Costa, Y., Speed, R., Ollinger, R., Alsheimer, M., Semple, C. A., Gautier, P., et al. (2005). Two novel proteins recruited by synaptonemal complex protein 1 (SYCP1) are at the centre of meiosis. *Journal of Cell Science*, *118*, 2755–2762. https://doi.org/10.1242/jcs.02402.

Dahlet, T., Truss, M., Frede, U., Al Adhami, H., Bardet, A. F., Dumas, M., et al. (2021). E2F6 initiates stable epigenetic silencing of germline genes during embryonic development. *Nature Communications*, *12*, 3582. https://doi.org/10.1038/s41467-021-23596-w.

de Vries, S. S., Baart, E. B., Dekker, M., Siezen, A., de Rooij, D. G., de Boer, P., et al. (1999). Mouse MutS-like protein Msh5 is required for proper chromosome synapsis in male and female meiosis. *Genes & Development*, *13*, 523–531.

de Vries, F. A., de Boer, E., van den Bosch, M., Baarends, W. M., Ooms, M., Yuan, L., et al. (2005). Mouse Sycp1 functions in synaptonemal complex assembly, meiotic recombination, and XY body formation. *Genes & Development*, *19*, 1376–1389. https://doi.org/10.1101/gad.329705.

Ding, X., Xu, R., Yu, J., Xu, T., Zhuang, Y., & Han, M. (2007). SUN1 is required for telomere attachment to nuclear envelope and gametogenesis in mice. *Developmental Cell*, *12*, 863–872. https://doi.org/10.1016/j.devcel.2007.03.018.

Dokshin, G. A., Baltus, A. E., Eppig, J. J., & Page, D. C. (2013). Oocyte differentiation is genetically dissociable from meiosis in mice. *Nature Genetics*, *45*, 877–883. https://doi.org/10.1038/ng.2672.

Endo, T., Romer, K. A., Anderson, E. L., Baltus, A. E., de Rooij, D. G., & Page, D. C. (2015). Periodic retinoic acid-STRA8 signaling intersects with periodic germ-cell competencies to regulate spermatogenesis. *Proceedings of the National Academy of Sciences of the United States of America*, *112*, E2347–E2356. https://doi.org/10.1073/pnas.1505683112.

Endoh, M., Endo, T. A., Shinga, J., Hayashi, K., Farcas, A., Ma, K. W., et al. (2017). PCGF6-PRC1 suppresses premature differentiation of mouse embryonic stem cells by regulating germ cell-related genes. *eLife*, *6*. https://doi.org/10.7554/eLife.21064.

Evans, E., Hogarth, C., Mitchell, D., & Griswold, M. (2014). Riding the spermatogenic wave: Profiling gene expression within neonatal germ and sertoli cells during a synchronized initial wave of spermatogenesis in mice. *Biology of Reproduction*, *90*, 108. https://doi.org/10.1095/biolreprod.114.118034.

Feng, C.-W., Burnet, G., Spiller, C. M., Cheung, F. K. M., Chawengsaksophak, K., Koopman, P., et al. (2021). Identification of regulatory elements required for Stra8 expression in fetal ovarian germ cells of the mouse. *Development*, *148*, dev.194977. https://doi.org/10.1242/dev.194977.

Fukuda, T., Fukuda, N., Agostinho, A., Hernandez-Hernandez, A., Kouznetsova, A., & Hoog, C. (2014). STAG3-mediated stabilization of REC8 cohesin complexes promotes chromosome synapsis during meiosis. *The EMBO Journal*, *33*, 1243–1255. https://doi.org/10.1002/embj.201387329.

Hamazaki, N., Kyogoku, H., Araki, H., Miura, F., Horikawa, C., Hamada, N., et al. (2021). Reconstitution of the oocyte transcriptional network with transcription factors. *Nature*, *589*, 264–269. https://doi.org/10.1038/s41586-020-3027-9.

Hamer, G., Gell, K., Kouznetsova, A., Novak, I., Benavente, R., & Hoog, C. (2006). Characterization of a novel meiosis-specific protein within the central element of the synaptonemal complex. *Journal of Cell Science*, *119*, 4025–4032. https://doi.org/10.1242/jcs.03182.

Herran, Y., Gutierrez-Caballero, C., Sanchez-Martin, M., Hernandez, T., Viera, A., Barbero, J. L., et al. (2011). The cohesin subunit RAD21L functions in meiotic synapsis and exhibits sexual dimorphism in fertility. *The EMBO Journal*, *30*, 3091–3105. https://doi.org/10.1038/emboj.2011.222.

Horisawa-Takada, Y., Kodera, C., Takemoto, K., Sakashita, A., Horisawa, K., Maeda, R., et al. (2021). Meiosis-specific ZFP541 repressor complex promotes developmental progression of meiotic prophase towards completion during mouse spermatogenesis. *Nature Communications*, *12*, 3184. https://doi.org/10.1038/s41467-021-23378-4.

Ishiguro, K. I. (2019). The cohesin complex in mammalian meiosis. *Genes to Cells*, *24*, 6–30. https://doi.org/10.1111/gtc.12652.

Ishiguro, K., Kim, J., Fujiyama-Nakamura, S., Kato, S., & Watanabe, Y. (2011). A new meiosis-specific cohesin complex implicated in the cohesin code for homologous pairing. *EMBO Reports*, *12*, 267–275. https://doi.org/10.1038/embor.2011.2.

Ishiguro, K., Kim, J., Shibuya, H., Hernandez-Hernandez, A., Suzuki, A., Fukagawa, T., et al. (2014). Meiosis-specific cohesin mediates homolog recognition in mouse spermatocytes. *Genes & Development*, *28*, 594–607. https://doi.org/10.1101/gad.237313.113.

Ishiguro, K. I., Matsuura, K., Tani, N., Takeda, N., Usuki, S., Yamane, M., et al. (2020). MEIOSIN directs the switch from mitosis to meiosis in mammalian germ cells. *Developmental Cell*, *52*(4), 429–445. https://doi.org/10.1016/j.devcel.2020.01.010.

Jain, D., Puno, M. R., Meydan, C., Lailler, N., Mason, C. E., Lima, C. D., et al. (2018). ketu mutant mice uncover an essential meiotic function for the ancient RNA helicase YTHDC2. *eLife*, 7. https://doi.org/10.7554/eLife.30919.

Jan, S. Z., Vormer, T. L., Jongejan, A., Roling, M. D., Silber, S. J., de Rooij, D. G., et al. (2017). Unraveling transcriptome dynamics in human spermatogenesis. *Development*, *144*, 3659–3673. https://doi.org/10.1242/dev.152413.

Kanatsu-Shinohara, M., Ogonuki, N., Inoue, K., Miki, H., Ogura, A., Toyokuni, S., et al. (2003). Long-term proliferation in culture and germline transmission of mouse male germline stem cells. *Biology of Reproduction*, *69*, 612–616. https://doi.org/10.1095/biolreprod.103.017012.

Keeney, S., Lange, J., & Mohibullah, N. (2014). Self-organization of meiotic recombination initiation: General principles and molecular pathways. *Annual Review of Genetics*, *48*, 187–214. https://doi.org/10.1146/annurev-genet-120213-092304.

Kehoe, S. M., Oka, M., Hankowski, K. E., Reichert, N., Garcia, S., McCarrey, J. R., et al. (2008). A conserved E2F6-binding element in murine meiosis-specific gene promoters. *Biology of Reproduction*, *79*, 921–930. https://doi.org/10.1095/biolreprod.108.067645.

Kimble, J. (2011). Molecular regulation of the mitosis/meiosis decision in multicellular organisms. *Cold Spring Harbor Perspectives in Biology*, *3*, a002683. https://doi.org/10.1101/cshperspect.a002683.

Kojima, M. L., de Rooij, D. G., & Page, D. C. (2019). Amplification of a broad transcriptional program by a common factor triggers the meiotic cell cycle in mice. *eLife*, *8*. https://doi.org/10.7554/eLife.43738.

Koubova, J., Hu, Y. C., Bhattacharyya, T., Soh, Y. Q., Gill, M. E., Goodheart, M. L., et al. (2014). Retinoic acid activates two pathways required for meiosis in mice. *PLoS Genetics*, *10*, e1004541. https://doi.org/10.1371/journal.pgen.1004541.

Koubova, J., Menke, D. B., Zhou, Q., Capel, B., Griswold, M. D., & Page, D. C. (2006). Retinoic acid regulates sex-specific timing of meiotic initiation in mice. *Proceedings of the National Academy of Sciences of the United States of America, 103*, 2474–2479. https://doi.org/10.1073/pnas.0510813103.

Kumar, S., Chatzi, C., Brade, T., Cunningham, T. J., Zhao, X., & Duester, G. (2011). Sex-specific timing of meiotic initiation is regulated by Cyp26b1 independent of retinoic acid signalling. *Nature Communications, 2*, 151. https://doi.org/10.1038/ncomms1136.

Le Bouffant, R., Souquet, B., Duval, N., Duquenne, C., Herve, R., Frydman, N., et al. (2011). Msx1 and Msx2 promote meiosis initiation. *Development, 138*, 5393–5402. https://doi.org/10.1242/dev.068452.

Le Rolle, M., Massa, F., Siggers, P., Turchi, L., Loubat, A., Koo, B. K., et al. (2021). Arrest of WNT/beta-catenin signaling enables the transition from pluripotent to differentiated germ cells in mouse ovaries. *Proceedings of the National Academy of Sciences of the United States of America, 118*. https://doi.org/10.1073/pnas.2023376118.

Lee, J., & Hirano, T. (2011). RAD21L, a novel cohesin subunit implicated in linking homologous chromosomes in mammalian meiosis. *The Journal of Cell Biology, 192*, 263–276. https://doi.org/10.1083/jcb.201008005.

Leseva, M., Santostefano, K. E., Rosenbluth, A. L., Hamazaki, T., & Terada, N. (2013). E2f6-mediated repression of the meiotic Stag3 and Smc1beta genes during early embryonic development requires Ezh2 and not the de novo methyltransferase Dnmt3b. *Epigenetics, 8*, 873–884. https://doi.org/10.4161/epi.25522.

Li, X. Z., Roy, C. K., Dong, X., Bolcun-Filas, E., Wang, J., Han, B. W., et al. (2013). An ancient transcription factor initiates the burst of piRNA production during early meiosis in mouse testes. *Molecular Cell, 50*, 67–81. https://doi.org/10.1016/j.molcel.2013.02.016.

Lin, Y., Gill, M. E., Koubova, J., & Page, D. C. (2008). Germ cell-intrinsic and -extrinsic factors govern meiotic initiation in mouse embryos. *Science, 322*, 1685–1687. https://doi.org/10.1126/science.1166340.

Luo, M., Yang, F., Leu, N. A., Landaiche, J., Handel, M. A., Benavente, R., et al. (2013). MEIOB exhibits single-stranded DNA-binding and exonuclease activities and is essential for meiotic recombination. *Nature Communications, 4*, 2788. https://doi.org/10.1038/ncomms3788.

MacLean, G., Li, H., Metzger, D., Chambon, P., & Petkovich, M. (2007). Apoptotic extinction of germ cells in testes of Cyp26b1 knockout mice. *Endocrinology, 148*, 4560–4567. https://doi.org/10.1210/en.2007-0492.

Mark, M., Jacobs, H., Oulad-Abdelghani, M., Dennefeld, C., Feret, B., Vernet, N., et al. (2008). STRA8-deficient spermatocytes initiate, but fail to complete, meiosis and undergo premature chromosome condensation. *Journal of Cell Science, 121*, 3233–3242. https://doi.org/10.1242/jcs.035071.

Matson, C. K., Murphy, M. W., Griswold, M. D., Yoshida, S., Bardwell, V. J., & Zarkower, D. (2010). The mammalian doublesex homolog DMRT1 is a transcriptional gatekeeper that controls the mitosis versus meiosis decision in male germ cells. *Developmental Cell, 19*, 612–624. https://doi.org/10.1016/j.devcel.2010.09.010.

Menke, D. B., Koubova, J., & Page, D. C. (2003). Sexual differentiation of germ cells in XX mouse gonads occurs in an anterior-to-posterior wave. *Developmental Biology, 262*, 303–312.

Miyauchi, H., Ohta, H., Nagaoka, S., Nakaki, F., Sasaki, K., Hayashi, K., et al. (2017). Bone morphogenetic protein and retinoic acid synergistically specify female germ-cell fate in mice. *The EMBO Journal, 36*, 3100–3119. https://doi.org/10.15252/embj.201796875.

Morimoto, A., Shibuya, H., Zhu, X., Kim, J., Ishiguro, K., Han, M., et al. (2012). A conserved KASH domain protein associates with telomeres, SUN1, and dynactin during mammalian meiosis. *The Journal of Cell Biology, 198*, 165–172. https://doi.org/10.1083/jcb.201204085.

Murphy, M. W., Lee, J. K., Rojo, S., Gearhart, M. D., Kurahashi, K., Banerjee, S., et al. (2015). An ancient protein-DNA interaction underlying metazoan sex determination. *Nature Structural & Molecular Biology*, *22*, 442–451. https://doi.org/10.1038/nsmb.3032.

Nagaoka, S. I., Nakaki, F., Miyauchi, H., Nosaka, Y., Ohta, H., Yabuta, Y., et al. (2020). ZGLP1 is a determinant for the oogenic fate in mice. *Science*, *367*(6482), eaaw4115. https://doi.org/10.1126/science.aaw4115.

Nelson, M. A. (1996). Mating systems in ascomycetes: A romp in the sac. *Trends in Genetics*, *12*, 69–74.

Oulad-Abdelghani, M., Bouillet, P., Decimo, D., Gansmuller, A., Heyberger, S., Dolle, P., et al. (1996). Characterization of a premeiotic germ cell-specific cytoplasmic protein encoded by Stra8, a novel retinoic acid-responsive gene. *The Journal of Cell Biology*, *135*, 469–477.

Papanikos, F., Clement, J. A. J., Testa, E., Ravindranathan, R., Grey, C., Dereli, I., et al. (2019). Mouse ANKRD31 regulates spatiotemporal patterning of meiotic recombination initiation and ensures recombination between X and Y Sex chromosomes. *Molecular Cell*, *74*, 1069–1085 e1011. https://doi.org/10.1016/j.molcel.2019.03.022.

Pittman, D. L., Cobb, J., Schimenti, K. J., Wilson, L. A., Cooper, D. M., Brignull, E., et al. (1998). Meiotic prophase arrest with failure of chromosome synapsis in mice deficient for Dmc1, a germline-specific RecA homolog. *Molecular Cell*, *1*, 697–705.

Pohlers, M., Truss, M., Frede, U., Scholz, A., Strehle, M., Kuban, R. J., et al. (2005). A role for E2F6 in the restriction of male-germ-cell-specific gene expression. *Current Biology*, *15*, 1051–1057. https://doi.org/10.1016/j.cub.2005.04.060.

Pratto, F., Brick, K., Cheng, G., Lam, K. G., Cloutier, J. M., Dahiya, D., et al. (2021). Meiotic recombination mirrors patterns of germline replication in mice and humans. *Cell*, *184*(16), 4251–4267. https://doi.org/10.1016/j.cell.2021.06.025.

Revenkova, E., Eijpe, M., Heyting, C., Gross, B., & Jessberger, R. (2001). Novel meiosis-specific isoform of mammalian SMC1. *Molecular and Cellular Biology*, *21*, 6984–6998.

Robert, T., Nore, A., Brun, C., Maffre, C., Crimi, B., Bourbon, H. M., et al. (2016). The TopoVIB-Like protein family is required for meiotic DNA double-strand break formation. *Science*, *351*, 943–949. https://doi.org/10.1126/science.aad5309.

Romanienko, P. J., & Camerini-Otero, R. D. (2000). The mouse Spo11 gene is required for meiotic chromosome synapsis. *Molecular Cell*, *6*, 975–987.

Saba, R., Wu, Q., & Saga, Y. (2014). CYP26B1 promotes male germ cell differentiation by suppressing STRA8-dependent meiotic and STRA8-independent mitotic pathways. *Developmental Biology*, *389*, 173–181. https://doi.org/10.1016/j.ydbio.2014.02.013.

Saitou, M., & Yamaji, M. (2012). Primordial germ cells in mice. *Cold Spring Harbor Perspectives in Biology*, *4*(11), a008375. https://doi.org/10.1101/cshperspect.a008375.

Schramm, S., Fraune, J., Naumann, R., Hernandez-Hernandez, A., Hoog, C., Cooke, H. J., et al. (2011). A novel mouse synaptonemal complex protein is essential for loading of central element proteins, recombination, and fertility. *PLoS Genetics*, *7*, e1002088. https://doi.org/10.1371/journal.pgen.1002088.

Shibuya, H., Hernandez-Hernandez, A., Morimoto, A., Negishi, L., Hoog, C., & Watanabe, Y. (2015). MAJIN links telomeric DNA to the nuclear membrane by exchanging telomere cap. *Cell*, *163*, 1252–1266. https://doi.org/10.1016/j.cell.2015.10.030.

Shibuya, H., Ishiguro, K., & Watanabe, Y. (2014). The TRF1-binding protein TERB1 promotes chromosome movement and telomere rigidity in meiosis. *Nature Cell Biology*, *16*, 145–156. https://doi.org/10.1038/ncb2896.

Shibuya, H., & Watanabe, Y. (2014). The meiosis-specific modification of mammalian telomeres. *Cell Cycle*, *13*, 2024–2028. https://doi.org/10.4161/cc.29350.

Shin, Y. H., Choi, Y., Erdin, S. U., Yatsenko, S. A., Kloc, M., Yang, F., et al. (2010). Hormad1 mutation disrupts synaptonemal complex formation, recombination, and chromosome segregation in mammalian meiosis. *PLoS Genetics*, *6*, e1001190. https://doi.org/10.1371/journal.pgen.1001190.

Sin, H. S., Kartashov, A. V., Hasegawa, K., Barski, A., & Namekawa, S. H. (2015). Poised chromatin and bivalent domains facilitate the mitosis-to-meiosis transition in the male germline. *BMC Biology*, *13*, 53. https://doi.org/10.1186/s12915-015-0159-8.

Soh, Y. Q., Junker, J. P., Gill, M. E., Mueller, J. L., van Oudenaarden, A., & Page, D. C. (2015). A gene regulatory program for meiotic prophase in the fetal ovary. *PLoS Genetics*, *11*, e1005531. https://doi.org/10.1371/journal.pgen.1005531.

Soh, Y. Q. S., Mikedis, M. M., Kojima, M., Godfrey, A. K., de Rooij, D. G., & Page, D. C. (2017). Meioc maintains an extended meiotic prophase I in mice. *PLoS Genetics*, *13*, e1006704. https://doi.org/10.1371/journal.pgen.1006704.

Souquet, B., Abby, E., Herve, R., Finsterbusch, F., Tourpin, S., Le Bouffant, R., et al. (2013). MEIOB targets single-strand DNA and is necessary for meiotic recombination. *PLoS Genetics*, *9*, e1003784. https://doi.org/10.1371/journal.pgen.1003784.

Sugimoto, A., Iino, Y., Maeda, T., Watanabe, Y., & Yamamoto, M. (1991). Schizosaccharomyces pombe ste11+ encodes a transcription factor with an HMG motif that is a critical regulator of sexual development. *Genes & Development*, *5*, 1990–1999.

Sun, S., et al. (2022). Znhit1 controls meiotic initiation in male germ cells by coordinating with Stra8 to activate meiotic gene expression. *Developmental Cell*, *57*, 901–913. https://doi.org/10.1016/j.devcel.2022.03.006.

Suzuki, A., Hirasaki, M., Hishida, T., Wu, J., Okamura, D., Ueda, A., et al. (2016). Loss of MAX results in meiotic entry in mouse embryonic and germline stem cells. *Nature Communications*, *7*, 11056. https://doi.org/10.1038/ncomms11056.

Suzuki, A., & Saga, Y. (2008). Nanos2 suppresses meiosis and promotes male germ cell differentiation. *Genes & Development*, *22*, 430–435. https://doi.org/10.1101/gad.1612708.

Takemoto, K., Tani, N., Takada-Horisawa, Y., Fujimura, S., Tanno, N., Yamane, M., et al. (2020). Meiosis-specific C19orf57/4930432K21Rik/BRME1 modulates localization of RAD51 and DMC1 to DSBs in mouse meiotic recombination. *Cell Reports*, *31*. https://doi.org/10.1016/j.celrep.2020.107686, 107686.

Tanno, N., Takemoto, K., Takada-Horisawa, Y., Shimada, R., Fujimura, S., Tani, N., et al. (2022). FBXO47 is essential for preventing the synaptonemal complex from premature disassembly in mouse male meiosis. *iScience*, *25*, 104008. https://doi.org/10.1016/j.isci.2022.104008.

Tedesco, M., La Sala, G., Barbagallo, F., De Felici, M., & Farini, D. (2009). STRA8 shuttles between nucleus and cytoplasm and displays transcriptional activity. *The Journal of Biological Chemistry*, *284*, 35781–35793. https://doi.org/10.1074/jbc.M109.056481.

van Werven, F. J., & Amon, A. (2011). Regulation of entry into gametogenesis. *Philosophical Transactions of the Royal Society of London. Series B, Biological Sciences*, *366*, 3521–3531. https://doi.org/10.1098/rstb.2011.0081.

Vernet, N., Condrea, D., Mayere, C., Feret, B., Klopfenstein, M., Magnant, W., et al. (2020). Meiosis occurs normally in the fetal ovary of mice lacking all retinoic acid receptors. *Science Advances*, *6*(21), eaaz1139. https://doi.org/10.1126/sciadv.aaz1139.

Vrielynck, N., Chambon, A., Vezon, D., Pereira, L., Chelysheva, L., De Muyt, A., et al. (2016). A DNA topoisomerase VI-like complex initiates meiotic recombination. *Science*, *351*, 939–943. https://doi.org/10.1126/science.aad5196.

Wojtasz, L., Daniel, K., Roig, I., Bolcun-Filas, E., Xu, H., Boonsanay, V., et al. (2009). Mouse HORMAD1 and HORMAD2, two conserved meiotic chromosomal proteins, are depleted from synapsed chromosome axes with the help of TRIP13 AAA-ATPase. *PLoS Genetics*, *5*, e1000702. https://doi.org/10.1371/journal.pgen.1000702.

Xu, Y., Greenberg, R. A., Schonbrunn, E., & Wang, P. J. (2017). Meiosis-specific proteins MEIOB and SPATA22 cooperatively associate with the single-stranded DNA-binding replication protein A complex and DNA double-strand breaks. *Biology of Reproduction*, *96*, 1096–1104. https://doi.org/10.1093/biolre/iox040.

Yamaguchi, S., Hong, K., Liu, R., Shen, L., Inoue, A., Diep, D., et al. (2012). Tet1 controls meiosis by regulating meiotic gene expression. *Nature*, *492*, 443–447. https://doi.org/10.1038/nature11709.

Yang, F., De La Fuente, R., Leu, N. A., Baumann, C., McLaughlin, K. J., & Wang, P. J. (2006). Mouse SYCP2 is required for synaptonemal complex assembly and chromosomal synapsis during male meiosis. *The Journal of Cell Biology*, *173*, 497–507. https://doi.org/10.1083/jcb.200603063.

Yao, H. H., DiNapoli, L., & Capel, B. (2003). Meiotic germ cells antagonize mesonephric cell migration and testis cord formation in mouse gonads. *Development*, *130*, 5895–5902. https://doi.org/10.1242/dev.00836.

Yokobayashi, S., Liang, C. Y., Kohler, H., Nestorov, P., Liu, Z., Vidal, M., et al. (2013). PRC1 coordinates timing of sexual differentiation of female primordial germ cells. *Nature*, *495*, 236–240. https://doi.org/10.1038/nature11918.

Yoshida, K., Kondoh, G., Matsuda, Y., Habu, T., Nishimune, Y., & Morita, T. (1998). The mouse RecA-like gene Dmc1 is required for homologous chromosome synapsis during meiosis. *Molecular Cell*, *1*, 707–718.

Yuan, L., Liu, J. G., Zhao, J., Brundell, E., Daneholt, B., & Hoog, C. (2000). The murine SCP3 gene is required for synaptonemal complex assembly, chromosome synapsis, and male fertility. *Molecular Cell*, *5*, 73–83. https://doi.org/10.1016/s1097-2765(00)80404-9.

Zhang, J., Fujiwara, Y., Yamamoto, S., & Shibuya, H. (2019). A meiosis-specific BRCA2 binding protein recruits recombinases to DNA double-strand breaks to ensure homologous recombination. *Nature Communications*, *10*, 722. https://doi.org/10.1038/s41467-019-08676-2.

Zhou, Q., Li, Y., Nie, R., Friel, P., Mitchell, D., Evanoff, R. M., et al. (2008). Expression of stimulated by retinoic acid gene 8 (Stra8) and maturation of murine gonocytes and spermatogonia induced by retinoic acid in vitro. *Biology of Reproduction*, *78*, 537–545. https://doi.org/10.1095/biolreprod.107.064337.

Zhou, Q., Nie, R., Li, Y., Friel, P., Mitchell, D., Hess, R. A., et al. (2008). Expression of stimulated by retinoic acid gene 8 (Stra8) in spermatogenic cells induced by retinoic acid: An in vivo study in vitamin A-sufficient postnatal murine testes. *Biology of Reproduction*, *79*, 35–42. https://doi.org/10.1095/biolreprod.107.066795.

Zickler, D., & Kleckner, N. (2015). Recombination, pairing, and synapsis of homologs during meiosis. *Cold Spring Harbor Perspectives in Biology*, 7. https://doi.org/10.1101/cshperspect.a016626.

CHAPTER TWO

# Orchestrating recombination initiation in mice and men

**Elena Damm and Linda Odenthal-Hesse***
Department Evolutionary Genetics, Research Group Meiotic Recombination and Genome Instability, Max Planck Institute for Evolutionary Biology, Plön, Germany
*Corresponding author: odenthalhesse@evolbio.mpg.de

## Contents



## Abstract

Recent discoveries have advanced our understanding of recombination initiation beyond the placement of double-stranded DNA breaks (DSBs) from **germline replication timing** to the dynamic **reorganization of chromatin,** and defined critical players of **recombination initiation**. This article focuses on recombination initiation in mammals utilizing the PRDM9 protein to orchestrate crucial stages of meiotic recombination initiation by interacting with the local DNA environment and several protein complexes. The **Pioneer Complex** with the SNF2-type chromatin remodeling enzyme HELLS, exposes PRDM9-bound DNA. At the same time, a **Compass-Complex** containing EWSR1, CXXC1, CDYL, EHMT2 and PRDM9 facilitates the association of putative hotspot sites in DNA loops with the chromosomal axis where **DSB-promoting complexes** are located, and DSBs are catalyzed by the **SPO11/TOPOVIBL complex**. Finally, homology search is facilitated at PRDM9-directed sites by ANKRD31. The **Reader-Writer system** consists of PRDM9 writing characteristic histone methylation signatures, which are read by ZCWPW1, promoting efficient homology engagement.

*Current Topics in Developmental Biology*, Volume 151
ISSN 0070-2153
https://doi.org/10.1016/bs.ctdb.2022.05.001

## 1. Introduction—Orchestrating the critical stages of recombination initiation

Meiosis is the critical phase in the development of sexually reproducing species, where a single round of DNA replication precedes two rounds of cell division in meiocyte nuclei. Engaging the homologous chromosome ensures correct segregation at the meiotic spindle. Both DNA replication and molecular recombination are tightly orchestrated processes that fulfill critical tasks. Preceding recombination, the parental DNA must be faithfully replicated before chromatin condenses into chromosomes along the chromosomal axis. Locally accessible putative recombination sites in DNA loops need to associate with the chromosomal axis, where DNA double-stranded breaks (DSBs) are catalyzed by topoisomerase-like enzyme SPO11 (Keeney et al., 1999; Metzler-Guillemain & Massy, 1999; Romanienko & Camerini-Otero, 1999). Faithful DSB repair follows homology search, as only interhomolog repair provides the necessary physical connections to prevent chromosomal non-disjunction. At the Megabase-scale, the patterning of meiotic recombination is conserved across large evolutionary timescales and on average, smaller chromosomes have higher recombination activity than larger chromosomes and higher numbers of meiotic DSBs in the sub-telomeric regions are seen in males of distinct species (Stapley, Feulner, Johnston, Santure, & Smadja, 2017). Recombination varies extensively at the fine scale, and DSBs cluster into discrete recombination hotspots in many organisms. Meiotic DSBs are repaired either as reciprocal crossovers (CO) or non-reciprocal non-crossovers (NCO), which may result in gene conversions. Ultimately, meiotic recombination between homologous chromosomes reshuffles genetic information, thus increasing genetic diversity.

## 2. Germline replication timing affects broad-scale recombination patterning

A single round of DNA replication in precursor nuclei duplicates paternal chromatin before two rounds of successive cell divisions create haploid germ-cells. Parental chromatids are replicated, forming two identical sister chromatids, connected by the centromere. Replication timing—the temporal order of chromatin replication and assembly—is slower in meiosis than mitosis, probably due to reduced origin firing (Pratto et al., 2021). High-density replication origins are found near transcription start sites and in GC-rich sequences, including CpG islands. In males, the CG density alone indicates broad replicating regions, containing multiple origins of replication at high density. Together with chromosome size, these can sufficiently explain variation in crossover density between chromosomes (Pratto et al., 2021). Interphase chromosomes can be categorized into two distinct compartments using high-throughput chromatin conformation capture analyses (Hi-C). Actively transcribed chromatin is loosely packed, defined as A-compartment. While inactive heterochromatin is densely-packed and defined as B-compartment. Meiotic recombination shows strong bias towards A-compartments, which have dramatically increased DSBs and crossover densities (Jin, Fudenberg, & Pollard, 2021; Patel et al., 2019; Zuo et al., 2021). Early replicating regions are found in A-compartments, while late replicating regions most often occur in B-compartments (Pratto et al., 2021). A/B compartment identity is maintained at least from interphase throughout meiotic Prophase I (Alavattam et al., 2019; Luo et al., 2020; Patel et al., 2019; Vara et al., 2019; Wang et al., 2019; Zuo et al., 2021).

## 3. Chromatin loop array formation and loop extrusion

After replication, both a maternal and a paternal chromosome are present, these homologous chromosomes are located in distinct chromosomal territories. During meiosis, individualized homologous chromosomes have to come into contact, eventually pair and finally synapse for meiotic recombination to complete successfully. During Prophase I, extensive chromatin reorganization takes place, when chromatin is compacted into chromosomes. DNA is first organized into linear arrays and then helical arrays of brush-loops, which protrude from a central chromosomal axis of meiosis-specific Cohesin complexes (Llano et al., 2012; Pelttari et al., 2001). A 3D organizational structure, analogous to mitotic chromatin, where loops are fed through the ring-like Cohesins attached to the rod-like core affects spatial accessibility of chromatin, may also occur in meiosis.

The ring-shaped Cohesin-complexes contain structural maintenance-of-chromosome protein (SMC), a Kleisin-subunit (RAD21L or Rec8) and a Stromalin-subunit, such as STAG3, which is essential for proper axis formation in meiosis (Bhattacharyya et al., 2019; Winters, McNicoll, & Jessberger, 2014). In mice, loop sizes increase from 0.8–1 Mb to 1.5–2 Mb as Prophase progresses, suggesting that loops continue to be extruded from the condensing axis (Patel et al., 2019). While consistent physical loop sizes are most likely maintained at each stage, gene-dense A-compartment loops contain 3-time less base pairs than B-compartment loops – probably because of lower chromatin compaction (Jin et al., 2021). Chromosome-associated Cohesin complexes can extrude loops from the chromosomal axis, forming topologically associated domains (TADs), a dynamic organizational structure of interacting DNA sequences that constrains genic and regulatory regions. The 11-ZNF CTCF protein forms TAD boundaries by interacting with CTCF binding motifs (Luo et al., 2020; Oomen, Hansen, Liu, Darzacq, & Dekker, 2019; Vara et al., 2019). While CTCF proteins remain attached, the loop-axis configuration of TADs are lost throughout Pachytene, likely abolishing previously constrained promoter *cis*-regulatory interactions (Luo et al., 2020; Wang et al., 2019; Zuo et al., 2021). A and B compartment loops remain distinct but can associate with their homologous counterpart, where transcription clustering can occur between homologous loci (Patel et al., 2019).

## 4. Creating an accessible local chromatin environment

In some species, the meiotic recombination machinery utilizes already open chromatin, such as in functional elements and actively transcribed promoters. All mammals, with the exception of *Canidae,* possess the KRAB-zinc-finger-methyltransferase protein "PR-domain-containing 9" (PRDM9) (Baker et al., 2017), that directs recombination away from functional elements (Brick, Smagulova, Khil, Camerini-Otero, & Petukhova, 2012). PRDM9-guided recombination initiation designates hotspots by creating an accessible local chromatin environment and associating DNA in loops with DSB-promoting proteins located at the chromosomal axis. A number of different proteins are necessary to perform these steps outside of functional elements ( Fig. 1). PRDM9 possesses four functional domains, the N-terminal Kruppel-associated box (KRAB) domain,

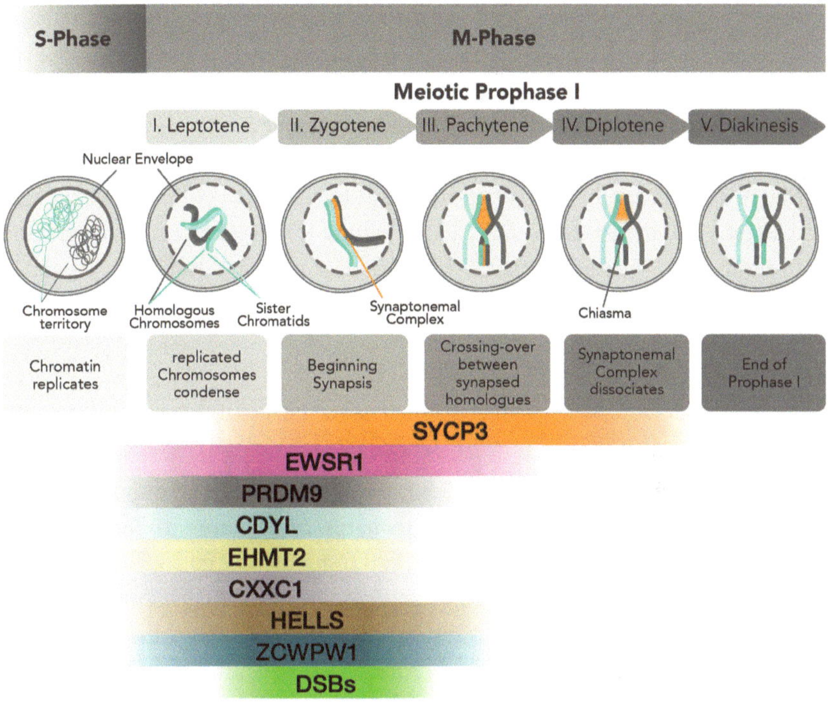


**Fig. 1** The Initiation of Meiosis and Meiotic Recombination (top) In the S-phase of Interphase parental chromatids are located far apart from each other in distinct chromosomal territories, where they are replicated. After replication individualized homologous chromosomes, are formed by two identical sister chromatids from each parental progenitor connected by the centromere. In Leptotene meiotic recombination is initiated by double-stranded DNA breaks. Homologs have to find each other, and pair to then synapse in zygotene. Crossing-over between synapsed homologs in Pachytene creates physical tension, visible as Chiasma in Diplotene, when the synaptonemal complex dissociates in preparation for Diakinesis (bottom). A cartoon shows expression times of critical enzymes that regulate meiotic recombination initiation in mammals *via* hotspot designation, or by promoting DSB formation or by fostering homolog engagement necessary for efficient downstream DSB repair.

necessary for protein-protein interactions (Imai et al., 2017), (i) an SSX-repression-domain (SSXRD) with nuclear localization signal (ii) the central PR/SET domain has methyltransferase activity (Hayashi, Yoshida, & Matsui, 2005) for Lysine 4 (Wu et al., 2013)and Lysine 36 of Histone H3 (Diagouraga et al., 2018), and auto-methylation ability for its Lysine sequences (Koh-Stenta et al., 2017), finally (iii) the C-terminal, highly diverse zinc finger domain of Cystin$_2$Histidin$_2$-type that specifically binds DNA-motifs (Baudat et al., 2010; Brick et al., 2012; Grey et al., 2011; Myers et al., 2010). PRDM9 proteins can occur as multimers *in-vitro*, or when overexpressed in cell-culture (Altemose et al., 2017; C. L. Baker et al., 2015; Schwarz et al., 2019), however PRDM9 multimerization has not been confirmed *in vivo* (Diagouraga et al., 2018). PRDM9 is expressed in the testis and ovary (Sun et al., 2015) where it first localizes to the nucleus at Preleptotene, persisting until the end of Zygotene (Grey et al., 2017; Parvanov et al., 2017; Sun et al., 2015).

## 5. Chromatin opening meiosis-specific Pioneer Complex

Opening of chromatin is facilitated by HELLS, a member of the SNF-2 family of ATP-dependent chromatin-remodeling enzymes, which remove and reposition nucleosomes making chromatin locally accessible (Kasten, Clapier, & Cairns, 2011) (as depicted in Fig. 2B). HELLS is strongly co-expressed with PRDM9 independent of DNA-binding- suggesting that a complex is formed even before PRDM9 attaches to the DNA (Spruce et al., 2020) and promotes the stable anchoring of PRDM9 to its specific DNA motif (Imai et al., 2020). HELLS promotes efficient placement of characteristic histone modifications by PRDM9, which are Histone 3 trimethylation marks on lysine-4 (H3K4) and lysine-36 (H3K36) (C. L. Baker, Walker, Kajita, Petkov, & Paigen, 2014; Brick et al., 2012; Diagouraga et al., 2018; Powers et al., 2016) as shown in Fig. 2A. Here, HELLS facilitates 5-hydroxymethylcytosine (5hmc) enrichment near the PRDM9 binding motif, most likely by conversion of preexisting 5-methylcytosine (5mC) (Imai et al., 2020). Meiosis-specific Ewing sarcoma RNA binding protein 1 (EWSR1) binds to PRDM9 anchored to its DNA motifs forming PRDM9-EWSR1 complexes (Parvanov et al., 2017) affecting the efficiency of the SET domain of PRDM9 thus increasing the H3K4 trimethylation signal (Tian, Billings, & Petkov, 2021). During this time, PRDM9 also interacts directly with two other proteins, CDYL, and EHMT2, to form a PRDM9-CDYL-EHMT2 complex that persists until Leptonema when both CDYL and EHMT2 disappear from the nucleus (Parvanov et al., 2017) (fig. 4B).

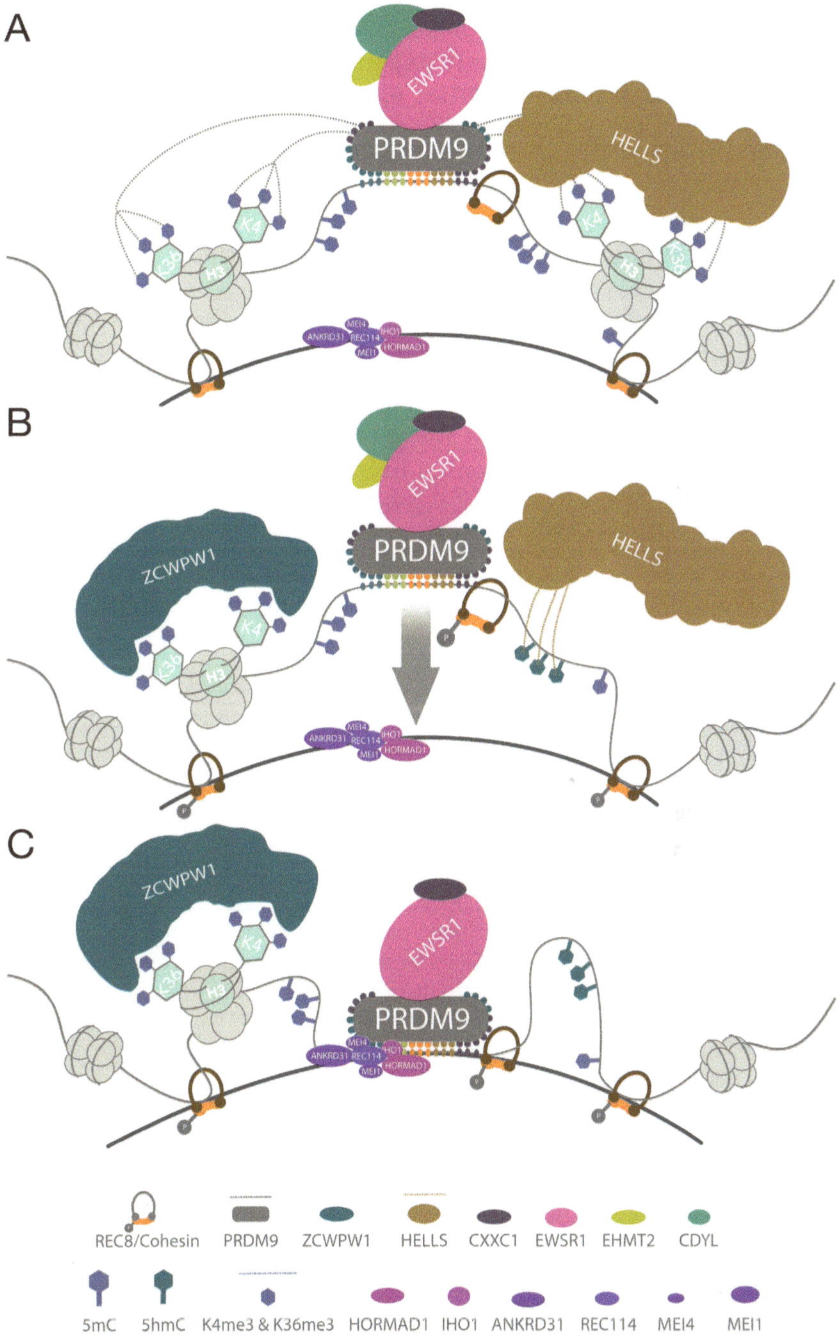


**Fig. 2** See figure legend on next page.

## 6. Associate putative hotspot sites in the DNA loops with the chromosomal axis

PRDM9-EWSR1 interactions were also observed with CXXC1 (Parvanov et al., 2017). The CXXC1 protein is an ortholog of *S. cerevisiae Spp1*. Yeast *Spp1* is a histone methyl reader that links DSB sites, located at promoters through its interaction with Mer2 and H3K4 trimethylation marks with DSB-promoting complexes located at the axis (Acquaviva et al., 2013; Sommermeyer, Beneut, Chaplais, Serrentino, & Borde, 2013). The importance for CXXC1 remains unclear in mouse meiosis, as conditional CXXC1 knockout mice on the same genetic backgrounds show conflicting results. While one study reports deems CXXC1 as essential, as knockouts are sterile, have significant decrease of hotspot trimethylation and severe meiotic defects (Jiang et al., 2020), another study reports normal H3K4 enrichment and healthy meiosis progression – deeming CXXC1 non-essential for mouse meiosis (Tian, Billings, & Petkov, 2018). The PRDM9-EWSR1-**COMPASS complex** interacts with phosphorylated REC8 (pREC8), as well as with synaptonemal complex proteins, thus it is likely that this complex in involved in translocating hotspot DNA to the chromosomal axis (fig. 4C) (Parvanov et al., 2017; Tian et al., 2021).

**Fig. 2** PRDM9 as mediator of the initiation stages of meiotic recombination during Leptotene (A) PRDM9 interacts with DNA, by recognizing specific sequence motifs by the zinc-finger array. Stable anchoring to the motif is facilitated by HELLS, which forms a pioneer complex with PRDM9. PRDM9 places tri-methylation marks on both Lysine-4 and lysine-36 on Histones H3 on two to four adjacent nucleosomes. This process is promoted by HELLS but also EWSR1 binding to PRDM9 (B) HELLS facilitates the remodeling of nucleosomes, forming a nucleosome depleted region, and converts pre-methylated cytosines (5mC) to 5-hydroxymethylcytosine (5hmC) in close proximity to the PRDM9 binding motif. The characteristic epigenetic signature placed by PRDM9 is recognized by histone-methyl reader ZCWPW1, which binds to nucleosomes with high affinity. The EWSR1-PRDM9 Complex is joined by CDYL and EHMT2 which then act together to bring the motif to the chromosomal axis. (C) At the chromosomal axis, PRDM9 interacts with cohesin, facilitating localization with important DSB-promoting proteins MEI4, MEI1, HORMAD1, IHO1, REC114 and ANKRD31 for DSB break formation. DSB-promoting protein ANKRD31 as well as methyl-reader ZCWPW1 also facilitate efficient homology search and DSB processing.

## 6.1 DSB formation

PRDM9 interacts with three Cohesin subunits, REC8, STAG3 and SMC3 at the chromosomal axis *in vivo* (Bhattacharyya et al., 2019). Important DSB-promoting proteins are located at the axis including Rec114 (Kumar et al., 2018), MEI1 and MEI4 (ortholog of *S. cerevisiae* meiosis 4) (Kumar et al., 2015), as well as HORMAD1 (HORMA Domain Containing 1) (Daniel et al., 2011) and IHO1 (Interactor Of HORMAD1) (Stanzione et al., 2016) and ANKRD31 (Papanikos et al., 2019). SPO11-mediated double-stranded DNA break activity is promoted particularly at PRDM9-dependent DSB sites by ANKRD31 (Boekhout et al., 2019; Papanikos et al., 2019).

The DSB repair machinery is recruited after PRDM9 has "written" H3K4me3/H3K36me3 double-positive marks on histones, to which the co-expressed "reader" ZCWPW1 was then bound (Wells et al., 2020). Functionally, ZCWPW1 consists of two domains i) CW with H3K4me3 binding capacity (He et al., 2010) and ii) PWWP with H3K36me3 binding capacity (Rona, Eleutherio, & Pinheiro, 2016). In humans and mice, ZCWPW1 binds H3K4me3/H3K36me3 double-positive sites with higher affinity than H3K4me3 present in open chromatin at promoter regions and other functional elements (Wells et al., 2020). Thus, ZCWPW1 reads epigenetic marks written by PRDM9 as part of the PRDM9/ ZCWPW1 Reader-Writer-System (Fig. 2B).

After DSBs begin to form in pre-Leptotene, soluble unbound PRDM9 disappears first, followed by the PRDM9 proteins associated with the axis (Parvanov et al., 2017). PRDM9 remains detectable on chromatin until the end of Zygotene (Grey et al., 2017; Sun et al., 2015). At PRDM9 bound hotspots, a SPO11-bound meiotic recombination intermediate (SPO11-RI) is present, presumably because persisting PRDM9 blocks DSB repair nuclease MRE11 from releasing SPO11 asymmetrically (Paiano et al., 2020).

# 7. Facilitating efficient homology engagement and DSB repair

After SPO11 is released, DMC1 binds to single-stranded overhangs and the recombinases DMC1 and RAD51 conduct homology search. Homology search is also aided by ZCWPW1, which not only reads PRDM9-dependent epigenetic trimethylation marks, but guides the DSB repair machinery towards these sites (as shown in Fig. 2C). DSB-formation is not impaired in *Zcwpw1*$^{-/-}$ male mice, but DSBs are repaired inefficiently and cells arrest at the zygotene to pachytene stage, with persisting DMC1 foci and high levels of asynapsis (Huang et al., 2020; Li et al., 2019; Mahgoub et al., 2020). Not all DSBs are repaired in a homolog-templated manner – particularly very early DSBs linked to early replication timing are apparently repaired using sister-chromatid templated repair (Pratto et al., 2021). DSBs that are fastest to engage with their homolog are more likely to be repaired as crossovers, suggesting that efficient homology search is an essential mechanism in meiotic recombination regulation (Hinch et al., 2019). Homologs are quicker to engage if the DSB is near the telomere and if the local GC content is high (Hinch et al., 2019). Symmetric PRDM9-binding sites on both homologs apparently also facilitates homology search, as more interhomolog-templated repair and a higher frequency of COs are seen (Hinch et al., 2019; Li et al., 2019). Until late pachytene, most breaks are repaired, and DMC1 levels remain elevated only at sites with obstructed repair (Chen et al., 2020; Davies et al., 2016; Hinch et al., 2019; Li, Bitoun, et al., 2019), such DSBs might instead be repaired as NCO (Chen et al., 2020; Wells et al., 2020). After SPO11 DSBs are formed, and PRDM9 has disappeared, EWSR1 expression increases dramatically (Parvanov et al., 2017), and EWSR1 physically facilitates the completion of genetic recombination through interactions with pREC8-positive meiosis-specific Cohesin complexes, participating further in the formation and resolution of holiday junctions (Parvanov et al., 2017; Tian et al., 2021).

## 8. PRDM9 redundant meiotic recombination initiation

PRDM9 function likely evolved early in metazoans but was also and has been lost multiple times within vertebrates (Baker et al., 2017). Organisms that lack PRDM9 also lack ZCWPW1 and its paralog ZCWPW2 (Cavassim et al., 2022), and at least dogs and birds successfully utilize existing H3K4-trimethylation marks in functional elements ads hotspot designation signal (Berglund, Quilez, Arndt, & Webster, 2015; Singhal et al., 2015). Snakes apparently also place meiotic DSBs in functional elements despite the presence of a complete genomic *Prdm9* sequence, though snake PRDM9 functionality remains unknown

(Schield et al., 2020). In these PRDM9-deficient species, historical recombination appears static. Mice genetically engineered to lack functional PRDM9 show DSBs formation in functional elements with default open chromatin (Brick et al., 2012). Two PRDM9 deficient mouse strains exist, while *Mus musculus domesticus* strain C57|Bl6/J is infertile (Brick et al., 2012), the *Mus musculus* PWD/Ph strain is fertile and apparently able to repair DSBs in functional elements (Mihola et al., 2019).

## 9. Summary

The *de-facto* initiation signal for meiotic recombination are programmed double-stranded DNA breaks catalyzed by the SPO11/TOPOVIBL complex in mammals (Robert, Vrielynck, Mezard, de Massy, & Grelon, 2016). Recent findings in meiosis research have highlighted critical stages occurring upstream and during mammalian DSB formation, that guide DSBs precisely to specific target sites in the genome and facilitate homolog engagement necessary for efficient DSB repair. Meiotic replication origin density, replication timing, and the size of individual chromosomes can also influences meiotic recombination (Pratto et al., 2021). Extensive reorganization of chromatin takes place in meiotic Prophase I, with chromatin loops elongating while maintaining compartment identity (Patel et al., 2019). A-compartments that are spatially accessible preferentially receive DSBs (Jin et al., 2021; Patel et al., 2019), but these DSBs may later be disfavored for crossover formation (Jin et al., 2021) and DSB placement in "default" open chromatin may reflect those seen in PRDM9-deficient organisms. In contrast, PRDM9-guided meiotic recombination, PRDM9 binding to specific DNA motifs outside of functional elements promotes recombination placement outside of functional elements. The SET domain of PRDM9 places characteristic H3K4 and H3K36-trimethylation marks aided by HELLS. HELLS also labels cytosines with Hydroxymethyl-groups and promotes local chromatin remodeling. PRDM9 furthermore interacts with EWSR1, CDYL, and EHMT2 and facilitates the translocation of initiation motifs in the DNA loops with the chromosomal axis, where DSB-promoting complexes are located and DSBs are catalyzed by SPO11/TOPOVIBL. The characteristic double-positive histone modification signal is recognized by ZCWPW1, which mediates homology search necessary for efficient DSB repair.

## References

Acquaviva, L., Székvölgyi, L. R. N., Dichtl, B., Dichtl, B. S., André, C. L. R. S., Nicolas, A., et al. (2013). The COMPASS subunit Spp1 links histone methylation to initiation of meiotic recombination. *Science, 339*, 215–218. https://doi.org/10.1126/science.1225739.

Alavattam, K. G., Maezawa, S., Sakashita, A., Khoury, H., Barski, A., Kaplan, N., et al. (2019). Attenuated chromatin compartmentalization in meiosis and its maturation in sperm development. *Nature Structural & Molecular Biology*, *26*(3), 175–184. https://doi.org/10.1038/s41594-019-0189-y.

Altemose, N., Noor, N., Bitoun, E., Tumian, A., Imbeault, M., Chapman, J. R., et al. (2017). A map of human PRDM9 binding provides evidence for novel behaviors of PRDM9 and other zinc-finger proteins in meiosis. *eLife, 6*. https://doi.org/10.7554/eLife.28383.

Baker, C. L., Petkova, P., Walker, M., Flachs, P., Mihola, O., Trachtulec, Z., et al. (2015). Multimer formation explains allelic suppression of PRDM9 recombination hotspots. *PLoS Genetics*, *11*(9), e1005512. https://doi.org/10.1371/journal.pgen.1005512.

Baker, Z., Schumer, M., Haba, Y., Bashkirova, L., Holland, C., Rosenthal, G. G., et al. (2017). Repeated losses of PRDM9-directed recombination despite the conservation of PRDM9 across vertebrates. *eLife, 6*. https://doi.org/10.7554/eLife.24133.

Baker, C. L., Walker, M., Kajita, S., Petkov, P. M., & Paigen, K. (2014). PRDM9 binding organizes hotspot nucleosomes and limits Holliday junction migration. *Genome Research*, *24*(5), 724–732. https://doi.org/10.1101/gr.170167.113.

Baudat, F., Buard, J., Grey, C., Fledel-Alon, A., Ober, C., Przeworski, M., et al. (2010). PRDM9 is a major determinant of meiotic recombination hotspots in humans and mice. *Science*, *327*(5967), 836–840. https://doi.org/10.1126/science.1183439.

Berglund, J., Quilez, J., Arndt, P. F., & Webster, M. T. (2015). Germline methylation patterns determine the distribution of recombination events in the dog genome. *Genome Biology and Evolution*, 7(2), 522–530. https://doi.org/10.1093/gbe/evu282.

Bhattacharyya, T., Walker, M., Powers, N. R., Brunton, C., Fine, A. D., Petkov, P. M., et al. (2019). Prdm9 and meiotic Cohesin proteins cooperatively promote DNA double-Strand break formation in mammalian spermatocytes. *Current Biology*, *29*(6), 1002–1018 e1007. https://doi.org/10.1016/j.cub.2019.02.007.

Boekhout, M., Karasu, M. E., Wang, J., Acquaviva, L., Pratto, F., Brick, K., et al. (2019). REC114 partner ANKRD31 controls number, timing, and location of meiotic DNA breaks. *Molecular Cell*, *74*(5), 1053–1068 e1058. https://doi.org/10.1016/j.molcel.2019.03.023.

Brick, K., Smagulova, F., Khil, P., Camerini-Otero, R. D., & Petukhova, G. V. (2012). Genetic recombination is directed away from functional genomic elements in mice. *Nature*, *485*(7400), 642–645. https://doi.org/10.1038/nature11089.

Cavassim, M. I. A., Baker, Z., Hoge, C., Schierup, M. H., Schumer, M., & Przeworski, M. (2022). PRDM9 losses in vertebrates are coupled to those of paralogs ZCWPW1 and ZCWPW2. *Proceedings of the National Academy of Sciences of the United States of America*, *119*(9). https://doi.org/10.1073/pnas.2114401119.

Chen, Y., Lyu, R., Rong, B., Zheng, Y., Lin, Z., Dai, R., et al. (2020). Refined spatial temporal epigenomic profiling reveals intrinsic connection between PRDM9-mediated H3K4me3 and the fate of double-stranded breaks. *Cell Research*, *30*(3), 256–268. https://doi.org/10.1038/s41422-020-0281-1.

Daniel, K., Lange, J., Hached, K., Fu, J., Anastassiadis, K., Roig, I., et al. (2011). Meiotic homologue alignment and its quality surveillance are controlled by mouse HORMAD1. *Nature Cell Biology*, *13*(5), 599–610. https://doi.org/10.1038/ncb2213.

Davies, B., Hatton, E., Altemose, N., Hussin, J. G., Pratto, F., Zhang, G., et al. (2016). Re-engineering the zinc fingers of PRDM9 reverses hybrid sterility in mice. *Nature, 530*(7589), 171–176. https://doi.org/10.1038/nature16931.

Diagouraga, B., Clement, J. A. J., Duret, L., Kadlec, J., de Massy, B., & Baudat, F. (2018). PRDM9 methyltransferase activity is essential for meiotic DNA double-Strand break formation at its binding sites. *Molecular Cell, 69*(5), 853–865 e856. https://doi.org/10.1016/j.molcel.2018.01.033.

Grey, C., Barthes, P., Chauveau-Le Friec, G., Langa, F., Baudat, F., & de Massy, B. (2011). Mouse PRDM9 DNA-binding specificity determines sites of histone H3 lysine 4 trimethylation for initiation of meiotic recombination. *PLoS Biology, 9*(10), e1001176. https://doi.org/10.1371/journal.pbio.1001176.

Grey, C., Clement, J. A., Buard, J., Leblanc, B., Gut, I., Gut, M., et al. (2017). In vivo binding of PRDM9 reveals interactions with noncanonical genomic sites. *Genome Research, 27*(4), 580–590. https://doi.org/10.1101/gr.217240.116.

Hayashi, K., Yoshida, K., & Matsui, Y. (2005). A histone H3 methyltransferase controls epigenetic events required for meiotic prophase. *Nature, 438*(7066), 374–378. https://doi.org/10.1038/nature04112.

He, F., Umehara, T., Saito, K., Harada, T., Watanabe, S., Yabuki, T., et al. (2010). Structural insight into the zinc finger CW domain as a histone modification reader. *Structure, 18*(9), 1127–1139. https://doi.org/10.1016/j.str.2010.06.012.

Hinch, A. G., Zhang, G., Becker, P. W., Moralli, D., Hinch, R., Davies, B., et al. (2019). Factors influencing meiotic recombination revealed by whole-genome sequencing of single sperm. *Science, 363*(6433). https://doi.org/10.1126/science.aau8861.

Huang, T., Yuan, S., Gao, L., Li, M., Yu, X., Zhang, J., et al. (2020). The histone modification reader ZCWPW1 links histone methylation to PRDM9-induced double-strand break repair. *eLife, 9*. https://doi.org/10.7554/eLife.53459.

Imai, Y., Baudat, F., Taillepierre, M., Stanzione, M., Toth, A., & de Massy, B. (2017). The PRDM9 KRAB domain is required for meiosis and involved in protein interactions. *Chromosoma, 126*(6), 681–695. https://doi.org/10.1007/s00412-017-0631-z.

Imai, Y., Biot, M., Clement, J. A., Teragaki, M., Urbach, S., Robert, T., et al. (2020). PRDM9 activity depends on HELLS and promotes local 5-hydroxymethylcytosine enrichment. *eLife, 9*. https://doi.org/10.7554/eLife.57117.

Jiang, Y., Zhang, H. Y., Lin, Z., Zhu, Y. Z., Yu, C., Sha, Q. Q., et al. (2020). CXXC finger protein 1-mediated histone H3 lysine-4 trimethylation is essential for proper meiotic crossover formation in mice. *Development, 147*(6). https://doi.org/10.1242/dev.183764.

Jin, X., Fudenberg, G., & Pollard, K. S. (2021). Genome-wide variability in recombination activity is associated with meiotic chromatin organization. *Genome Research, 31*(9), 1561–1572. https://doi.org/10.1101/gr.275358.121.

Kasten, M. M., Clapier, C. R., & Cairns, B. R. (2011). SnapShot: Chromatin remodeling: SWI/SNF. *Cell, 144*(2), 310 e311. https://doi.org/10.1016/j.cell.2011.01.007.

Keeney, S., Baudat, F., Angeles, M., Zhou, Z.-H., Copeland, N. G., Jenkins, N. A., et al. (1999). A Mouse Homolog of the *Saccharomyces cerevisiae* Meiotic Recombination DNA Transesterase Spo11p. *Genomics, 61*, 170–182.

Koh-Stenta, X., Poulsen, A., Li, R., Wee, J. L., Kwek, P. Z., Chew, S. Y., et al. (2017). Discovery and characterisation of the automethylation properties of PRDM9. *The Biochemical Journal, 474*(6), 971–982. https://doi.org/10.1042/BCJ20161067.

Kumar, R., Ghyselinck, N., Ishiguro, K., Watanabe, Y., Kouznetsova, A., Hoog, C., et al. (2015). MEI4 - a central player in the regulation of meiotic DNA double-strand break formation in the mouse. *Journal of Cell Science, 128*(9), 1800–1811. https://doi.org/10.1242/jcs.165464.

Kumar, R., Oliver, C., Brun, C., Juarez-Martinez, A. B., Tarabay, Y., Kadlec, J., et al. (2018). Mouse REC114 is essential for meiotic DNA double-strand break formation and forms a complex with MEI4. *Life Sci Alliance*, *1*(6), e201800259. https://doi.org/10.26508/lsa.201800259.

Li, R., Bitoun, E., Altemose, N., Davies, R. W., Davies, B., & Myers, S. R. (2019). A high-resolution map of non-crossover events reveals impacts of genetic diversity on mammalian meiotic recombination. *Nature Communications*, *10*(1), 3900. https://doi.org/10.1038/s41467-019-11675-y.

Li, M., Huang, T., Li, M.-J., Zhang, C.-X., Yu, X.-C., Yin, Y.-Y., et al. (2019). The histone modification reader ZCWPW1 is required for meiosis prophase I in male but not in female mice. *Science Advances*, *5*(eaax1101), 13. https://doi.org/10.1126/sciadv.aax1101.

Llano, E., Herran, Y., Garcia-Tunon, I., Gutierrez-Caballero, C., de Alava, E., Barbero, J. L., et al. (2012). Meiotic cohesin complexes are essential for the formation of the axial element in mice. *The Journal of Cell Biology*, *197*(7), 877–885. https://doi.org/10.1083/jcb.201201100.

Luo, Z., Wang, X., Jiang, H., Wang, R., Chen, J., Chen, Y., et al. (2020). Reorganized 3D genome structures support transcriptional regulation in mouse spermatogenesis. *iScience*, *23*(4), 101034. https://doi.org/10.1016/j.isci.2020.101034.

Mahgoub, M., Paiano, J., Bruno, M., Wu, W., Pathuri, S., Zhang, X., et al. (2020). Dual histone methyl reader ZCWPW1 facilitates repair of meiotic double strand breaks in male mice. *eLife*, *9*. https://doi.org/10.7554/eLife.53360.

Metzler-Guillemain, C., & Massy, B. D. (1999). Identification and characterization of an SPO11 homolog in the mouse. *Chromosoma*, *109*, 133–138.

Mihola, O., Pratto, F., Brick, K., Linhartova, E., Kobets, T., Flachs, P., et al. (2019). Histone methyltransferase PRDM9 is not essential for meiosis in male mice. *Genome Research*, *29*(7), 1078–1086. https://doi.org/10.1101/gr.244426.118.

Myers, S., Bowden, R., Tumian, A., Bontrop, R. E., Freeman, C., MacFie, T. S., et al. (2010). Drive against hotspot motifs in primates implicates the PRDM9 gene in meiotic recombination. *Science*, *327*(5967), 876–879. https://doi.org/10.1126/science.1182363.

Oomen, M. E., Hansen, A. S., Liu, Y., Darzacq, X., & Dekker, J. (2019). CTCF sites display cell cycle-dependent dynamics in factor binding and nucleosome positioning. *Genome Research*, *29*(2), 236–249. https://doi.org/10.1101/gr.241547.118.

Paiano, J., Wu, W., Yamada, S., Sciascia, N., Callen, E., Paola Cotrim, A., et al. (2020). ATM and PRDM9 regulate SPO11-bound recombination intermediates during meiosis. *Nature Communications*, *11*(1), 857. https://doi.org/10.1038/s41467-020-14654-w.

Papanikos, F., Clement, J. A. J., Testa, E., Ravindranathan, R., Grey, C., Dereli, I., et al. (2019). Mouse ANKRD31 regulates spatiotemporal patterning of meiotic recombination initiation and ensures recombination between X and Y sex chromosomes. *Molecular Cell*, *74*(5), 1069–1085 e1011. https://doi.org/10.1016/j.molcel.2019.03.022.

Parvanov, E. D., Tian, H., Billings, T., Saxl, R. L., Spruce, C., Aithal, R., et al. (2017). PRDM9 interactions with other proteins provide a link between recombination hotspots and the chromosomal axis in meiosis. *Molecular Biology of the Cell*, *28*(3), 488–499. https://doi.org/10.1091/mbc.E16-09-0686.

Patel, L., Kang, R., Rosenberg, S. C., Qiu, Y., Raviram, R., Chee, S., et al. (2019). Dynamic reorganization of the genome shapes the recombination landscape in meiotic prophase. *Nature Structural & Molecular Biology*. https://doi.org/10.1038/s41594-019-0187-0.

Pelttari, J., Hoja, M. R., Yuan, L., Liu, J. G., Brundell, E., Moens, P., et al. (2001). A meiotic chromosomal core consisting of cohesin complex proteins recruits DNA recombination proteins and promotes synapsis in the absence of an axial element in mammalian meiotic cells. *Molecular and Cellular Biology*, *21*(16), 5667–5677. https://doi.org/10.1128/MCB.21.16.5667-5677.2001.

Powers, N. R., Parvanov, E. D., Baker, C. L., Walker, M., Petkov, P. M., & Paigen, K. (2016). The meiotic recombination activator PRDM9 Trimethylates both H3K36 and H3K4 at recombination hotspots in vivo. *PLoS Genetics*, *12*(6), e1006146. https://doi.org/10.1371/journal.pgen.1006146.

Pratto, F., Brick, K., Cheng, G., Lam, K. G., Cloutier, J. M., Dahiya, D., et al. (2021). Meiotic recombination mirrors patterns of germline replication in mice and humans. *Cell*, *184*(16), 4251–4267 e4220. https://doi.org/10.1016/j.cell.2021.06.025.

Robert, T., Vrielynck, N., Mezard, C., de Massy, B., & Grelon, M. (2016). A new light on the meiotic DSB catalytic complex. *Seminars in Cell & Developmental Biology*, *54*, 165–176. https://doi.org/10.1016/j.semcdb.2016.02.025.

Romanienko, P. J., & Camerini-Otero, R. D. (1999). Cloning, characterization, and localization of mouse and human SPO11. *Genomics*, *61*, 156–169.

Rona, G. B., Eleutherio, E. C. A., & Pinheiro, A. S. (2016). PWWP domains and their modes of sensing DNA and histone methylated lysines. *Biophysical Reviews*, *8*(1), 63–74. https://doi.org/10.1007/s12551-015-0190-6.

Schield, D. R., Pasquesi, G. I. M., Perry, B. W., Adams, R. H., Nikolakis, Z. L., Westfall, A. K., et al. (2020). Snake recombination landscapes are concentrated in functional regions despite PRDM9. *Molecular Biology and Evolution*, *37*(5), 1272–1294. https://doi.org/10.1093/molbev/msaa003.

Schwarz, T., Striedner, Y., Horner, A., Haase, K., Kemptner, J., Zeppezauer, N., et al. (2019). PRDM9 forms a trimer by interactions within the zinc finger array. *Life Sci Alliance*, *2*(4). https://doi.org/10.26508/lsa.201800291.

Singhal, S., Leffler, E., Sannareddy, K., Turner, I., Venn, O., Hooper, D., et al. (2015). *Stable recombination hotspots in birds*. https://doi.org/10.1101/023101.

Sommermeyer, V., Beneut, C., Chaplais, E., Serrentino, M. E., & Borde, V. (2013). Spp1, a member of the Set1 complex, promotes meiotic DSB formation in promoters by tethering histone H3K4 methylation sites to chromosome axes. *Molecular Cell*, *49*(1), 43–54. https://doi.org/10.1016/j.molcel.2012.11.008.

Spruce, C., Dlamini, S., Ananda, G., Bronkema, N., Tian, H., Paigen, K., et al. (2020). HELLS and PRDM9 form a pioneer complex to open chromatin at meiotic recombination hot spots. *Genes & Development*, *34*(5–6), 398–412. https://doi.org/10.1101/gad.333542.119.

Stanzione, M., Baumann, M., Papanikos, F., Dereli, I., Lange, J., Ramlal, A., et al. (2016). Meiotic DNA break formation requires the unsynapsed chromosome axis-binding protein IHO1 (CCDC36) in mice. *Nature Cell Biology*, *18*(11), 1208–1220. https://doi.org/10.1038/ncb3417.

Stapley, J., Feulner, P. G. D., Johnston, S. E., Santure, A. W., & Smadja, C. M. (2017). Variation in recombination frequency and distribution across eukaryotes: Patterns and processes. *Philosophical Transactions of the Royal Society of London. Series B, Biological Sciences*, *372*(1736). https://doi.org/10.1098/rstb.2016.0455.

Sun, F., Fujiwara, Y., Reinholdt, L. G., Hu, J., Saxl, R. L., Baker, C. L., et al. (2015). Nuclear localization of PRDM9 and its role in meiotic chromatin modifications and homologous synapsis. *Chromosoma*, *124*(3), 397–415. https://doi.org/10.1007/s00412-015-0511-3.

Tian, H., Billings, T., & Petkov, P. M. (2018). CXXC1 is not essential for normal DNA double-strand break formation and meiotic recombination in mouse. *PLOS Genetics*, *14*(10). https://doi.org/10.1371/journal.pgen.1007657. 2. https://doi.org/10.1371/journal.pgen.1007657; (2018).

Tian, H., Billings, T., & Petkov, P. M. (2021). EWSR1 affects PRDM9-dependent histone 3 methylation and provides a link between recombination hotspots and the chromosome axis protein REC8. *Molecular Biology of the Cell*, *32*(1), 1–14. https://doi.org/10.1091/mbc.E20-09-0604.

Vara, C., Paytuvi-Gallart, A., Cuartero, Y., Le Dily, F., Garcia, F., Salva-Castro, J., et al. (2019). Three-dimensional genomic structure and Cohesin occupancy correlate with transcriptional activity during spermatogenesis. *Cell Reports*, *28*(2), 352–367 e359. https://doi.org/10.1016/j.celrep.2019.06.037.

Wang, Y., Wang, H., Zhang, Y., Du, Z., Si, W., Fan, S., et al. (2019). Reprogramming of meiotic chromatin architecture during spermatogenesis. *Molecular Cell*, *73*(3), 547–561 e546. https://doi.org/10.1016/j.molcel.2018.11.019.

Wells, D., Bitoun, E., Moralli, D., Zhang, G., Hinch, A., Jankowska, J., et al. (2020). ZCWPW1 is recruited to recombination hotspots by PRDM9 and is essential for meiotic double strand break repair. *eLife*, *9*. https://doi.org/10.7554/eLife.53392.

Winters, T., McNicoll, F., & Jessberger, R. (2014). Meiotic cohesin STAG3 is required for chromosome axis formation and sister chromatid cohesion. *The EMBO Journal*, *33*(11), 1256–1270. https://doi.org/10.1002/embj.201387330.

Wu, H., Mathioudakis, N., Diagouraga, B., Dong, A., Dombrovski, L., Baudat, F., et al. (2013). Molecular basis for the regulation of the H3K4 methyltransferase activity of PRDM9. *Cell Reports*, *5*(1), 13–20. https://doi.org/10.1016/j.celrep.2013.08.035.

Zuo, W., Chen, G., Gao, Z., Li, S., Chen, Y., Huang, C., et al. (2021). Stage-resolved hi-C analyses reveal meiotic chromosome organizational features influencing homolog alignment. *Nature Communications*, *12*(1), 5827. https://doi.org/10.1038/s41467-021-26033-0.

CHAPTER THREE

# Cancer and meiotic gene expression: Two sides of the same coin?

**Ieng Fong Sou[a,b], Geert Hamer[c], Wee-Wei Tee[b,d], Gerben Vader[c,e,f], and Urszula Lucja McClurg[a,*]**

[a]Institute of Systems, Molecular and Integrative Biology, University of Liverpool, Liverpool, United Kingdom
[b]Chromatin Dynamics and Disease Epigenetics Group, Institute of Molecular and Cell Biology, Agency for Science, Technology and Research (A*STAR), Singapore, Singapore
[c]Center for Reproductive Medicine, Reproductive Biology Laboratory, Amsterdam Reproduction and Development Research Institute, Amsterdam UMC, University of Amsterdam, Amsterdam, The Netherlands
[d]Department of Physiology, Yong Loo Lin School of Medicine, National University of Singapore, Singapore, Singapore
[e]Section of Oncogenetics, Department of Human Genetics, Amsterdam UMC, Vrije Universiteit, Amsterdam, The Netherlands
[f]Cancer Center Amsterdam, Cancer Biology and Immunology, Amsterdam, The Netherlands
*Corresponding author: e-mail address: urszula.mcclurg@liverpool.ac.uk

## Contents



## Abstract

Meiosis increases genetic diversity in offspring by generating genetically unique haploid gametes with reshuffled chromosomes. This process requires a specialized set of meiotic proteins, which facilitate chromosome recombination and segregation. However, re-expression of meiotic proteins in mitosis can have catastrophic oncogenic consequences and aberrant expression of meiotic proteins is a common occurrence in human tumors. Mechanistically, re-activation of meiotic genes in cancer promotes oncogenesis

*Current Topics in Developmental Biology*, Volume 151
ISSN 0070-2153
https://doi.org/10.1016/bs.ctdb.2022.06.002

likely because cancers—conversely to healthy mitosis—are fueled by genetic instability which promotes tumor evolution, and evasion of immune response and treatment pressure. In this review, we explore similarities between meiotic and cancer cells with a particular focus on the oncogenic activation of meiotic genes in cancer. We emphasize the role of histones and their modifications, DNA methylation, genome organization, R-loops and the availability of distal enhancers.

## 1. Introduction

The ability to replicate DNA and divide replicated genomes is essential for life. Mitosis is the somatic division where diploid cells faithfully replicate and segregate DNA to generate two identical daughter progenies. Alternatively, in meiosis, diploid cells produce four genetically diverse haploid gametes. This involves introducing numerous, highly controlled DNA double-strand breaks to recombine and segregate homologous chromosomes. However, in mitosis both aneuploidy and chromosomal recombination events can have catastrophic, oncogenic outcomes (Helleday, 2010). Consequently, for healthy somatic cell divisions it is critical that meiotic genes are accurately and specifically silenced following the transition of haploid reproductive cells into diploid mitotic cell cycles after fertilization (Sou et al., 2021). However, failure of meiotic gene silencing is commonly reported in cancer (Hosoya & Miyagawa, 2021; Hosoya, Ono, & Miyagawa, 2018; Hosoya et al., 2011; Masterson et al., 2015; Medrano et al., 2016; Mobasheri, Shirkoohi, & Modarressi, 2016; Neumann et al., 2005; Sandhu et al., 2019, 2021) where the majority of cancer patients present with global reactivation of the meiotic transcriptome and proteome (Bruggeman et al., 2018, 2020). In this review we will explore the similarities between meiotic and cancer chromatin which might promote reactivation of meiotic genes in cancer.

We will discuss gene expression regulation at multiple levels including (i) DNA methylation (ii) histone modifications and DNA accessibility (iii) higher-order chromatin structure. Chromatin structure includes resolution from Mbp to bp and is organized by chromosomal territories, "open" (A) and "closed" (B) compartments inside chromosomal territories, topologically associated domains (TADs) and DNA-looping interactions in that order of magnitude. These modes of expression regulation play a critical role in meiotic gene activation and during oncogenesis (Fig. 1).

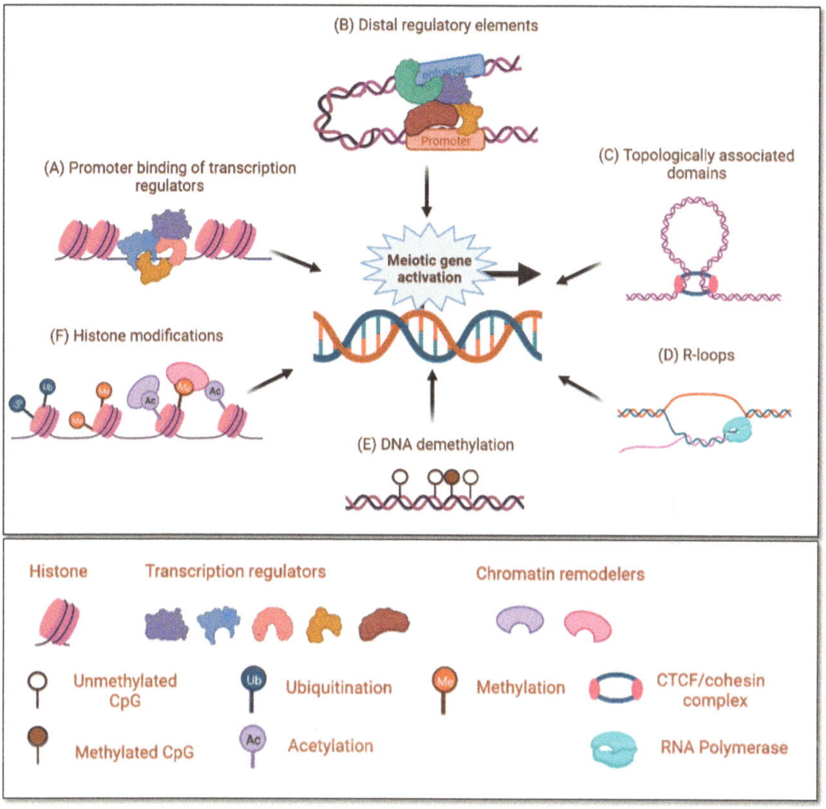


**Fig. 1** Similarities between DNA and chromatin during physiological meiotic gene activation and their aberrant activation in cancer. In this review we discuss a range of parallels between meiotic and cancer chromatin which are likely to contribute to the meiotic gene re-activation such as; (A) Large scale DNA demethylation induced by specific demethylases such as TETs characterizes cancer and meiotic DNA. (B) Histone modifications such as acetylation, methylation, and alternations in bivalency within the meiotic gene promoter regions that contain active (H3K4me3) and repressive (H3K27me3) marks can make meiotic and tumor chromatin akin. (C) Cancers lack many meiotic transcription factors and it remains to be understood what are meiotic promoter binding proteins in cancer and how similar are they to the meiotic machinery? (D) At a higher level of genome compaction large topologically associated domains (TADs) become the units of chromatin and a position of the gene within the TADs can affect its expression as well as promoter-distal enhancer interactions (E). (E) higher genome architecture such as promoter-enhancer interactions that control accessibility and consequently recruitment of transcription regulating machinery. Furthermore, the availability of promoter/enhancer binding proteins (C) is affected by a network of specific cofactors and modifying proteins including kinases, phosphatases, E3 ligases and deubiquitinating enzymes. Identification of the complete network of these meiotic expression master regulators and the distal enhancer regions essential for meiotic gene activation in cancer is needed. Both meiosis and cancer are characterized by disruption of TADs while it is unclear what role does the TAD positioning play in regulating the expression of meiotic genes? (F) Co-transcriptionally generated R-loops can be a major obstacle to the replication forks impacting gene expression. Both meiosis and cancer are characterized by high R-loop levels.

## 1.1 Role of meiotic genes in cancer

Despite the mechanisms of re-expression remaining an enigma, the roles of many meiotic genes in cancer have been well established and are covered by following reviews (Gantchev et al., 2020; Jay et al., 2021; Nielsen & Gjerstorff, 2016). Meiotic gene re-expression does not appear to lead to a complete conversion to a fully meiotic division in somatic cells. Rather it seems to interfere with error-free division as meiotic proteins serve functions that are detrimental outside of meiosis (Hosoya & Miyagawa, 2021; Hosoya et al., 2018, 2011; Masterson et al., 2015; Medrano et al., 2016; Mobasheri et al., 2016; Neumann et al., 2005; Sandhu et al., 2019, 2021). For instance, re-expression of synaptonemal complex (SC) proteins in cancer is frequent (Fig. 2) and has been reported to cause genome instability as a result of chromosome mis-segregation and thus aneuploidy (Collins et al., 1998; Shannon et al., 1999). When re-activated in cancer SC genes assume roles similar to those served in meiosis. They also exploit the same cellular niches. Previous findings from our lab demonstrated that SC protein TEX12 additionally localizes to meiotic microtubule organizing centres, similarly the aberrantly expressed TEX12 localizes to the cancer centrosomes leading to centrosome

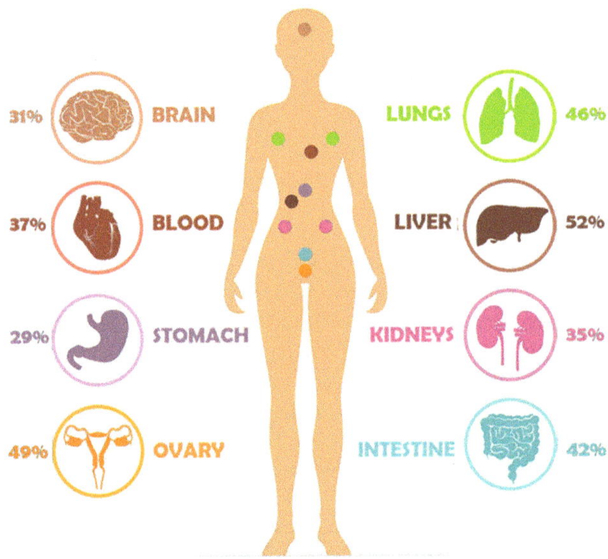


**Fig. 2** Percentage of cancer patients with re-activation of at least one meiotic synaptonemal complex gene. SC mRNA levels in patient samples from each cancer site which have been deposited to TCGA https://www.cancer.gov/tcga have been analyzed and percentage of patients re-expressing at least one SC gene has been determined.

amplification (Sandhu et al., 2021). Consequently, increased TEX12 levels correspond with poor prognosis (Sandhu et al., 2021). Furthermore, SC component SYCP3 has been shown to impair the BRCA2 modulated homologous recombination during DNA damage repair upon DSB leading to chromosomal instability (Hosoya et al., 2011), while a recent study showed that re-expression of SYCE2 in cancer alters ATM-mediated DNA damage response (Hosoya et al., 2018). Therefore, the re-expression of meiotic genes appears to be critical for cancer hallmarks responsible for both cancer initiation and maintenance. Moreover, it has previously been demonstrated that the pseudo-meiotic process can equip cancer cells with the ability to survive genotoxic treatments and thus leads to drug resistance (Erenpreisa et al., 2009). By virtue of their inherently restricted expression profiles in non-diseased somatic tissues, therapeutic targeting of such factors would be attractive with potentially limited side-effect restricted to infertility. Meiotic proteins thus potentially offer exciting novel prognostic biomarkers and drug targets in oncology. Hence, it is critical to understand the underlying mechanisms regulating meiotic gene re-expression in cancer.

## 2. Global DNA demethylation

DNA methylation is a chromatin remodeling mechanism where a methyl group from S-Adenosyl-Methionine (SAM) is added to the cytosine of CpG dinucleotides within the genome. A high level of methylated CpG islands, regions of the genome that contain a large number of CpG dinucleotide repeats, in a promoter sequence is commonly associated with transcriptional silencing (Fig. 1A). *De novo* DNA methylation patterns are established by DNA methyltransferases DNMT3A and DNMT3B during embryonic development (Hajkova et al., 2002; Okano et al., 1999). However, early during mammalian embryonic development a small group of cells, the so-called primordial germ cells (PGCs) are specified in the posterior epiblast (De Felici, 2016; Hamer & de Rooij, 2018). These PGCs migrate through the hindgut to the developing gonad, during which a global genome wide demethylation process starts. Upon arrival of the PGCs in the developing gonads, further demethylation occurs to enable either male or female imprinting and facilitate future gametogenesis (von Meyenn & Reik, 2015). In male mice this effect may be further enhanced by the difference between meiotic DNA replication compared to somatic cells in that during the meiotic S phase premeiotic DNA methylation levels are not restored in a timely manner (Gaysinskaya et al., 2018). Besides in PGCs, such massive global hypomethylation is only seen in tumors (Feinberg & Vogelstein, 1983a, 1983b; Gama-Sosa et al., 1983).

Germline gene expression can be induced through the oxidation of 5mC by Ten Eleven Translocation enzymes (TETs), in particular TET1 and 2 (Hackett et al., 2013). TETs are dioxygenases that add a hydroxyl group to 5-methylcytosine and give rise to several products such as 5-carboxylcytosine (5caC), 5-formylcytosine (5fC) and 5-hydroxymethylctyosine (5hmC) (Pastor, Aravind, & Rao, 2013). 5hmC is an intermediate toward DNA demethylation observed at high level in gonadal PGCs (Hackett et al., 2013). In mice, TET1 and 2 are expressed at E9.5–11.5, which is sufficient to induce imprint erasure and global DNA demethylation of germline genes (Hackett et al., 2013). Of note, TET1 is expressed predominantly in female PGCs and mediates DNA demethylation that drives the expression of a subset of meiotic prophase genes, including *SYCP1*, *SYCP3*, *PRDM9*, *STRA8* and *MAEl* (Yamaguchi et al., 2012). TET1 deficiency can cause significant reduction in female germ cells due to the absence of meiotic gene expression. However, it does not affect the development of male germ cells (Yamaguchi et al., 2012), most likely because male meiosis does not initiate during embryonic development. Instead, in a male gonad PGCs will become quiescent and meiosis will not start until the onset of puberty (Feng, Bowles, & Koopman, 2014; Hamer & de Rooij, 2018; Vergouwen et al., 1991). Interestingly, in gonadal PGCs, TET1 is reported to safeguard instead of acting as a driver of global DNA demethylation and germline demethylation can still proceed in the absence of TET1. TET1 is also responsible for removing *de novo*, or residual, DNA methylation (Hill et al., 2018). It is thus reasonable to propose that TET1-driven 5hmC protects against DNA methylation and primes germline genes for transcriptional activation. TETs play equally critical roles in the soma with malfunctions of TET family proteins responsible for a wide range of hematological cancers and solid tumors (Rasmussen & Helin, 2016).

Global DNA demethylation is also associated with DNA repair mechanisms. The base excision repair (BER) components APE1, PARP1 and XCRR1, but not nucleotide excision repair (NER), are enriched in PGCs at E10.5–12.5 coinciding with the large scale DNA demethylation (Hajkova et al., 2010). Unmodified cytosine can either be restored by the dilution of modified cytosine (5mC, 5hmC, 5fC and 5caC) over DNA replication, or through BER, which involves mismatch or damaged base repair catalyzed by DNA glycosylases. 5hmC can be recognized by the BER machinery during global DNA demethylation. Furthermore, high PARP1 expression correlates with global DNA demethylation in mouse

PGCs where histone demethylase UTX plays a role in safeguarding demethylation of H3K27me3 in PGCs at E10.5–11 (Mansour et al., 2012). Altered methylation has also been shown to contribute to cancer initiation by changing the borders of normally conserved undermethylated "canyon" regions, leading to demethylation of promoters and gene bodies and subsequent oncogene expression (Su et al., 2018).

It will be important to consider the similarities in the role of global demethylation between cancer and meiosis. They could be simply an outcome of the wide-scale structural remodeling that occurs in both processes but it remains a possibility that the permissive conditions generated by demethylation are critical to facilitate recombination events in meiosis and oncogenesis. Furthermore, demethylated chromatin allows for addition of new epigenetic marks, potentially facilitating asymmetric modifications that could be directing meiotic division and loss of heterozygosity common in cancer.

## 3. Histone modifications

Gene expression is partly regulated by chromatin architecture which, in turn, is largely determined by wrapping of DNA around nucleosomes that are built of histone molecules. In addition to CpG dinucleotides histone methylation plays a big role in chromatin organization and gene expression (Black, Van Rechem, & Whetstine, 2012) (Fig. 1B). To facilitate global changes to the histone methylation pattern, two histone methyltransferases (HMTs), G9a and GL9, form a complex associated with catalyzing histone 3 methylation on lysine 9 (forming H3K9me2/3), which is involved in the transcriptional silencing of developmental genes (Tachibana et al., 2002). Migratory PGCs downregulate G9a, resulting in global loss of H3K9me2, which contributes to global DNA demethylation. In addition, TET1 can drive transcription in a CpG demethylation independent manner by recruiting O-GlcNAc transferase OGT and SER1/COMPASS to promoters, leading to deposition of H3K4me3 and thus transcriptional activation (Deplus et al., 2013).

Deposition of specific histones on the DNA is facilitated by histone chaperones, which thereby direct cell type specific chromatin states in a context dependent manner (Hammond et al., 2017). Chromatin restructuring by reshuffling of repressive histone marks and histone variants is crucial for germline gene expression (Sou et al., 2021). Recent studies indicate that histone displacements establish adapted chromatin states and histone modifications prior to meiotic entry. Histone replacement occurs actively and specifically in PGCs prior to meiotic entry (Hajkova et al., 2008). Upon BER-driven demethylation, histone variants are displaced by the histone chaperones HIRA and NAP1 to facilitate chromatin rearrangement (Hajkova et al., 2008). NAP1 has a role in nucleosome disassembly *in vivo* as it disassembles and removes linker histone H1 in PGCs at E11.5 (Walfridsson et al., 2007), followed by the extraction of histone H2A.Z in PGCs at E11.5–12.5. At the same time the repressive marks H2A/H4R3me2s are lost. Furthermore, prior to H2A.Z disappearance, H3 modifications such as H3K9ac, H3K9me3 and H3K27me3 are diminished at E11.5, while HIRA-deposited H3.3 is enriched in the nucleus of PGCs (Hajkova et al., 2008). During spermatogenesis, new differential histone modifications are established to activate meiosis-specific gene expression, while histone variants in sperm are characterized by enrichment of H3K27cr, and to some extent of H3K27ac, on the variant H3.3 (Crespo et al., 2020). Furthermore, histone displacement prior to meiotic entry is also likely to be sex-specific. During spermatogenesis, replacement of H3 variant H3.1 by H3t is essential for spermatogonial differentiation and ensures meiotic entry by inducing an open chromatin structure *in vitro* (Ueda et al., 2017). Also alterations of H1 play a role in meiosis with three testis-specific variants (H1t, HILS1, and H1T2) and one oocyte-specific variant (H1oo) identified so far (Pérez-Montero, Carbonell, & Azorín, 2016). In pachytene spermatocytes, H1t is the major linker histone variant which ensures that retrotransposable repeat elements are repressed through association with methylated promoters and the repressive histone marks H3K9me3 and H4K20me3 (Mahadevan, Kumar, & Rao, 2020). Similarly, high levels of PAR (a product of PARP1) are detected at E10.5 PGCs where auto-modified PARP1 can act transiently as a histone chaperone performing similar activity to H1 (Muthurajan et al., 2014). Furthermore, disturbance in PAR metabolism causes retention of H1t, HILS1 and core histones in the mature spermatozoa (Meyer-Ficca et al., 2011).

Several histone variants play critical roles in both meiosis and cancer. For instance, H1t, which is normally restricted to testis, has also been observed in cancer where it represses rDNA transcription by condensing chromatin structure similarly to its role in sperm development (Tani et al., 2016). Also H2A.Z, besides its role during meiosis where it promotes recombination and nuclear relaxation (Yamada et al., 2017), has been linked to several cancer types (Giaimo et al., 2019). Furthermore, histone replacements in meiosis resemble the occurrence of oncohistones, which are caused by missense mutations common in cancer (Nacev et al., 2019). Both testis specific histones and oncohistones decrease nucleosome stability, which may lead to enhanced cell reprogramming (Wang et al., 2019a). Oncohistones have in some cases been identified as the founding oncogenic mutations (Nikbakht et al., 2016). Similar to meiosis, an altered chromatin landscape driven by oncohistones has wide-ranging implications for chromatin structure and function. For example, oncohistone H3.3-G34R/V causes aberrant chromatin looping promoting a permissive chromatin environment which leads to transcriptional activation of the oncogene Platelet-Derived Growth Factor Receptor-α (*PDGFRA*) (Chen et al., 2020). Other oncohistones capable of rewiring the enhancer network and driving oncogene expression include H3.3K27M, H3.1K27M, H3K36M and H3G34V/R/W/L (Ghiraldini, Filipescu, & Bernstein, 2021). H3K27M mutation in glioma causes inhibition of the H3K27me3 methyltransferase complex PRC2 by disrupting its ability to recognize targets. Furthermore, mutations of H3K27 demonstrate that, even in the case of the same K27M mutation, if histone variant or cellular context changes, the resulting enhancer-promoter interacting landscape will differ and so does the histone genomic location. Consequently, different PRC2 targets become disrupted by H3.3K27M compered to H3.1K27M (Nagaraja et al., 2019). Similarly, H2A.Z variants roles seems to depend on cell context with H2A.Z.2 reported important in malignant melanoma (Vardabasso et al., 2015), while H2A.Z.1 was proposed to play a role in liver tumorigenesis through EMT (Yang et al., 2016). However, methodical assessment of the roles for H2A.Z variants throughout cancer models will need to be performed to functionally confirm this.

In summary, similar to oncohistones, aberrant expression of germline specific histone variants can disturb healthy somatic chromatin architecture, which can lead to altered gene expression and, in the worst case, a cancer phenotype.

## 4. DNA accessibility

The modulation of chromatin accessibility is critical for DNA-based transactions such as transcription (Fig. 1C), replication, recombination and DNA damage repair. Histone modifications can influence these processes in different ways: (i) indirectly by recruiting trans-acting proteins such as chromatin remodeling complexes, and/or (ii) altering the binding affinity between DNA and histones, leading to changes in chromatin accessibility.

In general, trimethylations of H3 (H3K4me3, H3K36me3 and H3K79me3) and acetylation of lysines (Kac) at histones H3 and H4 are associated with active "open" chromatin whereas histone modifications such as H3K27me3 and H3K9me3 mark "closed" repressed chromatin. Histone lysine acetylation is arguably one of the best studied PTMs that can directly modulate chromatin accessibility. The covalent addition of an acetyl group neutralizes the positively charged lysine residue, hence reducing the binding between histones and DNA, leading to a more open structure with enhanced accessibility to the transcriptional machinery. As a case in point, the placement of H4K16ac inhibits the formation of the 30 nm chromatin fibre (Shogren-Knaak et al., 2006).

More recently, with the advent of advanced mass spectrometry tools, short-chain acyl modifications of histone lysine residues other than acetylation have been identified. These modifications include propionylation, butyrylation, isobutyration, hydroxybutyrylation, benzoylation, crotonylation, malonylation, succinylation, lactylation and glutarylation (Zhao, Zhang, & Li, 2018). These modifications have a structural similarity with acetylation and are regulated by known HATs and HDACs. Importantly, several of these modifications have been implicated in cancer and germ cell development. For example, H3K23 propionylation and histone crotonylation (Kcr) are implicated in leukemia cell differentiation (Liu et al., 2009) and spermatogenesis (Tan et al., 2011) respectively. *In vitro* transcription studies demonstrated that Kcr stimulates gene activation, to a greater extent than Kac (Sabari et al., 2015). The *in vivo* function of Kcr was originally described in the context of mouse spermatogenesis, a process that entails global changes in the chromatin structure, including genome-wide histone-protamine replacement. Consistent with a role in gene activation, ChIP-seq studies showed that Kcr is enriched at the promoters of testis-specific genes that are activated in the post-meiotic germ cells. Additionally, a significant proportion of X-linked genes that escape sex chromosome inactivation in the round spermatids are also marked by Kcr, indicating an epigenetic priming function of Kcr. Furthermore, in the elongating spermatids, the presence of Kcr facilitates histone-protamine replacement, key to the formation of a highly compacted sperm chromatin (Liu et al., 2017). Taken together, these results highlight the importance of histone acylation in enforcing specific gene expression programs through modulating chromatin accessibility.

Similar to Kac, the gene regulatory function of Kcr is likely achieved through specific reader proteins. For example, YEATS2, a scaffolding subunit of the Ada-two-A-containing (ATAC) HAT complex, is a histone crotonylation reader that shows preference for H3K27cr (Zhao et al., 2016). More recently, YEATS domain was also found to act as reader of histone benzoylation (Ren et al., 2021), paving the way for further mechanistic and functional dissections of histone benzoylation in development and disease. In spermatogenesis, H3K27cr also promotes the recruitment of other factors including BRD4, SOX30, SLY, CTCF and its germ paralogue BORIS (Crespo et al., 2020) that establishes a transcriptional active state. Conversely, factors can negative regulate histone Kcr, as in the case of histone deacetylation. For example, CDYL acts as a crotonyl-CoA hydratase to downregulate Kcr, leading to transcription repression during spermatogenesis (Liu et al., 2017).

Similarly, to what happens in germ cells, widespread chromatin changes take place during carcinogenesis. For example, genome-wide epigenetic reprogramming occurs during epithelial-to-mesenchymal transition, characterized by increased chromatin accessibility and gain of active histone modification marks (Gomes et al., 2019; McDonald et al., 2011). As discussed earlier, histone variant replacement plays an important role during germ cell development, and chromatin accessibility is heavily influenced by histone variants. Histone variant, H3.3, plays a major role in spermatogenesis through dynamic regulation of chromatin states (Yuen et al., 2014). Strikingly, recent study reported that selective incorporation of H3.3 into chromatin promotes pro-metastatic transcriptional reprogramming and is essential for tumor progression (Gomes et al., 2019). Additionally, NFIB, a member of the Nuclear Factor I family (NFI) of proteins, promotes metastasis through a widespread increase in chromatin accessibility (Denny et al., 2016). In germ cells, NFIB is highly expressed in round spermatids, highlighting a possible function in spermatogenesis, possibly through regulation of chromatin dynamics (Davila et al., 2022). In summary, many of the factors involved in meiotic gene expression are also implicated in cancer pointing to a possible mechanistic convergence.

One of the key features of meiosis and cancer is the reliance on genomic instability with introduction of DNA breaks to allow for genomic rearrangements. Consequently, it is key that both cancer and meiotic chromatin is open and DNA is accessible. Notably, aberrant expression of meiosis-specific genes is associated with genomic instability in cancer. For example, a meiosis-specific methyltransferase, PRDM9, is ectopically expressed in a range of cancer cell lines (Feichtinger et al., 2012). PRDM9 has a major role in DSB induction at recombination hotspots during meiotic prophase I and repairs SPO11-induced DSB by accurate chromatin remodeling for homologous recombination (Paiano et al., 2020). These meiotic recombination hotspots are characterized by opened chromatin that is enriched with two histone marks, H3K4me3 and H3K36me3, catalyzed by PRDM9 (Eram et al., 2014). This enables ATM to initiate and extend DSB resection, allowing admittance of DNA repair machinery recruited by ATM to the damage sites. Of note, aberrant deposit of H3K36me3 caused by PDRM9 expression is associated with favoring DNA damage response in cancer, which is observed in B-ALL patients (Woodward et al., 2014). In addition, PRDM9 binds to promoter regions of meiotic genes for activation. Therefore, its expression in cancer might result in aberrant transcriptional activation and chromatin state changes. In fact, PRDM9 was reported to induce expression of testis specific genes such as BORIS (CTCFL) and VCX by binding to their regulatory regions (Altemose et al., 2017). Collectively, these findings demonstrate that cancer may harness some of the DNA damage repair mechanisms in meiosis, resulting in genome instability and aberrant meiotic gene expression.

## 5. Higher-order of chromatin structure

In addition to the organization of the genome into chromatin through the assembly of nucleosomes, genomes are also dynamically organized into higher-order structures. On different scales, genomes are organized into domain/structures of increasing size. Nucleosomal DNA is organized into loops (or topologically associating domains—TADs, see also below) sized in range of the hundreds of kilobases. Over sizes of megabases, chromatin is further organized into what are referred to as A/B compartments (Serizay & Ahringer, 2018). These A/B compartments correspond to transcriptionally active/inactive regions, respectively, and cytologically manifest as euchromatin and heterochromatin.

The organization of genomes into large chromatin loops emanating from a proteinaceous axial structure was first shown well over 100 years ago in cells undergoing meiosis (Flemming, n.d.). During meiotic development—and especially in meiotic prophase—chromosomes organize into typical loop-axis structures, and this organization is key to enable efficient meiotic DSB formation and recombination. Over the recent years it has become clear that chromosome organization, especially at the level of loops/TADs is mediated to a large extent by the action of large Structural Maintenance of Chromosomes (SMC) complexes, of which the cohesin and condensin complexes are most well-known and studied. This organization being increasingly recognized as important for transcriptional regulation (especially during development), V(*D*)J recombination and chromosome individualization and segregation (Davidson & Peters, 2021). Cohesin as well as condensin complexes consist of two SMC subunits, in association with a kleisin protein, resulting in the formation of ring-shaped complex (Bauer et al., 2021). The kleisin subunit of these complexes in turn can interact with additional dedicated factors, termed HAWK proteins (for HEAT repeat proteins associated with kleisins). The identity of these factors differs between cohesin or condensin complexes, and these proteins serve different functions. We will focus on the cohesin complex, mostly because meiosis-specific subunits of the cohesin complex can dictate differentiated functions, and their mis-expression in non-germ cells is associated with cancer.

SMC complexes can use energy provided by ATP hydrolysis to reel in linear DNA molecules, *via* a process called loop-extrusion (Davidson & Peters, 2021). By still incompletely understood mechanisms, and in combination with other factors, such as the CTCF insulator protein, the action of these chromosomal enzymes leads to specific chromosome folding and organization that dynamically changes during cell cycle progression (Davidson & Peters, 2021). In addition to chromosome folding, the cohesin and condensin complexes are also important for orchestrating chromosome segregation in mitosis and meiosis (Yatskevich, Rhodes, & Nasmyth, 2019). Cohesin is responsible for establishing sister chromatid cohesion, and cleavage of its kleisin subunit SCC1 is sufficient, and required, to initiate chromosome disjunction at the metaphase-anaphase transition (Watanabe, 2004). During meiosis, cohesin and condensin also play key roles in mediating DNA break formation and recombination, and the proper organization of meiotic chromosomes into a stereotypical structure at the hands of SMCs plays a central role in these processes (Alavattam et al., 2019; Ke et al., 2017; Mitter et al., 2020; Muller et al., 2018; Patel et al., 2019; Schalbetter et al., 2019; Wang et al., 2019b). As already eluded to above, some subunits of SMCs have meiosis-specific counterparts, such as REC8 (for SCC1) and STAG3 (for STAG2/3) in the case of cohesin. In addition, a germ cell-specific orthologue of CTCF, BORIS/CTCFL (Nishana et al., 2020), exhibits differential functionality as compared to CTCF (Bhan et al., 2011; Cheema et al., 2014; D'Arcy et al., 2008; Link et al., 2013). All these germ cell-specific factors are regularly expressed in human cancers, and specific phenotypic effects have been ascribed to their somatic expression (Choi et al., 2022; Folco et al., 2017; Gao et al., 2018;

Han et al., 2021; Hosoya et al., 2018, 2011; Hou et al., 2021; Masterson et al., 2015; Sandhu et al., 2021; Sasaki et al., 2021; Strunnikov, 2013). It is currently poorly understood how these meiotic factors might impact tumorigenesis. Potentially, their expression might lead to alterations in chromosome organization—leading to widespread transcriptional effects. Alternatively, in light of the close connection of SMC complexes to meiosis-specific DNA repair, their aberrant expression could have an impact on genome stability, by for example interfering with DNA repair or chromosome segregation. Future research is needed to fill this knowledge gap. In this light it is interesting to note that components of the meiosis-specific synaptonemal complex, most prominently SYCP2/3, have been suggested to co-localize and directly interact with REC8-containing cohesin during meiosis (Panizza et al., 2011; Ur & Corbett, 2021). This association with chromosomes, in combination with meiotic HORMADs, is required to establish a meiosis-specific loop-axis structure (Gu, Desai, & Corbett, 2022). This organization is important to facilitate execution of several key meiotic processes such as DNA break formation and recombination (Ur & Corbett, 2021). Aforementioned factors are suggested to directly associate with, and potentially influence the function of cohesin complexes, such as for example the role of this complex in chromosome folding and organization. Thus, it will be interesting to consider whether, under pathological conditions, they interfere with cohesin-mediated chromosome organization.

## 5.1 Topologically associated domains

Genomes are often arranged by grouping sets of neighboring genes that are positioned on a linear chromosome which tend to preferentially interact with each other in three-dimensional structures, as defined by TAD establishment (Fig. 1D). TADs play a critical role in regulating gene expression with genes within the same TAD characterized by similar expression and histone-modification profiles. Chromosome folding and organization is largely driven by a process called loop-extrusion, and involves activity of cohesin and condensin complexes. During oncogenesis, spatial changes to genome organization often happen at the chromatin loop level with insulated neighborhoods defined by chromatin loops formed by CTCF–CTCF homodimers, the major anchoring components that defines TAD establishment. In cancer, TAD rearrangements can drive gene expression by promoting novel enhancer-promoter interactions and chromatin opening.

One such example of meiosis-cancer TAD relationship is the aberrant activation of a germ cell specific paralogue of CTCF (BORIS/CTCFL) in cancer (Bhan et al., 2011; Cheema et al., 2014; D'Arcy et al., 2008; Link et al., 2013). In cancer, BORIS promotes novel chromatin interactions which create super-enhancers that drive expression of genes responsible for hallmarks of cancer and treatment resistance (Debruyne et al., 2019), while it remains to be established if this is driven by H3K27cr as in meiosis. It might be beneficial for cancers to promote "meiosis-like" chromatin organization, which in turn will lead to activation of meiotic genes. BORIS re-expression is one such example of a circuit where cancer evolution drives re-expression of a meiotic factor which in turn promotes further instability and meiotic-like chromatin.

In cancer, TAD disruptions can be caused by mutations or epigenetic inactivation in just a single TAD boundary resulting in disruption of CTCF binding and, consequently, altered gene regulation at both sides of the TAD. TAD destruction can also be caused by genome rearrangements which lead to disruptions and/or fusions of TADs. This has no effect on TAD boundaries but instead generates new regulatory sub-domains (subTADs) which can activate oncogenes (Du et al., 2021). Similarly, CTCF binding loss causes TAD disruption leading to misregulation of gene expression such as oncogene overexpression and activation of proto-oncogenes (Akdemir et al., 2020; Sauerwald & Kingsford, 2018). Consequently, alterations in the TAD positioning of genes relate to patients' prognosis (Du et al., 2021). Furthermore, a TAD nesting score can distinguish healthy cells from early tumors and even from advanced disease (Du et al., 2021). TAD disassembly in tumors, which is not dependent on CTCF binding loss as it is in mitosis, resembles the dramatic changes in chromatin architecture during meiosis. TAD boundaries are weakened in differentiating spermatogonia (Zheng et al., 2022) and are largely lost by zygotene (Patel et al., 2019; Vara et al., 2019) while this is not due to CTCF binding loss (Zuo et al., 2021). Progression through meiosis is characterized by a significant increase in chromatin loop size potentially in a cohesion-driven fashion. Changes to chromatin loop size and architecture are likely needed to direct the distribution of meiotic recombination events to the gene rich regions. In cancer cells these alterations might conceivably target genome instability toward oncogene-rich regions. As new technologies allow us to define chromatin architecture and 3D organization with increased precision, also at a single cell resolution, it will be interesting to further explore this similarity.

## 5.2 Distal regulatory elements

Exceptional plasticity of transcriptional programs allows and coordinates a wide range of biological processes, from cellular growth and differentiation to specialized functions in distinct tissues. Cancers exhibit behaviors that exploit this plasticity. Beyond the interactions at gene promoters, distal *cis*-acting regulatory elements have emerged as prominent determinants of cell type-specific transcription (Fig. 1E). Chromatin allows the packaging of DNA into discreet states which can either be "open" for transcriptional machinery to access (euchromatin) or "closed" and generally repressed (heterochromatin). Although all cells of a multicellular organism share the same DNA sequence, the relative timing and levels of gene expression differ between cell types. The spatiotemporal control of gene expression is influenced by distal DNA sequences known as enhancers which are often characterized by H3 acetylation and H3K4me1 outside of promoter regions. Gene-specific enhancer–promoter interactions can be required to control the temporal and spatial transcriptional activity of individual genes. In a cell where an enhancer is "active," it can recruit transcriptional machinery and form a loop in the DNA so that the enhancer can directly interact with the target gene promoter, thus increasing the concentration of transcriptional machinery at the promoter and inducing transcription (Spitz & Furlong, 2012). Physiological expression of meiotic genes is carefully regulated and limited to gametocytes while silenced in healthy somatic cells. It is unclear how gametocyte specific expression is regulated, as many meiotic genes are autosomal and positioned between two somatically expressed genes. Local chromatin environment pre-conditions transcriptional responses as access for the transcription machinery is key however, the precise chromatin configuration and the exact mechanistic control that underlies the transcriptional outcome at meiotic genes and their distal regulatory elements facilitating re-expression in cancer are not known. Notably, enhancer activity has recently been shown crucial in controlling gene expression during the mitosis-meiosis transition (Maezawa et al., 2020). Aggregates of enhancers known as super-enhancers regulate expression of lineage specific genes as they concentrate transcription factors and co-activators. Sperm super-enhancers, including those that control expression of round spermatid (RS) specific genes, are marked by H3K27cr and H3K27ac where H3K27cr exists in association with H3K36me2 (Crespo et al., 2020). Activation of the meiotic expression programme is driven by a switch form mitotic to meiotic super-enhancers established in differentiating mitotic spermatogonia prior to meiotic entry through the MYBL1 transcription factor (Maezawa et al., 2020). This is possible thanks to the resolution of mitotic super-enhancers *via* the germline specific polycomb protein SCML2 (Maezawa et al., 2020).

Chromosomal architecture, and consequently the pattern of promoter-enhancer interactions, are controlled by cohesin complexes and regulatory proteins (such as CTCF) that affect cohesin-DNA interactions (Ishiguro, 2019). Consequently, cohesin depletion leads to defects in the activation of inducible genes and promoter interactions are impaired following cohesin and CTCF silencing (Thiecke et al., 2020). However, some interactions are not affected and additional new interactions can be generated (Thiecke et al., 2020). It is the position within the TAD, interaction distance and identity of associated proteins that determine what happens to an individual interaction (Thiecke et al., 2020). Whereas cohesin depletion results in the loss of multiple promoter-enhancer interactions, it can also lead to superenhancer rewiring of subsets of short-range promoter interactions that emerge *de novo* (Thiecke et al., *2020).* These novel contacts often cross TAD boundaries, suggesting general reshaping of promoter-enhancer interactions with active enhancers themselves able to function as TAD boundary elements (Thiecke et al., 2020). Consequently, contacts between super-enhancers recover more rapidly upon cohesin reconstitution compared with other chromosomal loops (Thiecke et al., 2020).

Cancer-specific super-enhancers are often found at oncogenic drivers (Hnisz et al., 2013) and are regulated by the BET protein BRD4 with BRD4 suppression selectively inhibiting the MYC oncogene (Lovén et al., 2013). Similarly, the germ cell specific BET protein BRDT is required in spermatogonia (Gaucher et al., 2012). BRD4 and BRDT are highly similar and BRD4s has a role in post-meiotic spermatids (Bryant et al., 2015). Therefore, BRD4 might play a supportive role in meiosis, while BRDT re-expression in cancer could drive meiosis-like chromatin architecture. As discussed above, altered chromatin looping can bring distal enhancers and promoters together leading to transcription. This has been demonstrated when forced chromatin looping reactivated silenced foetal globin genes in adult erythroid cells (Deng et al., 2014). Oncogenic stress can also cause reshaping of the regulatory element landscape in stressed cells. Heat shock has been shown to cause novel patterning of enhancers (Vihervaara et al., 2017). As reshaping of a single enhancer can be sufficient for oncogenesis (Gröschel et al., 2014) it is not inconceivable that alternative chromatin architecture in cancer driven by the oncohistones and changes to TAD domains could lead to distal enhancers activating meiotic genes. For example oncogenic gain-of-function mutation of EZH2, the H3K27me3 writer, can induce structural changes and long-range chromatin interactions in cancer (Barsotti et al., 2015). Cancers can also have unique oncogenic super-enhancers. Such changes could lead to rearrangements allowing for interactions of meiotic distal enhancers with meiotic gene promoters, especially as at least some sperm super-enhancers are already present in post-meiotic cells (Crespo et al., 2020). As our ability to accurately identify RNA-producing enhancers across the whole genome expands, comparing the enhancer network between cancers and meiosis could explain the mechanisms behind meiotic gene activation in cancer. Because enhancer functions are often distributed and latent enhancer sequences are pervasive in the genome, effects of non-coding mutations in cancer may be especially challenging to characterize and require careful functional assessment.

## 5.3 R-loops

R-loops are DNA-RNA hybrids that occur co-transcriptionally when the nascent RNA binds the template DNA (*cis*), leading to the formation of a three-stranded RNA–DNA structure displacing the non-template single-stranded DNA (Fig. 1F). R-loops can also be formed in *trans* by reannealing of RNA transcripts, which is likely the mechanism responsible for R-loop formation during meiosis, especially at DSB sites (Yang et al., 2021). Unresolved R-loops can present a genotoxic obstacle to the replication forks. R-loop levels are particularly prevalent at distinct regions including promoter CpG islands, terminators and transposable elements (Rinaldi et al., 2021). R-loops at sites of DNA damage facilitate repair by utilizing RNA as a template which regulates resection of DSB ends, while they can also impair DSB repair (Ohle et al., 2016). These effects of R-loops on the fidelity of DNA repair can ostensibly impact on genome stability, and thus might be relevant for meiosis as well as cancer development. Indeed, many cancer cells are characterized by high levels of R-loops, which contribute to their increased genome instability necessitating compensatory factors like INO80 (Prendergast et al., 2020) or ADAR1p110 (Shiromoto et al., 2021). Furthermore, R-loops themselves can be oncogenic (Chiang et al., 2019; Vohhodina et al., 2021). For example, R-loops regulate gene expression by preventing methylation and transcriptional silencing at CpG promoters (Ginno et al., 2012) which can lead to aberrant re-expression of genes in cancer. Predictably, chromatin structure plays a role in regulating R-loops with mutations in H3, H4 and oncohistones, leading to increased

R-loop formation, while topoisomerase I (Tuduri et al., 2009) and histone chaperone FACT complex (Herrera-Moyano et al., 2014) are responsible for preventing this. Similarly, to cancer, meiosis is characterized by high R-loop prevalence, particularly in regions of condensed chromatin and at ssDNA tracts that are present adjacent to meiotic DSBs (Castellano-Pozo et al., 2013). High levels of R-loops might be beneficial at the early stages of meiosis when DNA is undergoing Spo11-dependent DSB formation and transcription is active. Indeed, a recent report described an active role for RNA-DNA hybrids in the regulation of meiotic recombination, *via* currently poorly understood mechanisms. R-loops appear critical for meiosis to occur with fidelity, and overexpression of RNaseH1, which cleaves the RNA moiety, impairs meiotic recombination (Yang et al., 2021). However, excessive accumulation of R-loops at DSBs decreases crossover (CO) and non-CO (NCO) recombination (Yang et al., 2021), suggesting that R-loop formation needs to be carefully balanced. R-loop resolution by Senataxin is required for DNA damage repair to ensure genomic integrity during meiotic recombination and meiotic sex chromosome inactivation (Becherel et al., 2013). Consequently, loss of Senataxin abrogates CO formation, causing a pachytene-diplotene arrest and infertility (Becherel et al., 2013). These findings highlight the need for tight control of R-loops as meiosis progresses. The similarity between high levels of R-loops in early meiosis and cancer, both accompanied by DSBs, might contribute to similarities between chromatin in cancerous conditions and the typical chromosome architecture that is established in germ cells undergoing meiosis. Specifically, R-loops could play a role in cancer-associated activation of meiotic gene expression *via* de-silencing, especially as high levels of R-loops prevent transcriptional silencing.

## 6. Conclusions

*Re*-expression of meiotic genes is a common event during cancer development and progression. However, little is known about the events that permit meiotic genes to be activated and if the chromatin features of a gene ectopically expressed in a tumor are similar to the chromatin features of the same gene in meiosis. Chromatin structure is essential for maintaining cell identity, and it is highly remodelled during tumorigenesis. Multiple aspects of cancer and meiotic cell chromatin discussed in this review, are shared and we propose that it is likely that a meiosis-like chromatin structure and architecture support tumorigenesis. In order to fully appreciate the causal connections between the observed aberrant establishment of meiosis-like cellular characteristic and cancer development, comprehensive functional and mechanistic analysis of the dysfunctional effects of these genes on genomic and cellular behavior will be essential. Furthermore, it will be important to determine whether these meiosis-like changes, chromatin states and interactions can already be observed in precancerous conditions, such as ulceration or cirrhosis. Detailed molecular understanding of these processes and their origins will allow us to validate if they could be exploited as biomarkers and if future treatments should aim to harness cancer chromatin to return it to the soma-like state.

## Acknowledgments

The results presented in Fig. 2 are in whole based upon data generated by the TCGA Research Network: https://www.cancer.gov/tcga. U.L.M. was supported by University of Liverpool Tenure Track programme, the Royal Society Research Grant RG170342, North West Cancer Research RDG2021.15 and by the Wellcome Trustf 204822/Z/16/Z. I.F.S. was supported by a joint University of Liverpool—A* PhD programme. G.V. is financially supported by an Amsterdam UMC Research Fellowship.

## Author contributions

I.F.S. generated Fig. 1 and contributed some of the writing to the DNA methylation section. G.V., G.H. and W.-W.T. provided critical evaluation of the manuscript. G.V. contributed to the R-loop and TAD domain sections. U.L.M. wrote the manuscript.

## References

Akdemir, K. C., et al. (2020). Disruption of chromatin folding domains by somatic genomic rearrangements in human cancer. *Nature Genetics*, *52*(3), 294–305.

Alavattam, K. G., et al. (2019). Attenuated chromatin compartmentalization in meiosis and its maturation in sperm development. *Nature Structural & Molecular Biology*, *26*(3), 175–184.

Altemose, N., et al. (2017). A map of human PRDM9 binding provides evidence for novel behaviors of PRDM9 and other zinc-finger proteins in meiosis. *eLife*, *6*.

Barsotti, A. M., et al. (2015). Epigenetic reprogramming by tumor-derived EZH2 gain-of-function mutations promotes aggressive 3D cell morphologies and enhances melanoma tumor growth. *Oncotarget*, *6*(5), 2928–2938.

Bauer, B. W., et al. (2021). Cohesin mediates DNA loop extrusion by a "swing and clamp" mechanism. *Cell*, *184*(21). 5448–5464.e22.

Becherel, O. J., et al. (2013). Senataxin plays an essential role with DNA damage response proteins in meiotic recombination and gene silencing. *PLoS Genetics*, *9*(4), e1003435.

Bhan, S., et al. (2011). BORIS binding to the promoters of cancer testis antigens, MAGEA2, MAGEA3, and MAGEA4, is associated with their transcriptional activation in lung cancer. *Clinical Cancer Research*, *17*(13), 4267–4276.

Black, J. C., Van Rechem, C., & Whetstine, J. R. (2012). Histone lysine methylation dynamics: Establishment, regulation, and biological impact. *Molecular Cell*, *48*(4), 491–507.

Bruggeman, J. W., et al. (2018). Massive expression of germ cell-specific genes is a hallmark of cancer and a potential target for novel treatment development. *Oncogene*, *37*(42), 5694–5700.
Bruggeman, J. W., Irie, N., Lodder, P., van Pelt, A. M. M., Koster, J., & Hamer, G. (2020). Tumors widely express hundreds of embryonic germline genes. *Cancers*, *12*(12), 3812. https://doi.org/10.3390/cancers12123812.
Bryant, J. M., et al. (2015). Characterization of BRD4 during mammalian postmeiotic sperm development. *Molecular and Cellular Biology*, *35*(8), 1433–1448.
Castellano-Pozo, M., et al. (2013). R loops are linked to histone H3 S10 phosphorylation and chromatin condensation. *Molecular Cell*, *52*(4), 583–590.
Cheema, Z., et al. (2014). Expression of the cancer-testis antigen BORIS correlates with prostate cancer. *Prostate*, *74*(2), 164–176.
Chen, C. C. L., et al. (2020). Histone H3.3G34-mutant interneuron progenitors co-opt PDGFRA for Gliomagenesis. *Cell*, *183*(6), 1617–1633.e22.
Chiang, H.-C., et al. (2019). BRCA1-associated R-loop affects transcription and differentiation in breast luminal epithelial cells. *Nucleic Acids Research*, *47*(10), 5086–5099.
Choi, E. H., et al. (2022). Meiosis-specific cohesin complexes display essential and distinct roles in mitotic embryonic stem cell chromosomes. *Genome Biology*, *23*(1), 70.
Collins, C., et al. (1998). Positional cloning of <i>ZNF</i><i>217</i> and <i>NABC</i><i>1</i>: Genes amplified at 20q13.2 and overexpressed in breast carcinoma. *Proceedings of the National Academy of Sciences of the United States of America*, *95*(15), 8703–8708.
Crespo, M., et al. (2020). Multi-omic analysis of gametogenesis reveals a novel signature at the promoters and distal enhancers of active genes. *Nucleic Acids Research*, *48*(8), 4115–4138.
D'Arcy, V., et al. (2008). BORIS, a paralogue of the transcription factor, CTCF, is aberrantly expressed in breast tumours. *British Journal of Cancer*, *98*(3), 571–579.
Davidson, I. F., & Peters, J. M. (2021). Genome folding through loop extrusion by SMC complexes. *Nature Reviews. Molecular Cell Biology*, *22*(7), 445–464.
Davila, R. A., et al. (2022). Deletion of NFIX results in defective progression through meiosis within the mouse testis. *Biology of Reproduction*.
De Felici, M. (2016). The formation and migration of primordial germ cells in mouse and man. In R. P. Piprek (Ed.), *Molecular mechanisms of cell differentiation in gonad development* (pp. 23–46). Cham: Springer International Publishing.
Debruyne, D. N., et al. (2019). BORIS promotes chromatin regulatory interactions in treatment-resistant cancer cells. *Nature*, *572*(7771), 676–680.
Deng, W., et al. (2014). Reactivation of developmentally silenced globin genes by forced chromatin looping. *Cell*, *158*(4), 849–860.
Denny, S. K., et al. (2016). Nfib promotes metastasis through a widespread increase in chromatin accessibility. *Cell*, *166*(2), 328–342.
Deplus, R., et al. (2013). TET2 and TET3 regulate GlcNAcylation and H3K4 methylation through OGT and SET1/COMPASS. *The EMBO Journal*, *32*(5), 645–655.
Du, G., et al. (2021). The hierarchical folding dynamics of topologically associating domains are closely related to transcriptional abnormalities in cancers. *Computational and Structural Biotechnology Journal*, *19*, 1684–1693.
Eram, M. S., et al. (2014). Trimethylation of histone H3 lysine 36 by human methyltransferase PRDM9 protein. *The Journal of Biological Chemistry*, *289*(17), 12177–12188.
Erenpreisa, J., et al. (2009). The role of meiotic cohesin REC8 in chromosome segregation in gamma irradiation-induced endopolyploid tumour cells. *Experimental Cell Research*, *315*(15), 2593–2603.
Feichtinger, J., et al. (2012). Meta-analysis of clinical data using human meiotic genes identifies a novel cohort of highly restricted cancer-specific marker genes. *Oncotarget*, *3*(8), 843–853.

Feinberg, A. P., & Vogelstein, B. (1983a). Hypomethylation distinguishes genes of some human cancers from their normal counterparts. *Nature, 301*(5895), 89–92.

Feinberg, A. P., & Vogelstein, B. (1983b). Hypomethylation of ras oncogenes in primary human cancers. *Biochemical and Biophysical Research Communications, 111*(1), 47–54.

Feng, C. W., Bowles, J., & Koopman, P. (2014). Control of mammalian germ cell entry into meiosis. *Molecular and Cellular Endocrinology, 382*(1), 488–497.

Flemming, W. (1882). *Zellsubstanz, kern und zelltheilung*. Vogel.

Folco, H. D., et al. (2017). Untimely expression of gametogenic genes in vegetative cells causes uniparental disomy. *Nature, 543*(7643), 126–130.

Gama-Sosa, M. A., et al. (1983). The 5-methylcytosine content of DNA from human tumors. *Nucleic Acids Research, 11*(19), 6883–6894.

Gantchev, J., et al. (2020). The ectopic expression of meiCT genes promotes meiomitosis and may facilitate carcinogenesis. *Cell Cycle, 19*(8), 837–854.

Gao, Y., et al. (2018). The cancer/testes (CT) antigen HORMAD1 promotes homologous recombinational DNA repair and radioresistance in lung adenocarcinoma cells. *Scientific Reports, 8*(1), 15304.

Gaucher, J., et al. (2012). Bromodomain-dependent stage-specific male genome programming by Brdt. *The EMBO Journal, 31*(19), 3809–3820.

Gaysinskaya, V., et al. (2018). Transient reduction of DNA methylation at the onset of meiosis in male mice. *Epigenetics & Chromatin, 11*(1), 15.

Ghiraldini, F. G., Filipescu, D., & Bernstein, E. (2021). Solid tumours hijack the histone variant network. *Nature Reviews Cancer, 21*(4), 257–275.

Giaimo, B. D., et al. (2019). The histone variant H2A.Z in gene regulation. *Epigenetics & Chromatin, 12*(1), 37.

Ginno, P. A., et al. (2012). R-loop formation is a distinctive characteristic of unmethylated human CpG island promoters. *Molecular Cell, 45*(6), 814–825.

Gomes, A. P., et al. (2019). Dynamic incorporation of histone H3 variants into chromatin is essential for acquisition of aggressive traits and metastatic colonization. *Cancer Cell, 36*(4), 402–417.e13.

Gröschel, S., et al. (2014). A single oncogenic enhancer rearrangement causes concomitant EVI1 and GATA2 deregulation in leukemia. *Cell, 157*(2), 369–381.

Gu, Y., Desai, A., & Corbett, K. D. (2022). Evolutionary dynamics and molecular mechanisms of HORMA domain protein signaling. *Annual Review of Biochemistry, 91*(1), 541–569.

Hackett, J. A., et al. (2013). Germline DNA demethylation dynamics and imprint erasure through 5-hydroxymethylcytosine. *Science, 339*(6118), 448–452.

Hajkova, P., et al. (2002). Epigenetic reprogramming in mouse primordial germ cells. *Mechanisms of Development, 117*(1–2), 15–23.

Hajkova, P., et al. (2008). Chromatin dynamics during epigenetic reprogramming in the mouse germ line. *Nature, 452*(7189), 877–881.

Hajkova, P., et al. (2010). Genome-wide reprogramming in the mouse germ line entails the base excision repair pathway. *Science, 329*(5987), 78–82.

Hamer, G., & de Rooij, D. G. (2018). Mutations causing specific arrests in the development of mouse primordial germ cells and gonocytes. *Biology of Reproduction, 99*(1), 75–86.

Hammond, C. M., et al. (2017). Histone chaperone networks shaping chromatin function. *Nature Reviews. Molecular Cell Biology, 18*(3), 141–158.

Han, J., et al. (2021). REC8 promotes tumor migration, invasion and angiogenesis by targeting the PKA pathway in hepatocellular carcinoma. *Clinical and Experimental Medicine, 21*(3), 479–492.

Helleday, T. (2010). Homologous recombination in cancer development, treatment and development of drug resistance. *Carcinogenesis, 31*(6), 955–960.

Herrera-Moyano, E., et al. (2014). The yeast and human FACT chromatin-reorganizing complexes solve R-loop-mediated transcription-replication conflicts. *Genes & Development, 28*(7), 735–748.

Hill, P. W. S., et al. (2018). Epigenetic reprogramming enables the transition from primordial germ cell to gonocyte. *Nature, 555*(7696), 392–396.
Hnisz, D., et al. (2013). Super-enhancers in the control of cell identity and disease. *Cell, 155*(4), 934–947.
Hosoya, N., & Miyagawa, K. (2021). Synaptonemal complex proteins modulate the level of genome integrity in cancers. *Cancer Science, 112*(3), 989–996.
Hosoya, N., Ono, M., & Miyagawa, K. (2018). Somatic role of SYCE2: An insulator that dissociates HP1α from H3K9me3 and potentiates DNA repair. *Life Science Alliance, 1*(3), e201800021. https://doi.org/10.26508/lsa.201800021.
Hosoya, N., et al. (2011). Synaptonemal complex protein SYCP3 impairs mitotic recombination by interfering with BRCA2. *EMBO Reports, 13*(1), 44–51.
Hou, H., et al. (2021). Centromeres are dismantled by foundational meiotic proteins Spo11 and Rec8. *Nature, 591*(7851), 671–676.
Ishiguro, K. I. (2019). The cohesin complex in mammalian meiosis. *Genes to Cells, 24*(1), 6–30.
Jay, A., et al. (2021). Cancer testis antigens and genomic instability: More than immunology. *DNA Repair, 108*, 103214.
Ke, Y., et al. (2017). 3D chromatin structures of mature gametes and structural reprogramming during mammalian embryogenesis. *Cell, 170*(2), 367–381.e20.
Link, P. A., et al. (2013). BORIS/CTCFL mRNA isoform expression and epigenetic regulation in epithelial ovarian cancer. *Cancer Immunity, 13*, 6.
Liu, B., et al. (2009). Identification and characterization of propionylation at histone H3 lysine 23 in mammalian cells. *The Journal of Biological Chemistry, 284*(47), 32288–32295.
Liu, S., et al. (2017). Chromodomain protein CDYL acts as a Crotonyl-CoA hydratase to regulate histone Crotonylation and spermatogenesis. *Molecular Cell, 67*(5), 853–866.e5.
Lovén, J., et al. (2013). Selective inhibition of tumor oncogenes by disruption of super-enhancers. *Cell, 153*(2), 320–334.
Maezawa, S., et al. (2020). Super-enhancer switching drives a burst in gene expression at the mitosis-to-meiosis transition. *Nature Structural & Molecular Biology, 27*(10), 978–988.
Mahadevan, I. A., Kumar, S., & Rao, M. R. S. (2020). Linker histone variant H1t is closely associated with repressed repeat-element chromatin domains in pachytene spermatocytes. *Epigenetics & Chromatin, 13*(1), 9.
Mansour, A. A., et al. (2012). The H3K27 demethylase Utx regulates somatic and germ cell epigenetic reprogramming. *Nature, 488*(7411), 409–413.
Masterson, L., et al. (2015). Deregulation of SYCP2 predicts early stage human papillomavirus-positive oropharyngeal carcinoma: A prospective whole transcriptome analysis. *Cancer Science, 106*(11), 1568–1575.
McDonald, O. G., et al. (2011). Genome-scale epigenetic reprogramming during epithelial-to-mesenchymal transition. *Nature Structural & Molecular Biology, 18*(8), 867–874.
Medrano, J. V., et al. (2016). Human somatic cells subjected to genetic induction with six germ line-related factors display meiotic germ cell-like features. *Scientific Reports, 6*, 24956.
Meyer-Ficca, M. L., et al. (2011). Poly(ADP-ribose) metabolism is essential for proper nucleoprotein exchange during mouse spermiogenesis. *Biology of Reproduction, 84*(2), 218–228.
Mitter, M., et al. (2020). Conformation of sister chromatids in the replicated human genome. *Nature, 586*(7827), 139–144.
Mobasheri, M. B., Shirkoohi, R., & Modarressi, M. H. (2016). Synaptonemal complex protein 3 transcript analysis in breast Cancer. *Iranian Journal of Public Health, 45*(12), 1618–1624.
Muller, H., et al. (2018). Characterizing meiotic chromosomes' structure and pairing using a designer sequence optimized for Hi-C. *Molecular Systems Biology, 14*(7), e8293.
Muthurajan, U. M., et al. (2014). Automodification switches PARP-1 function from chromatin architectural protein to histone chaperone. *Proceedings of the National Academy of Sciences of the United States of America, 111*(35), 12752–12757.

Nacev, B. A., et al. (2019). The expanding landscape of 'oncohistone' mutations in human cancers. *Nature*, *567*(7749), 473–478.

Nagaraja, S., et al. (2019). Histone variant and cell context determine H3K27M reprogramming of the enhancer landscape and oncogenic state. *Molecular Cell*, *76*(6). 965–980.e12.

Neumann, F., et al. (2005). Identification of an epitope derived from the cancer testis antigen HOM-TES-14/SCP1 and presented by dendritic cells to circulating CD4+ T cells. *Blood*, *106*(9), 3105–3113.

Nielsen, A. Y., & Gjerstorff, M. F. (2016). Ectopic expression of testis germ cell proteins in Cancer and its potential role in genomic instability. *International Journal of Molecular Sciences*, *17*(6), 890.

Nikbakht, H., et al. (2016). Spatial and temporal homogeneity of driver mutations in diffuse intrinsic pontine glioma. *Nature Communications*, 7, 11185.

Nishana, M., et al. (2020). Defining the relative and combined contribution of CTCF and CTCFL to genomic regulation. *Genome Biology*, *21*(1), 108.

Ohle, C., et al. (2016). Transient RNA-DNA hybrids are required for efficient double-Strand break repair. *Cell*, *167*(4). 1001–1013.e7.

Okano, M., et al. (1999). DNA methyltransferases Dnmt3a and Dnmt3b are essential for de novo methylation and mammalian development. *Cell*, *99*(3), 247–257.

Paiano, J., et al. (2020). ATM and PRDM9 regulate SPO11-bound recombination intermediates during meiosis. *Nature Communications*, *11*(1), 857.

Panizza, S., et al. (2011). Spo11-accessory proteins link double-strand break sites to the chromosome axis in early meiotic recombination. *Cell*, *146*(3), 372–383.

Pastor, W. A., Aravind, L., & Rao, A. (2013). TETonic shift: Biological roles of TET proteins in DNA demethylation and transcription. *Nature Reviews. Molecular Cell Biology*, *14*(6), 341–356.

Patel, L., et al. (2019). Dynamic reorganization of the genome shapes the recombination landscape in meiotic prophase. *Nature Structural & Molecular Biology*, *26*(3), 164–174.

Pérez-Montero, S., Carbonell, A., & Azorín, F. (2016). Germline-specific H1 variants: The "sexy" linker histones. *Chromosoma*, *125*(1), 1–13.

Prendergast, L., et al. (2020). Resolution of R-loops by INO80 promotes DNA replication and maintains cancer cell proliferation and viability. *Nature Communications*, *11*(1), 4534.

Rasmussen, K. D., & Helin, K. (2016). Role of TET enzymes in DNA methylation, development, and cancer. *Genes & Development*, *30*(7), 733–750.

Ren, X., et al. (2021). Histone benzoylation serves as an epigenetic mark for DPF and YEATS family proteins. *Nucleic Acids Research*, *49*(1), 114–126.

Rinaldi, C., et al. (2021). Sensing R-loop-associated DNA damage to safeguard genome stability. *Frontiers in Cell and Developmental Biology*, *8*(1657).

Sabari, B. R., et al. (2015). Intracellular crotonyl-CoA stimulates transcription through p300-catalyzed histone crotonylation. *Molecular Cell*, *58*(2), 203–215.

Sandhu, S., et al. (2019). A pseudo-meiotic centrosomal function of TEX12 in cancer. *bioRxiv*, 509869.

Sandhu, S., et al. (2021). Centrosome dysfunction associated with somatic expression of the synaptonemal complex protein TEX12. *Communications Biology*, *4*(1), 1371.

Sasaki, M., et al. (2021). The meiosis-specific cohesin component stromal antigen 3 promotes cell migration and chemotherapeutic resistance in colorectal cancer. *Cancer Letters*, *497*, 112–122.

Sauerwald, N., & Kingsford, C. (2018). Quantifying the similarity of topological domains across normal and cancer human cell types. *Bioinformatics*, *34*(13), i475–i483.

Schalbetter, S. A., et al. (2019). Principles of meiotic chromosome assembly revealed in S. cerevisiae. *Nature Communications*, *10*(1), 4795.

Serizay, J., & Ahringer, J. (2018). Genome organization at different scales: Nature, formation and function. *Current Opinion in Cell Biology*, *52*, 145–153.
Shannon, M., et al. (1999). Differential gene expression of mammalian SPO11/TOP6A homologs during meiosis. *FEBS Letters*, *462*(3), 329–334.
Shiromoto, Y., et al. (2021). ADAR1 RNA editing enzyme regulates R-loop formation and genome stability at telomeres in cancer cells. *Nature Communications*, *12*(1), 1654.
Shogren-Knaak, M., et al. (2006). Histone H4-K16 acetylation controls chromatin structure and protein interactions. *Science*, *311*(5762), 844–847.
Sou, I. F., et al. (2021). Meiosis initiation: A story of two sexes in all creatures great and small. *The Biochemical Journal*, *478*(20), 3791–3805.
Spitz, F., & Furlong, E. E. (2012). Transcription factors: From enhancer binding to developmental control. *Nature Reviews. Genetics*, *13*(9), 613–626.
Strunnikov, A. (2013). Cohesin complexes with a potential to link mammalian meiosis to cancer. *Cell Regeneration*, *2*(1), 4.
Su, J., et al. (2018). Homeobox oncogene activation by pan-cancer DNA hypermethylation. *Genome Biology*, *19*(1), 108.
Tachibana, M., et al. (2002). G9a histone methyltransferase plays a dominant role in euchromatic histone H3 lysine 9 methylation and is essential for early embryogenesis. *Genes & Development*, *16*(14), 1779–1791.
Tan, M., et al. (2011). Identification of 67 histone marks and histone lysine crotonylation as a new type of histone modification. *Cell*, *146*(6), 1016–1028.
Tani, R., et al. (2016). Linker histone variant H1T targets rDNA repeats. *Epigenetics*, *11*(4), 288–302.
Thiecke, M. J., et al. (2020). Cohesin-dependent and -independent mechanisms mediate chromosomal contacts between promoters and enhancers. *Cell Reports*, *32*(3), 107929.
Tuduri, S., et al. (2009). Topoisomerase I suppresses genomic instability by preventing interference between replication and transcription. *Nature Cell Biology*, *11*(11), 1315–1324.
Ueda, J., et al. (2017). Testis-specific histone variant H3t gene is essential for entry into spermatogenesis. *Cell Reports*, *18*(3), 593–600.
Ur, S. N., & Corbett, K. D. (2021). Architecture and dynamics of meiotic chromosomes. *Annual Review of Genetics*, *55*, 497–526.
Vara, C., et al. (2019). Three-dimensional genomic structure and Cohesin occupancy correlate with transcriptional activity during spermatogenesis. *Cell Reports*, *28*(2), 352–367.e9.
Vardabasso, C., et al. (2015). Histone variant H2A.Z.2 mediates proliferation and drug sensitivity of malignant melanoma. *Molecular Cell*, *59*(1), 75–88.
Vergouwen, R. P., et al. (1991). Proliferative activity of gonocytes, Sertoli cells and interstitial cells during testicular development in mice. *Journal of Reproduction and Fertility*, *93*(1), 233–243.
Vihervaara, A., et al. (2017). Transcriptional response to stress is pre-wired by promoter and enhancer architecture. *Nature Communications*, *8*(1), 255.
Vohhodina, J., et al. (2021). BRCA1 binds TERRA RNA and suppresses R-loop-based telomeric DNA damage. *Nature Communications*, *12*(1), 3542.
von Meyenn, F., & Reik, W. (2015). Forget the parents: Epigenetic reprogramming in human germ cells. *Cell*, *161*(6), 1248–1251.
Walfridsson, J., et al. (2007). A genome-wide role for CHD remodelling factors and Nap1 in nucleosome disassembly. *The EMBO Journal*, *26*(12), 2868–2879.
Wang, T., et al. (2019a). Histone variants: Critical determinants in tumour heterogeneity. *Frontiers in Medicine*, *13*(3), 289–297.
Wang, Y., et al. (2019b). Reprogramming of meiotic chromatin architecture during spermatogenesis. *Molecular Cell*, *73*(3). 547–561.e6.

Watanabe, Y. (2004). Modifying sister chromatid cohesion for meiosis. *Journal of Cell Science*, *117*(Pt. 18), 4017–4023.
Woodward, E. L., et al. (2014). Allelic variants of PRDM9 associated with high hyperdiploid childhood acute lymphoblastic leukaemia. *British Journal of Haematology*, *166*(6), 947–949.
Yamada, S., et al. (2017). The histone variant H2A.Z promotes initiation of meiotic recombination in fission yeast. *Nucleic Acids Research*, *46*(2), 609–620.
Yamaguchi, S., et al. (2012). Tet1 controls meiosis by regulating meiotic gene expression. *Nature*, *492*(7429), 443–447.
Yang, H. D., et al. (2016). Oncogenic potential of histone-variant H2A.Z.1 and its regulatory role in cell cycle and epithelial-mesenchymal transition in liver cancer. *Oncotarget*, 7(10), 11412–11423.
Yang, X., et al. (2021). RNA-DNA hybrids regulate meiotic recombination. *Cell Reports*, *37*(10), 110097.
Yatskevich, S., Rhodes, J., & Nasmyth, K. (2019). Organization of Chromosomal DNA by SMC complexes. *Annual Review of Genetics*, *53*, 445–482.
Yuen, B. T., et al. (2014). Histone H3.3 regulates dynamic chromatin states during spermatogenesis. *Development*, *141*(18), 3483–3494.
Zhao, S., Zhang, X., & Li, H. (2018). Beyond histone acetylation-writing and erasing histone acylations. *Current Opinion in Structural Biology*, *53*, 169–177.
Zhao, D., et al. (2016). YEATS2 is a selective histone crotonylation reader. *Cell Research*, *26*(5), 629–632.
Zheng, Y., et al. (2022). Unraveling three-dimensional chromatin structural dynamics during spermatogonial differentiation. *Journal of Biological Chemistry*, *298*(2), 101559.
Zuo, W., et al. (2021). Stage-resolved Hi-C analyses reveal meiotic chromosome organizational features influencing homolog alignment. *Nature Communications*, *12*(1), 5827.

CHAPTER FOUR

# Phase separation in controlling meiotic chromosome dynamics

**Ruirui Zhang†, Yuanyuan Liu†, and Jinmin Gao***
Institute of Biomedical Sciences, College of Life Sciences, Key Laboratory of Animal Resistance Biology of Shandong Province, Shandong Normal University, Jinan, China
*Corresponding author: e-mail address: jinmingao@sdnu.edu.cn

## Contents



## Abstract

Sexually reproducing organisms produce haploid gametes through meiotic cell division, during which a single round of DNA replication is followed by two consecutive chromosome segregation. A series of meiosis-specific events take place during the meiotic prophase to ensure successful chromosome segregation. These events include programmed DNA double-strand break formation, chromosome movement driven by cytoplasmic forces, homologous pairing, synaptonemal complex installation, and inter-homolog crossover formation. Phase separation has emerged as a key principle controlling cellular biomolecular material organization and biological processes. Recent studies have revealed the involvements of phase separation in assembling meiotic chromosome-associated structures. Here we review and discuss how phase separation may participate in meiotic chromosome dynamics and propose that it may provide opportunities to understand the mysteries in meiotic regulations.

† These authors contributed equally to this work.

*Current Topics in Developmental Biology*, Volume 151
ISSN 0070-2153
https://doi.org/10.1016/bs.ctdb.2022.04.004

Sexually reproducing organisms use meiosis to produce haploid gametes. During meiosis, a single round of DNA replication is followed by two consecutive chromosome segregation, meiosis I and meiosis II, reducing chromosomes number by half. Abnormal chromosome segregation during meiosis causes infertility, miscarriages, and birth defects (Hassold, Hall, & Hunt, 2007; Zickler & Kleckner, 2015). Understanding the molecular control of meiotic progression that ensures accurate chromosome segregation is pivotal for human reproductive health.

Liquid-liquid phase separation (simplified as phase separation) has been revealed as a key principle underlying cellular biomolecular material organization and biological process regulation. Various cellular structures have been revealed to assemble through phase separation. It is also associated with the pathological protein aggregates linked to multiple neurodegenerative diseases such as amyotrophic lateral sclerosis (ALS), frontotemporal dementia, and Alzheimer's disease (Zhang, Ji, et al., 2020). Furthermore, recent studies have revealed that some meiotic chromosome-associated structures form through phase separation (Claeys Bouuaert et al., 2021; Ding et al., 2019; Rog, Kohler, & Dernburg, 2017; Zhang, Xie, et al., 2020). Here we review and discuss how phase separation is involved in meiotic chromosome dynamics and propose that phase separation may help to understand the mysteries in meiotic regulations.

## 1. Chromosome dynamics during meiosis

Upon meiotic entry, chromatin is organized into a loop-axial structure. A group of meiosis-specific proteins and cohesin complexes mediate the assembly of the axis, which serves as the structural basis for the following meiotic events. A canonical meiotic program is featured with several meiosis-specific events, including programmed DNA double-strand break (DSB) formation, chromosome movement driven by cytoplasmic forces, DSB-mediated coalignment (pairing), synaptonemal complex installation (synapsis), and inter-homolog crossover (CO) formation (Fig. 1A) (Zickler & Kleckner, 2015). Organisms such as mammals, plants, budding yeast, and filamentous fungi use canonical meiotic programs to produce haploid gametes. Meiotic DSBs are generated by the conserved topoisomerase-like protein Spo11, whose activity is supported by a group of accessory proteins, and the catalysis process is spatially tethered to the meiotic chromosome axis (Lam & Keeney, 2014). After DSB formation, the DSB ends are resected to form 3′ single-strand DNA, and a group of proteins is

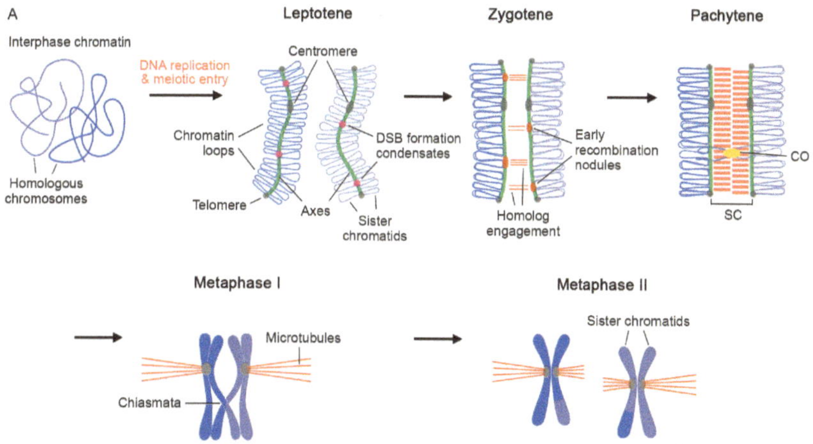


B

| Meiotic structure | Organism | Reference |
|---|---|---|
| DSB formation condensates | *Saccharomyces cerevisiae* | Claeys Bouuaert et al., 2021 |
| Pairing droplets | *Schizosaccharomyces pombe* | Ding et al., 2019 |
| The synaptonemal complex | *Caenorhabditis elegans* | Rog et al., 2017; Zhang et al., 2020b |
| Sex body (hypothetical) | Mammals | Alavattam et al., 2021; Xu and Qiao, 2021 |

**Fig. 1** Meiotic chromosome structures. (A) Chromosome dynamics during meiosis. Chromatin (blue) is organized into arrays of loops with their basis attached to a proteinaceous axis structure (green). Programmed DNA double-strand break (DSB) formation is induced by forming axis-associated condensates (DSB formation condensates). Early recombination nodules mediate homolog engagement during early prophase, following which a proteinaceous structure called synaptonemal complex (SC) is assembled between the paired homologs. A subset of early recombination nodules matures into crossovers (COs), which cause the exchange of genetic information between homologs. CO formation further provides cytological connections (Chiasma) between homologs, ensuring that the homologs and sister chromatids are sequentially segregated during meiosis I and meiosis II, respectively. (B) A table lists meiosis-specific structures in the indicated organisms that may be formed by phase separation.

recruited to the DSB sites to promote DSB repair (Hunter, 2015). A small subset of the DSBs is repaired via inter-homologous chromosomes to generate crossovers, which along with inter-sister cohesion, provide physical attachments between homologs required for proper metaphase alignment and accurate segregation (Fig. 1A). Failure to introduce the normal numbers of DSBs causes pairing and synapsis defects in these organisms (Zickler & Kleckner, 2015).

Meiotic recombination is differentially required for homologous pairing and SC installation in different organisms. In *Caenorhabditis elegans* (*C. elegans*) and female *Drosophila*, homologous pairing and synapsis occur independent of DSB formation, but DSBs are later required for CO formation and successful meiosis (Zickler & Kleckner, 2015). In *C. elegans*, pairing is mediated by pairing centers located toward one end of the chromosomes, which recruit a group of Zinc finger proteins (Phillips & Dernburg, 2006). In female *Drosophila*, the pairing of homologous chromosomes is established a few mitotic cycles before the onset of meiosis (Christophorou, Rubin, & Huynh, 2013; Joyce, Apostolopoulos, Beliveau, & Wu, 2013). Although the requirement of meiotic recombination in pairing and synapsis may be divergent, the role of SC in regulating meiotic recombination is conserved (Cahoon & Hawley, 2016; Zhang, Stauffer, Zwicker, & Dernburg, 2021). Consistently, the ultrastructure of the SC is also conserved, which consists of two lateral elements and a central region. Surprisingly, SC components are divergent and usually lack sequence similarities between organisms. However, SC central region proteins have similar secondary structural features, usually containing large fractions of coiled coils and intrinsically disordered regions (IDRs) (Gao & Colaiacovo, 2018). Moreover, the transverse filaments, an essential part of the SC central region, also exhibit a conserved organization within the SC, with their N-termini located in the middle and C-termini facing toward the lateral elements.

Albeit different programs might be applied to achieve proper chromosome segregation, dynamic chromosome movement during early prophase is a conserved feature for meiosis. Meiotic chromosome ends are recruited to a restricted area of the nuclear membrane during early prophase by trans-nuclear membrane complexes, which transduce cytoplasmic forces to drive rapid chromosome movements (Zetka, Paouneskou, & Jantsch, 2020). Such meiotic chromosome movements may help move chromosome pairs out of the way of one another to permit full-length synapsis, and/or promote the resolution of chromosome entanglements (Zickler & Kleckner, 2015).

## 2. Phase separation to form biomolecular condensates in cells

Biomaterials in the cells are organized by classical membrane-bounded organelles and membrane-less biomolecular condensates (or called membrane-less organelles). Classical membrane-bounded organelles include the nucleus, the endoplasmic reticulum, Golgi apparatus, mitochondria, lysosome, etc. Representative membrane-less biomolecular condensates include P granules, stress granules, nucleoli, Cajal bodies, nuclear speckles, etc., which are formed through phase separation (Brangwynne et al., 2009; Handwerger, Cordero, & Gall, 2005). Membrane-less organelles have higher permeability than membrane-bound organelles, and they are more dynamic in assembly and dissipation. The formation and regulation of biomolecular condensates are closely related to various biological functions, such as controlling reaction kinetics, sequestering molecules, buffering molecular concentration, and turning on and off biological processes (reviewed in Banani, Lee, Hyman, & Rosen, 2017; Zhang, Ji, et al., 2020).

From a physical perspective, phase separation occurs in cells when interactions between biomacromolecules (e.g., protein, DNA, and RNA) are more thermodynamically favorable than with the surrounding solvent. Consequently, they are highly condensed in the phase-separated volume than in the rest of the cell. Multivalent interactions between the molecules drive phase separation. Such multivalency can be achieved by forming multiple modular domains mediating protein-protein or protein-nucleic acid interactions or forming long intrinsically disordered regions (IDRs) (Banani et al., 2017). IDRs are highly flexible and can form many contacts with one another simultaneously, serving as an important indicator for phase separation potential (Uversky, Kuznetsova, Turoverov, & Zaslavsky, 2015). At the amino acid sequence level, phase separation can be driven by multiple types of interactions, including charge-charge interactions, π-Interactions, hydrophobic contacts, and hydrogen bonds (Zhang, Ji, et al., 2020). In addition, phase-separated structures usually can be disrupted by 1,6-hexanediol, a reagent that disrupts explicitly weak hydrophobic interactions (Kroschwald, Maharana, & Simon, 2017).

Phase-separated condensates may have different states, including the liquid, hydrogel, and amyloid states. Some condensates have highly dynamic properties, and their components exchange rapidly with the surrounding environment. It has also been observed that liquid-like condensates undergo liquid-to-gel transition over time (e.g., microtubule-binding protein Tau) (Wegmann et al., 2018). Physiological or pathological conditions can also cause similar transitions, and their dynamics and reversibility are compromised (Mateju et al., 2017; Murakami et al., 2015; Patel et al., 2015). While 1,6-hexanediol can usually disrupt biomolecular condensates with liquid properties, some condensates formed through LLPS cannot be completely disrupted by 1,6-hexanediol when transiting to hydrogel or amyloid states (So et al., 2019; Trivedi et al., 2019). Moreover, biomolecular condensates may have internal heterogeneities. For example, the nucleoli contain subcompartments with distinct viscosities, surface tension, and compositions (Feric et al., 2016), forming phase-in-phase or multi-phase organization.

Phase separation is actively regulated in cells to achieve various functions. Cells not only regulate the formation and dissolution of biomolecular condensates but, in many cases, also need to regulate their sizes, numbers, and locations through multiple mechanisms (Soding, Zwicker, Sohrabi-Jahromi, Boehning, & Kirschbaum, 2020; Zhang, Ji, et al., 2020). However, we are still at the early stages of understanding the mechanisms of such regulations. The current understanding of physical principles and regulations has been extensively reviewed (Dignon, Best, & Mittal, 2020; Snead & Gladfelter, 2019; Soding et al., 2020). Phase separation requires a certain concentration of its components; thus, the expression control of RNA and proteins involved in phase separation is essential. In many cases, phase separation occurs in response to specific stimuli, and enzymes and posttranslational modifications (PMTs) such as phosphorylation, methylation, and SUMOylation, are involved in controlling phase separation in a quick response (Soding et al., 2020). Moreover, many phase-separated condensates do not have random localization. They form at specific genome loci, chromosome subdomains, or attach to membrane surfaces. Such types of condensates may form around immobilized enzymes, DNA/RNA molecules, or other scaffold proteins, which have a role in triggering phase separation (Soding et al., 2020). Most of the chromatin or chromosome-associated condensates, if not all, likely belong to such localization induced condensates, as exemplified by the formation of DNA damage response (DDR) foci, DSB machinery condensates, fission yeast pairing foci, etc. (Claeys Bouuaert et al., 2021; Ding et al., 2019; Pessina et al., 2019).

A list of phase-separated biomolecular condensates present in germ cell development has been reviewed in (So, Cheng, & Schuh, 2021). We have previously summarized how phase separation controls interphase chromatin organization and chromosome behaviors during the cell cycle (Li, Gao, & Wang, 2021). In the current review, we focus on meiosis-specific structures that may be formed by phase separation (Fig. 1B) and discuss the potential functional relevance of such mechanism. We also pay special attention to the presence of IDRs in various meiotic structures, which is a prominent indicator of phase separation potential. However, modular domains and DNA/RNA sequences may have similarly important roles in phase separation of meiosis-specific structures, and they should also be considered for future investigations.

## 3. Phase separation in controlling chromatin organization and activities

Eukaryotic chromatin is organized into topologically associating domains. Analysis of reconstituted chromatin showed that nucleosome arrays have phase separation properties (Gibson et al., 2019). Moreover, nuclear RNA has an intriguing role in shaping large-scale chromatin structure and regulating genome function, and such regulation is at least partially regulated through phase separation (Michieletto & Gilbert, 2019). Following, we highlight a few examples of how phase separation controls chromatin-related structures and activities.

Studies have revealed the phase separation mechanism of heterochromatin domain formation. Two heterochromatin proteins, HP1 and CBX2, which recognize heterochromatin markers H3K9me3 and H3K27me3, respectively, exhibit IDR-driven phase separation properties (Larson et al., 2017; Plys et al., 2019; Tatavosian et al., 2019). In *Drosophila* embryos, HP1α-associated heterochromatin behaves as liquid-like condensates (Strom et al., 2017). Orthologs of HP1α and CBX2, e.g., HP1β, HP1γ, and CBX7, themselves do not have phase separation properties but have similar functions in heterochromatin organization (Larson et al., 2017; Plys et al., 2019; Tatavosian et al., 2019). However, accumulated evidence indicates that epigenetic markers and their binding proteins also control the formation of chromatin compartments via phase separation. Thus, IDR interactions, multivalent modification-reader interactions, intrinsic phase separation tendency of nucleosome arrays together drive phase separation of heterochromatin (reviewed in Zhang, Ji, et al., 2020). On the other hand, phase separation also controls euchromatin transcriptional activities. Most eukaryotic transcription factors contain IDRs in their activation domains, and the IDRs are more prevalent than in their DNA-binding domains (Liu et al., 2006). IDR regions of the transcription factors drive their phase separation and activate genes by forming high-concentration interaction hubs and recruiting RNA polymerase II (RNAPII) (Boija et al., 2018; Chong et al., 2018).

Moreover, phase separation has been revealed to be the mechanism for DNA damage response (DDR) focus formation. DDR detects DNA damage and facilitates repair by forming DDR foci at sites of DNA damage, and this response is evolutionarily conserved (Lanz, Dibitetto, & Smolka, 2019). Factors such as MDC1 and 53BP1 are recruited to the DDR foci to promote signaling and repair. DSB induced DDR also requires the recruitment of the sensor complex MRE11-RAD50-NBS1 (MRN) to the exposed DNA ends, where they further recruit the apical protein kinase ATM to phosphorylate the histone variant H2AX at serine 139 (i.e., γH2AX) (Lanz et al., 2019). Interestingly, RNAPII is also recruited to the DSB sites and induces the production of long non-coding RNA (lncRNA), which drives the phase separation of DDR factors in the DDR foci (Pessina et al., 2019). DDR foci may thus allow the very weak signals from a few DSB sites to be amplified through phase separation by recruiting a multitude of factors required to react to the signal robustly.

During mitosis, phase separation has also been suggested to underly chromosome compaction, formation of chromosome subdomains, and the special inner centromere territory (Liu et al., 2020; Ou et al., 2017; Trivedi et al., 2019). Chromosome passenger complex (CPC), the cell division master controller, localizes to the inner centromere during mitosis. A recent study showed that CPC has phase separation properties in vitro and in cells, and such properties are critical for mitotic regulation. It was proposed that phase separation of CPC at the inner centromere enable the formation of a chromatin body containing specific histone modifications and kinases, providing unique biochemical activities required for accurate chromosome segregation (Trivedi et al., 2019).

## 4. Phase separation in meiotic recombination initiation

In budding yeast, mammals, and plants, meiotic recombination is required for homologous pairing. Meiotic recombination is initiated by the conserved topoisomerase-like protein SPO11 (Lam & Keeney, 2014). The formed single-stranded DNA searches for a homologous sequence, thus facilitating the contacts and homologous recognition (Hunter, 2015). In budding yeast, DSB formation machinery subcomplex RMM (Rec114, Mei4, and Mer2) form foci and localize to the chromosome axis (Panizza et al., 2011). Rec114-Mei4 heterotrimer or Mer2 homotetramer can independently phase separate with DNA in vitro (Claeys Bouuaert et al., 2021). Mutations that weaken their DNA binding strongly disrupt both condensate formation in vitro and DSB formation in vivo. Moreover, Spo11 can be recruited to condensates of mixed RMM clusters. These observations revealed the self-assembly properties of the DSB machinery on chromosome axes, where it creates the centers of DSB activity (Fig. 2) (Claeys Bouuaert et al., 2021). This finding has several implications: each condensate is likely to recruit multiple Spo11 core complexes that may explain multi-cuts and

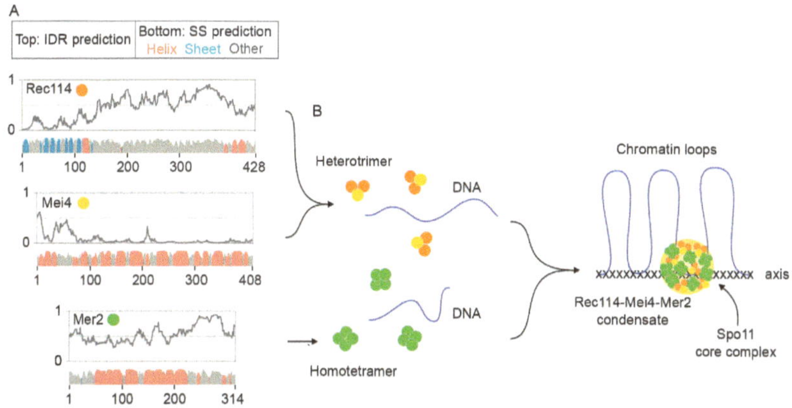


**Fig. 2** Assembly of DSB formation machinery by phase separation in *Saccharomyces cerevisiae*. (A) prediction of protein secondary structures (SS) and IDRs of RMM proteins, the essential and conserved components of the DSB machinery. The top graphs show the prediction of disorder tendencies along the length of the proteins, and the bottom graphs are predicated secondary structures (Helix, orange; Sheet, cyan-blue; Other structure, gray). The secondary structures were predicted by IUPred2 and Porter5, respectively. (B) Rec114-Mei4 forms 2:1 heterotrimers, and Mer2 forms homotetramers. RMM subcomplexes condense with DNA into reversible nucleoprotein aggregates on the chromosome axis, providing a platform to recruit the Spo11 core complex (Claeys Bouuaert et al., 2021).

periodic spacing in cuts; this model may also explain hotspot competition and DSB interference; moreover, the condensates may link DSB initiation and repair, and nucleate recombination nodules (Claeys Bouuaert et al., 2021). Given the conservation of the RMM complex in other organisms such as mice and *Arabidopsis*, assembly of DSB formation machinery through phase separation might be a conserved mechanism.

## 5. Phase separation in homologous chromosome pairing

Although distinct mechanisms of homologous pairing are used in different species (Zickler & Kleckner, 2015), homologous recognition may be achieved in common at the DNA or RNA level. However, how this operates is still a mystery. Evidence in fission yeast supports that phase separation is directly involved in DSB-independent pairing, providing insights into how RNA transcripts mediate pairing (Ding et al., 2019). The exciting findings that phase separation drives the formation of DSB machinery and DDR foci provide an intriguing possibility that DSB-mediated pairing may also involve phase separation.

## 5.1 DSB-independent pairing

The role of RNA in mediating pairing has been revealed in fission yeast *Schizosaccharomyces pombe*. A meiosis-specific lncRNA *sme2* RNA accumulates at its chromosome loci and greatly enhances pairing. RNA transcripts retained on the chromosome thus play an active role in homologs recognition and pairing (Ding et al., 2012). Further investigation identified a group of conserved *sme2* RNA-associated proteins (Smps) and additional meiosis-specific lncRNAs without sequence homology (e.g., *omt3* RNA and lncRNA584) that promote pairing. Interestingly, lncRNA-protein complexes exhibit phase separation properties and can be disrupted by 1,6-hexanediol treatment. lncRNA likely determines the pairing specificity, and it was proposed that lncRNA-protein complexes formed at specific chromosomal loci mediate recognition and subsequent pairing of homologous chromosomes (Fig. 3) (Ding et al., 2019). The loss of Smp-binding sites does not eliminate the pairing of the entire chromosomes, and the effect is local. The alignment of the entire chromosomes likely requires an array of pairing sites, which may be analogous to a bar-code model where an array of transcription sites juxtaposes a pair of homologous chromosomes (Xu & Cook, 2008). Interestingly, lncRNA-Smp droplets can fuse only when they are bearing the same RNA species, consistent with the idea that RNA sequence influences the physical properties of the droplets and determines the specificity of chromosomal loci for fusion (Ding et al., 2019). lncRNAs associate with their gene loci, acting as a recognition marker without the need for direct DNA sequence search. lncRNA-mediated condensate formation may stabilize components required for homologous recognition and increase interaction abundance, resulting in stable pairing. Such a mechanism is based on the physicochemical properties of RNA condensates and might evolve in primitive eukaryotes early in evolution and be conserved (Hiraoka, 2020). How RNA transcripts are retained on the chromosome loci is not clear. However, Some Smp proteins contain RNAPII CTD-interacting domain, and lncRNAs may be retained at the chromosomal loci through coupling with transcription and termination (Ding et al., 2019; Hiraoka, 2020).

In *Drosophila* male meiosis, meiotic recombination does not occur. However, pairing between homologous heterochromatic regions persists

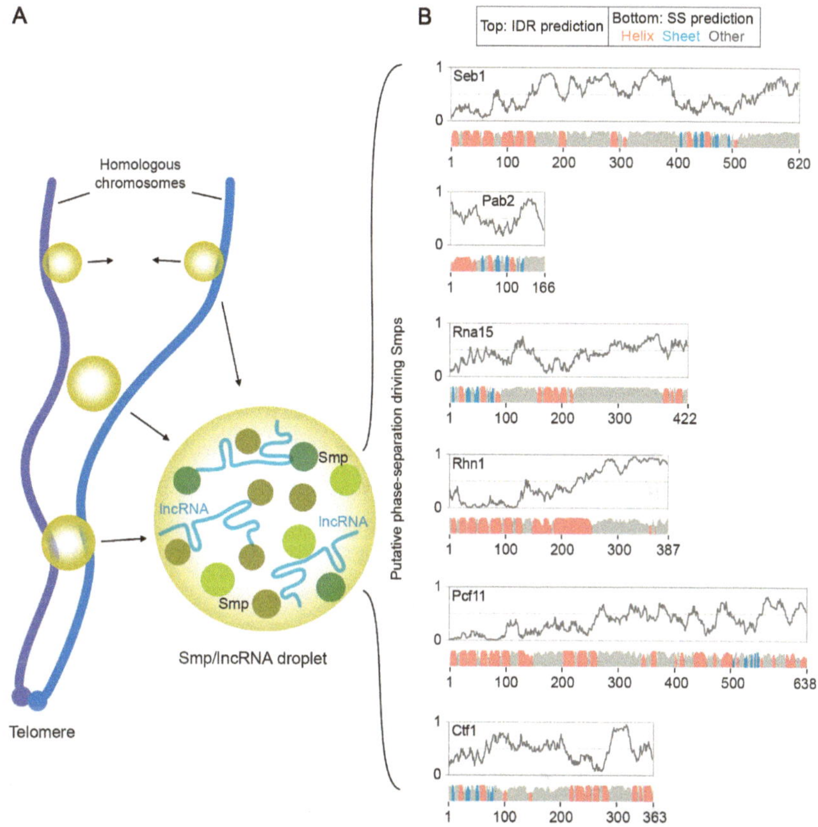


**Fig. 3** lncRNA-protein complexes promote homologous pairing in *S. pombe*. (A) A model of homologous pairing mediated by arrays of Smp/lncRNA droplets. Homologous droplets fuse together upon contact, promoting homologous pairing (Ding et al., 2019). (B) Prediction of protein secondary structures and IDRs of Smp proteins that may drive phase separation.

after prophase, taking the place of chiasmata to promote accurate segregation. Three chromosome-associated proteins, TEF, MNM, and SNM, are specifically required for homolog pairing and segregation in *Drosophila* male meiosis (McKee, Yan, & Tsai, 2012). X-Y pairing seems mediated by the tandemly repeated rRNA genes present on both the X and Y chromosomes but not the autosomes. MNM and SNM likely function together in pairing, and they colocalize to the nucleolus, where the rDNA is sequestered. For autosomes, euchromatic homology seems to be the major determinant of proper pairing and segregation. MNM and SNM first localize to multiple faint foci on autosomes. After the onset of condensation, these foci coalesce into large foci in small numbers and persist until anaphase I (Thomas et al., 2005). Both TEF and MNM have DNA-binding motifs and IDRs, and MNM also contains a BTB domain that usually mediates dimerization or multimerization. These proteins may thus have the multivalency to promote phase separation. It will be interesting to determine if phase separation is indeed involved in pairing in *Drosophila* male meiosis, which will expand our understanding of how phase separation controls homologous pairing across species. To address this, phase separation properties of the proteins involved in pairing can be tested in vitro with or without DNA. Moreover, in vivo examining the dynamic properties by experiments such as live-cell imaging and fluorescence recovery after photobleaching (FRAP) will be informative.

In *C. elegans*, homology pairing is mediated by pairing centers located near one end of each chromosome. Pairing centers comprise arrays of DNA repeats that can recruit specific zinc-finger proteins (Phillips et al., 2009; Phillips & Dernburg, 2006). Zinc-finger proteins themselves are not sufficient for homology discrimination, and sequences within or adjacent to the pairing centers likely have a role in homology recognition. However, it is unclear how the sequence homology is recognized and if RNA and phase separation might be involved. Interestingly, the Zinc finger proteins also contain large fractions of IDRs. Testing the phase separation properties of the pairing center proteins will help to understand the molecular mechanism of pairing in *C. elegans*.

## 5.2 DSB-dependent pairing

Although DSBs are required for homologous pairing in many organisms, how they promote pairing remains elusive. The numbers of DSBs generated vary a lot between organisms. They are high in organisms requiring DSBs for pairing (e.g., 200–300 per cell in mouse) and much less in organisms that do not require DSBs (12–14 in *Drosophila* and *C. elegans*). It was proposed that interactions between DSBs and their homolog partners may mediate coalignment via inter-axis bridges, where recombination proteins such as mouse RecA, human RPA, and *Sordaria* Mer3 can be observed (Zickler & Kleckner, 2015). In diverse organisms, ellipsoidal bodies with diameters up to ~100 nm can be observed closely attached to the chromosome axes by electron microscopy (von Wettstein, Rasmussen, & Holm, 1984). These ellipsoidal bodies observed during the early meiotic prophase are called early recombination nodules (ERNs), which are thought to promote homologous interaction and are often present at synaptic initiation sites. Similar bodies observed during mid-to-late pachynema are called late recombination nodules (LRNs). Recombination proteins involved in homology search (e.g., RAD51/DMC1) are present in at least some of these ERNs (Cohen & Pollard, 2001). Thus, the formation of ERNs is a potential mechanism for DSB-mediated pairing in some organisms. However, it is not clear how recombination nodules are assembled and if there is a connection between DSB formation condensates and ERNs. An in-depth dissection of the components of the recombination nodules will help determine if they may be formed through phase separation and if phase separation is involved in DSB-dependent pairing.

## 6. Phase separation in synapsis

For a long time, the SC was considered a stable structure and to play mainly a structural role, merely holding homologs together. However, recent studies suggest that the SC is dynamic, and its dynamic property can be modulated by the meiotic recombination process (Zhang, Stauffer, et al., 2021). The dynamic property of the SC may, in turn, regulate meiotic recombination (Liu et al., 2021; Nadarajan et al., 2017).

Several lines of evidence have suggested that the SC is formed through phase separation and has liquid crystalline properties in *C. elegans* (Rog et al., 2017; Zhang, Xie, et al., 2020). Both the linear SC structure assembled between the paired homologous chromosomes and round polycomplexes formed in axis-defective mutants exhibit dynamic properties and are highly sensitive to 1,6-hexanediol treatment. Moreover, SC proteins expressed in mammalian cells form droplet-like foci (Zhang, Xie, et al., 2020). Interaction analysis of SC central region proteins suggest that they form assembly units through stable interactions, and multivalent weak interactions between assembly units likely drive SC formation (Fig. 4) (Zhang, Xie, et al., 2020).

While the SC components are dynamic, they have a well-organized structure between the paired chromosomes. Transmission electron microscopy analysis revealed a multilaminar internal organization of the polycomplexes in various organisms (Hughes & Hawley, 2020), suggesting an ordered organization. SC structure is thus different from the canonical phase separation structures, which are usually small and round with unorganized internal molecules. A liquid-crystallin structure can explain the dynamics as well as the ordered organization of the SC (Rog et al., 2017).

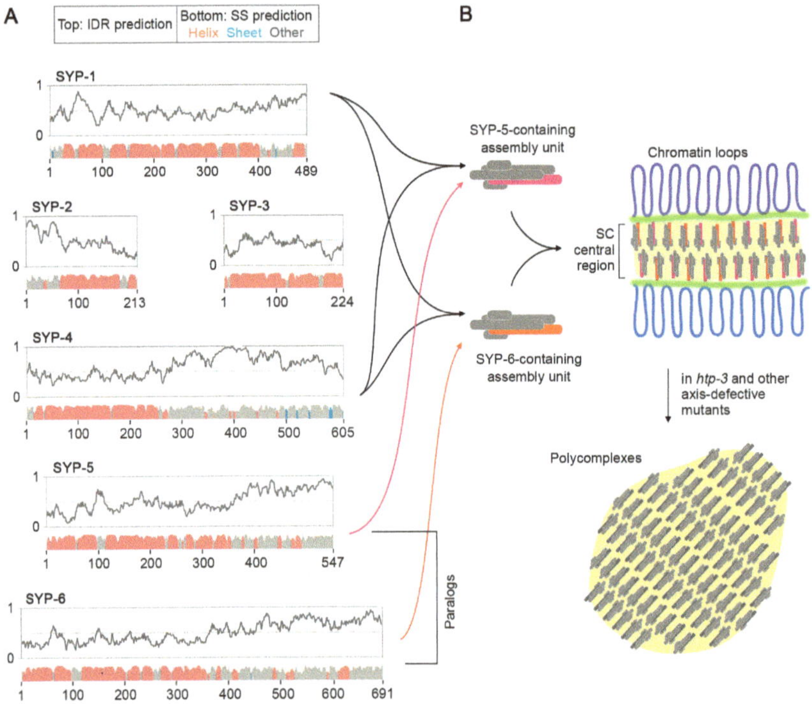


**Fig. 4** SC formation through phase separation in *C. elegans*. (A) Prediction of protein secondary structures and IDRs of SC central region proteins. SYP-5 and SYP-6 are encoded by paralog genes. (B) SYP proteins form assembly units through stable interactions, and SYP-5 and SYP-6 belong to distinct assembly units. Multivalent weak interactions between assembly units are suggested to drive SC formation between paired homologous chromosomes (Zhang, Xie, et al., 2020). In *htp-3* and other axis-defective mutants, SC central region proteins form polycomplexes, which have liquid-like properties as well as ordered internal organization (Rog et al., 2017).

Previous studies suggest that IDR regions of SC components are essential for SC assembly (Hurlock et al., 2020; Zhang, Xie, et al., 2020), suggesting phase separation of the SC may also be driven by IDRs, similar to many phase-separated structures. However, the ordered organization suggests that SC components or their N- and C-termini have interaction specificities in the middle or with the lateral elements, such that the assembly units do not randomly orient. Such properties also make the SC a unique system to investigate the principle of phase separation specificities.

SC central region proteins have been implicated in regulating CO formation across species (Zhang, Stauffer, et al., 2021). The liquid crystalline properties of the SC may allow the CO formation and interference to operate efficiently. Biochemical signals, including enzymes and PMTs involved in CO regulation, may diffuse within SC compartments. Supporting this, a group of RING finger proteins in *C. elegans*, whose homologs have a conserved role in CO regulation (Hunter, 2015), exhibit SC-dependent dynamics during meiotic prophase. They localize throughout the SCs during the early prophase and coacervate to CO sites or bivalent short arms during CO designation. These ZHP proteins might form a signaling network within the SC to control CO formation and interference (Zhang, Kohler, Rillo-Bohn, & Dernburg, 2018).

SC formation by phase separation might be conserved at least in some other organisms such as *Drosophila* and budding yeast, in which SCs are also sensitive to 1,6-hexanediol treatment (Rog et al., 2017). However, The SC might be formed through multiple types of interactions, which along with their densities, determine SC dynamic properties in distinct organisms. For example, in budding yeast, increasing the concentration of KCl potentiated the ability of 1,6-hexanediol to dissolve the SC (Rog et al., 2017), suggesting the involvement of electrostatic interactions in SC formation. In *C. elegans*, charge-interacting elements are also essential for SC formation and CO controls (Liu et al., 2021; Zhang, Xie, et al., 2020). Structural analysis of SC proteins in vitro has suggested oligomerization mechanisms of the SC proteins, and models of SC assembly were proposed (Dunce et al., 2018; Dunce, Salmon, & Davies, 2021; Dunne & Davies, 2019a, 2019b). Although some models were against the dynamic SC model, these analyses usually involve only a single or some of the SC central region proteins, thus may not reflect natural SC properties in vivo. Moreover, although the internal dynamic properties might be diverse between organisms, the idea that multivalent weak interactions mediate synapsis might be conserved. This is indirectly supported by the lack of reported interactions between SC central region proteins and lateral elements. Compared to stably paired interactions, multivalent weak interactions may allow the rapid evolve of SC components. Instead of SC central region per se, whether multivalent weak interactions between SC central region and lateral elements have a role in regulating CO interference is also a hypothesis that can be tested to explore mechanisms of CO interference.

## 7. Potential phase separation in mammalian sex body formation

In mammalian spermatocytes, XY chromosomes only pair at a small homologous region called the pseudoautosomal region (PAR). The rest of the unsynapsed chromosome regions undergo transcriptional silencing, i.e., meiotic sex chromosome inactivation (MSCI). The failure to induce MSCI causes a complete arrest and elimination of male germ cells during pachytene stage (Fernandez-Capetillo et al., 2003; Ichijima et al., 2011). During the early steps of MSCI, DDR factors including BRAC1, ATR, and γH2AX, are recruited, and repressive histone modifications including methylation, ubiquitination, SUMOylation, and acetylation are gradually established, resulting in the transcriptional silencing of the sex chromosomes, and they form a special nuclear territory, called sex body or XY body. Several lines of evidence suggest that phase separation may drive sex body formation (Alavattam, Maezawa, Andreassen, & Namekawa, 2021; Xu & Qiao, 2021). The 3D chromatin conformation of the X chromosome lack TADs, suggesting random conformation in pachytene spermatocyte population (Alavattam et al., 2019; Patel et al., 2019; Wang et al., 2019). XY body is a non-membrane-bound isolated structure located adjacent to the nucleolus, which is itself a phase-separated compartment (Frottin et al., 2019). Several key factors required for MSCI, including 53BP1, HP1, SUMOylation, and Poly(ADP-ribose), are involved in phase separation in other contexts; another two regulators, MDC1 and SCML2, contain large fractions of IDRs and are possible drivers of phase separation (Alavattam et al., 2021). Moreover, histone modifications enriched in the sex body such as γH2AX and other histone modifications may also affect phase separation of chromatin in the sex body.

The formation of the sex body through phase separation may have several functions. It may concentrate MSCI related factors, restrict specific signals within the XY body territory, exclude the transcription factors to the XY chromatin, and block the surveillance system to detect asynapsis. However, although the XY body is likely formed through phase separation, it remains to be determined what factors are the essential drivers for its phase separation and how its properties change over time.

## 8. Perspectives

Phase separation provides means for the cell to concentrate specific groups of biomolecules to perform various functions, such as sequestering biomolecules, increasing reaction rates, achieving robust signal transduction and switch-like regulations. Revealing the role of phase separation in the assembly of meiosis-specific structures provides novel insights into mechanisms coordinating various meiotic events. However, we are just starting to identify meiotic structures formed through phase separation, and how different phase-separated condensates are actively regulated during meiosis remains to be explored.

Although organized as a loop-axis structure, many chromatin structures formed through phase separation in interphase and mitotic cells might also exist on meiotic chromosomes, such as heterochromatin domains, transcriptionally active regions, and centromere-associated structures. In addition, chromosomes might condense through phase separation during the late meiotic prophase, as observed for mitotic metaphase chromosomes.

Several findings, including the phase separation properties observed for DSB formation machinery and DDR foci, recombination nodules observed under electron microscopes, trigger the proposal of whether phase separation is generally involved in the formation of meiotic recombination intermediates. With phase-separated condensates, the cells may be able to monitor the recombination process and provide proper feedback regulation, so that CO assurance and interference are operated robustly. In *C. elegans*, pro-crossover factors ZHP-3/4 were found to remain dynamic following their accumulation at recombination sites, together with other data supporting a model that recombination nodules arise through spatially scaffolded biomolecular condensates (Zhang, Zhang, & Gao, 2021). Thus, it needs to be determined what types of recombination intermediates might be formed through phase separation and what are their main drivers. Are CO-designated recombination intermediates cytologically inherited from the DSB formation condensates? If so, how are their compositions temporally regulated? Spatially, ERNs detach from the chromosome axis and relocate to the middle of the SC; what biological significance and molecular mechanism underly such spatial regulation? Moreover, what are the biochemical differences of different recombination nodules that have different repair fates?

Although the mechanisms of pairing and initiation of synapsis might be less conserved than the recombination process, could phase separation still be commonly involved? In fission yeast, DSB-independent homologous pairing is driven by the formation of protein-lncRNA droplets, requiring the transcriptional activity at specific gene loci. In *C. elegans*, a family of IDR-containing zinc finger proteins is responsible for pairing. Do these proteins have phase separation properties? In *Drosophila* male meiosis, do pairing factors such as TEF and MNM mediate pairing by forming condensates? In DSB-dependent pairing, do the chromosome axis-attached ERNs have phase separation properties? Addressing these questions will help understand if the pairing is mediated by multivalent weak interactions, making the correction of non-homologous pairing feasible. Moreover, phase separation properties of the pairing sites may have a role in locally enriching SC components to initiate synapsis. However, protein-only condensates likely lack the basis to exert homology recognition. Instead, RNAs transcribed from the specific gene loci or DNA sequences themselves may participate in this process. In DSB-dependent pairing, it will thus be important to confirm if DSB site-specific RNA products accumulated in ERNs during DSB-mediated pairing and if RNA molecules are involved in the pairing in *C. elegans* and *Drosophila*.

Various tools are now available to explore phase separation in vivo, such as applying light-controlled protein oligomerization to investigate phase separation in a spatially controlled manner in live cells (Bracha et al., 2018; Dine, Gil, Uribe, Brangwynne, & Toettcher, 2018; Shin et al., 2018). The application of photoconvertible fluorescent protein tags also helps to explore the dynamics of biomolecular assemblies. Moreover, it will be essential to dissect the compositions of various meiotic chromosome-associated condensates. Traditional proteomic approaches and recently developed techniques such as proximity labeling (Qin, Cho, Cavanagh, & Ting, 2021) will help to map the components assembled in biomolecular condensates. In addition, machine learning and phase separation prediction greatly help identify proteins driving phase separation. Finally, establishing in vitro systems will help determine the main drivers of phase separation and uncover regulation mechanisms.

## Acknowledgments

We thank the reviewers for their helpful comments. This work was supported by grants from the National Natural Science Foundation of China (32022018, 31871360, and 31701176).

## References

Alavattam, K. G., Maezawa, S., Andreassen, P. R., & Namekawa, S. H. (2021). Meiotic sex chromosome inactivation and the XY body: A phase separation hypothesis. *Cellular and Molecular Life Sciences*, *79*, 18.

Alavattam, K. G., Maezawa, S., Sakashita, A., Khoury, H., Barski, A., Kaplan, N., et al. (2019). Attenuated chromatin compartmentalization in meiosis and its maturation in sperm development. *Nature Structural & Molecular Biology*, *26*, 175–184.

Banani, S. F., Lee, H. O., Hyman, A. A., & Rosen, M. K. (2017). Biomolecular condensates: Organizers of cellular biochemistry. *Nature Reviews. Molecular Cell Biology*, *18*, 285–298.

Boija, A., Klein, I. A., Sabari, B. R., Dall'Agnese, A., Coffey, E. L., Zamudio, A. V., et al. (2018). Transcription factors activate genes through the phase-separation capacity of their activation domains. *Cell*, *175*, 1842–1855. e16.

Bracha, D., Walls, M. T., Wei, M. T., Zhu, L., Kurian, M., Avalos, J. L., et al. (2018). Mapping local and global liquid phase behavior in living cells using photo-oligomerizable seeds. *Cell, 175*, 1467–1480. e13.

Brangwynne, C. P., Eckmann, C. R., Courson, D. S., Rybarska, A., Hoege, C., Gharakhani, J., et al. (2009). Germline P granules are liquid droplets that localize by controlled dissolution/condensation. *Science, 324*, 1729–1732.

Cahoon, C. K., & Hawley, R. S. (2016). Regulating the construction and demolition of the synaptonemal complex. *Nature Structural & Molecular Biology, 23*, 369–377.

Chong, S., Dugast-Darzacq, C., Liu, Z., Dong, P., Dailey, G. M., Cattoglio, C., et al. (2018). Imaging dynamic and selective low-complexity domain interactions that control gene transcription. *Science, 361*, eaar2555.

Christophorou, N., Rubin, T., & Huynh, J. R. (2013). Synaptonemal complex components promote centromere pairing in pre-meiotic germ cells. *PLoS Genetics, 9*, e1004012.

Claeys Bouuaert, C., Pu, S., Wang, J., Oger, C., Daccache, D., Xie, W., et al. (2021). DNA-driven condensation assembles the meiotic DNA break machinery. *Nature, 592*, 144–149.

Cohen, P. E., & Pollard, J. W. (2001). Regulation of meiotic recombination and prophase I progression in mammals. *BioEssays, 23*, 996–1009.

Dignon, G. L., Best, R. B., & Mittal, J. (2020). Biomolecular phase separation: From molecular driving forces to macroscopic properties. *Annual Review of Physical Chemistry, 71*, 53–75.

Dine, E., Gil, A. A., Uribe, G., Brangwynne, C. P., & Toettcher, J. E. (2018). Protein phase separation provides long-term memory of transient spatial stimuli. *Cell Systems, 6*, 655–663. e5.

Ding, D. Q., Okamasa, K., Katou, Y., Oya, E., Nakayama, J. I., Chikashige, Y., et al. (2019). Chromosome-associated RNA-protein complexes promote pairing of homologous chromosomes during meiosis in *Schizosaccharomyces pombe*. *Nature Communications, 10*, 5598.

Ding, D. Q., Okamasa, K., Yamane, M., Tsutsumi, C., Haraguchi, T., Yamamoto, M., et al. (2012). Meiosis-specific noncoding RNA mediates robust pairing of homologous chromosomes in meiosis. *Science, 336*, 732–736.

Dunce, J. M., Dunne, O. M., Ratcliff, M., Millan, C., Madgwick, S., Uson, I., et al. (2018). Structural basis of meiotic chromosome synapsis through SYCP1 self-assembly. *Nature Structural & Molecular Biology, 25*, 557–569.

Dunce, J. M., Salmon, L. J., & Davies, O. R. (2021). Structural basis of meiotic chromosome synaptic elongation through hierarchical fibrous assembly of SYCE2-TEX12. *Nature Structural & Molecular Biology, 28*, 681–693.

Dunne, O. M., & Davies, O. R. (2019a). A molecular model for self-assembly of the synaptonemal complex protein SYCE3. *The Journal of Biological Chemistry, 294*, 9260–9275.

Dunne, O. M., & Davies, O. R. (2019b). Molecular structure of human synaptonemal complex protein SYCE1. *Chromosoma, 128*, 223–236.

Feric, M., Vaidya, N., Harmon, T. S., Mitrea, D. M., Zhu, L., Richardson, T. M., et al. (2016). Coexisting liquid phases underlie nucleolar subcompartments. *Cell, 165*, 1686–1697.

Fernandez-Capetillo, O., Mahadevaiah, S. K., Celeste, A., Romanienko, P. J., Camerini-Otero, R. D., Bonner, W. M., et al. (2003). H2AX is required for chromatin remodeling and inactivation of sex chromosomes in male mouse meiosis. *Developmental Cell, 4*, 497–508.

Frottin, F., Schueder, F., Tiwary, S., Gupta, R., Korner, R., Schlichthaerle, T., et al. (2019). The nucleolus functions as a phase-separated protein quality control compartment. *Science, 365*, 342–347.

Gao, J., & Colaiacovo, M. P. (2018). Zipping and unzipping: Protein modifications regulating synaptonemal complex dynamics. *Trends in Genetics, 34*, 232–245.

Gibson, B. A., Doolittle, L. K., Schneider, M. W. G., Jensen, L. E., Gamarra, N., Henry, L., et al. (2019). Organization of chromatin by intrinsic and regulated phase separation. *Cell, 179*, 470–484. e21.

Handwerger, K. E., Cordero, J. A., & Gall, J. G. (2005). Cajal bodies, nucleoli, and speckles in the *Xenopus* oocyte nucleus have a low-density, sponge-like structure. *Molecular Biology of the Cell, 16*, 202–211.

Hassold, T., Hall, H., & Hunt, P. (2007). The origin of human aneuploidy: Where we have been, where we are going. *Human Molecular Genetics, 16*, R203–R208. spec no. 2.

Hiraoka, Y. (2020). Phase separation drives pairing of homologous chromosomes. *Current Genetics, 66*, 881–887.

Hughes, S. E., & Hawley, R. S. (2020). Alternative synaptonemal complex structures: Too much of a good thing? *Trends in Genetics, 36*, 833–844.

Hunter, N. (2015). Meiotic recombination: The essence of heredity. *Cold Spring Harbor Perspectives in Biology*, 7, a016618.

Hurlock, M. E., Cavka, I., Kursel, L. E., Haversat, J., Wooten, M., Nizami, Z., et al. (2020). Identification of novel synaptonemal complex components in *C. elegans*. *The Journal of Cell Biology, 219*, e201910043.

Ichijima, Y., Ichijima, M., Lou, Z., Nussenzweig, A., Camerini-Otero, R. D., Chen, J., et al. (2011). MDC1 directs chromosome-wide silencing of the sex chromosomes in male germ cells. *Genes & Development, 25*, 959–971.

Joyce, E. F., Apostolopoulos, N., Beliveau, B. J., & Wu, C. T. (2013). Germline progenitors escape the widespread phenomenon of homolog pairing during *Drosophila* development. *PLoS Genetics, 9*, e1004013.

Kroschwald, S., Maharana, S., & Simon, A. (2017). Hexanediol: A chemical probe to investigate the material properties of membrane-less compartments. *Matters, 3*, e201702000010.

Lam, I., & Keeney, S. (2014). Mechanism and regulation of meiotic recombination initiation. *Cold Spring Harbor Perspectives in Biology*, 7, a016634.

Lanz, M. C., Dibitetto, D., & Smolka, M. B. (2019). DNA damage kinase signaling: Checkpoint and repair at 30 years. *The EMBO Journal, 38*, e101801.

Larson, A. G., Elnatan, D., Keenen, M. M., Trnka, M. J., Johnston, J. B., Burlingame, A. L., et al. (2017). Liquid droplet formation by HP1α suggests a role for phase separation in heterochromatin. *Nature, 547*, 236–240.

Li, J., Gao, J., & Wang, R. (2021). Control of chromatin organization and chromosome behavior during the cell cycle through phase separation. *International Journal of Molecular Sciences, 22*, 12271.

Liu, J., Perumal, N. B., Oldfield, C. J., Su, E. W., Uversky, V. N., & Dunker, A. K. (2006). Intrinsic disorder in transcription factors. *Biochemistry, 45*, 6873–6888.

Liu, X., Shen, J., Xie, L., Wei, Z., Wong, C., Li, Y., et al. (2020). Mitotic implantation of the transcription factor prospero via phase separation drives terminal neuronal differentiation. *Developmental Cell, 52*, 277–293. e8.

Liu, Y., Zhao, Q., Nie, H., Zhang, F., Fu, T., Zhang, Z., et al. (2021). SYP-5 regulates meiotic thermotolerance in *Caenorhabditis elegans*. *Journal of Molecular Cell Biology, 13*, 662–675.

Mateju, D., Franzmann, T. M., Patel, A., Kopach, A., Boczek, E. E., Maharana, S., et al. (2017). An aberrant phase transition of stress granules triggered by misfolded protein and prevented by chaperone function. *The EMBO Journal, 36*, 1669–1687.

McKee, B. D., Yan, R., & Tsai, J. H. (2012). Meiosis in male *Drosophila*. *Spermatogenesis, 2*, 167–184.

Michieletto, D., & Gilbert, N. (2019). Role of nuclear RNA in regulating chromatin structure and transcription. *Current Opinion in Cell Biology, 58*, 120–125.

Murakami, T., Qamar, S., Lin, J. Q., Schierle, G. S., Rees, E., Miyashita, A., et al. (2015). ALS/FTD mutation-induced phase transition of FUS liquid droplets and reversible hydrogels into irreversible hydrogels impairs RNP granule function. *Neuron*, *88*, 678–690.

Nadarajan, S., Lambert, T. J., Altendorfer, E., Gao, J., Blower, M. D., Waters, J. C., et al. (2017). Polo-like kinase-dependent phosphorylation of the synaptonemal complex protein SYP-4 regulates double-strand break formation through a negative feedback loop. *eLife*, *6*, e23437.

Ou, H. D., Phan, S., Deerinck, T. J., Thor, A., Ellisman, M. H., & O'Shea, C. C. (2017). ChromEMT: Visualizing 3D chromatin structure and compaction in interphase and mitotic cells. *Science*, *357*, eaag0025.

Panizza, S., Mendoza, M. A., Berlinger, M., Huang, L., Nicolas, A., Shirahige, K., et al. (2011). Spo11-accessory proteins link double-strand break sites to the chromosome axis in early meiotic recombination. *Cell*, *146*, 372–383.

Patel, A., Lee, H. O., Jawerth, L., Maharana, S., Jahnel, M., Hein, M. Y., et al. (2015). A liquid-to-solid phase transition of the ALS protein FUS accelerated by disease mutation. *Cell*, *162*, 1066–1077.

Patel, L., Kang, R., Rosenberg, S. C., Qiu, Y., Raviram, R., Chee, S., et al. (2019). Dynamic reorganization of the genome shapes the recombination landscape in meiotic prophase. *Nature Structural & Molecular Biology*, *26*, 164–174.

Pessina, F., Giavazzi, F., Yin, Y., Gioia, U., Vitelli, V., Galbiati, A., et al. (2019). Functional transcription promoters at DNA double-strand breaks mediate RNA-driven phase separation of damage-response factors. *Nature Cell Biology*, *21*, 1286–1299.

Phillips, C. M., & Dernburg, A. F. (2006). A family of zinc-finger proteins is required for chromosome-specific pairing and synapsis during meiosis in *C. elegans*. *Developmental Cell*, *11*, 817–829.

Phillips, C. M., Meng, X., Zhang, L., Chretien, J. H., Urnov, F. D., & Dernburg, A. F. (2009). Identification of chromosome sequence motifs that mediate meiotic pairing and synapsis in *C. elegans*. *Nature Cell Biology*, *11*, 934–942.

Plys, A. J., Davis, C. P., Kim, J., Rizki, G., Keenen, M. M., Marr, S. K., et al. (2019). Phase separation of Polycomb-repressive complex 1 is governed by a charged disordered region of CBX2. *Genes & Development*, *33*, 799–813.

Qin, W., Cho, K. F., Cavanagh, P. E., & Ting, A. Y. (2021). Deciphering molecular interactions by proximity labeling. *Nature Methods*, *18*, 133–143.

Rog, O., Kohler, S., & Dernburg, A. F. (2017). The synaptonemal complex has liquid crystalline properties and spatially regulates meiotic recombination factors. *eLife*, *6*, e21455.

Shin, Y., Chang, Y. C., Lee, D. S. W., Berry, J., Sanders, D. W., Ronceray, P., et al. (2018). Liquid nuclear condensates mechanically sense and restructure the genome. *Cell*, *175*, 1481–1491. e13.

Snead, W. T., & Gladfelter, A. S. (2019). The control centers of biomolecular phase separation: How membrane surfaces, PTMs, and active processes regulate condensation. *Molecular Cell*, *76*, 295–305.

So, C., Cheng, S., & Schuh, M. (2021). Phase separation during germline development. *Trends in Cell Biology*, *31*, 254–268.

So, C., Seres, K. B., Steyer, A. M., Mönnich, E., Clift, D., Pejkovska, A., et al. (2019). A liquid-like spindle domain promotes acentrosomal spindle assembly in mammalian oocytes. *Science*, *364*, eaat9557.

Soding, J., Zwicker, D., Sohrabi-Jahromi, S., Boehning, M., & Kirschbaum, J. (2020). Mechanisms for active regulation of biomolecular condensates. *Trends in Cell Biology*, *30*, 4–14.

Strom, A. R., Emelyanov, A. V., Mir, M., Fyodorov, D. V., Darzacq, X., & Karpen, G. H. (2017). Phase separation drives heterochromatin domain formation. *Nature*, *547*, 241–245.

Tatavosian, R., Kent, S., Brown, K., Yao, T., Duc, H. N., Huynh, T. N., et al. (2019). Nuclear condensates of the Polycomb protein chromobox 2 (CBX2) assemble through phase separation. *The Journal of Biological Chemistry*, *294*, 1451–1463.

Thomas, S. E., Soltani-Bejnood, M., Roth, P., Dorn, R., Logsdon, J. M., Jr., & McKee, B. D. (2005). Identification of two proteins required for conjunction and regular segregation of achiasmate homologs in *Drosophila* male meiosis. *Cell*, *123*, 555–568.

Trivedi, P., Palomba, F., Niedzialkowska, E., Digman, M. A., Gratton, E., & Stukenberg, P. T. (2019). The inner centromere is a biomolecular condensate scaffolded by the chromosomal passenger complex. *Nature Cell Biology*, *21*, 1127–1137.

Uversky, V. N., Kuznetsova, I. M., Turoverov, K. K., & Zaslavsky, B. (2015). Intrinsically disordered proteins as crucial constituents of cellular aqueous two phase systems and coacervates. *FEBS Letters*, *589*, 15–22.

von Wettstein, D., Rasmussen, S. W., & Holm, P. B. (1984). The synaptonemal complex in genetic segregation. *Annual Review of Genetics*, *18*, 331–413.

Wang, Y., Wang, H., Zhang, Y., Du, Z., Si, W., Fan, S., et al. (2019). Reprogramming of meiotic chromatin architecture during spermatogenesis. *Molecular Cell*, *73*, 547–561. e6.

Wegmann, S., Eftekharzadeh, B., Tepper, K., Zoltowska, K. M., Bennett, R. E., Dujardin, S., et al. (2018). Tau protein liquid-liquid phase separation can initiate tau aggregation. *The EMBO Journal*, *37*, e98049.

Xu, M., & Cook, P. R. (2008). Similar active genes cluster in specialized transcription factories. *The Journal of Cell Biology*, *181*, 615–623.

Xu, Y., & Qiao, H. (2021). A hypothesis: Linking phase separation to meiotic sex chromosome inactivation and sex-body formation. *Frontiers in Cell and Development Biology*, *9*, 674203.

Zetka, M., Paouneskou, D., & Jantsch, V. (2020). The nuclear envelope, a meiotic jack-of-all-trades. *Current Opinion in Cell Biology*, *64*, 34–42.

Zhang, H., Ji, X., Li, P., Liu, C., Lou, J., Wang, Z., et al. (2020). Liquid-liquid phase separation in biology: Mechanisms, physiological functions and human diseases. *Science China. Life Sciences*, *63*, 953–985.

Zhang, L., Kohler, S., Rillo-Bohn, R., & Dernburg, A. F. (2018). A compartmentalized signaling network mediates crossover control in meiosis. *eLife*, 7, e30789.

Zhang, L., Stauffer, W., Zwicker, D., & Dernburg, A. F. (2021). Crossover patterning through kinase-regulated condensation and coarsening of recombination nodules. *bioRxiv*.

Zhang, Z., Xie, S., Wang, R., Guo, S., Zhao, Q., Nie, H., et al. (2020). Multivalent weak interactions between assembly units drive synaptonemal complex formation. *The Journal of Cell Biology*, *219*, e201910086.

Zhang, F. G., Zhang, R. R., & Gao, J. M. (2021). The organization, regulation, and biological functions of the synaptonemal complex. *Asian Journal of Andrology*, *23*, 580–589.

Zickler, D., & Kleckner, N. (2015). Recombination, pairing, and synapsis of homologs during meiosis. *Cold Spring Harbor Perspectives in Biology*, 7, a016626.

CHAPTER FIVE

# Meiotic chromosome organization and its role in recombination and cancer

**Chris Morgan[a,†], Aditya Nayak[b,†], Noriko Hosoya[c], Gerald R. Smith[d], and Christophe Lambing[e,*]**

[a]John Innes Centre, Norwich, United Kingdom
[b]Department of Biology, Institute of Molecular Plant Biology, Swiss Federal Institute of Technology (ETH) Zurich, Zürich, Switzerland
[c]Laboratory of Molecular Radiology, Center for Disease Biology and Integrative Medicine, Graduate School of Medicine, The University of Tokyo, Tokyo, Japan
[d]Division of Basic Sciences, Fred Hutchinson Cancer Research Center, Seattle, WA, United States
[e]Plant Science Department, Rothamsted Research, Harpenden, United Kingdom
*Corresponding author: e-mail address: christophe.lambing@rothamsted.ac.uk

## Contents



## Abstract

Chromosomes adopt specific conformations to regulate various cellular processes. A well-documented chromosome configuration is the highly compacted chromosome structure during metaphase. More regional chromatin conformations have also been reported, including topologically associated domains encompassing mega-bases of DNA and local chromatin loops formed by kilo-bases of DNA. In this review, we discuss the changes in chromatin conformation taking place between somatic and meiotic cells, with a special focus on the establishment of a proteinaceous structure, called

† These authors contributed equally to this work.

*Current Topics in Developmental Biology*, Volume 151
ISSN 0070-2153
https://doi.org/10.1016/bs.ctdb.2022.04.008

the chromosome axis, at the beginning of meiosis. The chromosome axis is essential to support key meiotic processes such as chromosome pairing, homologous recombination, and balanced chromosome segregation to transition from a diploid to a haploid stage. We review the role of the chromosome axis in meiotic chromatin organization and provide a detailed description of its protein composition. We also review the conserved and distinct roles between species of axis proteins in meiotic recombination, which is a major factor contributing to the creation of genetic diversity and genome evolution. Finally, we discuss situations where the chromosome axis is deregulated and evaluate the effects on genome integrity and the consequences from protein deregulation in meiocytes exposed to heat stress, and aberrant expression of genes encoding axis proteins in mammalian somatic cells associated with certain types of cancers.

## 1. Chromatin and chromosome organization during meiosis

Meiosis is a specialized type of cell division essential for sexual reproduction. Meiosis consists of one round of DNA replication followed by two sequential events of chromosome segregation: the segregation of homologous chromosomes in meiosis I, and the segregation of sister centromeres and their attached chromosomal arms in meiosis II. The two rounds of chromosome segregation are central to transitioning from a diploid to a haploid stage. Proper chromosome segregation depends on the formation of a physical link between two homologous chromosomes (Kuo, Da Ines, & Lambing, 2021). In early meiosis, a topoisomerase-like complex forms DNA double-strand breaks (DSBs) that are repaired by homologous recombination. DSBs are resected to form single-strand DNA (ssDNA) molecules that can invade the sister chromatid or the homologous chromosome. ssDNAs anneal with the homologous chromosome to form heteroduplex structures. These structures can be resolved to form a crossover (CO) which consists of the reciprocal exchange of genetic information between two homologs (the nearly identical chromosomes from each parent). Alternatively, annealing can occur with the sister chromatid, which is genetically identical and cannot form a genetic recombinant (reviewed in Kuo et al., 2021). Gene conversion, which represents a unidirectional transfer of genetic information from one homolog to the other, is often associated with a CO, the reciprocal exchange of flanking DNA between homologs (reviewed in Berchowitz & Copenhaver, 2010). Gene conversion with or without a crossover (non-CO) can occur between homologous chromosomes. For most organisms, at least one CO is formed per chromosome pair to ensure correct segregation of the chromosomes during anaphase I

(Mercier, Mezard, Jenczewski, Macaisne, & Grelon, 2015). However, exceptions have been reported; male *Drosophila melanogaster* (fruit fly) and female *Bombyx mori* (moth) lack COs with no apparent defect in chromosome segregation (Morgan, 1910; Rasmussen, 1977).

Early cytological investigations of meiotic chromosomes revealed several distinct features between somatic and meiotic cells. The duration of meiotic S-phase appears to be significantly longer in several plant and mammalian species (2–6-fold depending on the species) compared to mitotic S-phase (Bennett & Smith, 1972; Callan, 1973; Cha, Weiner, Keeney, Dekker, & Kleckner, 2000; Holm, 1977). A longer duration of meiotic S-phase was also reported in *Mus musculus* (mouse) using genomic approaches, and a reduction in replication origin firing was suggested as the cause of this delay (Pratto et al., 2021). Moreover, the volume of chromatin and the nuclear size are larger in meiocytes (Fig. 1A and B), and meiotic chromosomes adopt a dense linear structure not observed with somatic chromosomes (Fig. 1C and D) (Bennett & Smith, 1972). In *Saccharomyces cerevisiae* (budding yeast) and *Schizosaccharomyces pombe* (fission yeast), Spo11, the protein forming meiotic DSBs, and the cohesin subunit Rec8, become associated with chromosomes at pre-meiotic G1 and S-phases (Cha et al., 2000; Kugou et al., 2009; Watanabe, Yokobayashi, Yamamoto, & Nurse, 2001). In these species, DNA replication is synchronized with the recruitment of meiotic proteins and is under the control of cell cycle regulatory kinases (Cdc7 in *S. cerevisiae*) (Murakami & Keeney, 2014). In the absence of *S. cerevisiae* Spo11, the duration of meiotic S-phase is decreased while the absence of Rec8 causes an increase in S-phase period (Cha et al., 2000). The coordination between DNA replication and the recruitment of meiotic proteins is a prerequisite for recombination and chromosomal processes taking place during meiosis. For instance, the pattern of meiotic recombination correlates with the pattern of germline replication in *M. musculus* and humans (Pratto et al., 2021), and meiosis initiated without a pre-meiotic S-phase exhibits recombination defects and a mitotic-like chromosome segregation pattern in *S. pombe* (Watanabe et al., 2001).

Following DNA replication, the chromatin expands along a dense linear structure, as observed under an electron microscope, and described as the chromosome axis (Fig. 1D) (Kleckner et al., 2004). The composition of the chromosome axis appears to be highly conserved between species and is composed of cohesin, coiled-coil proteins such as *S. cerevisiae* Red1, and HORMA (in reference to *H*op1, *R*ev7 and *MA*D2)-domain-containing proteins (Table 1). Genes coding for coiled-coil and HORMA-domain-containing axis proteins (HORMADs) are expressed specifically during meiosis (Fig. 1E).

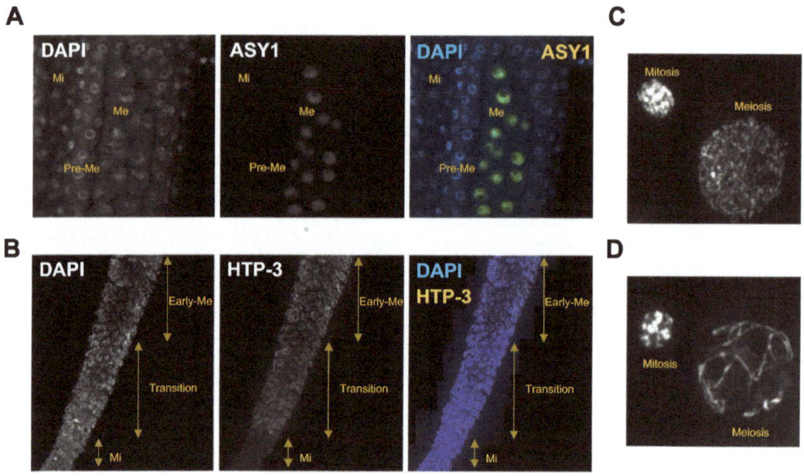


E

| | | | | | | | | | | | | |
|---|---|---|---|---|---|---|---|---|---|---|---|---|
| ***S. cerevisiae*** | *Hop1* 1030.0 / 1.4 | *Red1* 412.6 / 37.8 | *Rec8* 874.1 / 8.7 | *RAD61* (Wapl-like) 92.1 / 41.8 | *PDS5* 321.5 / 54.3 | | | | | | | |
| ***S. pombe*** | *Hop1* Ratio: 4.4 | *Rec10* Ratio: 14.9 | *Rec25* Ratio: 4.9 | *Rec27* Ratio: 1.1 | *Mug20* Ratio: 8.8 | *Rec8* Ratio: 16.4 | *Wpl1* Ratio: 1.6 | *Pds5* Ratio: 2.0 | | | | |
| ***A. thaliana*** | *ASY1* 37.9 / 0.0 | *ASY2* 0.4 / 0.2 | *ASY3* 30.2 / 0.2 | *ASY4* 77.1 / 0.1 | *REC8* 27.9 / 0.0 | *WAPL. 1* 0.9 / 0.4 | *WAPL. 2* 6.9 / 0.5 | *PDS5a* 14.2 / 3.8 | *PDS5b* 11.3 / 0.4 | *PDS5c* 10.4 / 1.9 | *PDS5d* 7.92 / 0.78 | *PDS5 e* 4.08 / 0.63 |
| ***Z. mays*** | *ASY1* 8569.1 / 920.4 | *ASY3* 876.6 / 88.7 | *ASY4* 2011.9 / 217.6 | *REC8* 13147.3 / 556.2 | *WAPL. 1* 165.1 / 349.1 | *WAPL. 2* 1009.6 / 937.9 | *WAPL. 3* 966.5 / 1468.0 | *PDS5* 2074.2 / 2876.4 | | | | |
| ***H. vulgare*** | *ASY1* 178.3 / 0.7 | *ASY3.1* 0.2 / 0.0 | *ASY3. 2* 29.9 / 0.1 | *ASY3.3* 15.6 / 5.4 | *ASY4.1* 10.9 / 1.2 | *ASY4. 2* 97.0 / 8.1 | *REC8.1* 69.7 / 6.0 | *REC8.2* 26.5 / 3.8 | *WAPL. 1* 0.8 / 0.3 | *WAPL. 2* 0.9 / 1.4 | *WAPL. 3* 36.1 / 22.9 | |
| ***D. melanogaster*** | *C(2)M* 5.1 / 0.1 | | | | | | | | | | | |
| ***C. elegans*** | *HTP-1* 133.5 / 3.9 | *HTP-2* 105.9 / 1.3 | *HTP-3* 41.1 / 2.4 | *HIM-3* 106.0 / 2.6 | *REC-8* 39.8 / 2.2 | *COH-3* 11.1 / 1.3 | *COH-4* 14.2 / 0.8 | | | | | |
| ***M. musculus*** | *HORMA D1* 91.0 / BC | *HORMA D2* 21.0 / BC | *SYCP2* 201.0 / BC | *SYCP3* 173.0 / 9.0 | *REC8* 127.0 / 0.7 | *RAD21* 77.0 / 149.0 | *RAD21 L* 6.0 / BC | *WAPL* 85.0 / 49.0 | *PDS5a* 10.0 / 10.0 | *PDS5b* 20.0 / 15.0 | | |

**Fig. 1** Comparison of gene expression and cell size between mitosis and meiosis. (A) Section of an *A. thaliana* bud. Chromatin is stained with DAPI (white or blue). ASY1-eYFP (white or yellow) is detected directly under a confocal microscope. Mi: mitosis; Pre-Me: pre-meiosis; Me: meiosis. Note the difference in cell and chromatin sizes between mitosis and meiosis. Image courtesy of Sebastien Andreuzza (B) Section of a *C. elegans* gonad. Chromatin is stained with DAPI (white or blue). HTP-3 is immunostained on chromosomes (white or yellow). Mi: mitosis; Transition: transition zone; Early-Me: early meiosis. Note the difference in chromatin size between mitosis and meiosis. Image courtesy of Chloé Girard. (C) Chromosome spread of tomato nuclei from inflorescence buds containing a mixture of somatic and meiotic cells. On the left side is a mitotic cell and on the right side is a meiotic cell in early prophase I. Chromatin is stained with DAPI. Note the difference in chromatin size between the mitotic and meiotic cells.

Cohesin is a multi-subunit complex with a ring structure essential to entrap DNA. The cohesin ring moves along the entrapped DNA to form a chromatin loop through a loop extrusion mechanism (reviewed in Davidson & Peters, 2021). The cohesin ring is composed of two Structural Maintenance of Chromosome (SMC) proteins, one kleisin and one stromal antigen (SA) protein (reviewed in Ishiguro, 2019). The SMC proteins are highly conserved across species, expressed ubiquitously, and consist of SMC1 and SMC3 in *Arabidopsis thaliana* (thale cress), *S. cerevisiae* and *D. melanogaster*; Psm1 and Psm3 in *S. pombe*; and SMC-1 and SMC-3 in *Caenorhabditis elegans* (nematode). In *M. musculus*, three SMC proteins are expressed: SMC1α, SMC1β and SMC3, among which only SMC1β is expressed specifically during meiosis (reviewed in Ishiguro, 2019). Plants and yeasts express the meiosis-specific kleisin Rec8. *C. elegans* has 3 meiotic kleisin proteins; COH-3, COH-4 and REC-8. COH-3 and COH-4 are functionally redundant and have functions distinct from those of REC-8. For instance, sister chromatin cohesion is more severely affected in *rec-8* than in *coh-3 coh-4*, and *rec-8 coh-3 coh-4* triple mutant has almost no sister chromatin cohesion (Crawley et al., 2016). *M. musculus* has three kleisins, with Rad21 being ubiquitously expressed, whereas Rad21L and Rec8 are meiosis-specific with a role in chromosome axis formation (Ward, Hopkins, Mckay, Murray, & Jordan, 2016). In contrast, *D. melanogaster* has two functionally distinct cohesin complexes; SOLO and the SA protein called SUNN are required for

(D) Chromosome spread of tomato nuclei from inflorescence buds containing a mixture of somatic and meiotic cells. On the left side is a mitotic cell and on the right side is a meiotic cell in mid prophase I. Chromatin is stained with DAPI. Note the change in chromatin conformation and the formation of a dense linear structure (synaptonemal complex). (E) Gene expression of axis proteins in meiotic and somatic cells. *S. cerevisiae* (meiosis vs vegetative stage; RNAseq; Reads Per Million reads (RPM) (Brar et al., 2012)), *S. pombe* (ratio is meiotic RNA level divided by vegetative RNA level; microarray data (http://www.bahlerlab.info/resources)), *A. thaliana* (isolated meiocyte vs leaf; RNAseq; Transcripts Per Million (TPM) (Walker et al., 2018)), *Z. mays* (isolated meiocyte vs seedling; RNAseq; RPM (Dukowic-Schulze et al., 2014)), *H. vulgare* (isolated meiocyte vs germinating embryo; RNAseq; TPM (Barakate et al., 2021)), *D. melanogaster* (ovary vs spermatheca; RNAseq; Fragments Per Kilobase of transcript per Million mapped reads (FPKM) (http://flyatlas.gla.ac.uk/FlyAtlas2/)), *C. elegans* (ovary vs larvae; RNAseq; FPKM (https://wormbase.org/)), *M. musculus* (meiosis vs kidney; RNAseq; TPM (http://www.ebi.ac.uk/gxa/)). Genes coding for HORMA-domain containing proteins are in blue, coiled-coil proteins in orange, kleisins in green, and cohesin regulators in purple.

**Table 1** Functions of chromosome axis proteins in the formation of DSBs and COs across species.

| Mutant | Species | DSB formation | CO formation | References |
|---|---|---|---|---|
| *Coiled-coil proteins* | | | | |
| *red1* | *S. cerevisiae* | – Strong reduction in DSBs<br>– Near abolishment of Rec114 and Mer2 binding to the chromatin genome-wide | – Reduction of interhomolog joint molecules and COs | Kim et al. (2010)<br>Panizza et al. (2011)<br>Schwacha and Kleckner (1997) |
| *rec10* | *S. pombe* | – Detectable DSBs at only one hotspot at approximately 1% of the wild type level | – Severe reduction in CO formation | Ellermeier and Smith (2005)<br>Fowler, Gutiérrez-Velasco, Martín-Castellanos, and Smith (2013) |
| *rec25* | *S. pombe* | – Differential DSB reduction | – Region-specific reduction in CO formation | Fowler et al. (2013)<br>Martín-Castellanos et al. (2005) |
| *rec27* | *S. pombe* | – Differential DSB reduction | – Region-specific reduction in CO formation | Fowler et al. (2013)<br>Martín-Castellanos et al. (2005) |
| *mug20* | *S. pombe* | – Differential DSB reduction | – Region-specific reduction in CO formation | Estreicher, Lorenz, and Loidl (2012)<br>Fowler et al. (2013) |
| *Sycp2* | *M. musculus* | – Slight reduction in RAD51 foci number in spermatocytes<br>– Less IHO1 foci co-localizing with SMC3-stained axis | – Unknown. Meiotic arrest and cellular apoptosis prevent the study of late stages with CO markers | Fujiwara et al. (2020) |
| *Sycp3* | *M. musculus* | – Reduction in RAD51 and DMC1 foci in spermatocytes | – Unknown. Meiotic arrests and cellular apoptosis prevent the study of late stages with CO markers | Yuan et al. (2000) |
| *asy3* | *A. thaliana* | – 29% less γH2AX foci<br>– 26% less DMC1 foci<br>– 29% less RAD51 foci<br>– 22% less MSH4 foci<br>– 77% less PRD3 foci | – 66% less chiasmata<br>– Influences class I and class II COs | Ferdous et al. (2012)<br>Vrielynck et al. (2021) |

| | | | | |
|---|---|---|---|---|
| *asy4* | *A. thaliana* | – No defect in DMC1 or MSH5 foci number | – 34% less chiasmata<br>– Influences class I and class II COs<br>– Regional effect: recombination frequency is decreased in 3 genetic intervals and increased in 1 genetic interval | Chambon et al. (2018) |
| *dsy2* | *Z. mays* | – 70% less RAD51 foci<br>– Reduction of Tunnel assay signal | – 76% less bivalent chromosomes.<br>– Unknown effect on class I or class II COs | Lee et al. (2015) |
| *HORMA proteins* | | | | |
| *hop1* | *S. cerevisiae* | – Strong reduction in DSBs<br>– Near abolishment of Rec114 binding to the chromatin genome-wide | – Reduction of interhomolog joint molecules and COs | Panizza et al. (2011)<br>Schwacha and Kleckner (1994) |
| *hop1* | *S. pombe* | – Reduction of DSBs<br>– Reduction of RAD51 foci | – Reduction of CO frequency | Latypov et al. (2010)<br>Lorenz, Estreicher, Kohli, and Loidl (2006) |
| *Hormad1* | *M. musculus* | – 91% less DMC1 foci in spermatocytes<br>– 63% less RAD51 foci in spermatocytes<br>– 64% less RPA foci in spermatocytes<br>– Reduction of MSH4 foci in spermatocytes<br>– Reduction to 2- to 4.8-fold in testis-weight-normalized SPO11-oligonucleotide level<br>– 62% less DMC1 foci in oocytes<br>– 57% less RAD51 foci in oocytes<br>– 84% less RPA foci in oocytes<br>– Reduced level of IHO1 | – Reduction in the number of MLH1 foci in spermatocytes<br>– 70% less MLH1 foci in oocytes | Daniel et al. (2011)<br>Shin et al. (2010)<br>Stanzione et al. (2016) |
| *Hormad2* | *M. musculus* | – Slight reduction in DMC1 and RAD51 foci in spermatocytes | – No change in MLH1 foci in oocytes | Wojtasz et al. (2012) |

*Continued*

**Table 1** Functions of chromosome axis proteins in the formation of DSBs and COs across species.—cont'd

| Mutant | Species | DSB formation | CO formation | References |
|---|---|---|---|---|
| *asy1* | *A. thaliana* | – No difference in γH2AX foci<br>– 66% less PRD3 foci | – 81% less chiasmata<br>– Influence class I and class II COs | Cuacos et al. (2021)<br>Sanchez-Moran, Santos, Jones, and Franklin (2007)<br>Vrielynck et al. (2021) |
| *asy1* | *Brassica rapa* | – Not reported | – 81% less chiasmata<br>– Influences class I COs. Unknown effect on class II COs | Cuacos et al. (2021) |
| *asy1* | *Triticum aestivum* (bread wheat) | – Not reported | – Partial loss of chiasmata and presence of multivalent chromosomes in *asy1* down-regulated line | Boden, Langridge, Spangenberg, and Able (2009) |
| *him-3* | *C. elegans* | – No defect in RAD51 foci count or localization | – Defect in inter-homolog recombination | Couteau, Nabeshima, Villeneuve, and Zetka (2004)<br>Couteau and Zetka (2005) |
| *htp-1* | *C. elegans* | – 70% less RAD-51 foci<br>– 23-fold increase in RAD-51 foci in *htp-1 him-3* compared to *htp-1*<br>– 12-fold increase in RAD-51 foci in *htp-1 htp-2* compared to *htp-1* | – 85% less bivalent chromosomes at diakinesis<br>– Absence of bivalent chromosomes in *htp-1 htp2*<br>– Reduction of 75% recombination frequency in a large genetic interval on the left arm of the X chromosome | Couteau and Zetka (2005)<br>Martinez-Perez and Villeneuve (2005) |
| *htp-2* | *C. elegans* | – 12-fold increase in RAD-51 foci in *htp-1 htp-2* compared to *htp-1* but below wild type level | – Absence of bivalent chromosomes in *htp-1 htp2* | Couteau and Zetka (2005) |
| *htp-3* | *C. elegans* | – Absence of RAD-51 foci<br>– Absence of RPA-1 foci | – Only univalent chromosomes | Goodyer et al. (2008) |

| | | | | |
|---|---|---|---|---|
| *Kleisins* | | | | |
| *rec8* | *S. cerevisiae* | – Redistribution of Rec114 genome-wide<br>– Differential localization of SPO11<br>– Region specific reduction in DSB formation | – No effect on inter-homolog bias of single-end invasions<br>– Inter-homolog bias is reduced on double Holliday junctions and CO rate is reduced | Kim et al. (2010)<br>Klein et al. (1999)<br>Kugou et al. (2009)<br>Panizza et al. (2011) |
| *rec8* | *S. pombe* | – Low level of DSBs at most hotspots | – Region specific reduction in CO formation | Ellermeier and Smith (2005)<br>Fowler et al. (2013) |
| *rec8* | *M. musculus* | – 23% less DMC1 foci number in spermatocytes<br>– IHO1 localization is restricted to the shorter SYCP3-stained axis | – Unknown. Meiotic arrests and cellular apoptosis prevent the study of late stages with CO markers | Bhattacharyya et al. (2019) |
| *rad21L* | *M. musculus* | 34% less DMC1 foci number in spermatocytes | – Unknown. Meiotic arrests and cellular apoptosis prevent the study of late stages with CO markers | Bhattacharyya et al. (2019) |
| *rec8/ syn1* | *A. thaliana* | – 74% less γH2AX foci<br>– 80% less RAD51 foci<br>– 77% less RPA1a foci<br>– 92% less DMC1 foci<br>– 92% less MSH4 foci | 53% less MLH1 foci | Lambing et al. (2020) |
| *afd1* | *Z. mays* | 89% less RAD51 foci | – Presence of univalent chromosomes | Pawlowski, Golubovskaya, and Cande (2003) |
| *rec-8* | *C. elegans* | – RAD51 present in *rec-8* but abolished in *rec-8 coh-3 coh-4* | – Presence of univalent chromosomes | Severson, Ling, van Zuylen, and Meyer (2009)<br>Severson and Meyer (2014) |

*Continued*

**Table 1** Functions of chromosome axis proteins in the formation of DSBs and COs across species.—cont'd

| Mutant | Species | DSB formation | CO formation | References |
|---|---|---|---|---|
| *coh-3* | *C. elegans* | – RAD51 present in *rec-8* but abolished in *rec-8 coh-3 coh-4* | – Presence of univalent chromosomes in *coh-3 coh-4* | Severson et al. (2009)<br>Severson and Meyer (2014) |
| *coh-4* | *C. elegans* | – RAD51 present in *rec-8* but abolished in *rec-8 coh-3 coh-4* | – Presence of univalent chromosomes in *coh-3 coh-4* | Severson et al. (2009)<br>Severson and Meyer (2014) |
| *c(2)m* | *D. melanogaster* | – Reduction of γH2Av foci | – Reduction in recombination frequency at several genetic intervals | Manheim and McKim (2003)<br>Mehrotra and McKim (2006) |
| *solo* | *D. melanogaster* | – No change in γH2Av foci | – Reduction in chiasma number | Yan and McKee (2013) |

sister-chromatid cohesion, and the kleisin C(2)M and Stromalin are required for homolog interactions (Gyuricza et al., 2016). In contrast, *A. thaliana* and *S. cerevisiae* express only one SA protein called SCC3. *S. pombe* has two SA proteins Psc3 and Rec11, and only Rec11 is meiosis-specific. Similarly, *M. musculus* has SA1, SA2 and SA3 (also called STAG3), with only SA3 being meiosis specific (reviewed in Ishiguro, 2019). These observations are representative of the similarities and differences of meiotic proteins and events among species, perhaps a reflection of the rapid evolution of meiosis.

Meiotic cohesin has a major role in organizing the chromatin in loop arrays along the chromosome axis (Fig. 1E) (Lambing, Tock, et al., 2020; Schalbetter, Fudenberg, Baxter, Pollard, & Neale, 2019). The role of cohesin is tightly regulated and is under the control of WAPL and PDS5. These two cohesin regulators are expressed in mitotic and meiotic cells and influence the association of cohesin with DNA to control chromatin organization. In *S. pombe* and *S. cerevisiae*, chromatin compaction is reduced in *rec8* mutants, whereas the chromosome axes are shortened and the chromatin is more compact in either *pds5* or *wapl* mutants (Challa, Lee, Shinohara, Kim, & Shinohara, 2016; Ding et al., 2016; Jin, Guacci, & Yu, 2009; Schalbetter et al., 2019). In *C. elegans wapl-1* mutant, COH-3 and COH-4, but not REC-8, accumulate on meiotic chromosomes, the chromosomes appear more compact, and the chromosome axis is shorter (Crawley et al., 2016). Intriguingly, the loss of Wapl in *M. musculus* leads to the accumulation of cohesin on the chromosomes and the formation of a dense meiotic-like axis structure in embryonic fibroblasts, revealing the propensity of somatic cohesin to form a dense chromosome axis in a deregulated environment (Tedeschi et al., 2013).

Meiotic recombination also influences chromatin interaction and the structure of the chromosome axis. The annealing of ssDNAs to the homologous chromosomes promotes novel chromatin interactions (Schalbetter et al., 2019; Zuo et al., 2021). In addition, homologous recombination facilitates the juxtaposition of homologous chromosomes leading to the formation of the synaptonemal complex (SC) in most studied organisms, with a few exceptions. For instance, *S. pombe* has an axial structure lacking all components of the central regions of the SC, while *C. elegans* and *D. melanogaster* form a SC independently of meiotic recombination (reviewed in Zickler & Kleckner, 1999). The SC has an evolutionarily conserved tripartite proteinaceous structure composed of two lateral elements, derived from axes of two homologous chromosomes, positioned in parallel orientation to each other and connected by transverse filaments located in the central region (Fig. 2).

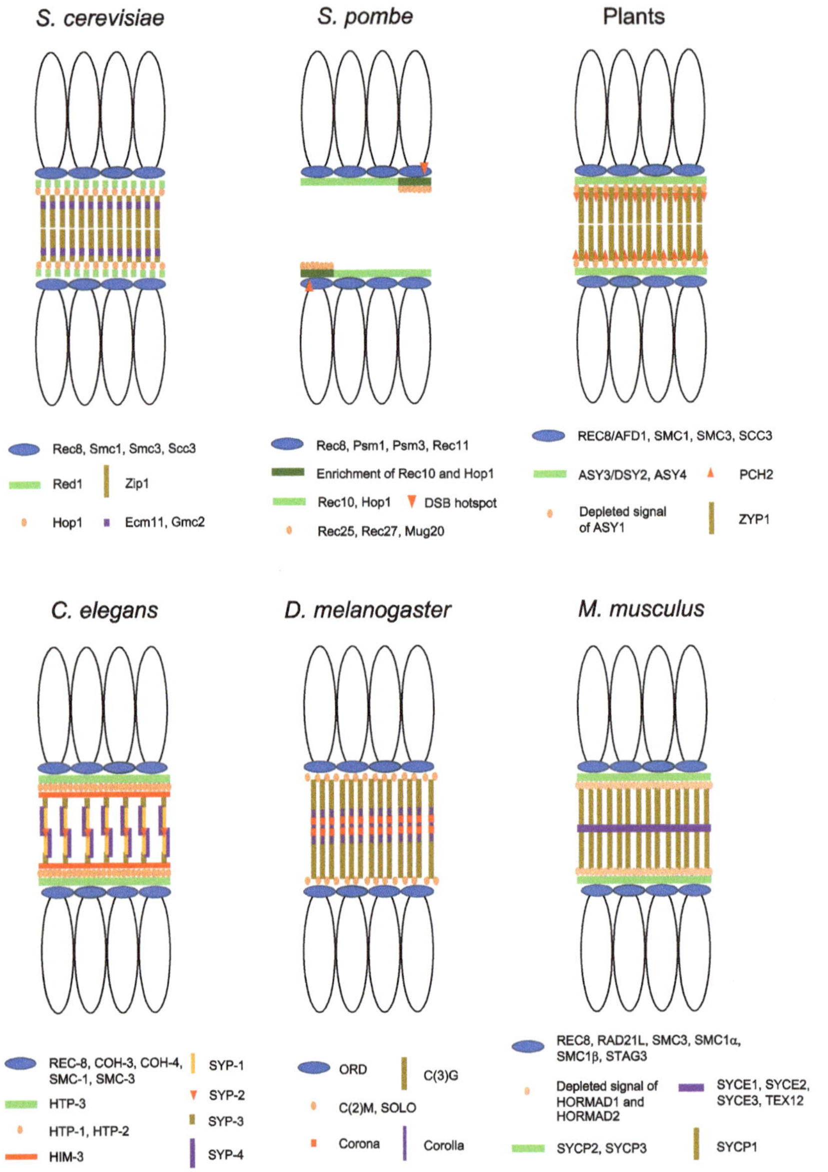


**Fig. 2** Composition of the meiotic chromosome axis across species. Schematic representation of the meiotic chromosome axis in *S. cerevisiae* (Humphryes et al., 2013; Panizza et al., 2011), *S. pombe* (Fowler et al., 2013; Kariyazono, Oda, Yamada, & Ohta, 2019), *C. elegans* (Köhler, Wojcik, Xu, & Dernburg, 2017; Schild-Prüfert et al., 2011), *D. melanogaster* (Anderson et al., 2005; Cahoon et al., 2017), *M. musculus* (Fujiwara et al., 2020; Roig et al., 2010; Yoon, Choi, Kim, & Kim, 2018) and plant (Lambing et al., 2015; Miao et al., 2013). For data leading to these models, see the indicated references.

## 2. Evolution of the core axis proteins and the synaptonemal complex

HORMADs and coiled-coil proteins belong to two protein families with structural domains that are conserved across species (Fig. 3). HORMADs are involved in several key meiotic processes, such as the formation of DSBs (a function not conserved across all orthologs) and COs (a function conserved across all orthologs) (Table 1). All HORMADs contain a short charged-hydrophobic amino acid patch called a closure motif. This closure motif allows HORMADs to undergo conformational changes between a closed and an unlocked state. These conformational changes influence protein interactions and the assembly and disassembly of the HORMADs on chromosomes, which is necessary for meiotic recombination and synapsis (West, Komives, & Corbett, 2018; Yang et al., 2020; Yang, Hu, Portheine, Chuenban, & Schnittger, 2020). Hop1/ASY1/HORMAD2 interacts with the HORMA-binding closure motifs of Red1/ASY3/SYCP2 and this interaction is necessary for their recruitment to the chromosome axis (West et al., 2019). The mode of recruitment of HORMADs to the chromosomes appears to be linked to the conserved core axis proteins in these species.

However, separation of function between HORMADs is apparent in some species. For instance, *A. thaliana ASY1* is highly expressed in meiosis and promotes CO formation, while *ASY2* shows limited expression; no functional redundancy has been reported between the two HORMADs (Table 1). A separation of function between *M. musculus* HORMAD1 and HORMAD2 in meiotic recombination has also been reported and HORMAD1 appears to have a dominant role in this process (Shin et al., 2010; Wojtasz et al., 2012) (Table 1). Hop1, ASY1 and HORMAD1 all share the common HORMA domain, but as these proteins evolved from yeast to humans, the proteins became smaller in size. *S. cerevisiae* Hop1 has 605 amino acids and *A. thaliana* ASY1 has 596 amino acid; these proteins are approximately 1.5 times longer than HORMAD1 in humans. In addition to the HORMA domain, the *S. cerevisiae* Hop1 protein contains a zinc finger motif, and this domain is replaced by a SWIRM domain in *A. thaliana* ASY1 protein. In contrast, HORMAD1 contains only the HORMA domain (Fig. 3). Both the zinc finger and SWIRM domains can bind DNA and might play a role in how Hop1 and ASY1 are loaded onto chromosomes. Although HORMAD1 lacks the DNA binding element, a recent

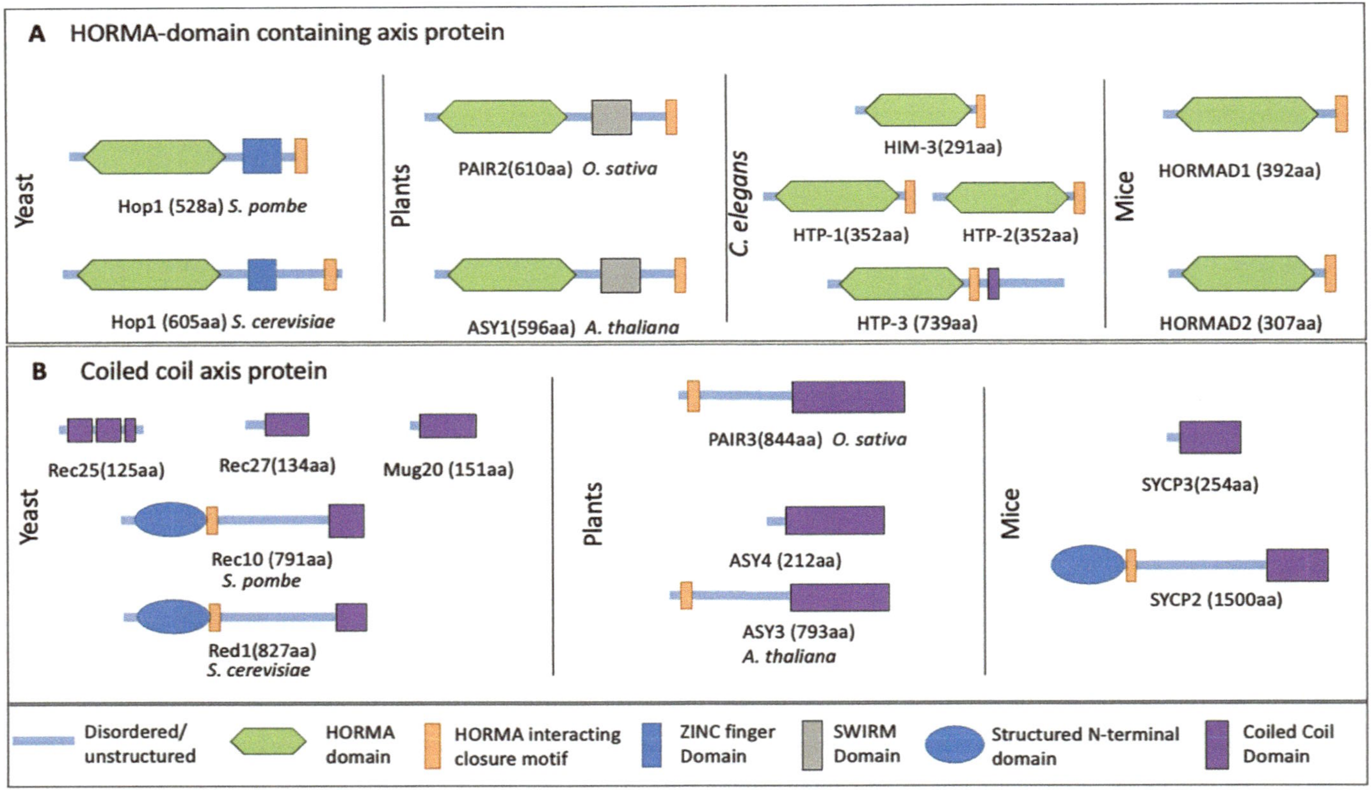


**Fig. 3** Domain architecture of meiotic axis proteins. Schematic representation depicting variation in length and domain architecture of (A) HORMA-domain containing proteins (HORMADs), (B) Coiled-coil axis proteins from species belonging to different kingdoms. Representations are made based on data from PDB and AlphaFold databases.

study showed that meiotic cohesins REC8 and RAD21L mediate initial loading of HORMAD1 to the chromosomes (Fujiwara et al., 2020).

In the yeast *Zygosaccharomyces rouxii* (budding yeast), the coiled-coil axis protein Red1 contains a 7 amino acid peptide stretch toward the C-terminal coiled-coil end that is responsible for axis formation (West et al., 2019). This property of the coiled-coil axis proteins to form an axis is conserved in mammals and plants. But as organisms evolve, the roles of axis formation and protein tetramerization become more distinct and evolve as separate protein functions. For example, in mammals, the SYCP2:SYCP3 antiparallel hetero-tetramer conformation is preferred over a homo-tetramer allowing for axis formation and bundling which leads to a stable core axis (West et al., 2019). In plants, ASY3 and ASY4 are orthologs of SYCP2 and SYCP3 and their functions are predicted to be conserved based on yeast two-hybrid interactions studies (West et al., 2019). Thus, the oligomerization and DNA interaction of axis proteins appears to be a crucial and evolutionarily conserved aspect of core axis formation.

## 3. Morphogenesis and remodeling of the chromosome axis

The chromosome axis results from a hierarchical assembly of the different axis proteins (Fig. 2). As noted above, the loading of cohesin on the chromosomes is a prerequisite for the recruitment of the other axis components (coiled-coil and HORMADs) (Fujiwara et al., 2020; Lambing, Tock, et al., 2020; Severson et al., 2009; Sun et al., 2015). In the absence of Rec8, shorter, and sometimes wider, axis-like structures or aggregates of axis proteins (polycomplexes) are formed in plants, yeasts, *M. musculus* and *C. elegans*. These aberrant structures often contain cohesin, coiled-coil proteins and HORMADs, such as in *M. musculus* and *A. thaliana* (Lambing, Tock, et al., 2020; Ward et al., 2016). Red1/ASY3 forms the second factor, after cohesin, to orchestrate axis morphogenesis, as it is required for the association of Hop1 with the chromosomes. In contrast, the role of Hop1/ASY1 in the recruitment of axis proteins is limited and varies between species. In *S. cerevisiae*, Hop1 modulates the distribution of Red1 genome-wide only when Rec8 is absent (Sun et al., 2015), while the localization of *S. pombe* Rec10 (Red1 ortholog) at DSB hotspots is dependent on Hop1 (Kariyazono et al., 2019). As opposed to *S. cerevisiae*, the localization of *A. thaliana* ASY3 (Red1 ortholog) does not require ASY1 (Hop1 ortholog) (Ferdous et al., 2012).

In *C. elegans*, no coiled-coil protein has been found associated with the chromosome axis. However, four HORMADs were identified, and their localizations are also hierarchical. All four HORMADs are dependent on the presence of the cohesin subunits COH-3, COH-4 and REC-8 (Severson et al., 2009). Following the loading of cohesin, HTP-3 is essential to recruit the other three HORMADs HTP-1, HTP-2 and HIM-3. Super-resolution imaging of meiotic chromosomes stained for axis proteins showed that HORMADs are located in the inner part while cohesins are located in the outer part of the lateral element in *C. elegans* (Fig. 2) (Köhler et al., 2017).

During chromosome pairing and the establishment of the SC, the chromosome axis is remodeled and HORMADs are dissociated from the axis by Pch2/Trip13 in *S. cerevisiae*, *A. thaliana* and *M. musculus* (Börner, Barot, & Kleckner, 2009; Lambing et al., 2015; Roig et al., 2010). Co-localization studies of Zip1/ZYP1/SYCP1 and Hop1/ASY1/HORMAD1/HORMAD2 suggest that the two proteins are mostly not co-localized in wild type conditions but show overlapping signal in *pch2/Trip13* mutants (Börner et al., 2009; Lambing et al., 2015; Roig et al., 2010). The exact role of the axis remodeling is not completely understood, but it is thought to be linked with the progression of meiotic recombination and the cell cycle, given the role of meiotic HORMADs in regulating cell cycle progression.

Genome-wide localization of axis proteins is influenced by transcription and epigenetic marks in several species. For example, *S. cerevisiae* Rec8 is enriched in the intergenic regions between convergent genes that are actively transcribed (Sun et al., 2015). Since Rec8 is a prerequisite for recruiting the other axis proteins on the DNA, a similar enrichment toward the 3′ end of convergent genes was also observed for Red1 (Sun et al., 2015). This pattern appears to be conserved between species, as *A. thaliana* REC8 occupancy is low and polarized toward the 3′ end of transcribed genes and transposons (Lambing, Tock, et al., 2020). In *S. pombe*, the pericentromeres are mostly composed of transposons that are transcriptionally repressed by chromatin marks and H3K9 methylation; the chromodomain protein Swi6 and the histone methyltransferase Clr4 are required for the recruitment of Rec8 to the pericentromeric regions (Kitajima, Yokobayashi, Yamamoto, & Watanabe, 2003; Nambiar & Smith, 2018). Swi6 has an additional role in meiosis as it prevents the recruitment of Rec11 cohesin subunit (Scc3 or STAG3 ortholog) at the pericentromeric regions to repress the formation of meiotic DSBs near centromeres (Nambiar & Smith, 2018). The *A. thaliana* genome contains an even larger region of heterochromatin compared to *S. pombe*. The loss of H3K9me2 in *A. thaliana suvh4 suvh5 suvh6* triple mutants is associated with a redistribution of REC8 over the pericentromeric regions. Certain transposons that become transcriptionally upregulated are associated with a reduction of REC8 occupancy, revealing a link between histone modification, transcription and REC8 localization (Lambing, Tock, et al., 2020). It is likely that other histone modifications influence the morphology of the axis. In *S. cerevisiae*, Esa1 is the catalytic subunit of the NuA4 complex and is responsible for histone acetylation. Esa1 regulates the length of the chromosome axis and the degree of chromosome compaction, and this is associated with a significant reduction in the level of histone acetylation during meiosis (Wang et al., 2021).

## 4. Roles of chromosomal axis proteins in DSB and crossover formation during meiosis

CO formation during meiosis appears to stem from DSB formation in all species examined. In the few species examined, DSBs are not uniformly distributed: in yeasts and *M. musculus*, they do not appear in regions with few or no COs, such as pericentric regions, and they appear at high frequency at special sites, called DSB hotspots, scattered across the rest of the genome (Cromie et al., 2007; Gerton et al., 2000; Lange et al., 2016). Where tested, DSB hotspots are also hotspots of gene conversion or crossing over or both (e.g., Cromie, Rubio, Hyppa, & Smith, 2005). In *A. thaliana* and *Zea mays* (maize), DSBs arise in regions even with low CO rates, such as the pericentric regions (Choi et al., 2018; He et al., 2017); presumably these DSBs are repaired as gene conversion or by repair with the sister chromatid (Shi et al., 2010), as discussed below.

Spo11 strictly requires several partner proteins for DSB formation (Keeney, 2007). Other proteins, such as those of the axis (Table 1), stimulate Spo11 activity or enhance Spo11's binding to the chromosome, sometimes in a region- or hotspot-specific manner. For example, the *S. cerevisiae* axis proteins Red1 and Hop1 are required nearly genome-wide for wild-type levels of DSB formation (Lam, 2016). These axis proteins recruit the DSB-forming complex or part of it. *S. cerevisiae* Hop1 binds to Mer2, an essential partner of the Spo11 complex (Panizza et al., 2011; Rousová et al., 2021). Similar interactions occur between HORMADs and Mer2 orthologs Rec15 in *S. pombe*, IHO1 in *M. musculus* and PRD3 in

*A. thaliana* (Kariyazono et al., 2019; Stanzione et al., 2016; Vrielynck et al., 2021). The recruitment of Mer2 and its orthologs to the axis appears conserved across species; PRD3 foci are reduced in *asy1* and *asy3* (Vrielynck et al., 2021), and axis-bound IHO1 localization is reduced in *Sycp2* (Fujiwara et al., 2020).

Hop1 is required for some but not all DSB formation in *S. cerevisiae* and *S. pombe* (Kariyazono et al., 2019; Schwacha & Kleckner, 1994), but its ortholog ASY1 is not required in *A. thaliana* (Sanchez-Moran et al., 2007). The assembly of the pre-DSB complex diverges between *S. cerevisiae* and *A. thaliana*. In *S. cerevisiae*, Mer2 physically interacts with the two DSB factors Rec114 and Mei4, whereas no interaction between PRD3 and PHS1 (Rec114 ortholog) or PRD2 (Mei4 ortholog) was detected in *A. thaliana* (Vrielynck et al., 2021). The non-canonical assembly of *A. thaliana* pre-DSB complex may relate to the lack of DSB defect in *asy1*.

*S. pombe* Rec10, which has limited homology with Red1, also binds Rec15 (Kariyazono et al., 2019) and is required for essentially all DSBs across the genome (Fowler et al., 2013). Three small linear element proteins with which Rec10 co-localizes (Rec25, Rec27, and Mug20) are required for DSBs at most but not all hotspots (Fowler et al., 2013). The three small proteins also bind to hotspot sites, even in Rec12's absence, with high specificity and thus are protein determinants of DSB hotspots nearly genome-wide (Fowler et al., 2013). Histone modification, such as trimethylation of histone H3, is associated with hotspot formation in *S. cerevisiae* and *M. musculus* (Baudat et al., 2010; Borde et al., 2009; Parvanov, Petkov, & Paigen, 2010). In *S. cerevisiae*, Mer2 interacts with Spp1, a histone H3K4 methyltransferase, and it was proposed that this interaction promotes DSB formation by tethering the region of a chromatin loop containing H3K4me3 with the chromosome axis (Sommermeyer, Béneut, Chaplais, Serrentino, & Borde, 2013). Other identified proteins, including several transcription factors, are hotspot determinants but at a more limited set of sites (e.g., Mieczkowski et al., 2006). The meiosis-specific cohesin subunit Rec8 is strongly required for DSB formation at most hotspots in *S. pombe* (Fowler et al., 2013) but is required to a lesser extent, in a region-specific manner, in *S. cerevisiae* (Kugou et al., 2009). In the mutants mentioned here, COs and DSBs are reduced co-ordinately, except that in *A. thaliana* using cytological assays, COs, but not DSBs, are reduced in *asy1* mutants, although they are both reduced in *rec8* and *asy3* mutants (Table 1) (Ferdous et al., 2012; Lambing, Tock, et al., 2020; Sanchez-Moran et al., 2007). Thus, some feature in addition to DSB frequency can govern CO frequency.

DSB repair with the homolog can produce a CO, a gene conversion, or both, but repair can also occur with the sister chromatid and produce (usually) no genetic recombinant, since the sisters are genetically identical. The choice of DNA for DSB repair is controlled in part by axis proteins and DNA strand-exchange proteins (see review by Humphryes & Hochwagen, 2014). Partner choice is most rigorously assayed as the relative frequency of intersister (IS) and interhomolog (IH) DNA joint molecules (Holliday junctions, or HJs). In *S. cerevisiae* the HJ assay has most frequently used the artificial *HIS4-LEU2* hotspot, which has insertions of *LEU2* and bacterial DNA into the *HIS4* locus. This assay shows that the axis proteins Red1 and Hop1; the cohesin subunit Rec8; and the strand-exchange proteins Dmc1 (meiosis-specific) and Rad51 affect the IS:IH ratio. For example, the IS:IH HJ ratio is about 1:5 in wild-type *S. cerevisiae* cells, 1:1 in *rec8Δ*, and 10:1 in *red1Δ* mutants (Kim et al., 2010). In *S. cerevisiae*, phosphorylation of Hop1 by Tel1 and Mec1 (homologs of ATM and ATR DNA damage response protein kinases) has been implicated in partner choice (Carballo, Johnson, Sedgwick, & Cha, 2008). The partner choice differs among DSB hotspots and between species. Indeed, at *S. pombe ade6* DSB hotspots, the IS:IH ratio ranges from ~6:1 with an array of *lacO* operator sequences activated by a Mug20-LacI fusion to ~3:1 with the native *mbs1* hotspot and the Atf1-Pcr1-activated hotspot *ade6-3049* (Hyppa, Cho, Nambiar, & Smith, 2022). Thus, axis proteins appear to have an important role in partner choice for DSB repair and recombination (Table 1).

The mechanism of partner choice for DSB repair is still unclear. One view, based on observations in *S. cerevisiae*, is that meiotic DSB repair is intrinsically with the homolog, but Rec8 cohesin switches repair to the sister; Red1 and Hop1, acting with strand exchange proteins Rad51 and Dmc1, counteract Rec8, returning repair to the homolog in meiosis (Kim et al., 2010). Further research may reveal additional roles for axis proteins and other factors involved in partner choice, which is critical for successful meiosis.

## 5. Does the chromosome axis play a role in CO interference?

CO interference describes the enigmatic phenomenon that when one CO forms at a particular chromosomal location it reduces the likelihood of additional COs forming nearby, influencing the eventual spacing and number of COs along individual chromosomes (Fig. 4A). Despite interference

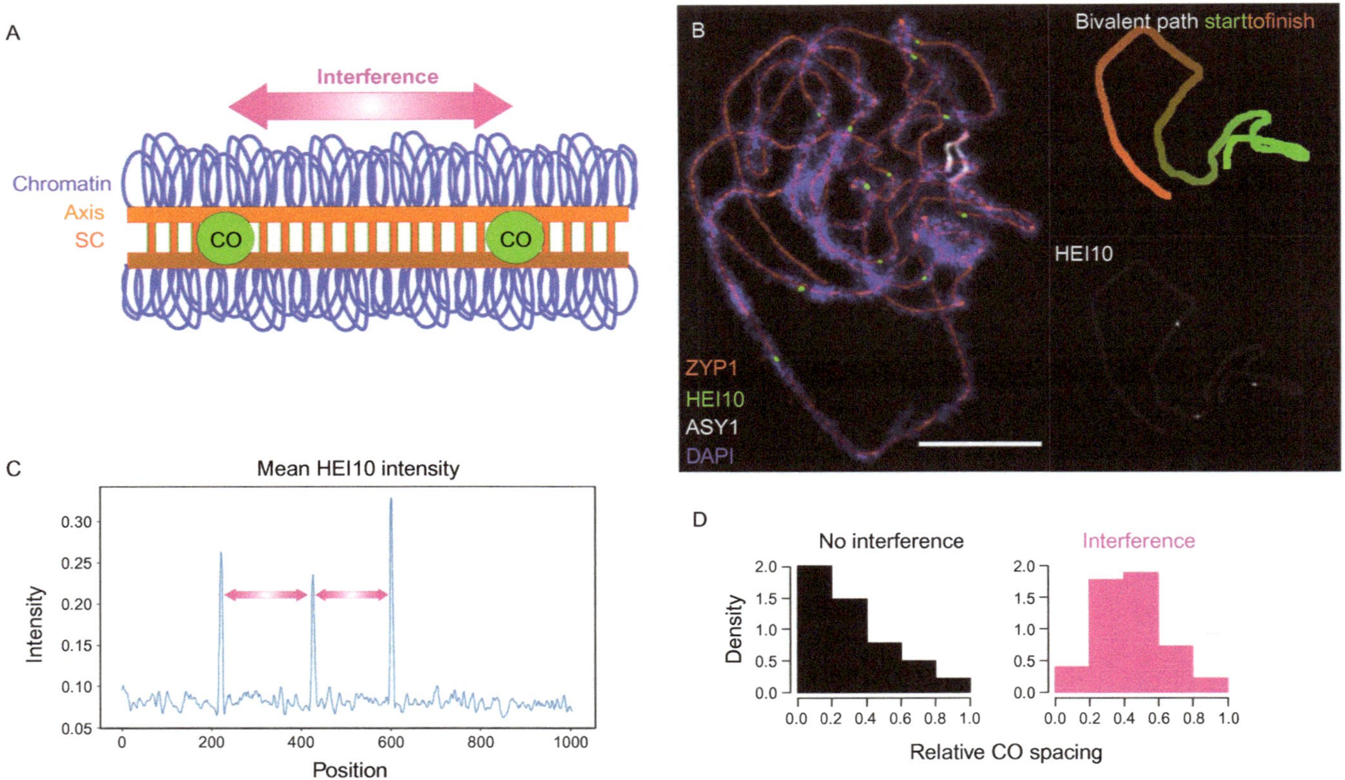


**Fig. 4** CO interference. (A) Cartoon showing the structure of a meiotic pachytene bivalent. CO interference inhibits COs from forming too close together along the length of the bivalent. (B) An example of an *A. thaliana* late-pachytene nucleus stained for ZYP1 (red), HEI10 (green), ASY1 (white) and DAPI (blue). Scale bar = 5 μm. Late HEI10 foci mark CO sites in *A. thaliana* and can be cytologically mapped along individual bivalents using the approach from Morgan et al. (2021). The orientation of a segmented bivalent from the example nucleus is shown along with the bivalent path's HEI10 channel. (C) Plot of the mean HEI10 intensity along the bivalent path from (B). Interference (signified by magenta arrows) prevents late-HEI10 foci from forming too close to one another. (D) Histograms comparing the expected spacing between adjacent COs if there was no interference (black) and late-HEI10 foci spacing along *A. thaliana* bivalents (magenta). The relative density is shown on the y-axis (so that each histogram has a total area of one) and the relative distance between adjacent COs/late-HEI10 foci (in units of SC length, normalized against total bivalent SC length in microns) is shown on the x-axis (Morgan et al., 2021).

being studied for over a century (Sturtevant, 1915), the functional role of the meiotic axis in mediating interference is still hotly debated.

Several mechanistic models have been proposed to explain interference (reviewed in: Chuang and Smith, this issue; Otto & Payseur, 2019) with some early models postulating that interference is mediated by assembly of, or transmission via, the SC (Egel, 1995). However, later studies questioned the role of the SC in interference. In *S. cerevisiae*, genetic interference is abolished in *zip1* mutants, which lack an SC (Sym & Roeder, 1994), but synapsis initiation complexes (SICs), which assemble in advance of the SC, maintain CO-like interference when assayed cytologically (Fung, Rockmill, Odell, & Roeder, 2004). This suggests that cytological interference is, at least in part, independent of the SC. *Sordaria* SICs also display cytological interference (Zhang, Espagne, de Muyt, Zickler, & Kleckner, 2014). Additionally, in *D. melanogaster* (Page & Hawley, 2001), a *c(3)G* mutant exhibits defective SC assembly but retains genetic interference.

The axis, which forms before SC assembly, then entered the spotlight as a prime candidate for transmitting interference. In support of this, perturbations of the axis can exert measurable effects on interference in a variety of organisms. For example, interference is abrogated in mutants of the HORMADs HIM-3 (Nabeshima, Villeneuve, & Hillers, 2004) and ASY1 (Lambing, Kuo, Tock, Topp, & Henderson, 2020) in *C. elegans* and *A. thaliana*, respectively. ASY1 (along with other axis and SC proteins) is also under strong selection in the model autotetraploid plant species *Arabidopsis arenosa* (Hollister et al., 2012), in which interference has evolved to stabilize polyploid meiosis (Morgan et al., 2021). By genetic analyses, interference is defective in *pch2* null-mutants in both *S. cerevisiae* and *A. thaliana* (Joshi, Barot, Jamison, & Börner, 2009; Lambing et al., 2015). However, neither of these mutants affects interference assayed cytologically (Lambing et al., 2015; Zhang et al., 2014). In *S. cerevisiae*, mutation of SUMOylation sites in axis proteins TopoII and Red1 weakens interference by cytological assay (Zhang, Wang, et al., 2014).

The reliance of interference in *S. cerevisiae* on the catalytic activity of TopoII lends support to a mechanical "stress and stress relief" model for interference (Kleckner et al., 2004), where interference is transmitted by the accumulation, relief and redistribution of mechanical stress along the meiotic axis. A mathematical model (the "beam-film" model) has also been formulated to quantitatively describe this mechanical process and has been successfully used to explain various aspects of CO patterning in several organisms (Zhang, Liang, Hutchinson, & Kleckner, 2014).

Despite this, a potential role for the SC, rather than the axis, in mediating interference has experienced a revival in recent years. In *A. thaliana* and *S. cerevisiae*, both synapsis and genetic interference are abolished in mutants lacking the SC transverse filament protein ZYP1 or ZIP1, while axis formation appears uncompromised (Capilla-Pérez et al., 2021; France et al., 2021; Sym & Roeder, 1994). In *C. elegans*, the liquid crystalline SC functions to spatially compartmentalize recombination proteins along paired chromosomes (Rog, Köhler, & Dernburg, 2017), and partial depletion of SC protein SYP-1 impairs interference (Libuda, Uzawa, Meyer, & Villeneuve, 2013). In some species that either lack or exhibit only weak interference, such as the fungi *S. pombe* and *Aspergillus nidulans* (ascomycetes), the SC is conspicuously absent (reviewed in: Chuang and Smith, this issue; Zickler & Kleckner, 1999). While *S. pombe* also lacks other ZMM proteins, such as the E3 ligase Zip3, the protist *Tetrahymena* lacks an SC and still requires Zip3 orthologues for CO formation (Shodhan, Kataoka, Mochizuki, Novatchkova, & Loidl, 2017). However, it remains unclear if *Tetrahymena thermophila* (ciliate) COs exhibit interference (Loidl, 2021).

The synthesis of these findings, combined with knowledge that interference strength is dependent upon the dosage of the Zip3 orthologue HEI10 in *A. thaliana* (Ziolkowski et al., 2017), has contributed to the development of an alternative "coarsening" model for interference. Here, CO patterning and interference are driven by the competitive coarsening of HEI10 protein clusters along pachytene bivalents (Morgan, Fozard, et al., 2021). This model is supported experimentally by quantitative cytological observations of HEI10 protein clusters and predictive mathematical simulations (Fig. 4B–D). As HEI10 is a member of a conserved family of RING-finger proteins with similar meiotic function, this coarsening paradigm may represent a conserved process explaining CO positioning in diverse species, although this has yet to be explicitly demonstrated. Thus, determining whether the axis plays a direct role (e.g., by modulating the coarsening dynamics of HEI10 or transmitting physical tension) or indirect role (e.g. by coordinating synapsis or determining bivalent length) in the mechanism of interference remains an exciting avenue for future study.

## 6. Temperature-based regulation of axis proteins and the impact on recombination

Fluctuations in temperature are known to affect recombination rates and patterns (reviewed in Morgan, Zhang, & Bomblies, 2017). Using electron microscopy and immunofluorescence, it has also been observed in various species that elevated temperatures can cause aggregation of the normally linear meiotic axis and the formation of SC polycomplexes (Fig. 5A and B) (Morgan et al., 2017). For example, in *A. thaliana* at 20 °C the axis proteins ASY1, ASY4, REC8, and the SC protein ZYP1, assemble linearly during prophase I, while at an elevated temperature of 37 °C the assembly of ASY1, ASY3 and ZYP1 (but not REC8) is disrupted and the proteins appear as punctate foci (Fu et al., 2022; Ning et al., 2021). Similar observations were also made for ASY1 and ZYP1 in *A. arenosa* grown at 22 °C and 33 °C (Morgan et al., 2017) (Fig. 3A) and, intriguingly, axis and SC proteins also appear to have undergone selection in *A. arenosa* populations adapted to warmer climates (Wright et al., 2015). In *A. thaliana*, heat-stress also affects meiotic DSB number (Ning et al., 2021), chromosomal segregation (de Storme & Geelen, 2020), meiotic duration (de Jaeger-Braet, Krause, Buchholz, & Schnittger, 2022) and the expression of numerous meiotic genes (Huang, Wang, Wang, & Copenhaver, 2021). For example, in *A. thaliana*, as well as *Hordeum vulgare* (barley), ASY1 expression is upregulated at high temperatures of 28 °C and 30 °C, respectively (Huang et al., 2021; Oshino et al., 2007).

With the advent of protein structure prediction, it is now possible to assess predicted structural elements of axis and SC proteins across kingdoms. In these predicted structures, several axis proteins contain intrinsically disordered regions which act as linkers between structured domains. For example, the predicted structure of the core axis proteins Red1 and Rec10 from *S. cerevisiae* and *S. pombe*, respectively, shows this feature, as do its plant (ASY3) and mammalian orthologs (SYCP2) (Fig. 5B). They all contain a coiled-coil domain which is surrounded by intrinsically disordered regions which are known to facilitate liquid-liquid phase separation to form bimolecular condensates. Furthermore, computational and experimental models show that temperature can be an important factor for driving liquid-liquid phase separation in disordered proteins (Dignon, Zheng, Kim, & Mittal, 2019). Intrinsically disordered regions in *C. elegans* axis and SC proteins are also enriched in charge-interacting elements (Liu et al., 2021; Zhang, Xie, et al., 2020), and mutations in these elements of SC proteins SYP-5 and SYP-4 lead to embryonic lethality at high temperature (Liu et al., 2021). Additionally, coiled-coil domains facilitate protein insolubility and aggregation at varying temperatures (Fiumara, Fioriti, Kandel, & Hendrickson, 2010) and are dominant structural elements of SC transverse filament proteins (Zip1 in *S. cerevisiae*, ZYP1a and ZYP1b in *A. thaliana*, and SYCP1 in *M. musculus*).

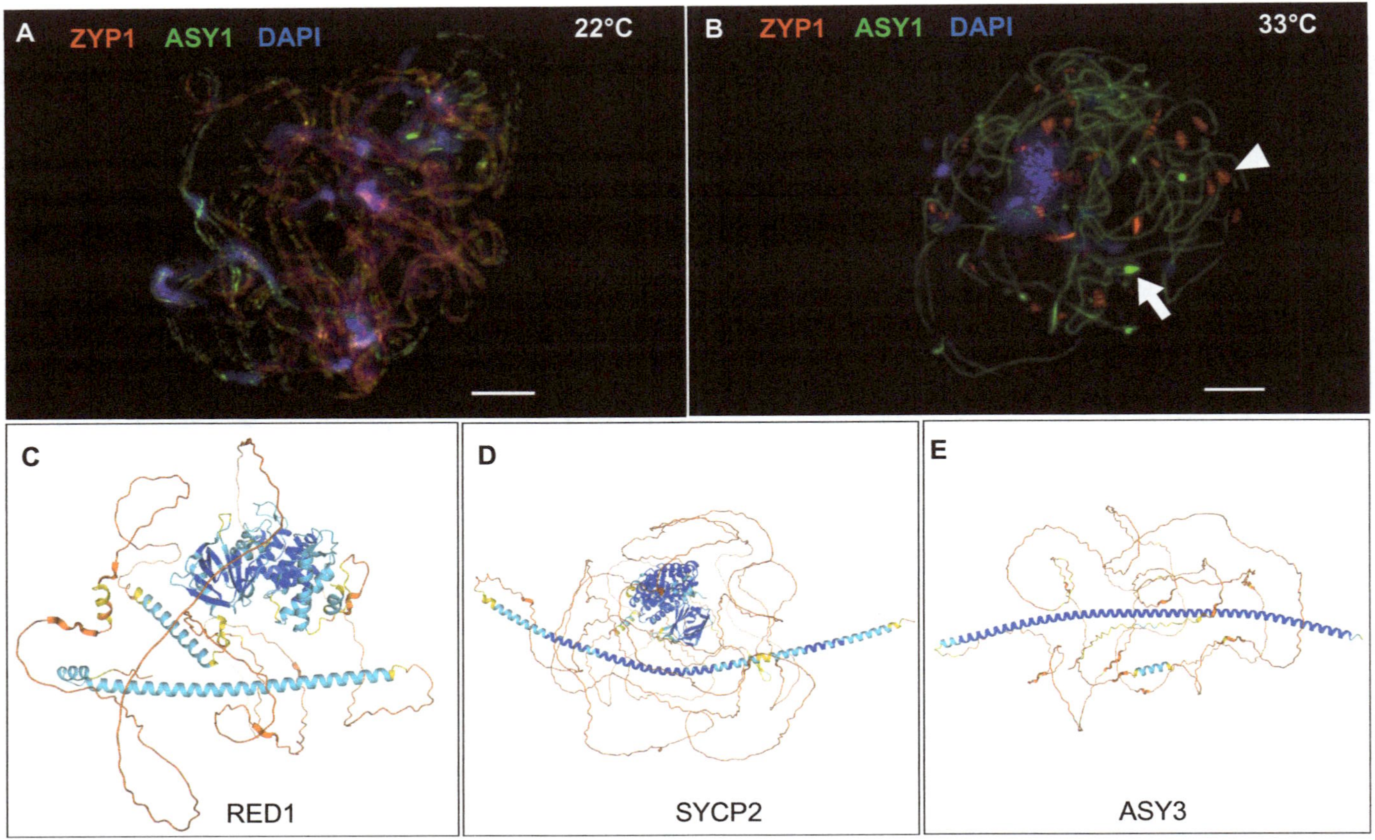


**Fig. 5** Temperature and axis proteins. (A, B) Examples of autotetraploid *Arabidopsis arenosa* prophase I nuclei from plants grown at 22 °C (A) and 33 °C (B). Nuclei are stained for ZYP1 (red), ASY1 (green) and DAPI (blue). The white arrow shows an example of an ASY1 aggregate and the white arrowhead shows an example of a ZYP1 polycomplex. Scale bars = 5 μm. (C–E) Structure of axis proteins predicted by AlphaFold (https://alphafold.ebi.ac.uk) for *Saccharomyces cerevisiae* Red1 (C), human SYCP2 (D) and *A. thaliana* ASY3 (E). Color code is used for depicting the per residue confidence score (pLDDT) which is an indicator of confidence in predicted secondary structure for each residue in a protein. Dark blue codes for confidence >90%, light blue for 70% to 90%, yellow for 50% to 70%, while orange for <50%.

Structural studies are essential for uncovering how and why individual meiotic proteins and protein complexes are affected by temperature, and combining these with molecular and cell biology approaches will be crucial for determining the underlying causes and consequences of meiotic thermal sensitivity.

## 7. Aberrant expression of the HORMADs and SC proteins regulates intrinsic DNA repair activities in somatic cancer cells

Although the SC proteins and HORMADs have long been considered to be expressed only in the germ cells, accumulating evidence has shown that these meiotic proteins are also aberrantly expressed in various somatic human cancer cells (Hosoya & Miyagawa, 2021a; Simpson, Caballero, Jungbluth, Chen, & Old, 2005). Such proteins have been called "the cancer/testis antigens" from their unique expression patterns and have been considered to be promising targets for cancer immunotherapy. Recently, the SC proteins SYCP3 and SYCE2, and the HORMAD1 protein have been reported to regulate intrinsic DSB repair activities in cancer, suggesting the roles of these proteins in the maintenance of genome integrity in somatic cells (Gao et al., 2018; Hosoya et al., 2012; Hosoya, Ono, & Miyagawa, 2018; Nichols et al., 2018).

SYCP3 is a component of the axial and lateral elements of the SC in meiotic cells. In somatic cells, SYCP3 expression has been documented in various cancers (Hosoya & Miyagawa, 2021a) and can be induced in SYCP3-nonexpressing cancer cells by treatment with the demethylating agent 5-azacytidine, indicating that a demethylation-dependent process is responsible for its ectopic expression (Hosoya et al., 2012). Clinical studies reveal that SYCP3 expression level may serve as a prognostic predictor for poor overall survival in cervical cancer and non-small cell lung cancer (Cho et al., 2014; Chung et al., 2013; Kitano et al., 2017). Mechanistically, SYCP3 inhibits intrinsic homologous recombination (HR) repair pathway for DSBs by interacting with BRCA2, a tumor suppressor whose mutations are responsible for hereditary breast and ovarian cancers (Hosoya et al., 2012). While BRCA2 binds to the meiosis-specific proteins MEILB2 and BRME1 and mediates strand invasion by RAD51 and DMC1 in meiotic recombination (Takemoto et al., 2020; Zhang, Gurusaran, et al., 2020; Zhang, Xie, et al., 2020), BRCA2 plays a mediator role at the early stages of HR by directly binding to RAD51 during mitotic recombination.

MEILB2 and BRME1 are also found activated in certain human cancers (Zhang, Gurusaran, et al., 2020; Zhang, Xie, et al., 2020). This inhibitory effect of SYCP3 on intrinsic HR in somatic cells is not only important as a cause of genomic instability but also provides an important clue in developing a novel therapeutic strategy for cancer. Cancer cells defective in HR are hypersensitive to poly(ADP-ribose) polymerase (PARP) inhibitors based on the principle of synthetic lethality, where the single-strand break repair pathway that compensates for the defects of HR in cancer cells is disrupted, leading to cancer-specific cell death (Hosoya & Miyagawa, 2014). While this concept is now being applied to BRCA1- or BRCA2-mutated cancers in cancer precision medicine (Hosoya & Miyagawa, 2021b), SYCP3-expressing cancers may also be sensitive to PARP inhibitors, even if they do not have BRCA mutations, which remains to be elucidated in the future.

SYCE2 is a component of the central elements of the SC in meiotic cells. It is also expressed at varying levels in somatic cancer cells, and its expression can be epigenetically induced by treatment with 5-azacytidine, like SYCP3 (Hosoya et al., 2018). Mechanistically, SYCE2 directly binds to heterochromatin-related protein HP1$\alpha$ through its N-terminal hydrophobic sequence and dissociates HP1$\alpha$ from trimethylated histone H3 lysine 9 (H3K9me3) to potentiate ATM-mediated DSB repair activity even in the absence of exogenous DNA damage (Hosoya et al., 2018). Among the DSB repair pathways, both HR and non-homologous end joining (NHEJ) are activated by SYCE2. These findings suggest that SYCE2 plays a role in the link between the nuclear microenvironment and the DNA damage response and repair when ectopically expressed in somatic cancer cells.

HORMAD1 expression is also observed in various cancers (Chen et al., 2005) and can be induced by treatment with 5-azacytidine (Nichols et al., 2018), like SYCP3 and SYCE2. Recent reports suggest that HORMAD1 promotes HR (Gao et al., 2018; Nichols et al., 2018). One report showed that generation of RPA foci, a protein binding to single-stranded DNA, was reduced by HORMAD1 depletion, suggesting that HORMAD1 promotes DSB end resection (Gao et al., 2018), whereas another report showed that HORMAD1 promoted RAD51-filament formation but not DNA resection (Nichols et al., 2018). Thus, the exact mechanisms and direct targets through which HORMAD1 regulates HR remain to be addressed. HORMAD1 expression in cancer cells correlates with resistance to DNA-damaging agents or PARP inhibitors (Nichols et al., 2018; Shahzad et al., 2013; Wang et al., 2018), in accord with the HR-promoting effect of HORMAD1.

These recent findings highlight the significance of meiosis-specific proteins in cancer biology and have great potential to impact the development of novel targeted cancer therapy. Thus, it would be worth investigating the currently unrecognized somatic roles of meiosis-related cancer/testis antigens to improve our understanding in mechanisms for cancer development and targets for cancer therapy.

## Acknowledgments

We thank Sebastien Andreuzza and Chloé Girard for providing images of *A. thaliana* buds and *C. elegans*, respectively; Abdellah Barakate, Changbin Chen and Gloria Brar for providing the expression of axis genes in *H. vulgare*, *Z. mays*, and *S. cerevisiae*, respectively; and Scott Keeney for fruitful discussions.

C.M. is supported by a BBSRC Discovery Fellowship (BB/V005774/1). N.H. research is funded by JSPS KAKENHI Grant Numbers JP23591836, JP25125705, JP26461881, JP16H01298, JP17K10471, and JP20K08127 and by grants from the Takeda Science Foundation and the Naito Foundation. G.R.S. research is funded by grant R35 GM118120 from the National Institutes of Health of the United States of America. C.L. research is funded by a BBSRC grant-aided support as part of the Institute Strategic Program Designing Future Wheat Grant (BB/P016855/1) and an Institutional Sponsorship Fund as part of the UKRI grant (BB/W510543/1).

## References

Anderson, L. K., Royer, S. M., Page, S. L., McKim, K. S., Lai, A., Lilly, M. A., et al. (2005). Juxtaposition of C(2)M and the transverse filament protein C(3)G within the central region of Drosophila synaptonemal complex. *Proceedings of the National Academy of Sciences of the United States of America*, *102*, 4482–4487.

Barakate, A., Orr, J., Schreiber, M., Colas, I., Lewandowska, D., McCallum, N., et al. (2021). Barley anther and meiocyte transcriptome dynamics in meiotic prophase I. *Frontiers in Plant Science*, *11*, 619404.

Baudat, F., Buard, J., Grey, C., Fledel-Alon, A., Ober, C., Przeworski, M., et al. (2010). PRDM9 is a major determinant of meiotic recombination hotspots in humans and mice. *Science*, *327*, 836–840.

Bennett, M. D., & Smith, J. B. (1972). The effect of polyploidy on meiotic duration and pollen development in cereal anthers. *Proceedings of the Royal Society of London*, *181*, 81–107.

Berchowitz, L. E., & Copenhaver, G. P. (2010). Genetic interference: Don't stand so close to me. *Current Genomics*, *11*, 91–102.

Bhattacharyya, T., Walker, M., Powers, N. R., Brunton, C., Fine, A. D., Petkov, P. M., et al. (2019). Prdm9 and meiotic cohesin proteins cooperatively promote DNA double-strand break formation in mammalian spermatocytes. *Current Biology*, *29*, 1002–1018.

Boden, S. A., Langridge, P., Spangenberg, G., & Able, J. A. (2009). TaASY1 promotes homologous chromosome interactions and is affected by deletion of *Ph1*. *The Plant Journal*, *57*, 487–497.

Borde, V., Robine, N., Lin, W., Bonfils, S., Géli, V., & Nicolas, A. (2009). Histone H3 lysine 4 trimethylation marks meiotic recombination initiation sites. *The EMBO Journal*, *28*, 99–111.

Börner, G. V., Barot, A., & Kleckner, N. (2009). Yeast Pch2 promotes domainal axis organization, timely recombination progression, and arrest of defective recombinosomes during meiosis. *Proceedings of the National Academy of Sciences of the United States of America*, *105*, 3327–3332.

Brar, G. A., Yassour, M., Friedman, N., Regev, A., Ingolia, N. T., & Weissman, J. S. (2012). High-resolution view of the yeast meiotic program revealed by ribosome profiling. *Science*, *335*, 552–557.

Cahoon, C. K., Yu, Z., Wang, Y., Guo, F., Unruh, J. R., Slaughter, B. D., et al. (2017). Superresolution expansion microscopy reveals the three-dimensional organization of the *Drosophila* synaptonemal complex. *Proceedings of the National Academy of Sciences of the United States of America*, *114*, E6857–E6866.

Callan, H. G. (1973). DNA replication in the chromosome of eukaryotes. *Cold Spring Harbor Symposia on Quantitative Biology*, *38*, 195–203.

Capilla-Pérez, L., Durand, S., Hurel, A., Lian, Q., Chambon, A., Taochy, C., et al. (2021). The synaptonemal complex imposes crossover interference and heterochiasmy in Arabidopsis. *Proceedings of the National Academy of Sciences*, *118*, e2023613118.

Carballo, J. A., Johnson, A. L., Sedgwick, S. G., & Cha, R. S. (2008). Phosphorylation of the axial element protein Hop1 by Mec1/Tel1 ensures meiotic interhomolog recombination. *Cell*, *132*, 758–770.

Cha, R. S., Weiner, B. M., Keeney, S., Dekker, J., & Kleckner, N. (2000). Progression of meiotic DNA replication is modulated by interchromosomal interaction proteins, negatively by Spo11p and positively by Rec8p. *Genes & Development*, *14*, 493–503.

Challa, K., Lee, M.-S., Shinohara, M., Kim, K. P., & Shinohara, A. (2016). Rad61/Wpl1 (Wapl), a cohesin regulator, controls chromosome compaction during meiosis. *Nucleic Acids Research*, *44*, 3190–3203.

Chambon, A., West, A., Vezon, D., Horlow, C., De Muyt, A., Chelysheva, L., et al. (2018). Identification of ASYNAPTIC4, a component of the meiotic chromosome axis. *Plant Physiology*, *178*, 233–246.

Chen, Y. T., Venditti, C. A., Theiler, G., Stevenson, B. J., Iseli, C., Gure, A. O., et al. (2005). Identification of CT46/HORMAD1, an immunogenic cancer/testis antigen encoding a putative meiosis-related protein. *Cancer Immunity*, *5*, 9.

Cho, H., Noh, K. H., Chung, J. Y., Takikita, M., Chung, E. J., Kim, B. W., et al. (2014). Synaptonemal complex protein 3 is a prognostic marker in cervical cancer. *PLoS One*, *9*, e98712.

Choi, K., Zhao, X., Tock, A. J., Lambing, C., Underwood, C. J., Hardcastle, T. J., et al. (2018). Nucleosomes and DNA methylation shape meiotic DSB frequency in *Arabidopsis thaliana* transposons and gene regulatory regions. *Genome Research*, *28*, 532–546.

Chung, J. Y., Kitano, H., Takikita, M., Cho, H., Noh, K. H., Kim, T. W., et al. (2013). Synaptonemal complex protein 3 as a novel prognostic marker in early stage non-small cell lung cancer. *Human Pathology*, *44*, 472–479.

Couteau, F., Nabeshima, K., Villeneuve, A., & Zetka, M. (2004). A component of *C. elegans* meiotic chromosome axes at the interface of homolog alignment, synapsis, nuclear reorganization, and recombination. *Current Biology*, *14*, 585–592.

Couteau, F., & Zetka, M. (2005). HTP-1 coordinates synaptonemal complex assembly with homolog alignment during meiosis in *C. elegans*. *Genes & Development*, *19*, 2744–2756.

Crawley, O., Barroso, C., Testori, S., Ferrandiz, N., Silva, N., Castellano-Pozo, M., et al. (2016). Cohesin-interacting protein WAPL-1 regulates meiotic chromosome structure and cohesion by antagonizing specific cohesin complexes. *eLife*, *5*, e10851.

Cromie, G. A., Hyppa, R. W., Cam, H. E., Farah, J. A., Grewal, S. H. I. S., & Smith, G. R. (2007). A discrete class of intergenic DNA dictates meiotic DNA break hotspots in fission yeast. *PLoS Genetics*, *3*, e141.

Cromie, G. A., Rubio, C. A., Hyppa, R. W., & Smith, G. R. (2005). A natural meiotic DNA break site in Schizosaccharomyces pombe is a hotspot of gene conversion, highly associated with crossing over. *Genetics*, *169*, 595–605.

Cuacos, M., Lambing, C., Pachon-Penalba, M., Osman, K., Armstrong, S. J., Henderson, I. R., et al. (2021). Meiotic chromosome axis remodelling is critical for meiotic recombination in *Brassica rapa*. *Journal of Experimental Botany*, *72*, 3012–3027.

Daniel, K., Lange, J., Hached, K., Fu, J., Anastassiadis, K., Roig, I., et al. (2011). Meiotic homologue alignment and its quality surveillance are controlled by mouse HORMAD1. *Nature Cell Biology*, *13*, 599–610.

Davidson, I. F., & Peters, J. M. (2021). Genome folding through loop extrusion by SMC complexes. *Nature Reviews. Molecular Cell Biology*, *22*, 445–464.

de Jaeger-Braet, J., Krause, L., Buchholz, A., & Schnittger, A. (2022). Heat stress reveals a specialized variant of the pachytene checkpoint in meiosis of Arabidopsis thaliana. *The Plant Cell*, *34*, 433–454.

de Storme, N., & Geelen, D. (2020). High temperatures alter cross-over distribution and induce male meiotic restitution in *Arabidopsis thaliana*. *Communications Biology*, *3*, 187.

Dignon, G. L., Zheng, W., Kim, Y. C., & Mittal, J. (2019). Temperature-Controlled Liquid–Liquid Phase Separation of Disordered Proteins. *ACS Central Science*, *5*, 821–830.

Ding, D.-Q., Matsuda, A., Okamasa, K., Nagahama, Y., Haraguchi, T., & Hiraoka, Y. (2016). Meiotic cohesin-based chromosome structure is essential for homologous chromosome pairing in Schizosaccharomyces pombe. *Chromosoma*, *125*, 205–214.

Dukowic-Schulze, S., Sundararajan, A., Mudge, J., Ramaraj, T., Farmer, A. D., Wang, M., et al. (2014). The transcriptome landscape of early maize meiosis. *BMC Plant Biology*, *14*, 118.

Egel, R. (1995). The synaptonemal complex and the distribution of meiotic recombination events. *Trends in Genetics*, *11*, 206–208.

Ellermeier, C., & Smith, G. R. (2005). Cohesins are required for meiotic DNA breakage and recombination in *Schizosaccharomyces pombe*. *Proceedings of the National Academy of Sciences of the United States of America*, *102*, 10952–10957.

Estreicher, A., Lorenz, A., & Loidl, J. (2012). Mug20, a novel protein associated with linear elements in fission yeast meiosis. *Current Genetics*, *58*, 119–127.

Ferdous, A., Higgins, J. D., Osman, K., Lambing, C., Roitinger, E., Mechtler, K., et al. (2012). Inter-homolog crossing-over and synapsis in Arabidopsis meiosis are dependent on the chromosome axis protein AtASY3. *PLoS Genetics*, *8*, e1002507.

Fiumara, F., Fioriti, L., Kandel, E. R., & Hendrickson, W. A. (2010). Essential role of coiled coils for aggregation and activity of Q/N-rich prions and PolyQ proteins. *Cell*, *143*, 1121–1135.

Fowler, K. R., Gutiérrez-Velasco, S., Martín-Castellanos, C., & Smith, G. R. (2013). Protein determinants of meiotic DNA break hot spots. *Molecular Cell*, *49*, 983–996.

France, M. G., Enderle, J., Röhrig, S., Puchta, H., Franklin, F. C. H., & Higgins, J. D. (2021). ZYP1 is required for obligate cross-over formation and cross-over interference in Arabidopsis. *Proceedings of the National Academy of Sciences of the United States of America*, *118*, e2021671118.

Fu, H., Zhao, J., Ren, Z., Yang, K., Wang, C., Zhang, X., et al. (2022). Interfered chromosome pairing at high temperature promotes meiotic instability in autotetraploid Arabidopsis. *Plant Physiology*, *188*, 1210–1228.

Fujiwara, Y., Horisawa-Takada, Y., Inoue, E., Tani, N., Shibuya, H., Fujimura, S., et al. (2020). Meiotic cohesins mediate initial loading of HORMAD1 to the chromosomes and coordinate SC formation during meiotic prophase. *PLoS Genetics*, *16*, e1009048.

Fung, J. C., Rockmill, B., Odell, M., & Roeder, G. S. (2004). Imposition of crossover interference through the nonrandom distribution of synapsis initiation complexes. *Cell*, *116*, 795–802.

Gao, Y., Kardos, J., Yang, Y., Tamir, T. Y., Mutter-Rottmayer, E., Weissman, B., et al. (2018). The cancer/testes (CT) antigen HORMAD1 promotes homologous recombinational DNA repair and radioresistance in lung adenocarcinoma cells. *Scientific Reports*, *8*, 15304.

Gerton, J. L., DeRisi, J., Shroff, R., Lichten, M., Brown, P. O., & Petes, T. D. (2000). Global mapping of meiotic recombination hotspots and coldspots in the yeast *Saccharomyces cerevisiae*. *Proceedings of the National Academy of Sciences of the United States of America*, *97*, 11383–11390.

Goodyer, W., Kaitna, S., Couteau, F., Ward, J. D., Boulton, S. J., & Zetka, M. (2008). HTP-3 links DSB formation with homolog pairing and crossing over during *C. elegans* meiosis. *Developmental Cell*, *14*, 263–274.

Gyuricza, M. R., Manheimer, K. B., Apte, V., Krishnan, B., Joyce, E. F., McKee, B. D., et al. (2016). Dynamic and stable cohesins regulate synaptonemal complex assembly and chromosome segregation. *Current Biology*, *26*, 1688–1698.

He, Y., Wang, M., Dukowic-Schulze, S., Zhou, A., Tiang, C.-L., Shilo, S., et al. (2017). Genomic features shaping the landscape of meiotic double-strand-break hotspots in maize. *Proceedings of the National Academy of Sciences of the United States of America*, *114*, 12231–12236.

Hollister, J. D., Arnold, B. J., Svedin, E., Xue, K. S., Dilkes, B. P., & Bomblies, K. (2012). Genetic adaptation associated with genome-doubling in autotetraploid *Arabidopsis arenosa*. *PLoS Genetics*, *8*, e1003093.

Holm, P. B. (1977). The premeiotic DNA replication of euchromatin and heterochromatin in *Lilium longiflorum* (Thunb.). *Carlsberg Research Communications*, *42*, 249–281.

Hosoya, N., & Miyagawa, K. (2014). Targeting DNA damage response in cancer therapy. *Cancer Science*, *105*, 370–388.

Hosoya, N., & Miyagawa, K. (2021a). Synaptonemal complex proteins modulate the level of genome integrity in cancers. *Cancer Science*, *112*, 989–996.

Hosoya, N., & Miyagawa, K. (2021b). Implications of the germline variants of DNA damage response genes detected by cancer precision medicine for radiological risk communication and cancer therapy decisions. *Journal of Radiation Research*, *62*, i44–i52.

Hosoya, N., Okajima, M., Kinomura, A., Fujii, Y., Hiyama, T., Sun, J., et al. (2012). Synaptonemal complex protein SYCP3 impairs mitotic recombination by interfering with BRCA2. *EMBO Reports*, *13*, 44–51.

Hosoya, N., Ono, M., & Miyagawa, K. (2018). Somatic role of SYCE2: An insulator that dissociates HP1α from H3K9me3 and potentiates DNA repair. *Life Sci Alliance*, *1*, e201800021.

Huang, J., Wang, H., Wang, Y., & Copenhaver, G. P. (2021). Comparative transcriptomic analysis of thermally stressed *Arabidopsis thaliana* meiotic recombination mutants. *BMC Genomics*, *22*, 181.

Humphryes, N., & Hochwagen, A. (2014). A non-sister act: Recombination template choice during meiosis. *Experimental Cell Research*, *329*, 53–60.

Humphryes, N., Leung, W.-K., Argunhan, B., Terentyev, Y., Dvorackova, M., & Tsubouchi, H. (2013). The Ecm11-Gmc2 complex promotes synaptonemal complex formation through assembly of transverse filaments in budding yeast. *PLoS Genetics, 9*, e1003194.

Hyppa, R. W., Cho, J. D., Nambiar, M., & Smith, G. R. (2022). Redirecting meiotic DNA break hotspot determinant proteins alters localized spatial control of DNA break formation and repair. *Nucleic Acids Research, 50*, 899–914.

Ishiguro, K.-I. (2019). The cohesin complex in mammalian meiosis. *Genes to Cells, 24*, 6–30.

Jin, H., Guacci, V., & Yu, H.-G. (2009). Pds5 is required for homologue pairing and inhibits synapsis of sister chromatids during yeast meiosis. *The Journal of Cell Biology, 186*, 713–725.

Joshi, N., Barot, A., Jamison, C., & Börner, G. V. (2009). Pch2 links chromosome axis remodeling at future crossover sites and crossover distribution during yeast meiosis. *PLoS Genetics, 5*, e1000557.

Kariyazono, R., Oda, A., Yamada, T., & Ohta, K. (2019). Conserved HORMA domain-containing protein Hop1 stabilizes interaction between proteins of meiotic DNA break hotspots and chromosome axis. *Nucleic Acids Research, 47*, 10166–10180.

Keeney, S. (2007). Spo11 and the formation of DNA double-strand breaks in meiosis. In R. Egel, & D.-H. Lankenau (Eds.), *Recombination and meiosis: Crossing-over and disjunction* (pp. 81–123). Berlin: Springer-Verlag.

Kim, K. P., Weiner, B. M., Zhang, L., Jordan, A., Dekker, J., & Kleckner, N. (2010). Sister cohesion and structural axis components mediate homolog bias of meiotic recombination. *Cell, 143*, 924–937.

Kitajima, T. S., Yokobayashi, S., Yamamoto, M., & Watanabe, Y. (2003). Distinct cohesin complexes organize meiotic chromosome domains. *Science, 300*, 1152–1155.

Kitano, H., Chung, J. Y., Noh, K. H., Lee, Y. H., Kim, T. W., Lee, S. H., et al. (2017). Synaptonemal complex protein 3 is associated with lymphangiogenesis in non-small cell lung cancer patients with lymph node metastasis. *Journal of Translational Medicine, 15*, 138.

Kleckner, N., Zickler, D., Jones, G. H., Dekker, J., Padmore, R., Henle, J., et al. (2004). A mechanical basis for chromosome function. *Proceedings of the National Academy of Sciences of the United States of America, 101*, 12592–12597.

Klein, F., Mahr, P., Galova, M., Buonomo, S. B., Michaelis, C., Nairz, K., et al. (1999). A central role for cohesins in sister chromatid cohesion, formation of axial elements, and recombination during yeast meiosis. *Cell, 98*, 91–103.

Köhler, S., Wojcik, M., Xu, K., & Dernburg, A. F. (2017). Superresolution microscopy reveals the three-dimensional organization of meiotic chromosome axes in intact *Caenorhabditis elegans* tissue. *Proceedings of the National Academy of Sciences of the United States of America, 114*, E4734–E4743.

Kugou, K., Fukuda, T., Yamada, S., Ito, M., Sasanuma, H., Mori, S., et al. (2009). Rec8 guides canonical Spo11 distribution along yeast meiotic chromosomes. *Molecular Biology of the Cell, 20*, 3064–3076.

Kuo, P., Da Ines, O., & Lambing, C. (2021). Rewiring meiosis for crop improvement. *Frontiers in Plant Science, 12*, 708948.

Lam, I. (2016). *The meiotic recombination initiation landscape in yeast: Evolutionary dynamics and factors that shape its distribution*. PhD thesis Memorial Sloan Kettering Cancer Center.

Lambing, C., Kuo, P. C., Tock, A. J., Topp, S. D., & Henderson, I. R. (2020). ASY1 acts as a dosage-dependent antagonist of telomere-led recombination and mediates crossover interference in *Arabidopsis*. *Proceedings of the National Academy of Sciences of the United States of America, 117*, 13647–13658.

Lambing, C., Osman, K., Nuntasoontorn, K., West, A., Higgins, J. D., Copenhaver, G. P., et al. (2015). Arabidopsis PCH2 mediates chromosome remodeling and maturation of crossovers. *PLoS Genetics*, *11*, e1005372.

Lambing, C., Tock, A. J., Topp, S. D., Choi, K., Kuo, P. C., Zhao, X., et al. (2020). Interacting genomic landscapes of REC8-cohesin, chromatin, and meiotic recombination in Arabidopsis. *The Plant Cell*, *32*, 1218–1239.

Lange, J., Yamada, S., Tischfield, S. E., Pan, J., Kim, S., Zhu, X., et al. (2016). The landscape of mouse meiotic double-strand break formation, processing, and repair. *Cell*, *167*, 695–708.

Latypov, V., Rothenberg, M., Lorenz, A., Octobre, G., Csutak, O., Lehmann, E., et al. (2010). Roles of Hop1 and Mek1 in meiotic chromosome pairing and recombination partner choice in *Schizosaccharomyces pombe*. *Molecular and Cellular Biology*, *30*, 1570–1581.

Lee, D. H., Kao, Y.-H., Ku, J.-C., Lin, C.-Y., Meeley, R., Jan, Y.-S., et al. (2015). The axial element protein DESYNAPTIC2 mediates meiotic double-strand break formation and synaptonemal complex assembly in maize. *The Plant Cell*, *27*, 2516–2529.

Libuda, D. E., Uzawa, S., Meyer, B. J., & Villeneuve, A. M. (2013). Meiotic chromosome structures constrain and respond to designation of crossover sites. *Nature*, *502*, 703–706.

Liu, Y., Zhao, Q., Nie, H., Zhang, F., Fu, T., Zhang, Z., et al. (2021). SYP-5 regulates meiotic thermotolerance in *Caenorhabditis elegans*. *Journal of Molecular Cell Biology*, *13*, 662–675.

Loidl, J. (2021). Tetrahymena meiosis: Simple yet ingenious. *PLoS Genetics*, *17*, e1009627.

Lorenz, A., Estreicher, A., Kohli, J., & Loidl, J. (2006). Meiotic recombination proteins localize to linear elements in *Schizosaccharomyces pombe*. *Chromosoma*, *115*, 330–340.

Manheim, E. A., & McKim, K. S. (2003). The synaptonemal complex component C(2)M regulates meiotic crossing over in Drosophila. *Current Biology*, *13*, 276–285.

Martín-Castellanos, C., Blanco, M., Rozalén, A. E., Pérez-Hidalgo, L., García, A. I., Conde, F., et al. (2005). A large-scale screen in S. pombe identifies seven novel genes required for critical meiotic events. *Current Biology*, *15*, 2056–2062.

Martinez-Perez, E., & Villeneuve, A. M. (2005). HTP-1 dependent constraints coordinate homolog pairing and synapsis and promote chiasma formation during *C. elegans* meiosis. *Genes & Development*, *19*, 2727–2743.

Mehrotra, S., & McKim, K. S. (2006). Temporal analysis of meiotic DNA double-strand break formation and repair in Drosophila females. *PLoS Genetics*, *2*, e200.

Mercier, R., Mezard, C., Jenczewski, E., Macaisne, N., & Grelon, M. (2015). The molecular biology of meiosis in plants. *Annual Review of Plant Biology*, *66*, 297–327.

Miao, C., Tang, D., Zhang, H., Wang, M., Li, Y., Tang, S., et al. (2013). Central region component1, a novel synaptonemal complex component, is essential for meiotic recombination initiation in rice. *The Plant Cell*, *25*, 2998–3009.

Mieczkowski, P. A., Dominska, M., Buck, M. J., Gerton, J. L., Lieb, J. D., & Petes, T. D. (2006). Global analysis of the relationship between the binding of the Bas1p transcription factor and meiosis-specific double-strand DNA breaks in *Saccharomyces cerevisiae*. *Molecular and Cellular Biology*, *26*, 1014–1027.

Morgan, T. H. (1910). Sex limited inheritance in Drosophila. *Science*, *32*, 120–122.

Morgan, C., Fozard, J. A., Hartley, M., Henderson, I. R., Bomblies, K., & Howard, M. (2021). Diffusion-mediated HEI10 coarsening can explain meiotic crossover positioning in Arabidopsis. *Nature Communications*, *12*, 4674.

Morgan, C., White, M. A., Franklin, F. C. H., Zickler, D., Kleckner, N., & Bomblies, K. (2021). Evolution of crossover interference enables stable autopolyploidy by ensuring pairwise partner connections in *Arabidopsis arenosa*. *Current Biology*, *31*, 4713–4726.

Morgan, C. H., Zhang, H., & Bomblies, K. (2017). Are the effects of elevated temperature on meiotic recombination and thermotolerance linked via the axis and synaptonemal complex? *Philosophical Transactions of the Royal Society of London. Series B, Biological Sciences*, *372*, 20160470.

Murakami, H., & Keeney, S. (2014). Temporospatial coordination of meiotic DNA replication and recombination via DDK recruitment to replisomes. *Cell*, *158*, 861–873.

Nabeshima, K., Villeneuve, A. M., & Hillers, K. J. (2004). Chromosome-wide regulation of meiotic crossover formation in *Caenorhabditis elegans* requires properly assembled chromosome axes. *Genetics*, *168*, 1275–1292.

Nambiar, M., & Smith, G. R. (2018). Pericentromere-specific cohesin complex prevents meiotic pericentric DNA double-strand breaks and lethal crossovers. *Molecular Cell*, *71*, 540–553.

Nichols, B. A., Oswald, N. W., McMillan, E. A., McGlynn, K., Yan, J., Kim, M. S., et al. (2018). HORMAD1 is a negative prognostic indicator in lung adenocarcinoma and specifies resistance to oxidative and genotoxic stress. *Cancer Research*, *78*, 6196–6208.

Ning, Y., Liu, Q., Wang, C., Qin, E., Wu, Z., Wang, M., et al. (2021). Heat stress interferes with formation of double-strand breaks and homolog synapsis. *Plant Physiology*, *185*, 1783–1797.

Oshino, T., Abiko, M., Saito, R., Ichiishi, E., Endo, M., Kawagishi-Kobayashi, M., et al. (2007). Premature progression of anther early developmental programs accompanied by comprehensive alterations in transcription during high-temperature injury in barley plants. *Molecular Genetics and Genomics*, *278*, 31–42.

Otto, S. P., & Payseur, B. A. (2019). Crossover interference: Shedding light on the evolution of recombination. *Annual Review of Genetics*, *53*, 19–44.

Page, S. L., & Hawley, R. S. (2001). C(3)G encodes a Drosophila synaptonemal complex protein. *Genes & Development*, *15*, 3130–3143.

Panizza, S., Mendoza, M. A., Berlinger, M., Hunag, L., Nicolas, A., Shirahige, K., et al. (2011). SPO11-accessory proteins link double strand break sites to the chromosome axis in early meiotic recombination. *Cell*, *146*, 372–383.

Parvanov, E. D., Petkov, P. M., & Paigen, K. (2010). Prdm9 controls activation of mammalian recombination hotspots. *Science*, *327*, 835.

Pawlowski, W. P., Golubovskaya, I. N., & Cande, W. Z. (2003). Altered nuclear distribution of recombination protein RAD51 in maize mutants suggests the involvement of RAD51 in meiotic homology recognition. *The Plant Cell*, *15*, 1807–1816.

Pratto, F., Brick, K., Cheng, G., Lam, K.-W. G., Cloutier, J. M., Dahiya, D., et al. (2021). Meiotic recombination mirrors patterns of germline replication in mice and humans. *Cell*, *184*, 4251–4267.

Rasmussen, S. W. (1977). Meiosis in *Bombyx mori* females. *Philosophical Transactions of the Royal Society of London. Series B, Biological Sciences*, *277*, 343–350.

Rog, O., Köhler, S., & Dernburg, A. F. (2017). The synaptonemal complex has liquid crystalline properties and spatially regulates meiotic recombination factors. *eLife*, *6*, e21455.

Roig, I., Dowdle, J. A., Toth, A., de Rooij, D. G., Jasin, M., & Keeney, S. (2010). Mouse TRIP13/PCH2 is required for recombination and normal higher-order chromosome structure during meiosis. *PLoS Genetics*, *6*, e1001062.

Rousová, D., Nivsarkar, V., Altmannova, V., Raina, V. B., Funk, S. K., Liedtke, D., et al. (2021). Novel mechanistic insights into the role of Mer2 as the keystone of meiotic DNA break formation. *eLife*, *10*, e72330.

Sanchez-Moran, E., Santos, J.-L., Jones, G. H., & Franklin, F. C. H. (2007). ASY1 mediates AtDMC1-dependent interhomolog recombination during meiosis in Arabidopsis. *Genes & Development*, *21*, 2220–2233.

Schalbetter, S. A., Fudenberg, G., Baxter, J., Pollard, K. S., & Neale, M. J. (2019). Principles of meiotic chromosome assembly revealed in *S. cerevisiae*. *Nature Communications*, *10*, 4795.

Schild-Prüfert, K., Saito, T. T., Smolikov, S., Gu, Y., Hincapie, M., Hill, D. E., et al. (2011). Organization of the synaptonemal complex during meiosis in *Caenorhabditis elegans*. *Genetics*, *189*, 411–421.

Schwacha, A., & Kleckner, N. (1994). Identification of joint molecules that form frequently between homologs but rarely between sister chromatids during yeast meiosis. *Cell*, *76*, 51–63.

Schwacha, A., & Kleckner, N. (1997). Interhomolog bias during meiotic recombination: Meiotic functions promote a highly differentiated interhomolog only pathway. *Cell*, *90*, 1123–1135.

Severson, A. F., Ling, L., van Zuylen, V., & Meyer, B. J. (2009). The axial element protein HTP-3 promotes cohesin loading and meiotic axis assembly in *C. elegans* to implement the meiotic program of chromosome segregation. *Genes & Development*, *23*, 1763–1778.

Severson, A. F., & Meyer, B. J. (2014). Divergent kleisin subunits of cohesin specify mechanisms to tether and release meiotic chromosomes. *eLife*, *3*, e03467.

Shahzad, M. M. K., Shin, Y. H., Matsuo, K., Lu, C., Nishimura, M., Shen, D. Y., et al. (2013). Biological significance of HORMA domain containing protein 1 (HORMAD1) in epithelial ovarian carcinoma. *Cancer Letters*, *330*, 123–129.

Shi, J., Wolf, S. E., Burke, J. M., Presting, G. G., Ross-Ibarra, J., & Dawe, R. K. (2010). Widespread gene conversion in centromere cores. *PLoS Biology*, *8*, e1000327.

Shin, Y.-H., Choi, Y., Erdin, S. U., Yatsenko, S. A., Kloc, M., Yang, F., et al. (2010). Hormad1 mutation disrupts synaptonemal complex formation, recombination, and chromosome segregation in mammalian meiosis. *PLoS Genetics*, *6*, e1001190.

Shodhan, A., Kataoka, K., Mochizuki, K., Novatchkova, M., & Loidl, J. (2017). A Zip3-like protein plays a role in crossover formation in the SC-less meiosis of the protist Tetrahymena. *Molecular Biology of the Cell*, *28*, 825–833.

Simpson, A. J., Caballero, O. L., Jungbluth, A., Chen, Y. T., & Old, L. J. (2005). Cancer/testis antigens, gametogenesis and cancer. *Nature Reviews. Cancer*, *5*, 615–625.

Sommermeyer, V., Béneut, C., Chaplais, E., Serrentino, M. E., & Borde, V. (2013). Spp1, a member of the Set1 complex, promotes meiotic DSB formation in promoters by tethering histone H3K4 methylation sites to chromosome axes. *Molecular Cell*, *49*, 43–54.

Stanzione, M., Baumann, M., Papanikos, F., Dereli, I., Lange, J., Ramlal, A., et al. (2016). Meiotic DNA break formation requires the unsynapsed chromosome axis-binding protein IHO1 (CCDC36) in mice. *Nature Cell Biology*, *18*, 1208–1220.

Sturtevant, A. H. (1915). The behavior of the chromosomes as studied through linkage. *Zeitschrift für Induktive Abstammungs- und Vererbungslehre*, *13*, 234–287.

Sun, X., Huang, L., Markowitz, T. E., Blitzblau, H. G., Chen, D., Klein, F., et al. (2015). Transcription dynamically patterns the meiotic chromosome-axis interface. *eLife*, *4*, e07424.

Sym, M., & Roeder, G. S. (1994). Crossover interference is abolished in the absence of a synaptonemal complex protein. *Cell*, *79*, 283–292.

Takemoto, K., Tani, N., Takada-Horisawa, Y., Fujimura, S., Tanno, N., Yamane, M., et al. (2020). Meiosis-specific C19orf57/4930432K21Rik/BRME1 modulates localization of RAD51 and DMC1 to DSBs in mouse meiotic recombination. *Cell Reports*, *31*, 107686.

Tedeschi, A., Wutz, G., Huet, S., Jaritz, M., Wuensche, A., Schirghuber, E., et al. (2013). Wapl is an essential regulator of chromatin structure and chromosome segregation. *Nature*, *501*, 564–568.

Vrielynck, N., Schneider, K., Rodriguez, M., Sims, J., Chambon, A., Hurel, A., et al. (2021). Conservation and divergence of meiotic DNA double strand break forming mechanisms in Arabidopsis thaliana. *Nucleic Acids Research*, *49*, 9821–9835.

Walker, J., Gao, H., Zhang, J., Aldridge, B., Vickers, M., Higgins, J. D., et al. (2018). Sexual-lineage-specific DNA methylation regulates meiosis in Arabidopsis. *Nature Genetics*, *50*, 130–137.

Wang, X., Tan, Y., Cao, X., Kim, J. A., Chen, T., Hu, Y., et al. (2018). Epigenetic activation of HORMAD1 in basal-like breast cancer: Role in Rucaparib sensitivity. *Oncotarget*, *9*, 30115–30127.

Wang, Y., Zhai, B., Tan, T., Yang, X., Zhang, J., Song, M., et al. (2021). ESA1 regulates meiotic chromosome axis and crossover frequency via acetylating histone H4. *Nucleic Acids Research*, *49*, 9353–9373.

Ward, A., Hopkins, J., Mckay, M., Murray, S., & Jordan, P. W. (2016). Genetic interactions between the meiosis-specific cohesin components, STAG3, REC8, and RAD21L. *G3-Genes Genomes Genetics*, *6*, 1713–1724.

Watanabe, Y., Yokobayashi, S., Yamamoto, M., & Nurse, P. (2001). Pre-meiotic S phase is linked to reductional chromosome segregation and recombination. *Nature*, *409*, 359–363.

West, A. M. V., Komives, E. A., & Corbett, K. D. (2018). Conformational dynamics of the Hop1 HORMA domain reveal a common mechanism with the spindle checkpoint protein Mad2. *Nucleic Acids Research*, *46*, 279–292.

West, A. M., Rosenberg, S. C., Ur, S. N., Lehmer, M. K., Ye, Q., Hagemann, G., et al. (2019). A conserved filamentous assembly underlies the structure of the meiotic chromosome axis. *eLife*, *8*, e40372.

Wojtasz, L., Cloutier, J. M., Baumann, M., Daniel, K., Varga, J., Fu, J., et al. (2012). Meiotic DNA double-strand breaks and chromosome asynapsis in mice are monitored by distinct HORMAD2-independent and -dependent mechanisms. *Genes & Development*, *26*, 958–973.

Wright, K. M., Arnold, B., Xue, K., Šurinová, M., O'Connell, J., & Bomblies, K. (2015). Selection on meiosis genes in diploid and tetraploid *Arabidopsis arenosa*. *Molecular Biology and Evolution*, *32*, 944–955.

Yan, R., & McKee, B. D. (2013). The cohesion protein SOLO associates with SMC1 and is required for synapsis, recombination, homolog bias and cohesion and pairing of centromeres in Drosophila meiosis. *PLoS Genetics*, *9*, e1003637.

Yang, C., Hu, B., Portheine, S. M., Chuenban, P., & Schnittger, A. (2020). State changes of the HORMA protein ASY1 are mediated by an interplay between its closure motif and PCH2. *Nucleic Acids Research*, *48*, 11521–11535.

Yang, C., Sofroni, K., Wijnker, E., Hamamura, Y., Carstens, L., Harashima, H., et al. (2020). The Arabidopsis Cdk1/Cdk2 homolog CDKA;1 controls chromosome axis assembly during plant meiosis. *The EMBO Journal*, *39*, e101625.

Yoon, S., Choi, E.-H., Kim, J.-W., & Kim, K. P. (2018). Structured illumination microscopy imaging reveals localization of replication protein A between chromosome lateral elements during mammalian meiosis. *Experimental & Molecular Medicine*, *50*, 1–12.

Yuan, L., Liu, J. G., Zhao, J., Brundell, E., Daneholt, B., & Höög, C. (2000). The murine SCP3 gene is required for synaptonemal complex assembly, chromosome synapsis, and male fertility. *Molecular Cell*, *5*, 73–83.

Zhang, L., Espagne, E., de Muyt, A., Zickler, D., & Kleckner, N. E. (2014). Interference-mediated synaptonemal complex formation with embedded crossover designation. *Proceedings of the National Academy of Sciences of the United States of America*, *111*, E5059–E5068.

Zhang, J., Gurusaran, M., Fujiwara, Y., Zhang, K., Echbarthi, M., Vorontsov, E., et al. (2020). The BRCA2-MEILB2-BRME1 complex governs meiotic recombination and impairs the mitotic BRCA2-RAD51 function in cancer cells. *Nature Communications*, *11*, 2055.

Zhang, L., Liang, Z., Hutchinson, J., & Kleckner, N. (2014). Crossover patterning by the beam-film model: Analysis and implications. *PLoS Genetics*, *10*, e1004042.

Zhang, L., Wang, S., Yin, S., Hong, S., Kim, K. P., & Kleckner, N. (2014). Topoisomerase II mediates meiotic crossover interference. *Nature*, *511*, 551–556.

Zhang, Z., Xie, S., Wang, R., Guo, S., Zhao, Q., Nie, H., et al. (2020). Multivalent weak interactions between assembly units drive synaptonemal complex formation. *The Journal of Cell Biology*, *219*, e201910086.

Zickler, D., & Kleckner, N. (1999). Meiotic chromosomes: Integrating structure and function. *Annual Review of Genetics*, *33*, 603–754.

Ziolkowski, P. A., Underwood, C. J., Lambing, C., Martinez-Garcia, M., Lawrence, E. J., Ziolkowska, L., et al. (2017). Natural variation and dosage of the HEI10 meiotic E3 ligase control Arabidopsis crossover recombination. *Genes & Development*, *31*, 306–317.

Zuo, W., Chen, G., Gao, Z., Li, S., Chen, Y., Huang, C., et al. (2021). Stage-resolved Hi-C analyses reveal meiotic chromosome organizational features influencing homolog alignment. *Nature Communications*, *12*, 5827.

CHAPTER SIX

# Chromosome-specific behaviors during early meiosis

**Katherine Kretovich Billmyre***
Stowers Institute for Medical Research, Kansas City, MO, United States
*Corresponding author: e-mail address: kbillmyre@stowers.org

## Contents



## Abstract

Inheriting the wrong number of chromosomes is one of the leading causes of infertility and birth defects in humans. However, in many organisms, individual chromosomes vary dramatically in both organization, sequence, and size. Chromosome segregation systems must be capable of accounting for these differences to reliably segregate chromosomes. During gametogenesis, meiosis ensures that all chromosomes segregate properly into gametes (i.e., egg or sperm). Interestingly, not all chromosomes exhibit the same dynamics during meiosis, which can lead to chromosome-specific behaviors and defects. This review will summarize some of the chromosome-specific meiotic events that are currently known and discuss their impact on meiotic outcomes.

*Current Topics in Developmental Biology*, Volume 151
ISSN 0070-2153
https://doi.org/10.1016/bs.ctdb.2022.05.002

## 1. Introduction

Meiosis is a specialized cell division that is necessary for sexual reproduction. The key role of meiosis is to take a diploid genome (containing two copies of the DNA) and segregate it equally into haploid gametes (containing one copy of the DNA). When chromosomes do not segregate properly, the wrong number of chromosomes can be inherited (called aneuploidy). In humans, aneuploidy is a leading cause of infertility, miscarriages, and birth defects (Hassold, Hall, & Hunt, 2007).

The genome is packaged into chromosomes that vary in shape, size, and composition. The variability that exists between chromosomes can be incredibly important in how chromosomes undergo many biological processes (e.g., chromosome segregation, imprinting, dosage compensation, gene expression, and replication). This review will focus on how differences between individual chromosomes can influence various aspects of meiotic and reproductive chromosome biology. After a brief overview of meiotic and chromosome biology, some key examples of chromosome-specific behaviors that occur during meiosis and their impact on meiotic outcomes will be discussed. There are two general categories of chromosome-specific behaviors that can occur between chromosomes: (1) *inter*chromsomal (between non-homologous chromosomes) and (2) *intra*-chromosomal (between two homologs). Most of this review will focus on interchromosomal meiotic events but some well-known examples of intrachromosomal differences will be discussed at the end. Lastly, this review will discuss how chromosomal differences extend outside of meiosis and their potential impact on human health. Gaining a better appreciation of how the meiotic machinery interacts with individual chromosomes is crucial to understanding what leads to chromosome-specific meiotic defects.

## 2. What is meiosis?

The key steps of meiosis have been extensively reviewed previously (Gerton & Hawley, 2005; Handel & Schimenti, 2010; Hughes, Miller, Miller, & Hawley, 2018; Ur & Corbett, 2021; Zickler & Kleckner, 2015). Therefore, this review will only summarize the basics of meiosis and how they relate to chromosomal biology. The main events that meiotic chromosomes must undergo are homolog pairing, double-strand break (DSB) formation, synapsis, crossing over, and segregation (Fig. 1A).

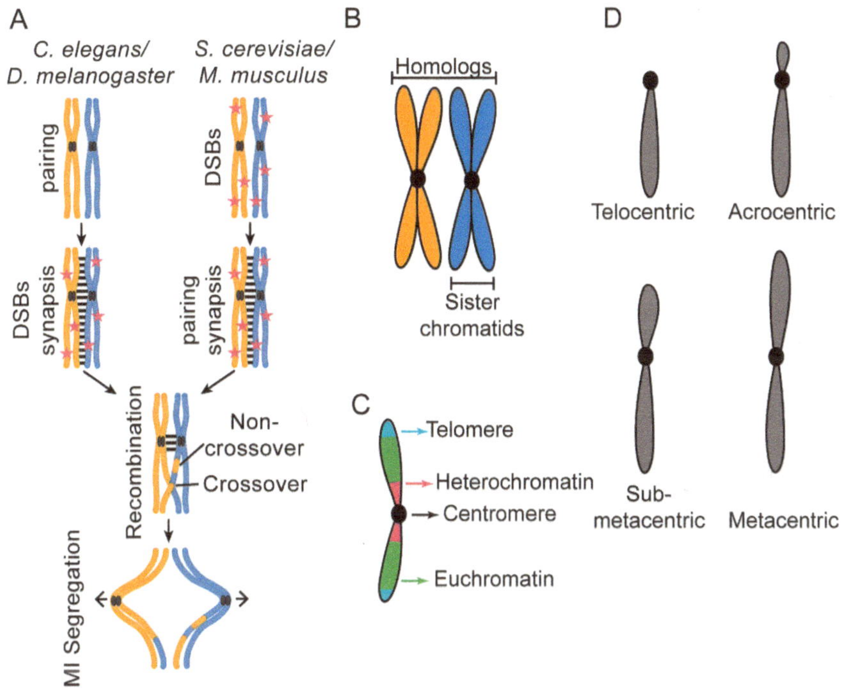


**Fig. 1** Different meiotic and chromosome features that contribute to chromosome-specific differences. (A) Key steps of meiosis in multiple organisms. In *C. elegans* and *D. melanogaster* homolog pairing occurs prior to DSB formation and synapsis. However, in mice and *S. cerevisiae* DSB formation helps facilitate pairing. (B) Diagram of homologs and sister chromatids. (C) Components that make up chromosomes (telomeres, centromeres, heterochromatin, and euchromatin). (D) Variations in chromosome structure due to centromere placement.

Currently, our understanding of the mechanisms of homolog pairing are relatively limited (Alleva & Smolikove, 2017; Hiraoka, 2020; Rubin, Macaisne, & Huynh, 2020). Interestingly, in many organisms (i.e., *Saccharomyces cerevisiae*, mice, humans), pairing of the homologous chromosomes depends on DSB formation which in these organisms occurs prior to synapsis (Keeney, 2001; Zickler & Kleckner, 2015). Synapsis is the formation of the synaptonemal complex (SC), a large protein structure present between homologs (Page & Hawley, 2004). However, in *Caenorhabditis elegans* and *Drosophila melanogaster*, pairing of the chromosomes is independent of DSB formation (Keeney, 2001; Zickler & Kleckner, 2015). In *D. melanogaster*, the mechanisms behind meiotic pairing are poorly understood in females but established pairing is maintained via synapsis (Gong, McKim, & Hawley, 2005; Page et al., 2008; Sherizen, Jang,

Bhagat, Kato, & McKim, 2005). In *C. elegans*, a set of proteins called pairing center proteins to bring homologs together and the SC helps stabilize these pairing interactions (Rog & Dernburg, 2013; Tsai & McKee, 2011).

In addition to differences in the requirement of DSBs for homolog pairing, species also exhibit differences in the timing and requirement of DSB formation with regard to synapsis. In both mice and *S. cerevisiae* DSB formation occurs prior to synapsis and full synapsis is partially dependent on normal levels of DSBs (Henderson & Keeney, 2004; Kauppi et al., 2013). In both organisms, a reduction in DSB levels results in delayed or abnormal synapsis. However, in *D. melanogaster*, synapsis is not dependent on DSBs (McKim & Hayashi-Hagihara, 1998) and in fact, SC mutants exhibit severely reduced numbers of DSBs suggesting that DSB formation is dependent on normal synapsis (Mehrotra & McKim, 2006). Lastly, in *C. elegans* DSB formation and synapsis are independent of one another. The loss of DSB formation does not affect synapsis (Dernburg et al., 1998; Rosu, Zawadzki, Stamper, Libuda, & Reese, 2013) and DSBs still form in SC mutants (Rog & Dernburg, 2013). This is in contrast to *Drosophila* where DSBs depend on the SC. It is unknown why organisms have evolved differences in the relationship between DSB formation and synapsis.

Regardless of these early meiotic differences, all sexually reproducing organisms must repair their DSBs into either crossovers or noncrossovers (also called gene conversions) (Youds & Boulton, 2011). Noncrossover gene conversions are small nonreciprocal exchanges of genetic material usually detected between two homologs. In contrast, the repair of DSBs into crossovers results in the reciprocal exchange of regions of the chromosomes between the crossover and the telomere (or a second crossover in the case of a double crossover). The crossover, along with sister chromatid cohesion, is then responsible for holding homologs together until the time of chromosome segregation. This review will focus on chiasmate segregation systems where crossing over is necessary for proper chromosome segregation (reviewed by Pazhayam, Turcotte, & Sekelsky, 2021). Additionally, a key aspect of meiosis is the occurrence of two divisions compared to one in mitosis. The first segregates homologous chromosomes (MI division) and the second separates sister chromatids (MII division). The examples presented here are related to MI meiotic events and divisions.

## 3. How do chromosomes differ from each other?

When looking at an individual's collection of chromosomes (karyotype) between different species it is immediately apparent that there is a wide variety in eukaryotic chromosome number, size, and structure. At the most basic level, all chromosomes have telomeres, centromeres, heterochromatic regions, and euchromatic regions (Fig. 1C). However, these building blocks are not identical between all the chromosomes within a species or even an individual (Britton-Davidian et al., 2000; Damas, Corbo, & Lewin, 2021; Deakin, 2018; Schubert, 2007; Wilson & Makova, 2009). There are multiple factors that can lead to variability in the components making up a chromosome. For example, telomeres are made up of repetitive units of DNA and repeat number can vary between chromosomes (Mason, Frydrychova, & Biessmann, 2008; Teixeira & Gilson, 2005). Similarly, the makeup of centromeres can be quite variable between chromosomes within a species. Centromeres can vary in repeat sequence, repeat number, overall size, and location (Fig. 1D) (Bloom, 2014; Miga & Alexandrov, 2021). Two examples that highlight the vast differences in centromere biology between species are found in *S. cerevisiae* and *C. elegans*. *S. cerevisiae* have point centromeres that are small (Furuyama & Biggins, 2007), while *C. elegans* are holocentric, meaning multiple kinetochores can form along the length of the chromosome (Howe, McDonald, Albertson, & Meyer, 2001). Further during meiosis, *C. elegans* chromosomes behave in a telocentric manner as the kinetochores assemble on ends of the chromosomes during meiotic segregation (Martinez-Perez et al., 2008). Additionally, chromosomes contain large regions of euchromatic (transcriptionally active) and heterochromatic (transcriptionally silent) DNA. The location, size, and composition of euchromatic and heterochromatic DNA can vary widely between chromosomes of an individual and can influence meiotic events (Babu & Verma, 1987; Morrison & Thakur, 2021).

An additional feature of meiotic chromosome biology is the organization of chromosomes within the nucleus. There are many drivers of nuclear organization but features such as genomic content and chromosome structure often play a role in how the chromosomes are packaged (Franklin & Cande, 1999; Grey & de Massy, 2021; van Bortle & Corces, 2012). One such genomic feature is the ribosomal DNA (rDNA) that is responsible for encoding ribosomal RNA, which is necessary for translation of mRNAs. rDNA is highly repetitive and localizes in a specialized compartment called the nucleolus, thereby bringing chromosomes containing rDNA to a similar location within the nucleus (Németh & Längst, 2011; Potapova & Gerton, 2019). The location of rDNA within the genome varies between species and depending on the organism it can be on one chromosome or split over many.

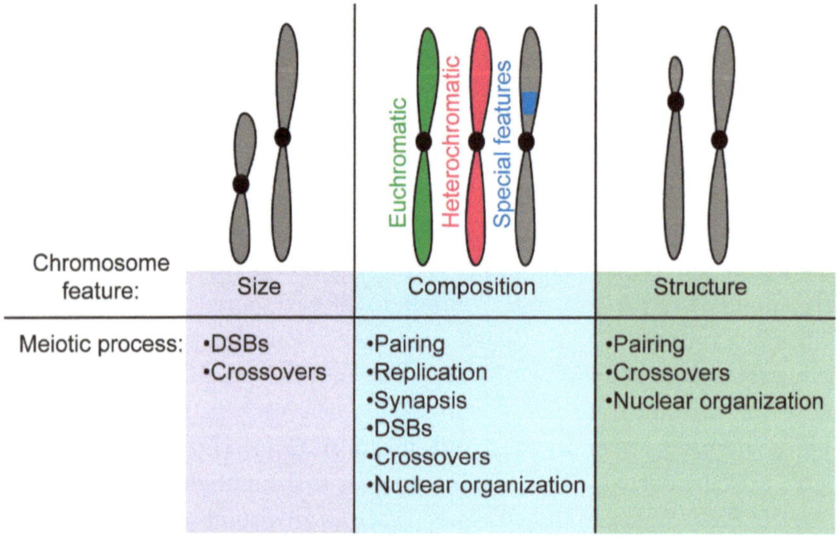


**Fig. 2** Relationship between chromosome-specific meiotic phenotypes and chromosome features. Summary of the meiotic events discussed in this review that have been shown to correlate with differences in chromosome features. Chromosome size, composition (chromatin environment and repetitive elements), and structure (centromere placement and translocations) have all been found to correlate with chromosome-specific meiotic phenotypes.

Variation in the building blocks discussed above result in chromosomes with different overall sizes, unique structures, and that are organized within the nucleus in unique ways. This review focuses on how chromosome-specific variability is linked to meiotic events such as synapsis, DSB formation, and crossing over, and their eventual effect on chromosome segregation (Fig. 2).

## 4. Association between meiotic errors and chromosome structure in humans

In humans, it is difficult to determine the causes of aneuploidy as the majority of aneuploidies lead to early miscarriages, resulting in a lack of genetic material to analyze. However, examination of oocytes and early embryos shows that aneuploidy is generally due to meiotic errors that can affect all chromosomes, groups of chromosomes, or individual chromosomes (Capalbo, Hoffmann, Cimadomo, Ubaldi, & Rienzi, 2017; Hassold et al., 2007).

Two factors that have been linked to aneuploidy are increased maternal age and disruption of chromosome segregation through various meiotic errors (Gruhn et al., 2019; Hassold, Merrill, Adkins, Freeman, & Sherman, 1995; Lamb, Yu, Shaffer, Feingold, & Sherman, 2005; Sherman, Lamb, & Feingold, 2006). It is well known that a complete loss of crossing over leads to high levels of random chromosome missegregation (Hassold & Hunt, 2001); however, even misplacement of a crossover can lead to improper chromosome segregation (Hassold, Hunt, & Nagaoka, 2012; Hou et al., 2013; Ottolini et al., 2015). For example, chromosome 21 trisomy has been associated with no crossovers, distal (telomeric) crossovers, and centromere proximal crossovers (Lamb et al., 2005). This contrasts with chromosome 16 missegregation where extreme distal crossovers are linked to trisomy (Hassold et al., 1995; Sherman et al., 2006) and chromosome 18 trisomy where only chromosomes lacking crossovers missegregate (Bugge et al., 1998; Fisher, Harvey, Morton, & Jacobs, 1995). Further, evidence suggests that acrocentric chromosomes (13, 14, 15, 21, 22) appear to exhibit similar patterns of recombination defects with the majority of nondisjunction arising due to a loss of crossing over (Hall et al., 2007). For nonacrocentric chromosomes there is greater variability in the underlying causes of nondisjunction making it difficult to identify a common mechanism in those cases.

The similar loss of crossing over seen in acrocentric chromosomes suggests that chromosome structure may play a role in MI chromosome segregation. The importance of chromosome structure in affecting aneuploidy rates is further confirmed in examination of human oocytes showing a correlation between chromosome size and structure with the differences in meiotic errors that occur in response to maternal age (Franasiak et al., 2014; Gruhn et al., 2019). This examination found aneuploidy was due to three different types of errors: (1) MI nondisjunction (the gain or loss of a whole chromosome), (2) precocious separation of sister chromatids (premature separation of the sister chromatids), and (3) reverse segregation (MI and MII are reversed, i.e., sister chromatids separate at MI and homologs separate at MII). Interestingly, these errors do not occur at similar rates as maternal age increases and individual chromosomes exhibit differing responses to maternal age. For example, larger chromosomes are more likely to exhibit MI nondisjunction in younger women compared to older women (Gruhn et al., 2019). On the other hand, acrocentric chromosomes are more likely to undergo precocious separation of sister chromatids and reverse segregation in older women (Gruhn et al., 2019). Little is known about the mechanisms that lead to the meiotic errors seen in humans and if there is a relationship between those errors and crossover loss or mispatterning. One attractive possibility is that sister chromatid cohesins weaken as maternal age increases, and therefore, might act as a type of clock predisposing specific chromosomes to separate prematurely (Duncan et al., 2012; Lagirand-Cantaloube et al., 2017; Zielinska, Holubcova, Blayney, Elder, & Schuh, 2015). However, how this mechanism would affect some chromosomes more severely than others is unknown.

Unfortunately, this research is incredibly difficult to undertake given the limitations on human research and scarcity of available samples. This makes establishing the types of chromosome-specific dynamics that occur in model systems (reviewed below) even more important for our ability to better understand what occurs in humans.

## 5. Regulation of crossing over on different chromosomes

Crossing over is necessary for proper chromosome segregation in organisms that undergo chiasmate chromosome segregation. Therefore, organisms have a phenomenon called crossover assurance that ensures every chromosome contains an obligate crossover regardless of differences between those chromosomes, such as chromosome length (reviewed by Pazhayam et al., 2021). This section will review how the meiotic recombination machinery works to ensure each chromosome receives a crossover and examples of meiotic errors that occur in a chromosome-specific pattern.

### 5.1 Regulation of DSBs and crossing over varies on different chromosomes

In *S. cerevisiae*, there are mutations that result in different crossover rates on different chromosomes (Chen et al., 2015; Lee et al., 2021; Murakami et al., 2020; Serrentino, Chaplais, Sommermeyer, & Borde, 2013; Zanders & Alani, 2009). If DSBs were randomly distributed across the genome, the smaller chromosomes would receive fewer DSBs per chromosome and are therefore less likely to obtain an obligatory crossover than a larger chromosome (Mancera, Bourgon, Brozzi, Huber, & Steinmetz, 2008; Pan et al., 2011; Thacker, Mohibullah, Zhu, & Keeney, 2014). Interestingly, a negative correlation between chromosome size and DSB/crossover density has been observed in *S. cerevisiae* (Chen et al., 2008; Kaback, Guacci, Barber, & Mahon, 1992). This correlation suggests that there is an active mechanism in place to ensure DSBs and crossovers are placed on smaller chromosomes. Short chromosomes are enriched for proteins necessary for DSB placement (Rec114 and Mer2) and retain those proteins for a longer period, allowing for greater numbers of DSBs to be made (Murakami et al., 2020). Additionally, the SC appears to be important in this process. The SC is composed of many different components including axis proteins that are localized along the DNA and central region proteins that are found in the middle of the structure (for a review of SC structure see Cahoon & Hawley, 2016; Page & Hawley, 2004; Zickler & Kleckner, 1999). In *S. cerevisiae,* the axis proteins (Hop1 and Red1) are overrepresented on short chromosomes compared to large chromosomes and are necessary for the increased concentration of Rec114 (Murakami et al., 2020). Interestingly, when short chromosomes are artificially lengthened, they still exhibit a boost in DSBs, and when a long chromosome is split into smaller chromosomes this boost does not occur (Murakami et al., 2020). This supports a hypothesis that the feature(s) identifying a short chromosome to the meiotic system is intrinsic to the chromosome itself, such as chromatin composition. How this intrinsic feature is established is of great interest because it will give insight into how chromosomes are specifically differentiated by the meiotic machinery.

Similar to how *S. cerevisiae* has mechanisms in place to ensure short chromosomes receive DSBs, male mammalian meiosis also needs to ensure that crossovers are placed between the *X* and *Y* chromosomes. *X* and *Y* chromosomes have different sequences but still need to pair and have a crossover to guarantee proper chromosome segregation. The region that pairs and receives a crossover is called the pseudoautosomal region (PAR) and in mice is only ~700 kb (Perry, Palmer, Gabriel, & Ashworth, 2001; Raudsepp, Das, Avila, & Chowdhary, 2012). There are multiple layers of mechanisms in place that work to place crossovers within the PAR. One of the key mechanisms is the accumulation of DSB promoting factors, such as REC114, MEI4, MEI1 and IHO1, in the PAR (Kumar et al., 2018; Kumar, Bourbon, & de Massy, 2010; Reinholdt & Schimenti, 2005; Stanzione et al., 2016). This accumulation is associated with heterochromatic *mo-2* minisatellite arrays, suggesting that the sequence of the PAR generates a unique chromatin environment that is necessary for crossing over (Acquaviva et al., 2020). Crossing over in the PAR region, along with the short chromosome DSB boost seen in *S. cerevisiae,* highlights the evolutionary importance of DSBs and crossovers in organisms that need chiasmata for proper segregation.

Another intriguing example of chromosome-specific crossover biology comes from *D. melanogaster* where the *4th* chromosome does not undergo crossing over while the rest of the chromosomes do. The *4th* chromosome has many unique properties that could potentially impair crossing over including that the *4th* chromosome is small, is highly heterochromatic, and has a high transposon content. However, the combination of multiple studies suggests it is unlikely that these physical properties of the *4th* chromosome are sufficient to prevent crossing over (reviewed by Hartmann & Sekelsky, 2017). Little work has investigated other potential mechanisms of crossover repression on the *4th* chromosome, but an attractive possibility is that the *4th* chromosome experiences a strong centromere effect (a phenomenon that prevents crossovers near the centromere). There is evidence that the centromere effect may play a large role in *4th* chromosome crossover suppression (Hatkevich et al., 2017), however, very little is known about the mechanism behind the centromere effect. In the future, more work needs to be done to better understand this phenomenon.

## 5.2 Chromosome-specific defects occur with defective synapsis

In addition to the connection between chromosome length and DSB regulation, SC central region proteins appear to play a role in how DSBs are repaired on different chromosomes. In *S. cerevisiae,* DSBs form prior to SC assembly and are key in initiating pairing and SC formation. Further, the assembly of the SC is important for efficient repair of DSBs into crossovers. The SC is made up of multiple components including axis proteins, transverse filament proteins, and central region proteins. In yeast, ZMM complex proteins that are involved in SC assembly, Zip1 (transverse filament protein) and Zip3 (E3 ligase), are important in establishing the recombination landscape (ZMM complex review: Lynn, Soucek, & Börner, 2007). Previous work has found that *zip1* and *zip3* mutants exhibit chromosome-specific defects in crossing over that are correlated with chromosome size (Chen et al., 2015; Serrentino et al., 2013). Moreover, mutants in *ecm11* and *gcm2* exhibit increases in crossing over on longer chromosomes versus shorter ones (Lee et al., 2021). Ecm11 and Gcm2 are central region SC proteins that are important for the assembly of Zip1 along the homologs. Interestingly, in late prophase, longer chromosomes accumulate increased numbers of DSBs in *ecm11* and *gcm2* mutants, suggesting that the regulation of homeostatic DSB formation requires Ecm11 and Gcm2 (Lee et al., 2021). This observation leads to the hypothesis that these extra DSBs are leading to higher levels of crossovers on longer chromosomes specifically due to late prophase accumulation of DSBs (the number of DSBs formed in early pachytene are not altered in *ecm11* and *gcm2* mutants) (Lee et al., 2021). The role of the SC in the regulation of meiotic DSBs and crossovers is poorly understood, but at least in *S. cerevisiae* the SC is clearly important for maintaining mechanisms that ensure equal DSBs and crossovers between chromosomes of different sizes.

Synapsis is also important in maintaining equal crossing over between different chromosomes in *D. melanogaster*. In a set of partial loss-of-function SC mutants, SC assembly and maintenance is disrupted, resulting in chromosome-specific crossing over and pairing defects (Billmyre et al., 2019). These mutants contain in-frame deletions within the SC transverse filament protein, C(3)G, and result in premature loss of the SC in early to mid-pachytene. The *X* chromosome appears to be more sensitive to early disassembly of the SC in early pachytene and has an almost complete loss of crossing over and a progressive loss of euchromatic pairing (Billmyre et al., 2019). This contrasts with the autosomes, which have almost normal euchromatic pairing and normal levels of total crossing over but a shift in the placement of crossovers proximally towards the centromere.

The reason for these dramatic differences between chromosomes is unknown but there are many attractive hypotheses. First, the *X* chromosome is smaller in size than the autosomes, and as in *S. cerevisiae* chromosome size may play a role in how meiotic DSBs and crossovers are placed (Chen et al., 2015; Lee et al., 2021; Murakami et al., 2020; Serrentino et al., 2013). Secondly, the *X* chromosome centromeres and heterochromatin pair earlier than those of the autosomes due to the presence of the rDNA and other large repeats that are unique to the *X* chromosome (Christophorou, Rubin, & Huynh, 2013). This could potentially change the timing of DSBs, synapsis, and/or crossing over on the *X* chromosome compared to the autosomes. Last, the *X* chromosome, in addition to being smaller, is acrocentric compared to the metacentric autosomes (Fig. 1D). Centromere placement is known to correlate with differences in human meiotic defects (Gruhn et al., 2019; Hassold et al., 2007). While it is unclear what mechanism underlies the chromosome-specific differences in these mutants, this model will help us to understand how different chromosomes undergo meiosis.

## 5.3 Chromosome-specific meiotic regulators

There are also regulators which localize specifically to different chromosomes and influence variation in meiotic events between chromosomes. In *C. elegans* there are two classes of these genes where mutations affect the autosomes or the *X* chromosome more severely. Examples of mutations which affect the autosomes more severely consist of *him-3* (hypomorphic allele), *htp-1* (removal of one paralog), and *cra-1* mutants, which all exhibit normal (or near normal) levels of *X* chromosome pairing and a partial loss of *X* chromosome crossing over but severe losses in pairing and crossing over on the autosomes (Couteau, Nabeshima, Villeneuve, & Zetka, 2004; Couteau & Zetka, 2005; Smolikov, Schild-Prüfert, & Colaiácovo, 2008). These phenotypes appear to be due in part to a more stable initial alignment between the *X* chromosomes than the autosomes in the absence of normal levels of HTP-1, as the *X* chromosomes are more aligned than the autosomes in *htp-1* mutants (Couteau & Zetka, 2005). One possible explanation for the initial difference in alignment of the *X* chromosome and autosomes is the presence of an unknown *X* chromosome specific stabilization mechanism early in meiosis. However, if this mechanism exists it has not yet been identified.

The second class includes mutations that cause severe losses of *X* chromosome DSB formation and crossing over, while the autosomes only experience some crossover patterning defects. The three genes that will be highlighted here are *xnd-1*, *him-5*, and *him-17* (Meneely, McGovern, Heinis, & Yanowitz, 2012; Reddy & Villeneuve, 2004; Wagner, Kuervers, Baillie, & Yanowitz, 2010). Both XND-1 and HIM-5 localize specifically to the autosomes (Meneely et al., 2012; Wagner et al., 2010), while HIM-17 localizes to all chromosomes and is required for the autosomal localization of HIM-5 (Reddy & Villeneuve, 2004). Interestingly, even though XND-1 and HIM-5 are localized to the autosomes, they are still necessary for DSB formation on the *X* chromosome (McClendon et al., 2016; Meneely et al., 2012; Wagner et al., 2010). Further they both regulate autosomal crossover positioning. Based on double mutant analysis, XND-1 and HIM-5 appear to function in the same pathway to guarantee DSB formation on the *X* chromosome and for some reason play a separate role in crossover patterning on the autosomes (Meneely et al., 2012).

The regulation of *X* chromosome DSBs may be due, in part, to the heterochromatic chromatin environment on the *X* chromosome (Kelly et al., 2002; Kelly & Fire, 1998). When *X* chromosome silencing is alleviated in histone methyltransferase mutant, *mes-2*, the loss of *X* chromosome DSBs in *xnd-1* and *him-5* mutants is also alleviated (Wagner et al., 2010). However, the relationship between these proteins is complex as they do not exhibit identical phenotypes in histone acetylation and DSB dynamics, suggesting that they likely function at different steps in *X* chromosome crossover regulation (McClendon et al., 2016). Why these genes are functioning to promote DSBs and crossover formation on the *X* chromosome specifically could be explained by the differences in chromatin environments between the autosomes and the *X* chromosome. It may be that these genes have evolved as a mechanism to ensure DSBs and crossovers are placed in the predominantly heterochromatic environment on the *X* chromosome, as heterochromatic regions usually exclude crossovers (Ohta, Shibata, & Nicolas, 1994; Wu & Lichten, 1994). An alternative possibility comes from a proposed model in which there is a window when chromosomes are competent to receive DSBs, and the window for DSB placement on the *X* chromosome could be shorter than on the autosomes (Reddy & Villeneuve, 2004). This model fits with some data from *D. melanogaster* suggesting that crossovers need to be placed earlier on the *X* chromosome than on the autosomes (Billmyre et al., 2019). The mechanisms underlying these chromosome-specific regulators are still poorly understood and gaining a better understanding of how and why chromosome-specific regulators have evolved is important to uncovering their impact on normal meiosis.

## 6. Chromatin composition can influence multiple aspects of meiosis

The chromatin landscape can play a key role in DSB and crossover location. For example, the heterochromatic, repetitive regions of the genome are less likely to receive DSBs than the euchromatic regions, likely as a mechanism to protect the genome from rearrangements within repeats (Sasaki, Lange, & Keeney, 2010). In addition to regulating DSB formation and repair, the chromatin environment also can serve to distinguish specific chromosomes.

### 6.1 Chromatin environment can alter timing of meiotic events between chromosomes

Variation in timing of meiotic events on different chromosomes can also drive chromosome-specific differences. A key example of this comes from work in *C. elegans* comparing the timing of replication and synapsis of the

*X* chromosomes to that of the autosomes. In *C. elegans* it is possible to monitor pre-meiotic DNA replication using a labelling technique that follows the incorporation of labeled nucleotides into replicating DNA (Jaramillo-Lambert, Ellefson, Villeneuve, & Engebrecht, 2007). This approach demonstrated that the *X* chromosome replicates later than the rest of the genome (Jaramillo-Lambert et al., 2007). In *C. elegans* the heterochromatic nature of the *X* chromosome (Kelly et al., 2002) is likely responsible for the late replication of the *X* chromosome. On the other hand, the autosomes are transcriptionally active and contain histone marks associated with active chromatin (Kelly et al., 2002). The late replication of the *X* chromosome led to a clever approach to specifically mark the *X* chromosome during meiosis and assess the timing and chromosome specificity of other meiotic events (Mlynarczyk-Evans & Villeneuve, 2017). These experiments showed that pairing of the *X* chromosome and two autosomes (*I* and *IV*) occurs at the same time during meiosis (Mlynarczyk-Evans & Villeneuve, 2017). Interestingly, even though the *X* chromosome and autosomes pair at the same time, synapsis of the *X* chromosome is significantly slower than synapsis of the autosomes (Mlynarczyk-Evans & Villeneuve, 2017). It is unknown if this later synapsis is due to a delay in synapsis initiation or a slower rate of SC assembly. These findings from *C. elegans* highlight the importance of chromatin composition in determining meiotic behavior.

## 6.2 Meiotic sex chromosome inactivation separates sex chromosomes from autosomes

Another key example of how chromatin marks can distinguish chromosomes from each other is the meiotic sex chromosome inactivation (MSCI) that occurs in many organisms with a heterogametic sex chromosome system (Ichijima, Sin, & Namekawa, 2012; Turner, 2015). MSCI is related to an older genome defense mechanism, meiotic silencing by unsynapsed chromatin/unpaired DNA (MSUC/MSUD) (Baarends et al., 2005; Schimenti, 2005; Turner et al., 2004). MSUC/MSUD silences DNA that does not have a pairing partner and likely evolved as a defense system against transposon movement in germline nuclei. In a *XY* (or *ZW* in chickens) sex determination system, the *X* and *Y* chromosomes only pair in the PAR. This means most of the chromosome is unpaired and is marked for silencing. Silencing occurs through the phosphorylation of serine 139 of histone H2AX by the DNA damage response (DDR) pathway member ATR (Fernandez-Capetillo et al., 2003; Ichijima et al., 2012;

Royo et al., 2013; Turner et al., 2004). This phosphorylation event results in the transcriptional silencing of the *XY* chromosomes and the formation of what is known as the *XY* body. The *XY* body sequesters the *XY* chromosomes away from the autosomes, which remain transcriptionally active during meiosis. Failure of MSCI results in the triggering of a meiotic checkpoint and complete meiotic arrest (Abe et al., 2020; Kogo et al., 2012; Mahadevaiah et al., 2008). Interestingly, silencing of the *XY* chromosomes continues after meiosis is complete, which could be a mechanism to achieve dosage compensation in the resulting offspring.

Dosage compensation is not itself a meiotic event but is an example of why the *X* chromosome may exhibit different behaviors and features during meiosis. There must be mechanisms to identify the *X* chromosome from the autosomes for dosage compensation to occur (reviewed by Disteche, 2012), some of which may be utilized or established during meiosis.

## 7. Aspects of chromosome biology that alter meiotic nuclear organization

Nuclear organization is critically important during meiosis because homologous chromosomes need to pair and synapse along their lengths to properly recombine (reviewed by Vara & Ruiz-Herrera, 2021). Additionally, nuclear localization can alter the meiotic environment of a chromosome. Unfortunately, much of the work regarding nuclear organization comes from non-meiotic nuclei in models of gene regulation and cancer. However, there are some clear examples in meiotic systems where nuclear organization and meiotic events are linked.

### 7.1 rDNA location affects pairing and nuclear organization

rDNA is known to be important in nucleolus and nuclear organization in multiple species and contexts (reviewed by (Cerqueira & Lemos, 2019; Potapova & Gerton, 2019; Shaw & McKeown, 2011). One example of how the rDNA location can contribute to chromosome specific behaviors is found in flies. In *D. melanogaster*, the chromosomes are always paired in somatic cells, but that is not the case in meiotic cells where chromosomes are unpaired in the adult germline stem cells (Joyce, Apostolopoulos, Beliveau, & Wu, 2013). Pairing of meiotic chromosomes in female meiosis must then be re-established during the pre-meiotic mitotic cycles starting with the centromeres and then euchromatic regions resulting in chromosomes being paired when they enter meiosis (Christophorou et al., 2013;

Joyce et al., 2013). There are differences between the *X* chromosome and autosomes in the timing of premeiotic pairing with the *X* chromosome centromeres pairing as early as the germline stem cells (Christophorou et al., 2013). In contrast, autosomal pairing is progressively established from the germline stems cells through the initiation of meiosis. The difference in timing is hypothesized to be due to the presence of either or both the rDNA and a large repeat called the 359 repeat on the *X* chromosome. The rDNA localizes to the nucleolus, thereby bringing the rDNA segments on both *X* chromosomes in close proximity. Therefore, it has been speculated that the rDNA may function as a pairing site as it does in *D. melanogaster* male meiosis (reviewed by McKee & Karpen, 1990; Tsai & McKee, 2011). This hypothesis is supported by the observation that autosomes require localization of SC proteins to the centromere for centromeric pairing, while *X* chromosome centromeric pairing is independent of SC proteins (Christophorou et al., 2013; Joyce et al., 2013). However, euchromatic pairing appears to be a different story as both the *X* chromosome and autosomes are dependent on the SC to maintain euchromatic pairing once a cell has fully entered meiosis (Takeo, Lake, Morais-de-Sá, Sunkel, & Hawley, 2011). rDNA location influencing pairing is a powerful example of how the rDNA locus may be influencing a chromosome's environment and behaviors.

Proper rDNA copy number is critical to survival and viability of an organism, and many studies have examined rDNA behaviors and the consequences of inappropriate numbers of rDNA repeats (reviewed by Lindström et al., 2018; Nelson, Watase, Warsinger-Pepe, & Yamashita, 2019). Because loss of the rDNA is detrimental to survival, there are mechanisms that protect the rDNA repeats from being lost due to recombination. One protective mechanism that has been found in *Arabidopsis* is that the nucleolus excludes meiotic DSB and repair machinery thereby protecting the rDNA from recombination events (Sims, Copenhaver, & Schlögelhofer, 2019). Additionally, ectopically placed rDNA arrays also localize to the nucleolus and are similarly protected from meiotic machinery (Sims et al., 2019). This confirms the importance of rDNA in altering the nuclear localization and environment of the specific chromosomes containing it.

## 7.2 Specialized chromosome-specific proteins pair homologs in *C. elegans*

How chromosomes are arranged and pair during meiosis is an important aspect of nuclear organization. Pairing mechanisms from other organisms are discussed above (see What is meiosis?), but *C. elegans* use a special set of chromosome-specific proteins called pairing center (PC) proteins. There are four PC proteins that are necessary for the pairing of the five autosomes and *X* chromosomes present in *C. elegans* (ZIM-1*: Chr II* and *III,* ZIM-2*: Chr V,* ZIM-3*: Chr I* and *IV,* HIM-8*: X Chr*) (Phillips et al., 2005; Phillips & Dernburg, 2006). PC proteins localize to the ends of chromosomes where they interact with a set of proteins that span the nuclear envelope, SUN-1 and ZYG-12. The interactions between SUN-1/ZYG-12 and PC proteins results in a set of rapid nuclear envelope movements causing the chromosomes to move around the nucleus until pairing and synapsis are complete (Baudrimont et al., 2010; Wynne, Rog, Carlton, & Dernburg, 2012). These movements are very similar to the telomere bouquet dynamics present in other organisms that are important for meiotic pairing (reviewed by Scherthan, 2007). The chromosome-specific nature of the PC proteins means it is easy to visualize specific chromosomes by marking individual PC proteins. Further, mutations in individual PC proteins can be used to investigate chromosome-specific disruptions in meiosis. As a result, *C. elegans* have a strong toolkit to better understand chromosome-specific behaviors in meiosis.

## 7.3 Changes in chromosome structure alter nuclear organization

Chromosomal rearrangements, such as Robertsonian translocations, are another key example of chromosome structure influencing meiotic events. Robertsonian translocations are end-to-end fusions of the long arms of acrocentric chromosomes that are seen in 1:1000 live births in humans and can cause miscarriages, infertility, and aneuploidy (Scriven, Flinter, Braude, & Ogilvie, 2001). The human acrocentric chromosomes (13, 14, 15, 21, 22) are associated with Robertsonian translocations. It is important to note that the short arms of these chromosomes contain mostly rDNA loci and repetitive DNA meaning that the loss of the short arms upon fusion is not deleterious to survival (Bandyopadhyay et al., 2002). These translocations cause fertility issues because during meiosis there is an increased risk for unbalanced gametes. Additionally, depending on the location of the breakpoint that formed the translocation, dicentric chromosomes can be formed (Sullivan, Jenkins, Karson, Leana-Cox, & Schwartz, 1996). In this case, one of the centromeres is usually inactivated leaving only one active centromere (reviewed by Stimpson, Matheny, & Sullivan, 2012).

Robertsonian translocations also occur naturally in other species, such as mice. Using the presence of many diverse Robertsonian fusions in wild mouse populations, it was found that genome rearrangements alter the nuclear architecture in germ cells (Vara et al., 2021). These alterations result in differences in chromosome synapsis and crossing over by inducing novel heterologous interactions (Vara et al., 2021). How cells handle the diversity in chromosome structures during meiosis is an important area of future research as alterations in structure can have major effects on genome organization and meiotic outcomes.

## 8. Other examples of chromosome-specific defects

In this review a few key examples of meiotic chromosome-specific defects have been highlighted. This is by no means an exhaustive list of chromosome-specific meiotic events. Further, intrachromosomal events can also occur between two homologs of the same chromosome. These events lead to meiotic and post-meiotic drive where one homolog is preferentially inherited over the other one. There are many mechanisms that can lead to meiotic drive and post-meiotic drive that have been previously reviewed (Bravo Núñez, Nuckolls, & Zanders, 2018; Courret, Chang, Wei, Montchamp-Moreau, & Larracuente, 2019; Kursel & Malik, 2018; Lampson & Black, 2017; Zanders & Johannesson, 2021).

Additionally, studies of chromosome-specific dynamics should not be limited to meiosis. For example, aneuploidy in humans is not only due to meiotic chromosome missegregation but also can be due to mitotic missegregation in the early embryonic divisions (Mantikou, Wong, Repping, & Mastenbroek, 2012; Schneider & Ellenberg, 2019). There is evidence to suggest that some chromosomes are more likely to missegregate during these initial divisions than others (Mantikou et al., 2012). However, this work is limited in humans and requires a better system to understand how chromosome-specific defects can occur in early embryonic divisions.

It is well known that cancerous cells are highly aneuploid and often not equally aneuploid across the entire genome (Ben-David & Amon, 2019; Dumont et al., 2020). Cancerous cells maintain aneuploidies of chromosomes that are beneficial, usually in promoting their growth and in evading treatment therapies (Ben-David & Amon, 2019; Lukow et al., 2021). Additionally, many studies that have induced chromosome missegregation events using chemicals or radiation have found that some chromosomes are more likely to missegregate than others (Balajee, Bertucci, Taveras, & Brenner, 2014; Caria, Chaveca, & Rueff, 1996; Dumont et al., 2020; Worrall et al., 2018). However, the underlying mitotic missegregation events that lead to chromosome-specific aneuploidies are poorly studied. There is evidence to suggest that centromere size plays a roll, with chromosomes bearing smaller centromeres missegregating more often (Drpic et al., 2018; Dumont et al., 2020). Additionally, cancer cell lines display aneuploidies of different chromosomes. This is likely due to specific chromosomes being beneficial to specific cancerous cells and therefore being maintained differently in distinct cancers. Both meiotic and mitotic chromosome-specific segregation defects are interesting areas of chromosome biology that are very relevant to human health.

## 9. Summary

Understanding why chromosome-specific dynamics arise in meiosis is a critical and often overlooked area of meiosis research. There are clear examples in humans and in many model systems that chromosome-specific defects occur and can lead to detrimental meiotic outcomes. In the future, attention needs to be paid to how all chromosomes, not just one or two representative ones, respond to meiotic defects and attempts need to be made to understand why chromosomes are behaving differently.

## Acknowledgments

Thank you to Hawley lab members and Blake Billmyre for critical reading and thoughtful comments. KEB is funded by a K99 from NIGMS (K99GM138759).

## References

Abe, H., Alavattam, K. G., Hu, Y. C., Pang, Q., Andreassen, P. R., Hegde, R. S., et al. (2020). The initiation of meiotic sex chromosome inactivation sequesters DNA damage signaling from autosomes in mouse spermatogenesis. *Current Biology*, *30*, 408–420. e5. https://doi.org/10.1016/J.CUB.2019.11.064.

Acquaviva, L., Boekhout, M., Karasu, M. E., Brick, K., Pratto, F., Li, T., et al. (2020). Ensuring meiotic DNA break formation in the mouse pseudoautosomal region. *Nature*, *582*(7812), 426–431. https://doi.org/10.1038/s41586-020-2327-4.

Alleva, B., & Smolikove, S. (2017). Moving and stopping: Regulation of chromosome movement to promote meiotic chromosome pairing and synapsis. *Nucleus*, *8*, 613–624. https://doi.org/10.1080/19491034.2017.1358329.

Baarends, W. M., Wassenaar, E., van der Laan, R., Hoogerbrugge, J., Sleddens-Linkels, E., Hoeijmakers, J. H. J., et al. (2005). Silencing of unpaired chromatin and histone H2A ubiquitination in mammalian meiosis. *Molecular and Cellular Biology*, *25*, 1041–1053. https://doi.org/10.1128/MCB.25.3.1041-1053.2005.

Babu, A., & Verma, R. S. (1987). Chromosome structure: Euchromatin and heterochromatin. *International Review of Cytology*, *108*, 1–60. https://doi.org/10.1016/S0074-7696(08)61435-7.

Balajee, A. S., Bertucci, A., Taveras, M., & Brenner, D. J. (2014). Multicolour FISH analysis of ionising radiation induced micronucleus formation in human lymphocytes. *Mutagenesis*, *29*, 447–455. https://doi.org/10.1093/MUTAGE/GEU041.

Bandyopadhyay, R., Heller, A., Knox-DuBois, C., McCaskill, C., Berend, S. A., Page, S. L., et al. (2002). Parental origin and timing of De novo Robertsonian translocation formation. *The American Journal of Human Genetics*, *71*, 1456–1462. https://doi.org/10.1086/344662.

Baudrimont, A., Penkner, A., Woglar, A., Machacek, T., Wegrostek, C., Gloggnitzer, J., et al. (2010). Leptotene/Zygotene chromosome movement via the SUN/KASH protein bridge in Caenorhabditis elegans. *PLoS Genetics*, *6*, e1001219. https://doi.org/10.1371/journal.pgen.1001219.

Ben-David, U., & Amon, A. (2019). Context is everything: Aneuploidy in cancer. *Nature Reviews Genetics*, *21*, 44–62. https://doi.org/10.1038/s41576-019-0171-x.

Billmyre, K. K., Cahoon, C. K., Matthew Heenan, G., Wesley, E. R., Yu, Z., Unruh, J. R., et al. (2019). X chromosome and autosomal recombination are differentially sensitive to disruptions in SC maintenance. *Proceedings of the National Academy of Sciences of the United States of America*, *116*, 21641–21650. https://doi.org/10.1073/pnas.1910840116.

Bloom, K. S. (2014). Centromeric Heterochromatin: The Primordial Segregation Machine. *Annual Review of Genetics*, *48*, 457–484. https://doi.org/10.1146/annurev-genet-120213-092033.

Bravo Núñez, M. A., Nuckolls, N. L., & Zanders, S. E. (2018). Genetic villains: Killer meiotic drivers. *Trends in Genetics*, *34*, 424. https://doi.org/10.1016/J.TIG.2018.02.003.

Britton-Davidian, J., Catalan, J., da Graça Ramalhinho, M., Ganem, G., Auffray, J. C., Capela, R., et al. (2000). Rapid chromosomal evolution in island mice. *Nature*, *403*(6766), 158. https://doi.org/10.1038/35003116.

Bugge, M., Collins, A., Petersen, M. B., Fisher, J., Brandt, G., Hertz, J. M., et al. (1998). Non-disjunction of chromosome 18. *Human Molecular Genetics*, 7, 661–669. https://doi.org/10.1093/HMG/7.4.661.

Cahoon, C. K., & Hawley, R. S. (2016). Regulating the construction and demolition of the synaptonemal complex. *Nature Structural & Molecular Biology*, *23*, 369–377. https://doi.org/10.1038/nsmb.3208.

Capalbo, A., Hoffmann, E. R., Cimadomo, D., Ubaldi, F. M., & Rienzi, L. (2017). Human female meiosis revised: New insights into the mechanisms of chromosome segregation and aneuploidies from advanced genomics and time-lapse imaging. *Human Reproduction Update*, *23*, 706–722. https://doi.org/10.1093/HUMUPD/DMX026.

Caria, H., Chaveca, T., & Rueff, J. (1996). Preferential sensitivity of acrocentric chromosomes to the aneugenic effect of colchicine. *Teratogenesis, Carcinogenesis, and Mutagenesis*. https://doi.org/10.1002/(SICI)1520-6866(1996)16:5<243::AID-TCM1>3.0.CO;2-I.

Cerqueira, A. V., & Lemos, B. (2019). Ribosomal DNA and the nucleolus as keystones of nuclear architecture, organization, and function. *Trends in Genetics*, *35*, 710–723. https://doi.org/10.1016/J.TIG.2019.07.011.

Chen, X., Suhandynata, R. T., Sandhu, R., Rockmill, B., Mohibullah, N., Niu, H., et al. (2015). Phosphorylation of the synaptonemal complex protein Zip1 regulates the crossover/noncrossover decision during yeast meiosis. *PLoS Biology*, *13*. https://doi.org/10.1371/journal.pbio.1002329.

Chen, S. Y., Tsubouchi, T., Rockmill, B., Sandler, J. S., Richards, D. R., Vader, G., et al. (2008). Global analysis of the meiotic crossover landscape. *Developmental Cell*, *15*, 401–415. https://doi.org/10.1016/j.devcel.2008.07.006.

Christophorou, N., Rubin, T., & Huynh, J. R. (2013). Synaptonemal complex components promote centromere pairing in pre-meiotic germ cells. *PLoS Genetics*, *9*. https://doi.org/10.1371/journal.pgen.1004012.

Courret, C., Chang, C. H., Wei, K. H. C., Montchamp-Moreau, C., & Larracuente, A. M. (2019). Meiotic drive mechanisms: Lessons from Drosophila. *Proceedings of the Royal Society B, 286*. https://doi.org/10.1098/RSPB.2019.1430.

Couteau, F., Nabeshima, K., Villeneuve, A., & Zetka, M. (2004). A component of C. elegans meiotic chromosome axes at the interface of homolog alignment, synapsis, nuclear reorganization, and recombination. *Current Biology, 14*, 585–592. https://doi.org/10.1016/J.CUB.2004.03.033.

Couteau, F., & Zetka, M. (2005). HTP-1 coordinates synaptonemal complex assembly with homolog alignment during meiosis in C. elegans. *Genes & Development, 19*, 2744–2756. https://doi.org/10.1101/GAD.1348205.

Damas, J., Corbo, M., & Lewin, H. A. (2021). Vertebrate chromosome. *Evolution, 9*, 1–27. https://doi.org/10.1146/annurev-animal-020518-114924.

Deakin, J. E. (2018). Chromosome evolution in marsupials. *Genes, 9*, 72. https://doi.org/10.3390/GENES9020072.

Dernburg, A. F., McDonald, K., Moulder, G., Barstead, R., Dresser, M., & Villeneuve, A. M. (1998). Meiotic recombination in C. elegans initiates by a conserved mechanism and is dispensable for homologous chromosome synapsis. *Cell, 94*, 387–398. https://doi.org/10.1016/S0092-8674(00)81481-6.

Disteche, C. M. (2012). *Dosage Compensation of the Sex Chromosomes. 46* (pp. 537–560). https://doi.org/10.1146/annurev-genet-110711-155454.

Drpic, D., Almeida, A. C., Aguiar, P., Renda, F., Damas, J., Lewin, H. A., et al. (2018). Chromosome segregation is biased by kinetochore size. *Current Biology, 28*, 1344–1356.e5. https://doi.org/10.1016/J.CUB.2018.03.023.

Dumont, M., Gamba, R., Gestraud, P., Klaasen, S., Worrall, J. T., Vries, S. G.d., et al. (2020). Human chromosome-specific aneuploidy is influenced by DNA-dependent centromeric features. *The EMBO Journal, 39*, e102924. https://doi.org/10.15252/EMBJ.2019102924.

Duncan, F. E., Hornick, J. E., Lampson, M. A., Schultz, R. M., Shea, L. D., & Woodruff, T. K. (2012). Chromosome cohesion decreases in human eggs with advanced maternal age. *Aging Cell, 11*, 1121–1124. https://doi.org/10.1111/J.1474-9726.2012.00866.X.

Fernandez-Capetillo, O., Mahadevaiah, S. K., Celeste, A., Romanienko, P. J., Camerini-Otero, R. D., Bonner, W. M., et al. (2003). H2AX is required for chromatin remodeling and inactivation of sex chromosomes in male mouse meiosis. *Developmental Cell, 4*, 497–508. https://doi.org/10.1016/S1534-5807(03)00093-5.

Fisher, J. M., Harvey, J. F., Morton, N. E., & Jacobs, P. A. (1995). Trisomy 18: Studies of the parent and cell division of origin and the effect of aberrant recombination on nondisjunction. *American Journal of Human Genetics, 56*, 669.

Franasiak, J. M., Forman, E. J., Hong, K. H., Werner, M. D., Upham, K. M., Treff, N. R., et al. (2014). Aneuploidy across individual chromosomes at the embryonic level in trophectoderm biopsies: Changes with patient age and chromosome structure. *Journal of Assisted Reproduction and Genetics, 31*(11), 1501–1509. https://doi.org/10.1007/S10815-014-0333-X.

Franklin, A. E., & Cande, W. Z. (1999). Nuclear organization and chromosome segregation. *The Plant Cell, 11*, 523–534. https://doi.org/10.1105/TPC.11.4.523.

Furuyama, S., & Biggins, S. (2007). Centromere identity is specified by a single centromeric nucleosome in budding yeast. *Proceedings of the National Academy of Sciences of the United States of America, 104*, 14706–14711. https://doi.org/10.1073/PNAS.0706985104.

Gerton, J. L., & Hawley, R. S. (2005). Homologous chromosome interactions in meiosis: Diversity amidst conservation. *Nature Reviews Genetics, 6*(6), 477–487. https://doi.org/10.1038/nrg1614.

Gong, W. J., McKim, K. S., & Hawley, R. S. (2005). All paired up with no place to go: Pairing, synapsis, and DSB formation in a balancer heterozygote. *PLoS Genetics*, *1*, e67. https://doi.org/10.1371/journal.pgen.0010067.

Grey, C., & de Massy, B. (2021). Chromosome organization in early meiotic prophase. *Frontiers in Cell and Development Biology*, *9*. https://doi.org/10.3389/FCELL.2021.688878.

Gruhn, J. R., Zielinska, A. P., Shukla, V., Blanshard, R., Capalbo, A., Cimadomo, D., et al. (2019). Chromosome errors in human eggs shape natural fertility over reproductive life span. *Science*, *1979*(365), 1466–1469. https://doi.org/10.1126/science.aav7321.

Hall, H. E., Surti, U., Hoffner, L., Shirley, S., Feingold, E., & Hassold, T. (2007). The origin of trisomy 22: Evidence for acrocentric chromosome-specific patterns of nondisjunction. *American Journal of Medical Genetics. Part A*, *143A*, 2249–2255. https://doi.org/10.1002/ajmg.a.31918.

Handel, M. A., & Schimenti, J. C. (2010). Genetics of mammalian meiosis: Regulation, dynamics and impact on fertility. *Nature Reviews Genetics*, *11*(2), 124–136. https://doi.org/10.1038/nrg2723.

Hartmann, M. A., & Sekelsky, J. (2017). The absence of crossovers on chromosome 4 in *Drosophila melanogaster*: Imperfection or interesting exception? *Fly*, *11*, 253–259. https://doi.org/10.1080/19336934.2017.1321181.

Hassold, T., Hall, H., & Hunt, P. (2007). The origin of human aneuploidy: Where we have been, where we are going. *Human Molecular Genetics*, *16*, R203–R208. https://doi.org/10.1093/hmg/ddm243.

Hassold, T., & Hunt, P. (2001). To err (meiotically) is human: The genesis of human aneuploidy. *Nature Reviews. Genetics*, *2*, 280–291. https://doi.org/10.1038/35066065.

Hassold, T. J., Hunt, P. A., & Nagaoka, S. I. (2012). Human aneuploidy: Mechanisms and new insights into an age-old problem. *Nature Reviews Genetics*, *13*, 493–504. https://doi.org/10.1038/nrg3245.

Hassold, T., Merrill, M., Adkins, K., Freeman, S., & Sherman, S. (1995). Recombination and maternal age-dependent nondisjunction: Molecular studies of trisomy 16. *American Journal of Human Genetics*, *57*, 867–874.

Hatkevich, T., Kohl, K. P., McMahan, S., Hartmann, M. A., Williams, A. M., & Sekelsky, J. (2017). Bloom syndrome helicase promotes meiotic crossover patterning and homolog disjunction. *Current Biology*, *27*, 96–102. https://doi.org/10.1016/J.CUB.2016.10.055.

Henderson, K. A., & Keeney, S. (2004). Tying synaptonemal complex initiation to the formation and programmed repair of DNA double-strand breaks. *Proceedings of the National Academy of Sciences of the United States of America*, *101*, 4519–4524. https://doi.org/10.1073/PNAS.0400843101.

Hiraoka, Y. (2020). Phase separation drives pairing of homologous chromosomes. *Current Genetics*, *66*(5), 881–887. https://doi.org/10.1007/S00294-020-01077-9.

Hou, Y., Fan, W., Yan, L., Li, R., Lian, Y., Huang, J., et al. (2013). Genome analyses of single human oocytes. *Cell*, *155*, 1492–1506. https://doi.org/10.1016/J.CELL.2013.11.040.

Howe, M., McDonald, K. L., Albertson, D. G., & Meyer, B. J. (2001). Him-10 is required for kinetochore structure and function on Caenorhabditis elegans holocentric chromosomes. *Journal of Cell Biology*, *153*, 1227–1238. https://doi.org/10.1083/JCB.153.6.1227.

Hughes, S. E., Miller, D. E., Miller, A. L., & Hawley, R. S. (2018). Female meiosis: Synapsis, recombination, and segregation in Drosophila melanogaster. *Genetics*, *208*, 875–908. https://doi.org/10.1534/genetics.117.300081.

Ichijima, Y., Sin, H. S., & Namekawa, S. H. (2012). Sex chromosome inactivation in germ cells: Emerging roles of DNA damage response pathways. *Cellular and Molecular Life Sciences*, *69*, 2559–2572. https://doi.org/10.1007/S00018-012-0941-5.

Jaramillo-Lambert, A., Ellefson, M., Villeneuve, A. M., & Engebrecht, J. A. (2007). Differential timing of S phases, X chromosome replication, and meiotic prophase in the C. elegans germ line. *Developmental Biology, 308*, 206–221. https://doi.org/10.1016/J.YDBIO.2007.05.019.

Joyce, E. F., Apostolopoulos, N., Beliveau, B. J., & Wu, C. T. (2013). Germline progenitors escape the widespread phenomenon of homolog pairing during Drosophila development. *PLoS Genetics, 9*, e1004013. https://doi.org/10.1371/JOURNAL.PGEN.1004013.

Kaback, D. B., Guacci, V., Barber, D., & Mahon, J. W. (1992). Chromosome size-dependent control of meiotic recombination. *Science, 1979*(256), 228–232. https://doi.org/10.1126/SCIENCE.1566070.

Kauppi, L., Barchi, M., Lange, J., Baudat, F., Jasin, M., & Keeney, S. (2013). Numerical constraints and feedback control of double-strand breaks in mouse meiosis. *Genes & Development, 27*, 873–886. https://doi.org/10.1101/GAD.213652.113.

Keeney, S. (2001). Mechanism and control of meiotic recombination initiation. *Current Topics in Developmental Biology, 52*, 1–53. https://doi.org/10.1016/S0070-2153(01)52008-6.

Kelly, W. G., & Fire, A. (1998). Chromatin silencing and the maintenance of a functional germline in Caenorhabditis elegans. *Development, 125*, 2451–2456. https://doi.org/10.1242/DEV.125.13.2451.

Kelly, W. G., Schaner, C. E., Dernburg, A. F., Lee, M. H., Kim, S. K., Villeneuve, A. M., et al. (2002). X-chromosome silencing in the germline of C. elegans. *Development, 129*, 479–492. https://doi.org/10.1242/DEV.129.2.479.

Kogo, H., Tsutsumi, M., Inagaki, H., Ohye, T., Kiyonari, H., & Kurahashi, H. (2012). HORMAD2 is essential for synapsis surveillance during meiotic prophase via the recruitment of ATR activity. *Genes to Cells, 17*, 897–912. https://doi.org/10.1111/GTC.12005.

Kumar, R., Bourbon, H. M., & de Massy, B. (2010). Functional conservation of Mei4 for meiotic DNA double-strand break formation from yeasts to mice. *Genes & Development, 24*, 1266–1280. https://doi.org/10.1101/gad.571710.

Kumar, R., Oliver, C., Brun, C., Juarez-Martinez, A. B., Tarabay, Y., Kadlec, J., et al. (2018). Mouse REC114 is essential for meiotic DNA double-strand break formation and forms a complex with MEI4. *Life Science Alliance, 1*. https://doi.org/10.26508/lsa.201800259.

Kursel, L. E., & Malik, H. S. (2018). The cellular mechanisms and consequences of centromere drive. *Current Opinion in Plant Biology, 52*, 58–65. https://doi.org/10.1016/j.pbi.2018.01.007.

Lagirand-Cantaloube, J., Ciabrini, C., Charrasse, S., Ferrieres, A., Castro, A., Anahory, T., et al. (2017). Loss of centromere cohesion in Aneuploid human oocytes correlates with decreased kinetochore localization of the sac proteins Bub1 and Bubr1. *Scientific Reports, 7*. https://doi.org/10.1038/SREP44001.

Lamb, N. E., Yu, K., Shaffer, J., Feingold, E., & Sherman, S. L. (2005). Association between maternal age and meiotic recombination for trisomy 21. *American Journal of Human Genetics, 76*, 91–99. https://doi.org/10.1086/427266.

Lampson, M. A., & Black, B. E. (2017). Cellular and molecular mechanisms of centromere drive. *Cold Spring Harbor Symposia on Quantitative Biology, 82*, 249. https://doi.org/10.1101/SQB.2017.82.034298.

Lee, M. S., Higashide, M. T., Choi, H., Li, K., Hong, S., Lee, K., et al. (2021). The synaptonemal complex central region modulates crossover pathways and feedback control of meiotic double-strand break formation. *Nucleic Acids Research, 49*, 7537–7553. https://doi.org/10.1093/NAR/GKAB566.

Lindström, M. S., Jurada, D., Bursac, S., Orsolic, I., Bartek, J., & Volarevic, S. (2018). Nucleolus as an emerging hub in maintenance of genome stability and cancer pathogenesis. *Oncogene, 37*(18), 2351–2366. https://doi.org/10.1038/s41388-017-0121-z.

Lukow, D. A., Sausville, E. L., Suri, P., Chunduri, N. K., Wieland, A., Leu, J., et al. (2021). Chromosomal instability accelerates the evolution of resistance to anti-cancer therapies. *Developmental Cell, 56*, 2427–2439.e4. https://doi.org/10.1016/s.devcel.2021.07.009.

Lynn, A., Soucek, R., & Börner, G. V. (2007). ZMM proteins during meiosis: Crossover artists at work. *Chromosome Research, 15*(5), 591–605. https://doi.org/10.1007/S10577-007-1150-1.

Mahadevaiah, S. K., Bourc'his, D., de Rooij, D. G., Bestor, T. H., Turner, J. M. A., & Burgoyne, P. S. (2008). Extensive meiotic asynapsis in mice antagonises meiotic silencing of unsynapsed chromatin and consequently disrupts meiotic sex chromosome inactivation. *Journal of Cell Biology, 182*, 263–276. https://doi.org/10.1083/JCB.200710195.

Mancera, E., Bourgon, R., Brozzi, A., Huber, W., & Steinmetz, L. M. (2008). High-resolution mapping of meiotic crossovers and non-crossovers in yeast. *Nature, 454*(7203), 479–485. https://doi.org/10.1038/nature07135.

Mantikou, E., Wong, K. M., Repping, S., & Mastenbroek, S. (2012). Molecular origin of mitotic aneuploidies in preimplantation embryos. *Biochimica et Biophysica Acta (BBA) - Molecular Basis of Disease, 1822*, 1921–1930. https://doi.org/10.1016/J.BBADIS.2012.06.013.

Martinez-Perez, E., Schvarzstein, M., Barroso, C., Lightfoot, J., Dernburg, A. F., & Villeneuve, A. M. (2008). Crossovers trigger a remodeling of meiotic chromosome axis composition that is linked to two-step loss of sister chromatid cohesion. *Genes & Development, 22*, 2886–2901. https://doi.org/10.1101/GAD.1694108.

Mason, J. M., Frydrychova, R. C., & Biessmann, H. (2008). Drosophila telomeres: An exception providing new insights. *BioEssays, 30*, 25–37. https://doi.org/10.1002/BIES.20688.

McClendon, T. B., Mainpal, R., Amrit, F. R. G., Krause, M. W., Ghazi, A., & Yanowitz, J. L. (2016). X Chromosome Crossover Formation and Genome Stability in *Caenorhabditis elegans* Are Independently Regulated by xnd-1. *G3 (Bethesda), 6*, 3913–3925. https://doi.org/10.1534/G3.116.035725.

McKee, B. D., & Karpen, G. H. (1990). Drosophila ribosomal RNA genes function as an X-Y pairing site during male meiosis. *Cell, 61*, 61–72. https://doi.org/10.1016/0092-8674(90)90215-Z.

McKim, K. S., & Hayashi-Hagihara, A. (1998). Mei-W68 in Drosophila melanogaster encodes a Spo11 homolog: Evidence that the mechanism for initiating meiotic recombination is conserved. *Genes & Development, 12*, 2932–2942. https://doi.org/10.1101/gad.12.18.2932.

Mehrotra, S., & McKim, K. S. (2006). Temporal analysis of meiotic DNA double-Strand break formation and repair in Drosophila females. *PLoS Genetics, 2*, e200. https://doi.org/10.1371/journal.pgen.0020200.

Meneely, P. M., McGovern, O. L., Heinis, F. I., & Yanowitz, J. L. (2012). Crossover distribution and frequency are regulated by him-5 in Caenorhabditis elegans. *Genetics, 190*, 1251–1266. https://doi.org/10.1534/Genetics.111.137463.

Miga, K. H., & Alexandrov, I. A. (2021). Variation and Evolution of Human Centromeres: A Field Guide and Perspective. *Annual Review of Genetics, 55*, 583–602. https://doi.org/10.1146/annurev-genet-071719-020519.

Mlynarczyk-Evans, S., & Villeneuve, A. M. (2017). Time-course analysis of early meiotic prophase events informs mechanisms of homolog pairing and synapsis in caenorhabditis elegans. *Genetics, 207*, 103–114. https://doi.org/10.1534/GENETICS.117.204172/-/DC1.

Morrison, O., & Thakur, J. (2021). Molecular complexes at Euchromatin, heterochromatin and Centromeric chromatin. *International Journal of Molecular Sciences, 22*, 6922. https://doi.org/10.3390/IJMS22136922.

Murakami, H., Lam, I., Huang, P. C., Song, J., van Overbeek, M., & Keeney, S. (2020). Multilayered mechanisms ensure that short chromosomes recombine in meiosis. *Nature*, *582*(7810), 124–128. https://doi.org/10.1038/s41586-020-2248-2.

Nelson, J. O., Watase, G. J., Warsinger-Pepe, N., & Yamashita, Y. M. (2019). Mechanisms of rDNA copy number maintenance. *Trends in Genetics*, *35*, 734–742. https://doi.org/10.1016/J.TIG.2019.07.006.

Németh, A., & Längst, G. (2011). Genome organization in and around the nucleolus. *Trends in Genetics*, *27*, 149–156. https://doi.org/10.1016/J.TIG.2011.01.002.

Ohta, K., Shibata, T., & Nicolas, A. (1994). Changes in chromatin structure at recombination initiation sites during yeast meiosis. *The EMBO Journal*, *13*, 5754–5763. https://doi.org/10.1002/J.1460-2075.1994.TB06913.X.

Ottolini, C. S., Newnham, L. J., Capalbo, A., Natesan, S. A., Joshi, H. A., Cimadomo, D., et al. (2015). Genome-wide maps of recombination and chromosome segregation in human oocytes and embryos show selection for maternal recombination rates. *Nature Genetics*, *47*(7), 727–735. https://doi.org/10.1038/ng.3306.

Page, S. L., & Hawley, R. S. (2004). The genetics and molecular biology of the synpatonemal complex. *Annual Review of Cell and Developmental Biology*, *20*, 525–558. https://doi.org/10.1146/annurev.cellbio.19.111301.155141.

Page, S. L., Khetani, R. S., Lake, C. M., Nielsen, R. J., Jeffress, J. K., Warren, W. D., et al. (2008). Corona is required for higher-order assembly of transverse filaments into full-length synaptonemal complex in Drosophila oocytes. *PLoS Genetics*, *4*, e1000194. https://doi.org/10.1371/journal.pgen.1000194.

Pan, J., Sasaki, M., Kniewel, R., Murakami, H., Blitzblau, H. G., Tischfield, S. E., et al. (2011). A hierarchical combination of factors shapes the genome-wide topography of yeast meiotic recombination initiation. *Cell*, *144*, 719–731. https://doi.org/10.1016/j.cell.2011.02.009.

Pazhayam, N. M., Turcotte, C. A., & Sekelsky, J. (2021). Meiotic crossover patterning. *Frontiers in Cell and Development Biology*, *9*, 1940. https://doi.org/10.3389/fcell.2021.681123/BIBTEX.

Perry, J., Palmer, S., Gabriel, A., & Ashworth, A. (2001). A short pseudoautosomal region in laboratory mice. *Genome Research*, *11*, 1826–1832. https://doi.org/10.1101/gr.203001.

Phillips, C. M., & Dernburg, A. F. (2006). A family of zinc-finger proteins is required for chromosome-specific pairing and synapsis during meiosis in C. elegans. *Developmental Cell*, *11*, 817–829. https://doi.org/10.1016/J.DEVCEL.2006.09.020.

Phillips, C. M., Wong, C., Bhalla, N., Carlton, P. M., Weiser, P., Meneely, P. M., et al. (2005). HIM-8 binds to the X chromosome pairing center and mediates chromosome-specific meiotic synapsis. *Cell*, *123*, 1051–1063. https://doi.org/10.1016/j.cell.2005.09.035.

Potapova, T. A., & Gerton, J. L. (2019). Ribosomal DNA and the nucleolus in the context of genome organization. *Chromosome Research*, *27*, 109–127. https://doi.org/10.1007/S10577-018-9600-5.

Raudsepp, T., Das, P. J., Avila, F., & Chowdhary, B. P. (2012). The pseudoautosomal region and sex chromosome aneuploidies in domestic species. *Sexual Development*, *6*, 72–83. https://doi.org/10.1159/000330627.

Reddy, K. C., & Villeneuve, A. M. (2004). C. elegans HIM-17 links chromatin modification and competence for initiation of meiotic recombination. *Cell*, *118*, 439–452. https://doi.org/10.1016/j.cell.2004.07.026.

Reinholdt, L. G., & Schimenti, J. C. (2005). Mei1 is epistatic to Dmc1 during mouse meiosis. *Chromosoma*, *114*, 127–134. https://doi.org/10.1007/s00412-005-0346-4.

Rog, O., & Dernburg, A. F. (2013). Chromosome pairing and synapsis during Caenorhabditis elegans meiosis. *Current Opinion in Cell Biology*, *25*, 349–356. https://doi.org/10.1016/j.ceb.2013.03.003.

Rosu, S., Zawadzki, K. A., Stamper, E. L., Libuda, D. E., & Reese, A. L. (2013). The C. elegans DSB-2 protein reveals a regulatory network that controls competence for meiotic DSB formation and promotes crossover assurance. *PLoS Genetics*, *9*, 1003674. https://doi.org/10.1371/journal.pgen.1003674.

Royo, H., Prosser, H., Ruzankina, Y., Mahadevaiah, S. K., Cloutier, J. M., Baumann, M., et al. (2013). ATR acts stage specifically to regulate multiple aspects of mammalian meiotic silencing. *Genes & Development*, *27*, 1484–1494. https://doi.org/10.1101/GAD.219477.113.

Rubin, T., Macaisne, N., & Huynh, J. R. (2020). Mixing and matching chromosomes during female meiosis. *Cell*, *9*, 696. https://doi.org/10.3390/CELLS9030696.

Sasaki, M., Lange, J., & Keeney, S. (2010). Genome destabilization by homologous recombination in the germ line. *Nature Reviews Molecular Cell Biology*, *11*(3), 182–195. https://doi.org/10.1038/nrm2849.

Scherthan, H. (2007). Telomere attachment and clustering during meiosis. *Cellular and Molecular Life Sciences*, *64*, 117–124. https://doi.org/10.1007/S00018-006-6463-2.

Schimenti, J. (2005). Synapsis or silence. *Nature Genetics*, *37*(1), 11–13. https://doi.org/10.1038/ng0105-11.

Schneider, I., & Ellenberg, J. (2019). Mysteries in embryonic development: How can errors arise so frequently at the beginning of mammalian life? *PLoS Biology*, *17*. https://doi.org/10.1371/JOURNAL.PBIO.3000173.

Schubert, I. (2007). Chromosome evolution. *Current Opinion in Plant Biology*, *10*, 109–115. https://doi.org/10.1016/J.PBI.2007.01.001.

Scriven, P. N., Flinter, F. A., Braude, P. R., & Ogilvie, C. M. (2001). Robertsonian translocations- -reproductive risks and indications for preimplantation genetic diagnosis. *Human Reproduction*, *16*, 2267–2273. https://doi.org/10.1093/HUMREP/16.11.2267.

Serrentino, M. E., Chaplais, E., Sommermeyer, V., & Borde, V. (2013). Differential association of the conserved SUMO ligase Zip3 with meiotic double-strand break sites reveals regional variations in the outcome of meiotic recombination. *PLoS Genetics*, *9*. https://doi.org/10.1371/JOURNAL.PGEN.1003416.

Shaw, P. J., & McKeown, P. C. (2011). The structure of rDNA chromatin. *Protein Review*, *15*, 43–55. https://doi.org/10.1007/978-1-4614-0514-6_3.

Sherizen, D., Jang, J. K., Bhagat, R., Kato, N., & McKim, K. S. (2005). Meiotic recombination in Drosophila females depends on chromosome continuity between genetically defined boundaries. *Genetics*, *169*, 767–781. https://doi.org/10.1534/genetics.104.035824.

Sherman, S. L., Lamb, N. E., & Feingold, E. (2006). Relationship of recombination patterns and maternal age among non-disjoined chromosomes 21. *Biochemical Society Transactions*, *34*, 578–580. https://doi.org/10.1042/BST0340578.

Sims, J., Copenhaver, G. P., & Schlögelhofer, P. (2019). Meiotic DNA repair in the nucleolus employs a nonhomologous end-joining mechanism. *The Plant Cell*, *31*, 2259–2275. https://doi.org/10.1105/tpc.19.00367.

Smolikov, S., Schild-Prüfert, K., & Colaiácovo, M. P. (2008). CRA-1 uncovers a double-Strand break-dependent pathway promoting the assembly of central region proteins on chromosome axes during C. elegans meiosis. *PLoS Genetics*, *4*, e1000088. https://doi.org/10.1371/JOURNAL.PGEN.1000088.

Stanzione, M., Baumann, M., Papanikos, F., Dereli, I., Lange, J., Ramlal, A., et al. (2016). Meiotic DNA break formation requires the unsynapsed chromosome axis-binding protein IHO1 (CCDC36) in mice. *Nature Cell Biology*, *18*, 1208–1220. https://doi.org/10.1038/ncb3417.

Stimpson, K. M., Matheny, J. E., & Sullivan, B. A. (2012). Dicentric chromosomes: Unique models to study centromere function and inactivation. *Chromosome Research*, *20*, 595–605. https://doi.org/10.1007/S10577-012-9302-3/FIGURES/2.

Sullivan, B. A., Jenkins, L. S., Karson, E. M., Leana-Cox, J., & Schwartz, S. (1996). Evidence for structural heterogeneity from molecular cytogenetic analysis of dicentric Robertsonian translocations. *American Journal of Human Genetics*, *59*, 167.

Takeo, S., Lake, C. M., Morais-de-Sá, E., Sunkel, C. E., & Hawley, R. S. (2011). Synaptonemal complex-dependent centromeric clustering and the initiation of synapsis in Drosophila oocytes. *Current Biology*, *21*, 1845–1851. https://doi.org/10.1016/j.cub.2011.09.044.

Teixeira, M. T., & Gilson, E. (2005). Telomere maintenance, function and evolution: The yeast paradigm. *Chromosome Research*, *13*(5), 535–548. https://doi.org/10.1007/S10577-005-0999-0.

Thacker, D., Mohibullah, N., Zhu, X., & Keeney, S. (2014). Homologue engagement controls meiotic DNA break number and distribution. *Nature*, *510*(7504), 241–246. https://doi.org/10.1038/nature13120.

Tsai, J. H., & McKee, B. D. (2011). Homologous pairing and the role of pairing centers in meiosis. *Journal of Cell Science*, *124*, 1955–1963. https://doi.org/10.1242/JCS.006387.

Turner, J. M. A. (2015). Meiotic silencing in mammals. *Annual Review of Genetics*, *49*, 395–412. https://doi.org/10.1146/ANNUREV-GENET-112414-055145.

Turner, J. M. A., Aprelikova, O., Xu, X., Wang, R., Kim, S., Chandramouli, G. V. R., et al. (2004). BRCA1, histone H2AX phosphorylation, and male meiotic sex chromosome inactivation. *Current Biology*, *14*, 2135–2142. https://doi.org/10.1016/J.CUB.2004.11.032.

Turner, J. M. A., Mahadevaiah, S. K., Fernandez-Capetillo, O., Nussenzweig, A., Xu, X., Deng, C. X., et al. (2004). Silencing of unsynapsed meiotic chromosomes in the mouse. *Nature Genetics*, *37*(1), 41–47. https://doi.org/10.1038/ng1484.

Ur, S. N., & Corbett, K. D. (2021). Architecture and Dynamics of Meiotic Chromosomes. *Annual Review of Genetics*, *55*, 497–526. https://doi.org/10.1146/annurev-genet-071719-020235.

van Bortle, K., & Corces, V. G. (2012). Nuclear Organization and Genome Function. *Annual Review of Cell and Developmental Biology*, *28*, 163–187. https://doi.org/10.1146/annurev-cellbio-101011-155824.

Vara, C., Paytuví-Gallart, A., Cuartero, Y., Álvarez-González, L., Marín-Gual, L., Garcia, F., et al. (2021). The impact of chromosomal fusions on 3D genome folding and recombination in the germ line. *Nature Communications*, *12*(1), 1–17. https://doi.org/10.1038/s41467-021-23270-1.

Vara, C., & Ruiz-Herrera, A. (2021). Unpacking chromatin remodelling in germ cells: Implications for development and evolution. *Trends in Genetics*. https://doi.org/10.1016/J.TIG.2021.10.007.

Wagner, C. R., Kuervers, L., Baillie, D. L., & Yanowitz, J. L. (2010). Xnd-1 regulates the global recombination landscape in Caenorhabditis elegans. *Nature*, *467*(7317), 839–843. https://doi.org/10.1038/nature09429.

Wilson, M. A., & Makova, K. D. (2009). Genomic analyses of sex chromosome. *Evolution*, *10*, 333–354. https://doi.org/10.1146/annurev-genom-082908-150105.

Worrall, J. T., Tamura, N., Mazzagatti, A., Shaikh, N., van Lingen, T., Bakker, B., et al. (2018). Non-random Mis-segregation of human chromosomes. *Cell Reports*, *23*, 3366–3380. https://doi.org/10.1016/J.CELREP.2018.05.047.

Wu, T. C., & Lichten, M. (1994). Meiosis-induced double-strand break sites determined by yeast chromatin structure. *Science*, *263*, 515–518. https://doi.org/10.1126/SCIENCE.8290959.

Wynne, D. J., Rog, O., Carlton, P. M., & Dernburg, A. F. (2012). Dynein-dependent processive chromosome motions promote homologous pairing in C. elegans meiosis. *Journal of Cell Biology*, *196*, 47–64. https://doi.org/10.1083/JCB.201106022/VIDEO-9.

Youds, J. L., & Boulton, S. J. (2011). The choice in meiosis – Defining the factors that influence crossover or non-crossover formation. *Journal of Cell Science, 124*, 501–513. https://doi.org/10.1242/JCS.074427.

Zanders, S., & Alani, E. (2009). The pch2Δ mutation in Baker's yeast alters meiotic crossover levels and confers a defect in crossover interference. *PLoS Genetics*, *5*, e1000571. https://doi.org/10.1371/JOURNAL.PGEN.1000571.

Zanders, S., & Johannesson, H. (2021). Molecular mechanisms and evolutionary consequences of spore killers in Ascomycetes. *Microbiology and Molecular Biology Reviews*. https://doi.org/10.1128/MMBR.00016-21.

Zickler, D., & Kleckner, N. (1999). Meiotic chromosomes: Integrating structure and function. *Annual Review of Genetics*, *33*, 603–754. https://doi.org/10.1146/annurev.genet.33.1.603.

Zickler, D., & Kleckner, N. (2015). Recombination, pairing, and synapsis of homologs during meiosis. *Cold Spring Harbor Perspectives in Biology*, 7, 1–28. https://doi.org/10.1101/CSHPERSPECT.A016626.

Zielinska, A. P., Holubcova, Z., Blayney, M., Elder, K., & Schuh, M. (2015). Sister kinetochore splitting and precocious disintegration of bivalents could explain the maternal age effect. *eLife*, *4*. https://doi.org/10.7554/ELIFE.11389.

CHAPTER SEVEN

# Emerging mechanisms and roles of meiotic crossover repression at centromeres

**Sucharita Sen, Ananya Dodamani, and Mridula Nambiar***
Department of Biology, Indian Institute of Science Education and Research, Pune, India
*Corresponding author: e-mail address: mridula@iiserpune.ac.in

## Contents



## Abstract

Crossover events during recombination in meiosis are essential for generating genetic diversity as well as crucial to allow accurate chromosomal segregation between homologous chromosomes. Spatial control for the distribution of crossover events along the chromosomes is largely a tightly regulated process and involves many facets such as interference, repression as well as assurance, to make sure that not too many or too few crossovers are generated. Repression of crossover events at the centromeres is a highly conserved process across all species tested. Failure to inhibit such recombination events can result in chromosomal mis-segregation during meiosis resulting in aneuploid gametes that are responsible for infertility or developmental disorders such as Down's syndrome and other trisomies in humans. In the past few decades, studies to understand the molecular mechanisms behind this repression have shown the involvement of a multitude of factors ranging from the centromere-specific proteins such as the kinetochore to the flanking pericentric heterochromatin as well as DNA

*Current Topics in Developmental Biology*, Volume 151
ISSN 0070-2153
https://doi.org/10.1016/bs.ctdb.2022.06.003

double-strand break repair pathways. In this chapter, we review the different mechanisms of pericentric repression mechanisms known till date as well as highlight the importance of understanding this regulation in the context of chromosomal segregation defects. We also discuss the clinical implications of dysregulation of this process, especially in human reproductive health and genetic diseases.

## 1. Introduction

A fundamental property of life is reproduction. Mitosis is a type of somatic cell division that aids unicellular organisms to grow in number and multicellular organisms to develop complex body plans. However, in order to evolve and adapt to changing environments, genetically diverse offspring are required that aid in survival of species. Cell division by meiosis offers an opportunity for such genetic variation via DNA recombination, that allows genetic exchange between the paired (parental) homologous chromosomes (Ohkura, 2015).

During meiosis, programmed double-strand breaks (DSBs) are generated that are repaired by the process of homologous recombination, which can give rise to either crossovers (COs) or non-crossovers (NCOs) (Whitby, 2005). The mechanistic steps of CO formation were initially proposed in the classic model by Robin Holliday and supplanted later with the DSB model that proposed initiation of recombination via DSBs (Holliday, 1964; Szostak, Orr-Weaver, Rothstein, & Stahl, 1983). The recombination-based DSB repair pathway mechanistically explains the formation of CO and NCO during meiosis (Fig. 1). DSBs are generated by the highly conserved, meiosis-specific Spo11 and partner proteins in a topoisomerase-like manner (Lam & Keeney, 2015). Spo11 remains covalently attached to the 5′-end of the DSB and is cleaved off by MRN complex and Ctp1, to release short Spo11-oligonucleotides (Ehmsen & Heyer, 2008). The now unbound DSB is further resected (5′ → 3′) leading to the formation of single-stranded (ss) DNA with 3′-overhangs. Strand exchange proteins Rad51 and the meiosis-specific paralog DMC1 bind to the ssDNA to aid in the homology search and strand-invasion into either a sister chromatid or the homolog, resulting in a D-loop (Ehmsen & Heyer, 2008). The subsequent steps can either lead to synthesis-dependent strand annealing (SDSA) or formation of a double Holliday Junction (dHJ) (Fig. 1). SDSA always gives rise to NCO products, whereas dHJ resolution via resolvases can be either symmetrical or asymmetrical, resulting in generation of NCOs and COs, respectively (Hunter, 2015; West et al., 2016). In an alternative dissolution model,

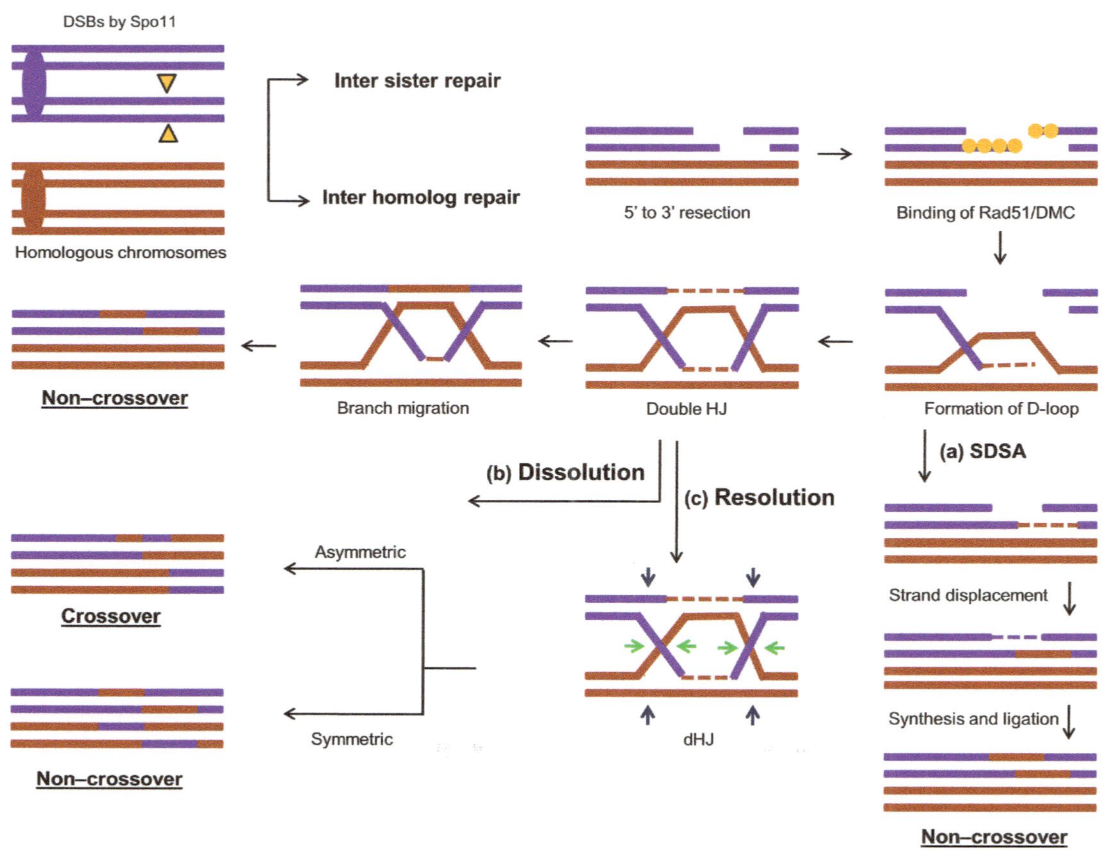


**Fig. 1** Homology-mediated DSB repair pathways in meiosis. After DNA DSB formation by Spo11 and partner proteins, the initial choice of repair template can result in two outcomes—inter-sister (IS) or inter-homolog (IH) repair. IS repair always results in a non-recombinant due to identical sequences of the sister-chromatids. IH repair on the other hand can be processed differentially to result in either non-crossover (NCO) or crossover (CO) events.

the topoisomerase II decatenates the migrated HJ and consequently results in NCOs (Bizard & Hickson, 2014; Swuec & Costa, 2014). Another specialized recombination event that can be associated with either a CO or NCO is known as gene conversion, whereupon a non-reciprocal exchange of genetic material from the donor to an acceptor DNA occurs at heteroduplex DNA. Apart from serving an evolutionary purpose, meiotic COs are also largely essential for proper segregation of chromosomes.

Cohesin complexes are ring-shaped proteins that hold the sister-chromatids together after replication (Peters, Tedeschi, & Schmitz, 2008). They are enriched around the centromeres as well as distributed along the chromosomal arms. During mitosis, cohesins help bi-orient the sister kinetochores during metaphase and are proteolytically cleaved in a cell-cycle dependent manner at the onset of anaphase to allow proper chromosomal segregation (Hauf, Waizenegger, & Peters, 2001; Tanaka, Fuchs, Loidl, & Nasmyth, 2000; Uhlmann, Lottspelch, & Nasmyth, 1999). However, meiosis involves two rounds of division, the first being reductional leading to separation of homologous chromosomes instead of the sister chromatids. The cohesins are removed in a step-wise manner, with those in the chromosomal arms removed at anaphase I to facilitate resolution of the chiasma, whereas the centromeric cohesins are protected till anaphase II (Ishiguro, 2019). The pericentric cohesins along with certain centromere-specific proteins—Monopolin complex in *Saccharomyces cerevisiae,* Moa I in *Schizosaccharomyces pombe* and Meikin in mice, help mono-orient the sister kinetochores to facilitate reductional division at meiosis I (MI) (Galander & Marston, 2020). The function of COs during MI is to act as physical links between the homologs, which mature into chiasma as the synaptonemal complex (SC) dissolves toward the end of prophase I. At the time of separation, they provide necessary tension to counter the pulling forces at the kinetochores and support mono-orientation of the sister kinetochores (Bascom-Slack, Ross, & Dawson, 1997). Absence of meiotic COs is a major cause for MI mis-segregation resulting in aneuploidy of gametes. Recombination-deficient mutants such as *rec12Δ* (Spo11-ortholog in *S. pombe*) lacking meiotic DSBs show defects in spore formation as well as errors in both MI and meiosis II (MII) suggesting gross defects in chromosomal segregation (Sharif, Glick, Davidson, & Wahls, 2002). However, a significant number of zygotes show proper MI segregation even in the absence of recombination and form dyads that are homozygous diploids in *S. pombe* (Davis & Smith, 2003). Such non-random homolog segregation is attributed to an achiasmate system dependent on dynein in *S. pombe* (Davis & Smith, 2005). A small fraction of achiasmate chromosomes is also tolerated in budding yeast via centromere-pairing, mediated by the residual SC proteins at the centromeres after SC disassembly, providing an alternative link between homologous chromosomes (Kurdzo, Chuong, Evatt, & Dawson, 2018). A robust achiasmate system is also prevalent in *Drosophila* females as well as males and their chromosome IV that are specifically devoid of COs (Hawley et al., 1992). Hence, presence of COs may not be a universal requirement for meiotic chromosomal segregation, as seen in these exceptions, but still remain one of the major factors determining the fidelity of separation of chromosomes in meiosis (Székvölgyi & Nicolas, 2010).

Meiotic crossovers are not uniformly distributed across the chromosomes and their patterning is tightly regulated (Nambiar, Chuang, & Smith, 2019; Pazhayam, Turcotte, & Sekelsky, 2021). Crossover assurance safeguards accurate chromosomal segregation by ensuring at least one crossover per bivalent during meiosis (Rosu, Libuda, & Villeneuve, 2011; Shinohara, Oh, Hunter, & Shinohara, 2008). Crossover interference limits the total number of COs by influencing the probability of occurrence of a crossover in the vicinity of another, as too many COs promote mis-segregation (Hollis et al., 2020; Smith & Nambiar, 2020; Sun et al., 2017; von Diezmann & Rog, 2021). Interestingly, a constant CO frequency could be explained via crossover invariance where the choice of partner for repair i.e., either the sister chromatid or the homolog, allows to modulate the outcome of recombination as in fission yeast (Hyppa & Smith, 2010). Spo11-mediated programmed DSBs also do not occur uniformly and are dependent on multiple molecular factors that dictate differential frequency of DSBs across chromosomal loci (Tock & Henderson, 2018; Yadav & Claeys Bouuaert, 2021). Chromosomal regions repressed for recombination during meiosis include the centromeres, telomeres, ribosomal RNA loci and the mating-type loci in yeast and fungi (Nambiar & Smith, 2016; Pazhayam et al., 2021). Among these, de-repression of recombination at centromeres is perhaps the most deleterious as it leads to chromosomal segregation defects resulting in aneuploidies, congenital disorders such as Down's syndrome (and other trisomies) as well as human infertility (Garcia-Cruz, Roig, & Caldés, 2009; Hassold & Hunt, 2001).

When Walther Flemming observed stained chromosomes for the first time under the microscope, he noticed singular constricted sites, that later became known as the centromeres (Flemming, 1882). The centromere is cytologically defined as the chromosomal region that mediates attachment to spindle microtubules through large protein complexes known as the kinetochores (McKinley & Cheeseman, 2016). The tension generated by the pulling force of the spindle fibers drives chromosomal segregation during both mitosis and meiosis. Centromeres were functionally characterized for the first time in *S. cerevisiae* as DNA sequences that conferred stable inheritance to the genetic material (Bloom & Carbon, 1982; Clarke & Carbon, 1980). *S. cerevisiae* has a point centromere, which is a well conserved ~125 bp sequence, containing three conserved DNA elements – CDEI, II and III, sequence-specificity of which is essential for kinetochore interaction (Saunders, Fitzgerald-Hayes, & Bloom, 1988) (Fig. 2). On the other

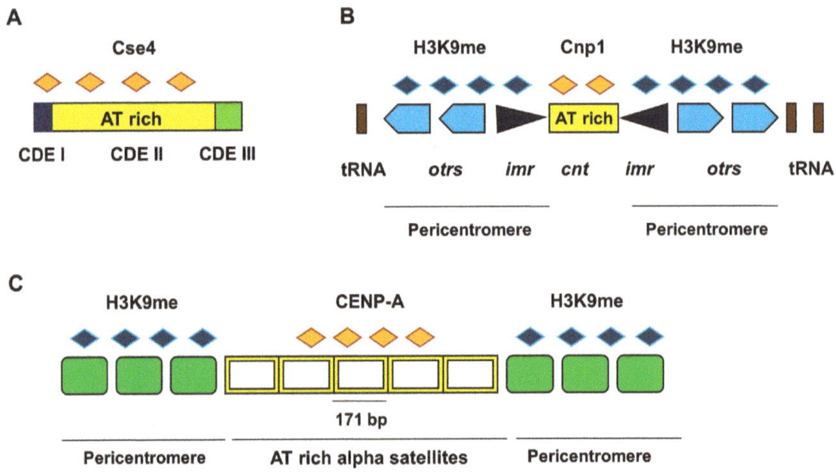


**Fig. 2** Centromere structures across species. (A) The point centromeres (125 bp) in *S. cerevisiae* has 3 conserved DNA elements (CDE). Sequence-specific protein binding sites (CDE I and CDE III) surrounds the core central region (CDE II). The histone variant Cse4 interacts with CDE I and CDE II. (B) In *S. pombe,* the centromeres are complex, regional in nature and around 30–125 kb. They contain a central core *cnt,* which is flanked by inverse innermost repeats (*imr*), multiple outer repeats (*otr*) and tRNA genes acting as the heterochromatin boundary elements. Cnp1, an ortholog of CENP-A interacts with the central core and the H3K9me epigenetic marks are found in the regions flanking the central core that are collectively called the pericentromere. (C) Mammals such as humans contain long, repetitive centromeres that are ~1–5 Mb in length. They have a central region consisting of tandem repeats of ~171 bp and interacts with CENP-A histone variant. Surrounding the alpha satellite is the heterochromatin region associated with H3K9me mark.

hand, *S. pombe* centromeres are complex, ranging from ~35 to 110 kb, with a central non-repetitive core flanked by repeat elements known as the pericentromere (Clarke & Baum, 1990; Murakami, Matsumoto, Niwa, & Yanagida, 1991; Steiner, Hahnenberger, & Clarke, 1993). However, these repetitive elements are organized into epigenetically modified heterochromatin containing H3K9 methylated histones that are transcriptionally repressive (Fig. 2) (Grewal & Jia, 2007). Furthermore, centromeres in animals and plants are even more complex spanning megabases in length (Aldrup-MacDonald & Sullivan, 2014). In humans, the centromeres are almost entirely made up of repetitive sequences; alpha-satellites, being the most common (Thakur, Packiaraj, & Henikoff, 2021) (Fig. 2). Centromeres are also markedly different from other chromosomal regions due to the presence of a specialized histone H3 variant known as CENP-A, essential for kinetochore assembly (Mellone & Fachinetti, 2021).

The pericentromeres have a major role during chromosomal segregation as they have an enriched deposition of cohesins, which are critical to maintain sister-chromatid cohesion. The role of these pericentric cohesins become all the more critical during meiosis, wherein the sister-chromatid cohesion needs to remain intact until anaphase II. Hence, removal of cohesins during mitosis and meiosis are regulated differentially in a spatial manner to cater to the needs of the segregating chromosomes in mitosis and meiosis. In yeasts, the cohesins are removed actively from the mitotic chromosomes during the transition from metaphase to anaphase by the action of Separase, which is kept in check till anaphase by Securin (Hornig, Knowles, McDonald, & Uhlmann, 2002). In vertebrates, there is differential removal of the cohesins from the arm and centromeric regions of chromosomes even during mitosis. During the prophase, the arm cohesins are removed by the concerted action of Wapl following phosphorylation of STAG1/STAG2 subunit of the cohesin complex by PLK1 (Polo like kinases) (Ishiguro, 2019). The centromeric cohesion is protected by the SGOL1, which recruits a phosphatase to dephosphorylate STAG1/STAG2. Finally, at anaphase, the centromeric cohesins are removed by Separase-mediated cleavage. The removal of the cohesins during meiosis also follows a two-step cleavage mechanism dependent on the genomic location. During anaphase I, the arm cohesins are removed but the centromeric cohesin containing REC8 (meiotic kleisin subunit of the cohesin complex) is protected by SGOL2 to retain sister-chromatid cohesion and facilitate only separation of homologous chromosomes (Rattani et al., 2013). The loss of SGOL2 at anaphase II leads to the separation of the sister chromatids. The centromeric cohesion is also important for proper attachment of the microtubule to the kinetochore of the chromosomes especially for mono-orientation during meiosis I. In the fission yeast, there are two Shugoshin proteins, Sgo1 (meiosis-specific) and Sgo2 (expressed both during mitosis and meiosis) (Ishiguro, 2019). Although Shugoshin deleted cells are viable, *sgo1Δ* cells exhibit premature loss of sister chromatid cohesion during meiosis, whereas *sgo2Δ* cells also exhibit mitotic chromosome segregation defects.

## 2. The "centromere" effect

Beadle was the first to report the "centromere effect" defined as the reduction in COs per unit cytological length near the centromeres in *Drosophila* (Beadle, 1932). Since then, there has been overwhelming evidence to demonstrate its conservation across species ranging from unicellular eukaryotes such as budding and fission yeasts, multicellular invertebrates such as fruit flies and worms, plant species as well as mammals including mice and humans (Choo, 1998; Nambiar & Smith, 2016; Talbert & Henikoff, 2010). This observation fit well with the idea of how centromeric regions are also genetically inert owing to the surrounding heterochromatin. However, Mather observed that heterochromatin when away from the centromere could show COs in a distance-dependent manner, concluding that the proximity to the centromeric regions itself was inhibitory (Mather, 1939). Among both budding and fission yeasts, the fold-reduction of centromeric COs varies. Early on, genetic analysis of *S. cerevisiae* centromere 3 (CEN3) reported ~5-fold reduction in CO frequency that was reversed upon introduction of a mutation in the CDEIII centromere element (Lambie & Shirleen Roeder, 1988). More recent analysis report repression in both COs and NCOs to be ~3–6 fold and extending over a span of 10 kb flanking the centromere, which is remarkably ~50× the length of the centromere itself (Chen et al., 2008; Mancera, Bourgon, Brozzi, Huber, & Steinmetz, 2008). In *S. pombe*, the first evidence came from *cen*2 with reduced recombination levels across a 50 kb centromeric region (Nakaseko, Adachi, Funahashi, Niwa, & Yanagida, 1986), whereas at *cen3* it was found to be ~200-fold lower than the genome-mean (Ellermeier et al., 2010). *Neurospora crassa* also exhibited high levels of repression across centromeres, similar to multicellular eukaryotes and extending well into the pericentric regions (Centola & Carbon, 1994). Analysis of large polymorphic restriction fragments at the centromeres in *Arabidopsis thaliana*, which possess several repeat arrays, revealed no non-parental fragments suggesting CO repression (Round, Flowers, & Richards, 1997). Interestingly, measurement of genetic and physical distances across the two pericentric regions (left and right) along with FISH data, revealed that the centromeric core region is repressed more than the pericentric regions and that there are differences in fold-repression between the right and the left pericentric regions as well (Haupt, Fischer, Winderl, Fransz, & Torres-Ruiz, 2001). The *CEN1* core region was ~220-fold repressed, whereas the left and right pericentric regions showed ~50-fold and 10-fold repression, respectively. There is substantial evidence that pericentric recombination is also repressed in other plant species such as tomato (Sherman & Stack, 1995), *Oryza sativa* (Harushima et al., 1998; Si et al., 2015), wheat (Saintenac et al., 2009), and maize (Anderson et al., 2003).

The first evidence for centromeric repression in humans was reported by Mahtani and Willard who used long range restriction maps across the alpha satellite array, DXZ1, followed by pulse-field gel electrophoresis to construct a physical map of the X chromosome pericentromeres (Mahtani & Willard, 1998). Upon constructing a genetic map using 27 loci within its pericentric region, the meiotic recombination frequency was found to be ~8 times lower than the genome-wide estimate for the X chromosome (Mahtani & Willard, 1998). Analysis of chromosome 10 centromere also showed similar repression (Jackson, See, Mulligan, & Lauffart, 1996).

Despite a tight control in repressing COs at centromeres, other recombination events such as gene conversions are found to be prevalent in several species such as maize (Shi et al., 2010). As mentioned above, gene conversion is a non-reciprocal unidirectional transfer of sequence information between donor and acceptor sites (Fig. 3). Such observations have also been seen in human alpha satellites, rice and mouse centromeres in somatic cells that lead to re-arrangements and expansions within the core centromeric regions (Jaco, Canela, Vera, & Blasco, 2008; Ma, Wing, Bennetzen, & Jackson, 2007; Talbert & Henikoff, 2010). The presence of gene conversion events at the centromeres may not be sufficient to trigger chromosomal mis-segregation errors and hence may not result in observable diseased phenotypes.

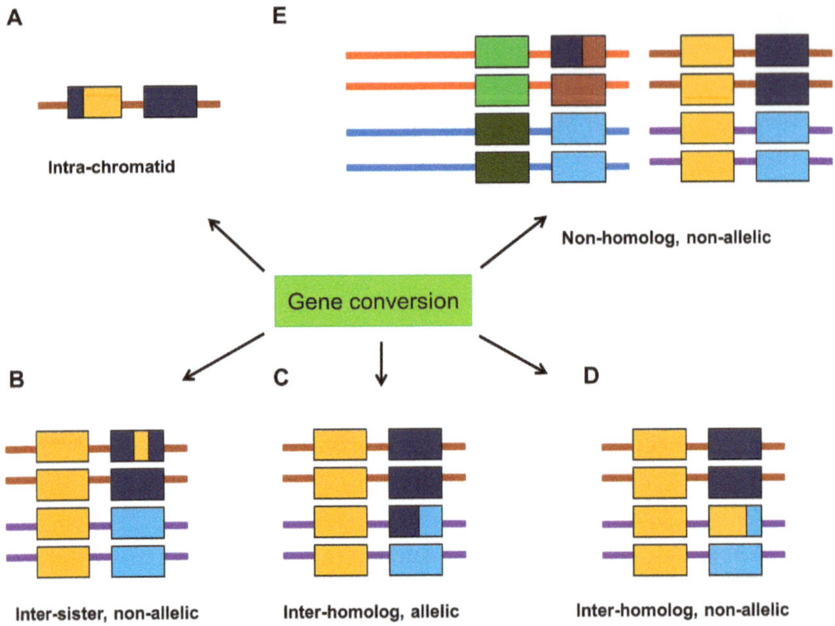


**Fig. 3** Various types of gene conversion events. Gene conversion, an unequal exchange of DNA at sites of heterology, requires donor and acceptor DNA sequences and therefore can take place in multiple ways. The various types of gene conversion events can be intra-chromatid (A), inter-sister and non-allelic (B), inter-homolog and allelic (C), inter-homolog and non-allelic (D) and non-homolog and non-allelic (E). Each line represents a single chromatid and the boxes represent different regions of the DNA. Differently coloured boxes on differently coloured lines represent the non-allelic regions. A pair of homologous chromosomes after DNA replication is shown. Only a few combinations of conversions are depicted here.

## 3. Mechanisms of meiotic crossover repression at centromeres

Despite the discovery of CO repression at centromeres for close to 90 years, the molecular mechanisms for such a tight control are not completely elucidated in any organism. Studies in the two yeasts, *S. cerevisiae* and *S. pombe* showcase the role of the kinetochore complex assembled on the centromeres as well as the flanking pericentric cohesins and heterochromatin to play crucial roles in this repression (Kuhl & Vader, 2019; Nambiar & Smith, 2016). The inaccessibility of the pericentric heterochromatin has been the most popular mechanism to explain the repressive nature of the centromeres, especially since it transcriptionally represses any genes inserted in the vicinity (Grewal & Jia, 2007). However, the pericentric heterochromatin (and other such regions) is accessible to the replication machinery and in fact has been reported to replicate early in S phase at least in *S. pombe* (Hayashi, Takahashi, Nakagawa, Nakayama, & Masukata, 2009; Kim, Dubey, & Huberman, 2003). They are also acted upon by a plethora of other proteins involved in the establishment and maintenance of heterochromatin such as the RNAi machinery and histone-modifying enzymes (Wang, Jia, & Jia, 2016). Since, DSBs are pre-requisites for initiation of recombination, the regulation of CO events could occur by inhibiting DSBs or later during partner choice for repair. In addition, as discussed previously, formation of NCOs is also a legitimate outcome during DSB repair, which can limit the CO frequency in a region. Different elements and properties of centromeres could potentially contribute toward any one or more of these possibilities in different species. We discuss below the various unique mechanisms for repression of DSBs and recombination that have been identified till date, across various species.

## 4. *Saccharomyces cerevisiae*

Genome-wide mapping for meiotic DSBs in *S. cerevisiae* showed the presence of centromere-proximal hotspots as close as 5 kb to the centromeres (Blitzblau, Bell, Rodriguez, Bell, & Hochwagen, 2007; Pan et al., 2011). The strength of these hotspots was found to be lower than genome-mean, but increased in *dmc1* mutants, suggesting that DSB initiation may not be the limiting factor to control centromere-proximal recombination, but could be dependent on the repair outcome. Mutations in *zip1*, needed for the assembly of SC, show significant reduction in total COs as well as interference. Interestingly, Zip1 has also been implicated in being a mediator of CO repression at the pericentromere by promoting inter-sister (IS) repair over inter-homolog (IH). In a genome-wide analysis, an increase in crossover density similar to the genome level, within 10 kb of the centromeres, was observed in *zip1* mutants (Chen et al., 2008). However, there was no difference in the CO/NCO ratio due to a concomitant increase in the level of the NCOs, suggesting the outcome of repair as a potential mechanism to explain low COs near centromeres.

Absence of pericentric heterochromatin in *S. cerevisiae*, prompted search for other factors to explain the centromere effect. Kinetochore proteins that bind to the spindle microtubules to mediate chromosomal segregation as well as play a role to deposit cohesins at the pericentric regions, fit the bill as potential candidates (Fernius et al., 2013; Hinshaw, Makrantoni, Harrison, & Marston, 2017). Using a fluorescence-based CO assay as well as genome-wide measurements it was shown that absence of the Ctf19 sub-complex increased both CO and NCO levels across centromeres and not the chromosomal arms, dramatically compared to a wild type (Thacker, Lam, Knop, & Keeney, 2011; Vincenten et al., 2015) (Table 1). An increase in the formation of centromeric DSBs, measured directly via

**Table 1** List of proteins involved in centromeric repression across species.

| Model organism | Defect in | Reference |
|---|---|---|
| *S. cerevisiae* | Sgs1 | Rockmill, Voelkel-Meiman, and Roeder (2006) |
| | Zip1 | Chen et al. (2008) |
| | Iml3 (CENP-L), Chl4 (CENP-N), Mcm21 (CENP-O) Ctf19 (CENP-P) Cnn1 (CENP-T), Wip1 (CENP-W), Nkp1, Nkp2 Mhf1 (CENP-S) Mhf2 (CENP-X) | Vincenten et al. (2015) |
| *S. pombe* | Dcr1 | Ellermeier et al. (2010) |
| | Rdp1 | |
| | Ago1 | |
| | Clr4 | |
| | Rik1 | |
| | Chp1 | |
| *Drosophila* | Blm | Hatkevich et al. (2017) |
| *C. elegans* | SLX-1 | Saito, Mohideen, Meyer, Harper, and Colaiácovo (2012) |
| *A. thaliana* | MET1 | Yelina et al. (2012) |
| | | Mirouze et al. (2012) |
| | | Yelina et al. (2015) |
| | CMT3 | Underwood et al. (2018) |
| | AXR1 | Christophorou et al. (2020) |

Southern blots and by using Spo11-oligos, was observed both at the centromere and the pericentric regions, suggesting direct blocking of DSBs (Vincenten et al., 2015) (Fig. 4). Although DSB formation increased upon depletion of Ctf19 complex, there was no concomitant increase in the level of COs. However, by using a cohesin-loading mutant that fails to establish cohesin deposition, they observed increased CO frequency with minimal changes in DSB frequency (Vincenten et al., 2015). This points toward regulation at both the level of DSB formation and repair, either directly or indirectly via recruitment of cohesins (Fig. 4). In addition, reduced Zip1 localization in Ctf19 and cohesin-loading mutants suggest a possible mechanism for repression via increased IS repair (Chen et al., 2008; Vincenten et al., 2015). More recent work targeting specific kinetochore subunits ectopically, using a dCas9/CRISPR system, led to the observation that Ctf19 forms the crux for kinetochore-mediated crossover repression observed in *S. cerevisiae* (Kuhl et al., 2020). DDK-mediated phosphorylation

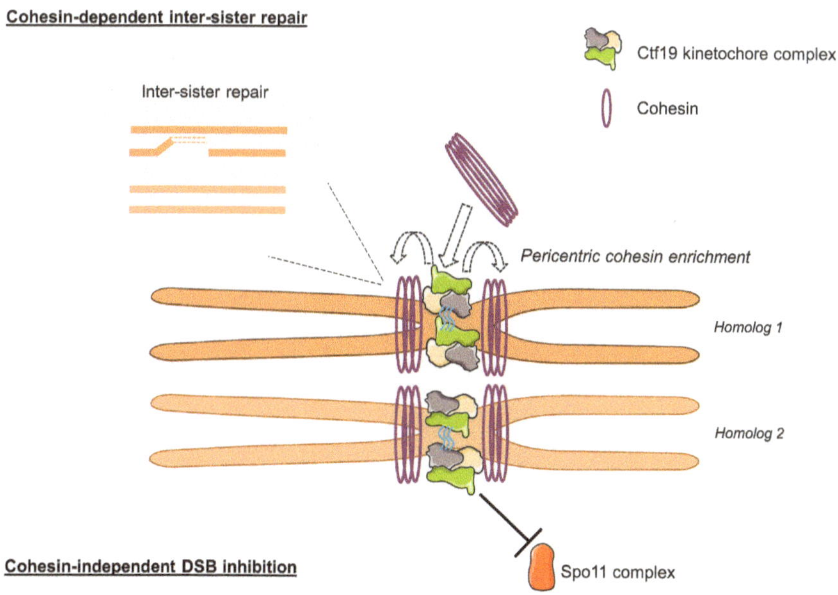


**Fig. 4** Role of the kinetochore in centromeric and pericentric repression. Crossovers near the centromeres are repressed in *S. cerevisiae*, mechanistically in a cohesin independent as well as dependent manner at two levels, DSB formation and DNA repair, respectively. Ctf19, the kinetochore protein complex, blocks DSB formation by Spo11 in an as yet unknown mechanism. In addition, it also recruits cohesins to the pericentric regions that can promote inter-sister DSB repair over inter- homolog, thereby resulting in a non-crossover outcome.

of Ctf19 plays a crucial role in this CO repression as Ctf19 phosphomutants have elevated recombination levels (Kuhl et al., 2020). However, a mechanistic understanding of how the IH to IS switch during repair is executed, still needs to be investigated.

## 5. *Schizosaccharomyces pombe*

Centromeres in fission yeast are much more elaborate than those in budding yeast, spanning around 35–125kb (Fig. 2). Understanding the mechanism of CO repression in *S. pombe* started with the observation that RNAi and heterochromatin mutants were significantly derepressed for both intragenic as well as intergenic recombination (Ellermeier et al., 2010) (Table 1). However, removing heterochromatin proteins such as Swi6 and Chp2 still maintained the repression at *cen3*, despite being derepressed for transcriptional silencing, suggesting a separation of regulation for the two processes that couldn't be just explained by presence of the heterochromatin. Recently, a more detailed molecular explanation for this repression has been put forward, which again puts focus on cohesins as the mediators of repression, as in *S. cerevisiae*. DSB formation in *S. pombe* depends on the meiosis-specific cohesin complex Rec8-Rec11 that helps recruit Rec10, an activator of the Rec12 complex to initiate break formation (Cromie & Smith, 2008; Ellermeier & Smith, 2005). Since, the levels of Rec12 at the pericentric repeats are significantly high, it rules out absence of Spo11-recruitment in the pericentromere as a potential mechanism (Ludin et al., 2008). However, the levels of Rec10, are significantly lower at the pericentric regions, suggesting missing Rec12-activators as a potential reason for absence of DSBs (Nambiar & Smith, 2018). Targeting such missing activating proteins to the pericentric regions genetically, elevated recombination levels over 50-fold as well as showed formation of DSBs (Nambiar & Smith, 2018). While the Rec8-Rec11 complex is loaded at the chromosomal arms, Rec8-Psc3 complex is deposited at the pericentric regions, in a heterochromatin binding protein Swi6-dependent manner (Kitajima, Yokobayashi, Yamamoto, & Watanabe, 2003; Nonaka et al., 2002). Replacement of Psc3 with Rec11 at the pericentric regions in the absence of Swi6, stimulated pericentric recombination, suggesting that exclusion of pro-recombinogenic Rec11 cohesin from the centromeres via specific cohesins and heterochromatin proteins is the key to repressing DSB initiation in *S. pombe* (Fig. 5).

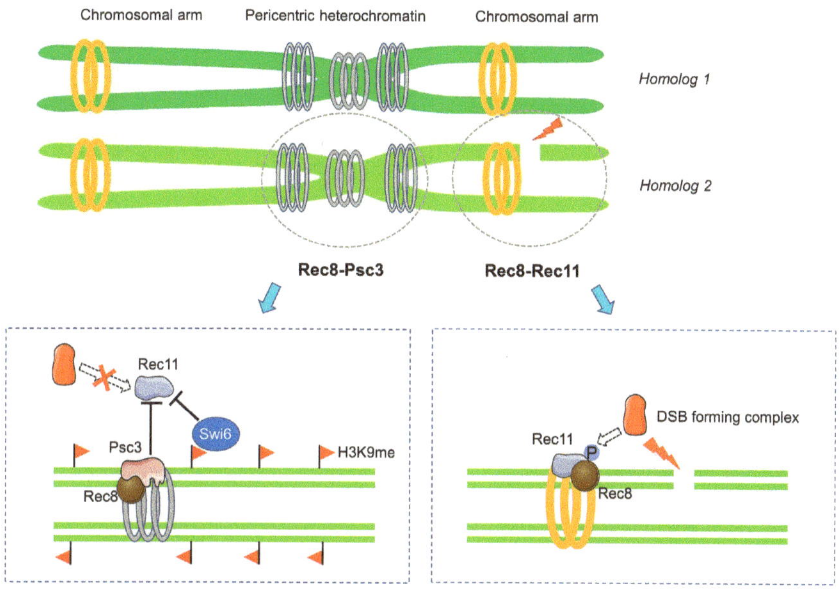


**Fig. 5** Role of pericentromere-specific cohesins in CO repression. In *S. pombe*, the mechanism of pericentric CO repression is mediated by cohesin and heterochromatin. There is differential loading of cohesins on meiotic chromosomes between chromosomal arms and the pericentric heterochromatin. In meiosis, the pericentric region is enriched in Rec8-Psc3 complex, whereas the arms regions are enriched in the pro-recombinogenic Rec8-Rec11 complex that activates the Rec12 complex to initiate DSB formation and consequently recombination.

## 6. *Caenorhabditis elegans*

The nematode has 6 holocentric chromosomes of similar lengths that exhibit a bias in distribution of recombination hotspots as seen in other eukaryotes (Barnes, Kohara, Coulson, & Hekimi, 1995). The distal parts of the chromosomes have significantly more COs than the central regions (Fig. 6) (Rockman & Kruglyak, 2009). Chromosomal segregation also follows a very different mechanism in *C. elegans,* where the centromere sequences are identified by the enzyme aurora kinase B and the nucleation and polymerization of spindle tubules pushes the two homologs away from each other (Duro & Marston, 2015). Interestingly, it has been observed that highly recombinogenic regions coincide with regions that are epigenetically marked as transcriptionally inactive (Liu et al., 2011). This appear to be counterintuitive to the conventional thought of heterochromatin being

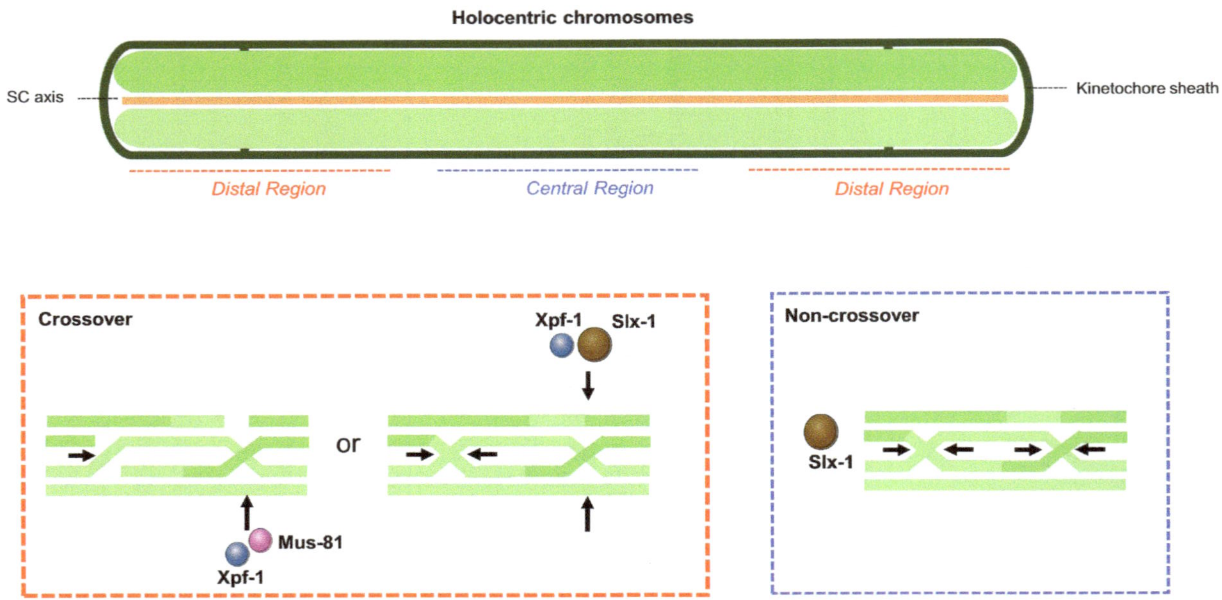


**Fig. 6** Mechanism of centromeric repression in worms. In *C. elegans*, the chromosomes are holocentric, that is, there is no one central centromeric structure present and the kinetochore assembles across the length of the chromosomes forming a sheath around it. The central regions observe CO repression, in contrast to the distal regions, which is mediated at the level of DSB repair. There are four structure-specific endonucleases/resolvases namely, XPF1, GEN1, SLX1 and MUS81, which aid in resolution of Holliday junctions. The proposed model hypothesizes that these structure-specific resolvases coordinate action at different chromosomal domains - MUS81, SLX-1 and XPF-1 at the arm regions, while SLX-1 functions at the central region. SLX-1 may promote NCO outcome during HJ resolution or facilitate DSB repair via the sister chromatids.

repressive in nature. However, since *C. elegans* does not possess large blocks of constitutive heterochromatin, the meiotic COs are most likely associated with transcriptionally active regions (Yu, Kim, & Dernburg, 2016). Studies show that the structure-specific endonuclease SLX-1 appears to reduce COs in the central regions, whereas the other three endonucleases MUS-81, SLX-1 and XPF-1 promote COs in the chromosomal arms, suggesting control at the level of DSB repair (Saito, Lui, Kim, Meyer, & Colaiácovo, 2013). Furthermore, mutations in XPF-1 and SLX-1 in oocytes also show impaired crossover maturation and repair, resulting in increased chromosomal aberrations.

## 7. *Drosophila melanogaster*

Pericentric heterochromatin in *Drosophila* has been the focus of many studies to explain the centromere effect, since its discovery (Beadle, 1932). Using 4 different X chromosome inversions $sc^8$, $sc^4$, $rst^3$ and $y^4$, Mather tested the roles of both the centromere and heterochromatin in influencing crossover repression, by moving the markers either distal or proximal to the centromere compared to their original position (Mather, 1939). He observed that movement of euchromatic regions closer to the centromere was more repressive for COs compared to moving it closer to a block of heterochromatin, which Mather interpreted as a strong effect of the centromere. This was supported by the observation that COs in the centromere-proximal euchromatin depended more on the distance from the centromere rather than the heterochromatin itself (Yamamoto & Miklos, 1978). Molecular differences in the nature of the heterochromatin at the pericentromeres, perhaps centromere-driven itself, as compared to those present elsewhere in the genome, could potentially lead to these differences as well. However, experiments in *D. virilis*, showed absolutely no COs within the heterochromatin even when shuffled between chromosomes, away from the centromere via translocation (Baker, 1958). Hence, heterochromatin-specific COs seen by Mather were attributed to a lack of sufficiently close markers to mark the boundary of the heterochromatin and presence of residual euchromatic regions that showed the COs. However, this propagated the long-held idea that pericentric heterochromatin regions block meiotic COs, by virtue of their inaccessibility to DSB repair proteins. In support of this, investigations of the dominant mutations in the suppressor-of-variegation (Su-Var) genes that disrupt heterochromatin were performed (Reuter & Wolff, 1981; Weiler & Wakimoto, 2003). An increase in recombination at the pericentric heterochromatin in mutant flies mapping to mutations in 9 Su(var) genes was observed, further giving fuel to the idea that heterochromatin was repressive for recombination (Westphal & Reuter, 2002). In contrast, insertion of a block of heterochromatin did not repress COs in the adjacent euchromatin, arguing against the role of heterochromatin (Hartmann, Umbanhowar, & Sekelsky, 2019). More recent studies performing whole-genome analysis of independent meiosis in *Drosophila* to measure recombination rates showed that NCOs are more uniformly distributed across the genome as compared to COs, supporting the idea DSBs may be formed at the centromeres but get repaired largely as NCOs (Comeron, Ratnappan, & Bailin, 2012; Miller et al., 2016).

Bloom Syndrome (Blm) helicase is an ATP-dependent helicase belonging to the RecQ family and plays a role in DNA replication and recombination (Kaur, Agrawal, & Sengupta, 2021). Recent studies describe that Blm mutants show loss of crossover assurance, centromeric effect as well as crossover interference (Hatkevich et al., 2017). This is in contrast to *mei-41* mutants that also disrupt meiotic patterning albeit selectively, as centromere-proximal CO repression is mostly intact in these mutants, while crossover interference is significantly decreased (Brady, McMahan, & Sekelsky, 2018). Furthermore, it was noted that although Blm mutants had a loss of meiotic patterning, repression at the highly repetitive heterochromatin was intact and hence indicated toward a separation of the centromere effect from that of the heterochromatin (Hartmann et al., 2019). This supports the studies described above in yeasts that regional centromeres (containing a central-kinetochore assembly region flanked by adjacent pericentric regions) may have a coupled (and hence stringent) mechanism for repression of COs dependent on both the kinetochore-binding "centromere" and the neighboring "pericentric" heterochromatin. An interesting aspect of *Drosophila* biology is the complete absence of COs on chromosome 4, making this an exception to the rule of crossover assurance (Comeron et al., 2012; Hartmann & Sekelsky, 2017). However, Blm mutants have COs on chromosome 4, in agreement with its role in disrupting meiotic patterning (Hatkevich et al., 2017). This suggests that perhaps the regulation of COs occurs during DSB repair, with the Blm helicase influencing the outcome and thereby restricting CO formation.

## 8. Plants

The sequence of meiotic events that happen in monocentric plants is very similar to how it happens in other species (Hofstatter, Thangavel, Castellani, & Marques, 2021). Centromeric repression, as discussed above, is seen in several plant species. Mechanistically, there is evidence suggesting that SPO11 recruitment is not the limiting step. Meiotic DSBs monitored across the length of *Arabidopsis* chromosomes show reduced DSBs at the pericentric regions, despite enrichment of SPO11-1 (Choi et al., 2018). Therefore, similar to *S. pombe,* mere SPO11 recruitment may not be sufficient to initiate DSB formation. In contrast, measurement of RAD51 foci as well as ChIP-seq data in maize confirm DSB generation and the absence of COs, arguing for regulation at the level of DNA repair as well (He et al., 2017; Shi et al., 2010). In addition, epigenetic modifications along the chromosomal length also play an important role in defining centromeric identity and repression of COs. In *Arabidopsis*, overexpression of a histone methyl transferase or inhibition of a deacetylase also result in changes in CO number as well as meiotic defects (Perrella et al., 2010). DNA methylation in plants is an RNA driven process (RNA directed DNA methylation), dependent on methyl transferase MET1 and is much higher at the pericentric heterochromatin, where the meiotic recombination rate is lower and meiotic COs increase as the distance from the centromeres increases (Cokus et al., 2008). *met1* mutants exhibited an increase in arm recombination and a stronger repression of COs at the pericentric region, without altering the total number of COs compared to wild type suggesting that CO homeostasis mechanisms are inclusive of pericentric COs and intact in *met1* mutants and other mechanisms may contribute to CO suppression at pericentromeres in the absence of MET1 (Mirouze et al., 2012; Yelina et al., 2012, 2015). In support of this, loss of DNA methylation at centromeric tandem repeats in mice led to improper chromosome pairing and synapsis as well as mutations in Suv39h methyl transferase in spermatocytes caused non-homologous chromosome pairing (Tachibana, Nozaki, Takeda, & Shinkai, 2007). Another study also found that increasing temperature led to an increase in proximal COs in barley, a species in which COs are primarily centromere-distal under typical temperature conditions (Phillips et al., 2015).

There are a number of plants that have holocentric chromosomal organization such as Cyperacaceae and Juncaceae. To maintain segregation fidelity, they adopt the process of inverted meiosis wherein the sister chromatids first separate and the homologs separate only in the second division (Cabral, Marques, Schubert, Pedrosa-Harand, & Schlögelhofer, 2014; Heckmann et al., 2013; Marques et al., 2015). There are even more drastic adaptations as in *Rhynchospora tenuis* (*Cyperacaceae*), which follows an achiasmatic inverted meiosis (Cabral et al., 2014). Although this contradicts the concept of crossover assurance, but even random segregation would have a significant chance of survival, since this species has only two pairs of chromosomes. Overall, our lack of understanding of the process and key players of crossover homeostasis in such plants have come in the way of the mechanistic understanding of how pericentric repression would work at holocentric centromeres. In holocentric animals such as *C. elegans*, centromere associated proteins CENH3 and CENP-C seem non-essential for meiotic recombination (Monen, Maddox, Hyndman, Oegema, & Desai, 2005). In case of holocentric plants, there is no study that documents the meiotic DSB or recombination profiles, till date.

## 9. Mammals

Mammalian centromeres are typically large stretches of tandem DNA repeat sequences. In mouse, there are major satellite repeats that span approximately 6 Mb of 230 bp repeats and minor satellites across a span of 600 kb of 120 bp repeats. The major satellite regions reside in the pericentric region, while the minor region is a part of the core centromere (Jaco et al., 2008). Meiotic CO repression has also been reported in humans, loss of which has implications for genetic disorders such as Down syndrome and other trisomies. In mammals, there is an additional level of complexity in gametogenesis, namely sexual dimorphism, which is both mechanistic as well as temporal in nature. Female gametogenesis or oogenesis involves the formation of only one oocyte post meiosis of an oogonium, while the meiosis of the male meiocyte or spermatogonium leads to the formation of four functional male gametes or sperms. Oogenesis begins during the fetal stage itself but is not completed until after the oocyte gets fertilized, while spermatogenesis begins in puberty and gets completed just weeks before fertilization. Lastly, oogenesis involves unequal division of the cytoplasm, wherein half of the chromosomes in both meiosis are segregated into polar bodies enabling the one viable oocyte to retain the major cytoplasmic content of the mother cell (Herbert, Kalleas, Cooney, Lamb, & Lister, 2015).

It has been implicated that COs too close to the centromeres may disrupt the cohesion at the sister chromatids that is important to keep them together during meiosis I while the homologous chromosomes separate. As a result of this, there is an increased risk of chromosome segregation errors. Ottolini et al. detected a novel segregation pattern, called the reverse segregation or RS (equivalent to the inverted meiosis observed in holocentric plants, as discussed in the previous section) by studying the SNPs at the pericentromere. They observed that in a population of oocytes derived from advanced maternal age females, trisomies occurred at a higher frequency. In contradiction to the hypothesis earlier that meiosis I non-disjunction (MI NDJ) is the most prevalent segregation error, they recorded higher incidences of precocious separation of sister chromatids (PSSC) in the age related trisomies. This is in accordance to the theory that centromeric crossovers could be leading to disturbed centromeric cohesion, which is important to maintain sister chromatid cohesion at the centromere. Failure to do so, might lead to premature separation of sister chromatids in MI itself. The RS pattern is not limited to just a scenario where there is no exchange between two homologs but also can be due to a CO close to the centromeres. There is high incidence of meiosis II non-disjunction (MII NDJ) in case of RS, which could be explained by the occurrence of COs proximal to the centromeres (Ottolini et al., 2015).

In mouse oocytes, univalents have been observed to predominantly segregate their sister chromatids in meiosis I itself and this could be happening in humans as well owing to the bi-orientation of the sister kinetochores in MI that evades the spindle assembly checkpoint. An important point to be noted here, is that whether or not the MII errors happen because of reverse segregation or centromeric crossovers, the MII errors can have origins during meiosis I in human females (Kouznetsova, Lister, Nordenskjöld, Herbert, & Höög, 2007). Another study in mice, using live-cell imaging revealed that TEV mediated cleavage of Rec8 promoted chiasma resolution in MI, whereas cleavage of Scc1 cohesin subunit triggered sister chromatid disjunction in the first round of mitosis of the embryo. Ectopically expressing Rec8 in the growing phase of the Rec8-TEV oocytes did not rescue or prevent the TEV mediated bivalent destruction, which signified that there is minimal or no cohesin turnover once the oocyte age crosses 2 weeks. This is critical since it points toward the possibility of oocytes being unable to re-establish cohesion giving rise to associated MI errors (Tachibana-Konwalski et al., 2010). This observation has also been reported from other similar studies, where Rec8 (activated by tamoxifen-inducible Cre) in mouse fetal oocytes established cohesion during DNA replication. However, when such a Rec8 activation was done in dictyate stage arrested oocytes (arrested at the diplotene stage of meiosis I in fetal stage), no new cohesion was observed, implying that cohesion once established in the fetal stage is maintained over months and that fertility would depend on the longevity of cohesin protein persistence (Burkhardt et al., 2016). Examination of cohesin levels in dictyate oocytes in humans as well as mice by immunofluorescence of ovarian sections revealed a significant drop in the levels of meiotic cohesins REC8 and SMC1β temporally. In humans, this decrease was observed in oocytes of women aged over 40 in comparison to 20-year-old females. An interesting observation was in fact the minimal expression of SMC1α (mitotic counterpart of SMC1β) in mouse oocytes that increased slightly with age. In human oocytes however, there was substantial expression of both SMC1α and SMC1β. This points toward the fact that these meiotic and mitotic cohesins might be working in coordination to promote the maintenance of sister chromatid cohesion across decades (Tsutsumi et al., 2014).

Evidence from studies on mouse oocytes also suggests the involvement of cohesins in maintaining segregation fidelity. There is a strong correlation between loss of cohesins in metaphase I leading to PSSC. REC8 levels in oocytes have been observed to progressively go down by 9 months and segregation errors increase significantly by 15 months. This difference could be due to discrepancies in detecting REC8 levels or uncertainty in knowing whether low levels of cohesins actually cause total functional loss. Thus, there is a need to figure out the local threshold level of meiotic cohesin to show a better correlation between the level of cohesins and susceptibility to chromosome segregation errors (Lee, 2019).

From a mechanistic perspective, not much is known about how pericentric COs give rise to chromosome segregation errors in mammals. However, similarities in the cohesin dynamics and centromere structure between mammals and other model organisms, strongly suggest a central role for cohesins in mediating centromeric recombination repression in humans and other mammals as well (Fig. 7).

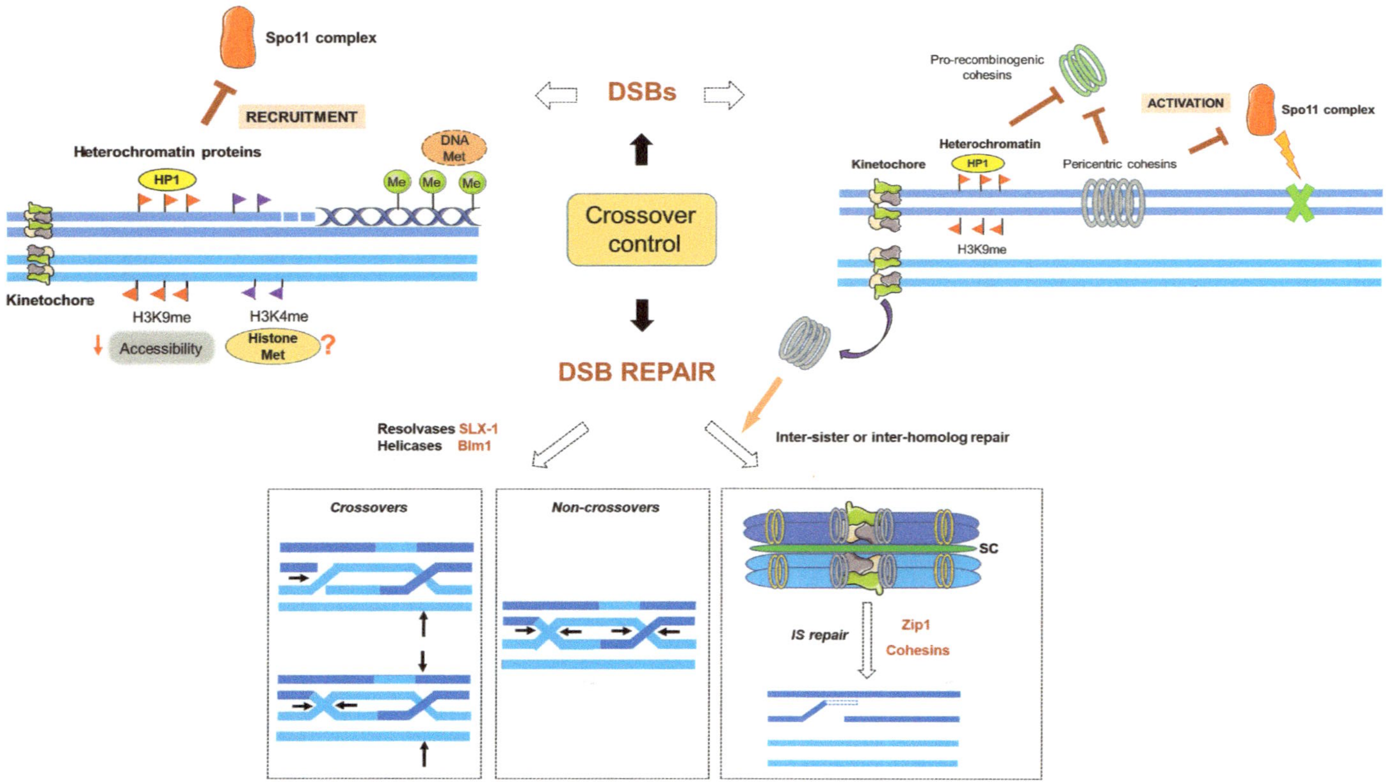


**Fig. 7** Key players in repression of recombination across centromeres. Schematic representation of the different levels of regulation of COs at the centromeric and pericentric regions known till date. HP1—heterochromatin protein 1; Met—methyl transferase.

## 10. Pericentric crossovers and chromosomal mis-segregation

Presence of COs near centromeres have been strongly correlated with an increased frequency of chromosomal segregation errors in humans (Hassold & Hunt, 2001). Studies in experimental model systems such as yeasts using mutants that are derepressed for centromeric COs or ectopic ways of stimulating centromere-proximal COs are crucial in understanding the type of segregation defects that are associated with them. Depending on which step of the meiotic program goes haywire, the segregation errors can differ (Fig. 8). MI mis-segregation broadly includes two kinds of errors—(a) MI non-disjunction that happens when the homologs fail to separate in meiosis I, (b) precocious separation of sister chromatids (PSSC), that occurs when the sister chromatids separate during MI to opposite spindle poles. Meiosis II errors primarily include MII NDJ, which signifies the failure of the sister chromatids to separate in anaphase II. Another type of aberrant segregation is reverse segregation, where MI becomes equational and MII becomes reductional (Ottolini et al., 2015). Reverse segregation is not a combination of two PSSC events, as equational division at MI due to reverse segregation is ~100 times higher than the expected frequency of a combination of two PSSC events (Ottolini et al., 2015).

In an earlier study involving YACs as marker chromosomes, it was shown that centromere-proximal gene conversion events were associated with high levels of MI NDJ and PSSC (Sears, Hegemann, Shero, & Hieter, 1995). Centromere-proximal recombination in *S. cerevisiae* is associated with 60% of MI segregation errors (Rockmill et al., 2006). These MI errors primarily appeared to be due to PSSC whose effects were seen in both MI and MII. In contrast, *S. pombe*, strains with elevated centromeric COs resulted in ~15-fold increase in MI NDJ compared to wild-type, while the PSSC levels were similar (Nambiar & Smith, 2018). Distinct mechanisms for pericentric repression in these two yeast species as well as differences in the centromere organization could account for this disparity. Characterization of spontaneous X chromosome non-disjunctions in *Drosophila* as well as human trisomy 21 cases showed that centromere-proximal exchanges were associated primarily with MII NDJ (Koehler et al., 1996; Lamb et al., 1996). A detailed mechanistic understanding of the way centromere-associated COs cause chromosomal mis-segregation during meiosis is essential to explain the clinical implications of the same.

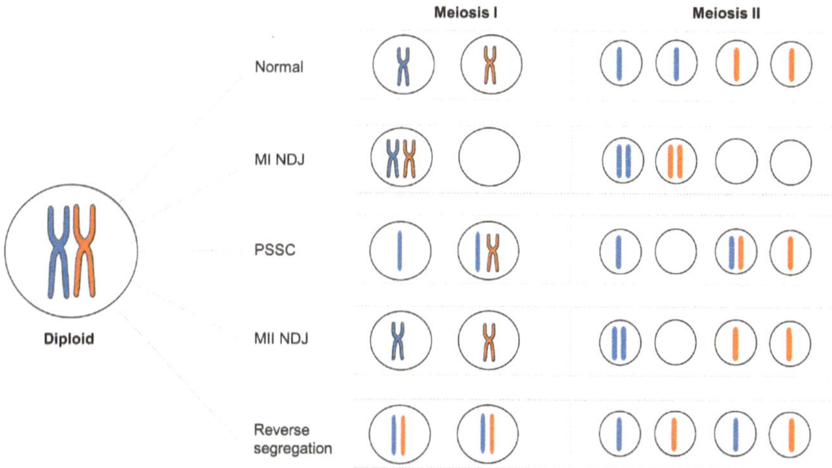


**Fig. 8** Different types of chromosomal segregation errors during meiosis. Homologous chromosomes pair and separate in meiosis I (MI) toward opposite poles, while the sister chromatids remain joined via centromeric cohesins. Meiosis II (MII) is equational division where the two sister chromatids separate into different nuclei, giving rise to four haploid nuclei. When the pair of homologous chromosomes fail to separate in meiosis I, it leads to MI non-disjunction (NDJ). Hence at MI, a nullisome is generated, and post-MII two disomes and two nullisomes are generated. During premature separation of sister chromatids (PSSC), sister chromatids separate at MI itself and therefore at the end of meiosis, a nullisome and a disome may be formed along with two normal gametes. During MII NDJ, while MI proceeds normally, the sister chromatids fail to separate at MII, thereby resulting in two normal nuclei, one nullisome and one disome. During reverse segregation, meiosis is inverted such that MI is equational, where the sister chromatids separate, while MII is reductional and the homologous, non-sister chromatids separate into daughter nuclei. While this may give rise to four normal nuclei, such a meiosis is achiasmate (without crossovers). The aberrations depicted here represent only one pair of homologs; such errors can arise in one or more pairs of homologous chromosomes. No recombinant chromatids are shown here for ease of representation.

## 11. Role of pericentric crossovers in human diseases

Chromosomal segregation errors result in aneuploidy, an imbalance in the amount of genetic content that can be at the level of an entire chromosome or only a segment (Pfau & Amon, 2012). The primary effect of aneuploidy is the imbalance in gene dosage that can result in proteotoxic stress due to disruption of the delicately balanced protein quality control systems (Oromendia, Dodgson, & Amon, 2012). All aneuploidies do not lead to inviability but can be detrimental enough to cause growth deficiencies, especially during development. Mitotic aneuploidies as those seen in cancers are largely an exception and can survive abnormal increases in chromosomal numbers (Sheltzer & Amon, 2011). On the other aneuploidies that arise due to segregation errors in meiosis are almost always detrimental, in the context of human health, and cause a varied range of diseases that include developmental disorders such as Down, Patau and Edward's syndrome (extra copy of chromosome 21, 15 and 18, respectively), miscarriages and infertility. Importantly, these chromosomal segregation errors mostly arise from defective oogenesis and have a strong correlation with an increase in age of females (Herbert et al., 2015; MacLennan, Crichton, Playfoot, & Adams, 2015). Loss of sister-chromatid cohesion due to poor turnover of the cohesin complexes in older oocytes is a major mechanistic explanation for the higher number of segregation defects observed in older oocytes (Handyside, 2012). In contrast, younger males also appear to have a higher risk of offspring with chromosomal aneuploidies (Steiner et al., 2015). Fewer COs in spermatocytes from juvenile mice and humans suggested weak CO maturation efficiency that could elevate the risk of segregation errors in the progeny (Zelazowski et al., 2017). Preferential activity of alternate repair pathways that potentially generate more NCOs compared to COs could explain the observed reduction in COs. As discussed above, a significant number of maternally derived Down's syndrome as well as sex chromosomes trisomies have reported meiotic exchanges close to the centromere. Interestingly, maternally-derived chromosome 21 was observed to have two CO events instead of one and yet, pericentric exchange was risk factor for its segregation (Oliver et al., 2012). Hence, recombination in the fetal stage of oocytes can confer a predisposition to segregation error that might get amplified as maternal age increases due to perhaps less robust and weakened surveillance mechanisms (MacLennan et al., 2015). In addition, maternal recombination rates in human oocytes seem to be higher than paternal rates by a fold of about 1.6, perhaps attributable to higher CO frequencies on the autosomes (Ottolini et al., 2015). Interestingly, maternal recombination appears to have higher incidences of centromeric recombination than paternal events, but the degree of such aberrant recombination can vary greatly within the same woman, due to which, some oocytes might be more pre-disposed to chromosome segregation errors by virtue of centromeric recombination (Ottolini et al., 2015).

A recent study provided more insights on the mechanistic details of how segregation errors might arise in human oocytes and the types of error that define the fertility curves in females, across a scale of time (age-dependency). Human oocytes from females aged between 9 and 43 years revealed that fertility in these women followed an inverted U-shaped curve (Gruhn et al., 2019). Females between the age groups 4–20 and 33–43 had low fertility, while the optimum fertile range was from 20 to 32 years. The advanced maternal age (AMA) group showed more susceptibility to MI PSSC, whereas the young age-group showed more susceptibility to MI NDJ errors. While weakening of sister chromatid cohesion in aged oocytes could lead to increased incidences of MI PSSC, the occurrence of abnormal recombination events (COs across centromere, telomeres or missing COs) in the younger oocytes could be the leading reason for increased MI NDJ events in them (Gruhn et al., 2019). Therefore, deciphering the molecular mechanisms behind loss of crossover patterning in meiosis is imperative to better understand human reproductive health and associated diseases.

## 12. Future perspectives

The phenomenon of crossover repression is conserved across various species and although we understand a lot more about the mechanistic basis of how this control is regulated, we are still a long way away from deciphering all the molecular players involved. Interestingly, all the species that have been used for understanding this process exhibit broadly two levels of regulation, at the level of DSB formation and its repair. The complexity of the centromere in itself varies among species and the epigenetic nature of the pericentromeres add another layer of control for crossover patterning during meiosis. However, the presence of such elaborate mechanisms to inhibit centromere-proximal COs across species point toward an evolutionary importance to keep recombination events away from these genomic regions in order to maintain proper chromosomal segregation, mechanisms for which we are yet to fully understand. It is possible that centromeric recombination intermediates could interfere with proper kinetochore orientation or due to poor resolution fail to separate in time at MI, resulting in nondisjunction. Owing to the high incidences of pericentric recombination-directed aneuploidies in humans and other species, deciphering the network of all the players involved in this regulation is essential to pave the way for better targeted prevention and treatment options for aneuploidy-associated diseases. More importantly, such studies might aid us in improving techniques for assisted reproductive strategies that would be crucial in managing complications in human reproductive health such as miscarriages and infertility.

## Acknowledgment

We thank MN lab members for their comments on the manuscript. Research in MN laboratory is supported by Science and Engineering Research Board (SERB) start-up research grant [SRG/2020/000434] and IISER, Pune. SS is supported by IISER Pune fellowship. AD is supported by KVPY fellowship.

## References

Aldrup-MacDonald, M. E., & Sullivan, B. A. (2014). The past, present, and future of human centromere genomics. *Genes*, *5*(1), 33. https://doi.org/10.3390/GENES5010033.

Anderson, L. K., Doyle, G. G., Brigham, B., Carter, J., Hooker, K. D., Lai, A., et al. (2003). High-resolution crossover maps for each bivalent of Zea mays using recombination nodules. *Genetics*, *165*(2), 849–865. https://doi.org/10.1093/GENETICS/165.2.849.

Baker, W. K. (1958). Crossing over in heterochromatin. *The American Naturalist*, *92*(862), 59–60. https://doi.org/10.1086/282010.

Barnes, T. M., Kohara, Y., Coulson, A., & Hekimi, S. (1995). Meiotic recombination, noncoding DNA and genomic organization in Caenorhabditis elegans. *Genetics*, *141*(1), 159–179. https://doi.org/10.1093/GENETICS/141.1.159.

Bascom-Slack, C. A., Ross, L. O., & Dawson, D. S. (1997). Chiasmata, crossovers, and meiotic chromosome segregation. *Advances in Genetics*, *35*(C), 253–284. https://doi.org/10.1016/S0065-2660(08)60452-6.

Beadle, G. W. (1932). A possible influence of the spindle fibre on crossing-over in Drosophila. *Proceedings of the National Academy of Sciences*, *18*(2), 160–165. https://doi.org/10.1073/PNAS.18.2.160.

Bizard, A. H., & Hickson, I. D. (2014). The dissolution of double Holliday junctions. *Cold Spring Harbor Perspectives in Biology*, *6*(7). https://doi.org/10.1101/cshperspect.a016477.

Blitzblau, H. G., Bell, G. W., Rodriguez, J., Bell, S. P., & Hochwagen, A. (2007). Mapping of meiotic single-stranded DNA reveals double-stranded-break hotspots near centromeres and telomeres. *Current Biology: CB*, *17*(23), 2003–2012. https://doi.org/10.1016/J.CUB.2007.10.066.

Bloom, K. S., & Carbon, J. (1982). Yeast centromere DNA is in a unique and highly ordered structure in chromosomes and small circular minichromosomes. *Cell*, *29*(2), 305–317. https://doi.org/10.1016/0092-8674(82)90147-7.

Brady, M. M., McMahan, S., & Sekelsky, J. (2018). Loss of drosophila Mei-41/ATR alters meiotic crossover patterning. *Genetics*, *208*(2), 579–588. https://doi.org/10.1534/GENETICS.117.300634/-/DC1.

Burkhardt, S., Borsos, M., Szydlowska, A., Godwin, J., Williams, S. A., Cohen, P. E., et al. (2016). Chromosome cohesion established by Rec8-cohesin in fetal oocytes is maintained without detectable turnover in oocytes arrested for months in mice. *Current Biology: CB*, *26*(5), 678–685. https://doi.org/10.1016/J.CUB.2015.12.073.

Cabral, G., Marques, A., Schubert, V., Pedrosa-Harand, A., & Schlögelhofer, P. (2014). Chiasmatic and achiasmatic inverted meiosis of plants with holocentric chromosomes. *Nature Communications*, *5*. https://doi.org/10.1038/NCOMMS6070.

Centola, M., & Carbon, J. (1994). Cloning and characterization of centromeric DNA from Neurospora crassa. *Molecular and Cellular Biology*, *14*(2), 1510–1519. https://doi.org/10.1128/MCB.14.2.1510-1519.1994.

Chen, S. Y., Tsubouchi, T., Rockmill, B., Sandler, J. S., Richards, D. R., Vader, G., et al. (2008). Global analysis of the meiotic crossover landscape. *Developmental Cell*, *15*(3), 401–415. https://doi.org/10.1016/J.DEVCEL.2008.07.006.

Choi, K., Zhao, X., Tock, A. J., Lambing, C., Underwood, C. J., Hardcastle, T. J., et al. (2018). Nucleosomes and DNA methylation shape meiotic DSB frequency in Arabidopsis thaliana transposons and gene regulatory regions. *Genome Research*, *28*(4), 532–546. https://doi.org/10.1101/GR.225599.117.

Choo, K. H. A. (1998). Why is the centromere so cold? *Genome Research, 8*(2), 81–82. https://doi.org/10.1101/GR.8.2.81.

Christophorou, N., She, W., Long, J., Hurel, A., Beaubiat, S., Idir, Y., et al. (2020). AXR1 affects DNA methylation independently of its role in regulating meiotic crossover localization. *PLOS Genetics 16*(6), e1008894. https://doi.org/10.1371/journal.pgen.1008894.

Clarke, L., & Baum, M. P. (1990). Functional analysis of a centromere from fission yeast: A role for centromere-specific repeated DNA sequences. *Molecular and Cellular Biology, 10*(5), 1863–1872. https://doi.org/10.1128/mcb.10.5.1863.

Clarke, L., & Carbon, J. (1980). Isolation of a yeast centromere and construction of functional small circular chromosomes. *Nature, 287*(5782), 504–509. https://doi.org/10.1038/287504a0.

Cokus, S. J., Feng, S., Zhang, X., Chen, Z., Merriman, B., Haudenschild, C. D., et al. (2008). Shotgun bisulphite sequencing of the Arabidopsis genome reveals DNA methylation patterning. *Nature, 452*(7184), 215–219. https://doi.org/10.1038/NATURE06745.

Comeron, J. M., Ratnappan, R., & Bailin, S. (2012). The many landscapes of recombination in Drosophila melanogaster. *PLoS Genetics, 8*(10). https://doi.org/10.1371/JOURNAL.PGEN.1002905.

Cromie, G., & Smith, G. R. (2008). Meiotic recombination in Schizosaccharomyces pombe: A paradigm for genetic and molecular analysis. *Genome Dynamics and Stability, 3*, 195–230. https://doi.org/10.1007/7050_2007_025.

Davis, L., & Smith, G. R. (2003). Nonrandom homolog segregation at meiosis I in Schizosaccharomyces pombe mutants lacking recombination. *Genetics, 163*(3), 857–874. https://doi.org/10.1093/genetics/163.3.857.

Davis, L., & Smith, G. R. (2005). Dynein promotes achiasmate segregation in Schizosaccharomyces pombe. *Genetics, 170*(2), 581–590. https://doi.org/10.1534/genetics.104.040253.

Duro, E., & Marston, A. L. (2015). From equator to pole: Splitting chromosomes in mitosis and meiosis. *Genes & Development, 29*(2), 109–122. https://doi.org/10.1101/GAD.255554.114.

Ehmsen, K. T., & Heyer, W. D. (2008). Biochemistry of meiotic recombination: Formation, processing, and resolution of recombination intermediates. *Genome Dynamics and Stability, 3*, 91–164. https://doi.org/10.1007/7050_2008_039.

Ellermeier, C., Higuchi, E. C., Phadnis, N., Holm, L., Geelhood, J. L., Thon, G., et al. (2010). RNAi and heterochromatin repress centromeric meiotic recombination. *Proceedings of the National Academy of Sciences of the United States of America, 107*(19), 8701–8705. https://doi.org/10.1073/pnas.0914160107.

Ellermeier, C., & Smith, G. R. (2005). Cohesins are required for meiotic DNA breakage and recombination in Schizosaccharomyces pombe. *Proceedings of the National Academy of Sciences of the United States of America, 102*(31), 10952–10957. https://doi.org/10.1073/pnas.0504805102.

Fernius, J., Nerusheva, O. O., Galander, S., Alves, F. D. L., Rappsilber, J., & Marston, A. L. (2013). Cohesin-dependent association of Scc2/4 with the centromere initiates pericentromeric cohesion establishment. *Current Biology, 23*(7), 599–606. https://doi.org/10.1016/j.cub.2013.02.022.

Flemming, W. (1882). *Zellsubstanz, Kern und Zelltheilung (in German)*. F. C. W. Vogel.

Galander, S., & Marston, A. L. (2020). Meiosis I kinase regulators: Conserved orchestrators of reductional chromosome segregation. *BioEssays: News and Reviews in Molecular, Cellular and Developmental Biology, 42*(10), e2000018. https://doi.org/10.1002/BIES.202000018.

Garcia-Cruz, R., Roig, I., & Caldés, M. G. (2009). Maternal origin of the human aneuploidies. Are homolog synapsis and recombination to blame? Notes (learned) from the underbelly. *Genome Dynamics, 5*, 128–136. https://doi.org/10.1159/000166638.

Grewal, S. I. S., & Jia, S. (2007). Heterochromatin revisited. *Nature Reviews Genetics*, *8*(1), 35–46. https://doi.org/10.1038/nrg2008.

Gruhn, J. R., Zielinska, A. P., Shukla, V., Blanshard, R., Capalbo, A., Cimadomo, D., et al. (2019). Chromosome errors in human eggs shape natural fertility over reproductive life span. *Science*, *365*(6460), 1466–1469. https://doi.org/10.1126/science.aav7321.

Handyside, A. H. (2012). Molecular origin of female meiotic aneuploidies. *Biochimica et Biophysica Acta*, *1822*(12), 1913–1920. https://doi.org/10.1016/J.BBADIS.2012.07.007.

Hartmann, M. A., & Sekelsky, J. (2017). The absence of crossovers on chromosome 4 in Drosophila melanogaster: Imperfection or interesting exception? *Fly*, *11*(4), 253–259. https://doi.org/10.1080/19336934.2017.1321181.

Hartmann, M., Umbanhowar, J., & Sekelsky, J. (2019). Centromere-proximal meiotic crossovers in Drosophila melanogaster are suppressed by both highly repetitive heterochromatin and proximity to the centromere. *Genetics*, *213*(1), 113–125. https://doi.org/10.1534/GENETICS.119.302509.

Harushima, Y., Yano, M., Shomura, A., Sato, M., Shimano, T., Kuboki, Y., et al. (1998). A high-density rice genetic linkage map with 2275 markers using a single F2 population. *Genetics*, *148*(1), 479–494. https://doi.org/10.1093/GENETICS/148.1.479.

Hassold, T., & Hunt, P. (2001). To err (meiotically) is human: The genesis of human aneuploidy. *Nature Reviews Genetics*, *2*(4), 280–291. https://doi.org/10.1038/35066065.

Hatkevich, T., Kohl, K. P., McMahan, S., Hartmann, M. A., Williams, A. M., & Sekelsky, J. (2017). Bloom syndrome helicase promotes meiotic crossover patterning and homolog disjunction. *Current Biology*, *27*(1), 96–102. https://doi.org/10.1016/J.CUB.2016.10.055.

Hauf, S., Waizenegger, I. C., & Peters, J. M. (2001). Cohesin cleavage by separase required for anaphase and cytokinesis in human cells. *Science (New York, N.Y.)*, *293*(5533), 1320–1323. https://doi.org/10.1126/SCIENCE.1061376.

Haupt, W., Fischer, T. C., Winderl, S., Fransz, P., & Torres-Ruiz, R. A. (2001). The centromere1 (CEN1) region of Arabidopsis thaliana: Architecture and functional impact of chromatin. *The Plant Journal for Cell and Molecular Biology*, *27*(4), 285–296. https://doi.org/10.1046/J.1365-313X.2001.01087.X.

Hawley, R. S., Irick, H., Haddox, D. A., Whitley, M. D., Arbel, T., Jang, J., et al. (1992). There are two mechanisms of achiasmate segregation in Drosophila females, one of which requires heterochromatic homology. *Developmental Genetics*, *13*(6), 440–467. https://doi.org/10.1002/DVG.1020130608.

Hayashi, M. T., Takahashi, T. S., Nakagawa, T., Nakayama, J. I., & Masukata, H. (2009). The heterochromatin protein Swi6/HP1 activates replication origins at the pericentromeric region and silent mating-type locus. *Nature Cell Biology*, *11*(3), 357–362. https://doi.org/10.1038/NCB1845.

He, Y., Wang, M., Dukowic-Schulze, S., Zhou, A., Tiang, C. L., Shilo, S., et al. (2017). Genomic features shaping the landscape of meiotic double-strand-break hotspots in maize. *Proceedings of the National Academy of Sciences of the United States of America*, *114*(46), 12231–12236. https://doi.org/10.1073/PNAS.1713225114.

Heckmann, S., MacAs, J., Kumke, K., Fuchs, J., Schubert, V., Ma, L., et al. (2013). The holocentric species Luzula elegans shows interplay between centromere and large-scale genome organization. *The Plant Journal: For Cell and Molecular Biology*, *73*(4), 555–565. https://doi.org/10.1111/TPJ.12054.

Herbert, M., Kalleas, D., Cooney, D., Lamb, M., & Lister, L. (2015). Meiosis and maternal aging: Insights from aneuploid oocytes and trisomy births. *Cold Spring Harbor Perspectives in Biology*, 7(4). https://doi.org/10.1101/CSHPERSPECT.A017970.

Hinshaw, S. M., Makrantoni, V., Harrison, S. C., & Marston, A. L. (2017). The kinetochore receptor for the cohesin loading complex. *Cell*, *171*(1), 72–84.e13. https://doi.org/10.1016/j.cell.2017.08.017.

Hofstatter, P. G., Thangavel, G., Castellani, M., & Marques, A. (2021). Meiosis progression and recombination in holocentric plants: What is known? *Frontiers in Plant Science*, *12*, 658296. https://doi.org/10.3389/FPLS.2021.658296.
Holliday, R. (1964). A mechanism for gene conversion in fungi. *Genetical Research*, *5*(2), 282–304. https://doi.org/10.1017/S0016672300001233.
Hollis, J. A., Glover, M. L., Schlientz, A. J., Cahoon, C. K., Bowerman, B., Wignall, S. M., et al. (2020). Excess crossovers impede faithful meiotic chromosome segregation in C. elegans. *PLoS Genetics*, *16*(9), e1009001. https://doi.org/10.1371/JOURNAL.PGEN.1009001.
Hornig, N. C. D., Knowles, P. P., McDonald, N. Q., & Uhlmann, F. (2002). The dual mechanism of separase regulation by securin. *Current Biology: CB*, *12*(12), 973–982. https://doi.org/10.1016/S0960-9822(02)00847-3.
Hunter, N. (2015). Meiotic recombination: The essence of heredity. *Cold Spring Harbor Perspectives in Biology*, 7(12), a016618. https://doi.org/10.1101/CSHPERSPECT.A016618.
Hyppa, R. W., & Smith, G. R. (2010). Crossover invariance determined by partner choice for meiotic DNA break repair. *Cell*, *142*(2), 243–255. https://doi.org/10.1016/j.cell.2010.05.041.
Ishiguro, K. I. (2019). The cohesin complex in mammalian meiosis. *Genes to Cells*, *24*(1), 6–30. https://doi.org/10.1111/GTC.12652.
Jackson, M. S., See, C. G., Mulligan, L. M., & Lauffart, B. F. (1996). A 9.75-Mb map across the centromere of human chromosome 10. *Genomics*, *33*(2), 258–270. https://doi.org/10.1006/geno.1996.0190.
Jaco, I., Canela, A., Vera, E., & Blasco, M. A. (2008). Centromere mitotic recombination in mammalian cells. *The Journal of Cell Biology*, *181*(6), 885–892. https://doi.org/10.1083/JCB.200803042.
Kaur, E., Agrawal, R., & Sengupta, S. (2021). Functions of BLM helicase in cells: Is it acting like a double-edged sword? *Frontiers in Genetics*, *12*, 277. https://doi.org/10.3389/FGENE.2021.634789/BIBTEX.
Kim, S. M., Dubey, D. D., & Huberman, J. A. (2003). Early-replicating heterochromatin. *Genes & Development*, *17*(3), 330–335. https://doi.org/10.1101/GAD.1046203.
Kitajima, T. S., Yokobayashi, S., Yamamoto, M., & Watanabe, Y. (2003). Distinct cohesin complexes organize meiotic chromosome domains. *Science*, *300*(5622), 1152–1155. https://doi.org/10.1126/SCIENCE.1083634.
Koehler, K. E., Boulton, C. L., Collins, H. E., French, R. L., Herman, K. C., Lacefield, S. M., et al. (1996). Spontaneous X chromosome MI and MII nondisjunction events in Drosophila melanogaster oocytes have different recombinational histories. *Nature Genetics*, *14*(4), 406–414. https://doi.org/10.1038/NG1296-406.
Kouznetsova, A., Lister, L., Nordenskjöld, M., Herbert, M., & Höög, C. (2007). Bi-orientation of achiasmatic chromosomes in meiosis I oocytes contributes to aneuploidy in mice. *Nature Genetics*, *39*(8), 966–968. https://doi.org/10.1038/NG2065.
Kuhl, L. M., Makrantoni, V., Recknagel, S., Vaze, A. N., Marston, A. L., & Vader, G. (2020). A dCas9-based system identifies a central role for Ctf 19 in kinetochore-derived suppression of meiotic recombination. *Genetics*, *216*(2), 395–408.
Kuhl, L. M., & Vader, G. (2019). Kinetochores, cohesin, and DNA breaks: Controlling meiotic recombination within pericentromeres. *Yeast*, *36*(3), 121–127. https://doi.org/10.1002/yea.3366.
Kurdzo, E. L., Chuong, H. H., Evatt, J. M., & Dawson, D. S. (2018). A ZIP1 separation-of-function allele reveals that centromere pairing drives meiotic segregation of achiasmate chromosomes in budding yeast. *PLoS Genetics*, *14*(8). https://doi.org/10.1371/JOURNAL.PGEN.1007513.

Lam, I., & Keeney, S. (2015). Mechanism and regulation of meiotic recombination initiation. *Cold Spring Harbor Perspectives in Biology*, *7*(1). https://doi.org/10.1101/CSHPERSPECT.A016634.

Lamb, N. E., Freeman, S. B., Savage-Austin, A., Pettay, D., Taft, L., Hersey, J., et al. (1996). Susceptible chiasmate configurations of chromosome 21 predispose to non-disjunction in both maternal meiosis I and meiosis II. *Nature Genetics*, *14*(4), 400–405. https://doi.org/10.1038/NG1296-400.

Lambie, E. J., & Shirleen Roeder, G. (1988). A yeast centromere acts in cis to inhibit meiotic gene conversion of adjacent sequences. *Cell*, *52*(6), 863–873. https://doi.org/10.1016/0092-8674(88)90428-X.

Lee, J. (2019). Is age-related increase of chromosome segregation errors in mammalian oocytes caused by cohesin deterioration? *Reproductive Medicine and Biology*, *19*(1), 32–41. https://doi.org/10.1002/RMB2.12299.

Liu, T., Rechtsteiner, A., Egelhofer, T. A., Vielle, A., Latorre, I., Cheung, M. S., et al. (2011). Broad chromosomal domains of histone modification patterns in C. elegans. *Genome Research*, *21*(2), 227–236. https://doi.org/10.1101/GR.115519.110.

Ludin, K., Mata, J., Watt, S., Lehmann, E., Bähler, J., & Kohli, J. (2008). Sites of strong Rec12/Spo11 binding in the fission yeast genome are associated with meiotic recombination and with centromeres. *Chromosoma*, *117*(5), 431–444. https://doi.org/10.1007/S00412-008-0159-3.

Ma, J., Wing, R. A., Bennetzen, J. L., & Jackson, S. A. (2007). Plant centromere organization: A dynamic structure with conserved functions. *Trends in Genetics: TIG*, *23*(3), 134–139. https://doi.org/10.1016/J.TIG.2007.01.004.

MacLennan, M., Crichton, J. H., Playfoot, C. J., & Adams, I. R. (2015). Oocyte development, meiosis and aneuploidy. *Seminars in Cell & Developmental Biology*, *45*, 68–76. https://doi.org/10.1016/J.SEMCDB.2015.10.005.

Mahtani, M. M., & Willard, H. F. (1998). Physical and genetic mapping of the human X chromosome centromere: Repression of recombination. *Genome Research*, *8*(2), 100–110. https://doi.org/10.1101/GR.8.2.100.

Mancera, E., Bourgon, R., Brozzi, A., Huber, W., & Steinmetz, L. M. (2008). High-resolution mapping of meiotic crossovers and non-crossovers in yeast. *Nature*, *454*(7203), 479–485. https://doi.org/10.1038/NATURE07135.

Marques, A., Ribeiro, T., Neumann, P., Macas, J., Novák, P., Schubert, V., et al. (2015). Holocentromeres in Rhynchospora are associated with genome-wide centromere-specific repeat arrays interspersed among euchromatin. *Proceedings of the National Academy of Sciences of the United States of America*, *112*(44), 13633–13638. https://doi.org/10.1073/PNAS.1512255112.

Mather, K. (1939). Crossing over and heterochromatin in the x chromosome of Drosophila melanogaster. *Genetics*, *24*(3), 413–435. https://doi.org/10.1093/genetics/24.3.413.

McKinley, K. L., & Cheeseman, I. M. (2016). The molecular basis for centromere identity and function. *Nature Reviews Molecular Cell Biology*, *17*(1), 16–29. https://doi.org/10.1038/nrm.2015.5.

Mellone, B. G., & Fachinetti, D. (2021). Diverse mechanisms of centromere specification. *Current Biology: CB*, *31*(22), R1491–R1504. https://doi.org/10.1016/j.cub.2021.09.083.

Miller, D. E., Smith, C. B., Kazemi, N. Y., Cockrell, A. J., Arvanitakis, A. V., Blumenstiel, J. P., et al. (2016). Whole-genome analysis of individual meiotic events in Drosophila melanogaster reveals that noncrossover gene conversions are insensitive to interference and the centromere effect. *Genetics*, *203*(1), 159–171. https://doi.org/10.1534/GENETICS.115.186486/-/DC1.

Mirouze, M., Lieberman-Lazarovich, M., Aversano, R., Bucher, E., Nicolet, J., Reinders, J., et al. (2012). Loss of DNA methylation affects the recombination landscape in Arabidopsis. *Proceedings of the National Academy of Sciences of the United States of America*, *109*(15), 5880–5885. https://doi.org/10.1073/PNAS.1120841109/-/DCSUPPLEMENTAL.

Monen, J., Maddox, P. S., Hyndman, F., Oegema, K., & Desai, A. (2005). Differential role of CENP-A in the segregation of holocentric C. elegans chromosomes during meiosis and mitosis. *Nature Cell Biology*, 7(12), 1148–1155. https://doi.org/10.1038/NCB1331.

Murakami, S., Matsumoto, T., Niwa, O., & Yanagida, M. (1991). Structure of the fission yeast centromere cen3: Direct analysis of the reiterated inverted region. *Chromosoma*, *101*(4), 214–221. https://doi.org/10.1007/BF00365153.

Nakaseko, Y., Adachi, Y., Funahashi, S., Niwa, O., & Yanagida, M. (1986). Chromosome walking shows a highly homologous repetitive sequence present in all the centromere regions of fission yeast. *The EMBO Journal*, *5*(5), 1011–1021. https://doi.org/10.1002/J.1460-2075.1986.TB04316.X/FORMAT/PDF.

Nambiar, M., Chuang, Y. C., & Smith, G. R. (2019). Distributing meiotic crossovers for optimal fertility and evolution. In *Vol. 81. DNA repair*, Elsevier B.V. https://doi.org/10.1016/j.dnarep.2019.102648.

Nambiar, M., & Smith, G. R. (2016). Repression of harmful meiotic recombination in centromeric regions. In *Vol. 54. Seminars in cell and developmental biology* (pp. 188–197). Academic Press. https://doi.org/10.1016/j.semcdb.2016.01.042.

Nambiar, M., & Smith, G. R. (2018). Pericentromere-specific cohesin complex prevents meiotic pericentric DNA double-strand breaks and lethal crossovers. *Molecular Cell*, *71*(4), 540–553.e4. https://doi.org/10.1016/j.molcel.2018.06.035.

Nonaka, N., Kitajima, T., Yokobayashi, S., Xiao, G., Yamamoto, M., Grewal, S. I. S., et al. (2002). Recruitment of cohesin to heterochromatic regions by Swi6/HP1 in fission yeast. *Nature Cell Biology*, *4*(1), 89–93. https://doi.org/10.1038/NCB739.

Ohkura, H. (2015). Meiosis: An overview of key differences from mitosis. *Cold Spring Harbor Perspectives in Biology*, 7(5), 1–15. https://doi.org/10.1101/cshperspect.a015859.

Oliver, T. R., Tinker, S. W., Allen, E. G., Hollis, N., Locke, A. E., Bean, L. J. H., et al. (2012). Altered patterns of multiple recombinant events are associated with nondisjunction of chromosome 21. *Human Genetics*, *131*(7), 1039–1046. https://doi.org/10.1007/S00439-011-1121-7.

Oromendia, A. B., Dodgson, S. E., & Amon, A. (2012). Aneuploidy causes proteotoxic stress in yeast. *Genes & Development*, *26*(24), 2696–2708. https://doi.org/10.1101/GAD.207407.112.

Ottolini, C. S., Newnham, L. J., Capalbo, A., Natesan, S. A., Joshi, H. A., Cimadomo, D., et al. (2015). Genome-wide maps of recombination and chromosome segregation in human oocytes and embryos show selection for maternal recombination rates. *Nature Genetics*, *47*(7), 727–735. https://doi.org/10.1038/NG.3306.

Pan, J., Sasaki, M., Kniewel, R., Murakami, H., Blitzblau, H. G., Tischfield, S. E., et al. (2011). A hierarchical combination of factors shapes the genome-wide topography of yeast meiotic recombination initiation. *Cell*, *144*(5), 719–731. https://doi.org/10.1016/J.CELL.2011.02.009.

Pazhayam, N. M., Turcotte, C. A., & Sekelsky, J. (2021). Meiotic crossover patterning. *Frontiers in Cell and Development Biology*, *9*, 681123. https://doi.org/10.3389/fcell.2021.681123.

Perrella, G., Consiglio, M. F., Aiese-Cigliano, R., Cremona, G., Sanchez-Moran, E., Barra, L., et al. (2010). Histone hyperacetylation affects meiotic recombination and chromosome segregation in Arabidopsis. *The Plant Journal for Cell and Molecular Biology*, *62*(5), 796–806. https://doi.org/10.1111/J.1365-313X.2010.04191.X.

Peters, J. M., Tedeschi, A., & Schmitz, J. (2008). The cohesin complex and its roles in chromosome biology. *Genes & Development*, *22*(22), 3089–3114. https://doi.org/10.1101/GAD.1724308.

Pfau, S. J., & Amon, A. (2012). Chromosomal instability and aneuploidy in cancer: From yeast to man. *EMBO Reports*, *13*(6), 515–527. https://doi.org/10.1038/EMBOR.2012.65.

Phillips, D., Jenkins, G., Macaulay, M., Nibau, C., Wnetrzak, J., Fallding, D., et al. (2015). The effect of temperature on the male and female recombination landscape of barley. *The New Phytologist*, *208*(2), 421–429. https://doi.org/10.1111/NPH.13548.

Rattani, A., Wolna, M., Ploquin, M., Helmhart, W., Morrone, S., Mayer, B., et al. (2013). Sgol2 provides a regulatory platform that coordinates essential cell cycle processes during meiosis I in oocytes. *eLife*, *2013*(2). https://doi.org/10.7554/ELIFE.01133.001.

Reuter, G., & Wolff, I. (1981). Isolation of dominant suppressor mutations for position-effect variegation in Drosophila melanogaster. *Molecular and General Genetics MGG*, *182*(3), 516–519. https://doi.org/10.1007/BF00293947.

Rockman, M. V., & Kruglyak, L. (2009). Recombinational landscape and population genomics of Caenorhabditis elegans. *PLoS Genetics*, *5*(3). https://doi.org/10.1371/JOURNAL.PGEN.1000419.

Rockmill, B., Voelkel-Meiman, K., & Roeder, G. S. (2006). Centromere-proximal crossovers are associated with precocious separation of sister chromatids during meiosis in Saccharomyces cerevisiae. *Genetics*, *174*(4), 1745–1754. https://doi.org/10.1534/GENETICS.106.058933.

Rosu, S., Libuda, D. E., & Villeneuve, A. M. (2011). Robust crossover assurance and regulated interhomolog access maintain meiotic crossover number. *Science (New York, N.Y.)*, *334*(6060), 1286–1289. https://doi.org/10.1126/science.1212424.

Round, E. K., Flowers, S. K., & Richards, E. J. (1997). Arabidopsis thaliana centromere regions: Genetic map positions and repetitive DNA structure. *Genome Research*, 7(11), 1045–1053. https://doi.org/10.1101/GR.7.11.1045.

Saintenac, C., Falque, M., Martin, O. C., Paux, E., Feuillet, C., & Sourdille, P. (2009). Detailed recombination studies along chromosome 3B provide new insights on crossover distribution in wheat (Triticum aestivum L.). *Genetics*, *181*(2), 393–403. https://doi.org/10.1534/GENETICS.108.097469.

Saito, T. T., Lui, D. Y., Kim, H. M., Meyer, K., & Colaiácovo, M. P. (2013). Interplay between structure-specific endonucleases for crossover control during Caenorhabditis elegans meiosis. *PLoS Genetics*, *9*(7). https://doi.org/10.1371/JOURNAL.PGEN.1003586.

Saito, T. T., Mohideen, F., Meyer, K., Harper, J. W., Colaiácovo, M. P. (2012). SLX-1 is required for maintaining genomic integrity and promoting meiotic noncrossovers in the Caenorhabditis elegans germline. *PLoS Genetics*, *8*(8), e1002888. https://doi.org/10.1371/journal.pgen.1002888.

Saunders, M., Fitzgerald-Hayes, M., & Bloom, K. (1988). Chromatin structure of altered yeast centromeres. *Proceedings of the National Academy of Sciences of the United States of America*, *85*(1), 175–179. https://doi.org/10.1073/PNAS.85.1.175.

Sears, D. D., Hegemann, J. H., Shero, J. H., & Hieter, P. (1995). Cis-acting determinants affecting centromere function, sister-chromatid cohesion and reciprocal recombination during meiosis in Saccharomyces cerevisiae. *Genetics*, *139*(3), 1159–1173. https://doi.org/10.1093/GENETICS/139.3.1159.

Sharif, W. D., Glick, G. G., Davidson, M. K., & Wahls, W. P. (2002). Distinct functions of S. pombe Rec12 (Spo11) protein and Rec12-dependent crossover recombination (chiasmata) in meiosis I; and a requirement for Rec12 in meiosis II. *Cell & Chromosome*, *1*(1). https://doi.org/10.1186/1475-9268-1-1.

Sheltzer, J. M., & Amon, A. (2011). The aneuploidy paradox: Costs and benefits of an incorrect karyotype. *Trends in Genetics: TIG*, *27*(11), 446–453. https://doi.org/10.1016/J.TIG.2011.07.003.

Sherman, J. D., & Stack, S. M. (1995). Two-dimensional spreads of synaptonemal complexes from solanaceous plants. VI. High-resolution recombination nodule map for tomato (Lycopersicon esculentum). *Genetics*, *141*(2), 683–708. https://doi.org/10.1093/GENETICS/141.2.683.

Shi, J., Wolf, S. E., Burke, J. M., Presting, G. G., Ross-Ibarra, J., & Dawe, R. K. (2010). Widespread gene conversion in centromere cores. *PLoS Biology*, *8*(3). https://doi.org/10.1371/JOURNAL.PBIO.1000327.

Shinohara, M., Oh, S. D., Hunter, N., & Shinohara, A. (2008). Crossover assurance and crossover interference are distinctly regulated by the ZMM proteins during yeast meiosis. *Nature Genetics*, *40*(3), 299–309. https://doi.org/10.1038/ng.83.

Si, W., Yuan, Y., Huang, J., Zhang, X., Zhang, Y., Zhang, Y., et al. (2015). Widely distributed hot and cold spots in meiotic recombination as shown by the sequencing of rice F2 plants. *The New Phytologist*, *206*(4), 1491–1502. https://doi.org/10.1111/NPH.13319.

Smith, G. R., & Nambiar, M. (2020). New solutions to old problems: Molecular mechanisms of meiotic crossover control. *Trends in Genetics*, *36*(5), 337–346. https://doi.org/10.1016/j.tig.2020.02.002.

Steiner, N. C., Hahnenberger, K. M., & Clarke, L. (1993). Centromeres of the fission yeast Schizosaccharomyces pombe are highly variable genetic loci. *Molecular and Cellular Biology*, *13*(8), 4578–4587. https://doi.org/10.1128/MCB.13.8.4578-4587.1993.

Steiner, B., Masood, R., Rufibach, K., Niedrist, D., Kundert, O., Riegel, M., et al. (2015). An unexpected finding: Younger fathers have a higher risk for offspring with chromosomal aneuploidies. *European Journal of Human Genetics: EJHG*, *23*(4), 466–472. https://doi.org/10.1038/EJHG.2014.122.

Sun, L., Wang, J., Sang, M., Jiang, L., Zhao, B., Cheng, T., et al. (2017). Landscaping crossover interference across a genome. *Trends in Plant Science*, *22*(10), 894–907. https://doi.org/10.1016/j.tplants.2017.06.008.

Swuec, P., & Costa, A. (2014). Molecular mechanism of double Holliday junction dissolution. *Cell & Bioscience*, *4*(1), 36. https://doi.org/10.1186/2045-3701-4-36.

Székvölgyi, L., & Nicolas, A. (2010). From meiosis to postmeiotic events: Homologous recombination is obligatory but flexible. *The FEBS Journal*, *277*(3), 571–589. https://doi.org/10.1111/J.1742-4658.2009.07502.X.

Szostak, J. W., Orr-Weaver, T. L., Rothstein, R. J., & Stahl, F. W. (1983). The double-strand-break repair model for recombination. *Cell*, *33*(1), 25–35. https://doi.org/10.1016/0092-8674(83)90331-8.

Tachibana, M., Nozaki, M., Takeda, N., & Shinkai, Y. (2007). Functional dynamics of H3K9 methylation during meiotic prophase progression. *The EMBO Journal*, *26*(14), 3346–3359. https://doi.org/10.1038/SJ.EMBOJ.7601767.

Tachibana-Konwalski, K., Godwin, J., van der Weyden, L., Champion, L., Kudo, N. R., Adams, D. J., et al. (2010). Rec8-containing cohesin maintains bivalents without turnover during the growing phase of mouse oocytes. *Genes & Development*, *24*(22), 2505–2516. https://doi.org/10.1101/GAD.605910.

Talbert, P. B., & Henikoff, S. (2010). Centromeres convert but don't cross. *PLoS Biology*, *8*(3), e1000326. https://doi.org/10.1371/JOURNAL.PBIO.1000326.

Tanaka, T., Fuchs, J., Loidl, J., & Nasmyth, K. (2000). Cohesin ensures bipolar attachment of microtubules to sister centromeres and resists their precocious separation. *Nature Cell Biology*, *2*(8), 492–499. https://doi.org/10.1038/35019529.

Thacker, D., Lam, I., Knop, M., & Keeney, S. (2011). Exploiting spore-autonomous fluorescent protein expression to quantify meiotic chromosome behaviors in Saccharomyces cerevisiae. *Genetics*, *189*(2), 423–439. https://doi.org/10.1534/GENETICS.111.131326.

Thakur, J., Packiaraj, J., & Henikoff, S. (2021). Sequence, chromatin and evolution of satellite DNA. *International Journal of Molecular Sciences*, *22*(9). https://doi.org/10.3390/IJMS22094309.

Tock, A. J., & Henderson, I. R. (2018). Hotspots for initiation of meiotic recombination. *Frontiers in Genetics*, *9*, 521. https://doi.org/10.3389/fgene.2018.00521.

Tsutsumi, M., Fujiwara, R., Nishizawa, H., Ito, M., Kogo, H., Inagaki, H., et al. (2014). Age-related decrease of meiotic cohesins in human oocytes. *PLoS One*, *9*(5). https://doi.org/10.1371/JOURNAL.PONE.0096710.

Uhlmann, F., Lottspelch, F., & Nasmyth, K. (1999). Sister-chromatid separation at anaphase onset is promoted by cleavage of the cohesin subunit Scc1. *Nature*, *400*(6739), 37–42. https://doi.org/10.1038/21831.

Underwood, C. J., Choi, K., Lambing, C., Zhao, X., Serra, H., Borges, F., et al. (2018). Epigenetic activation of meiotic recombination near *Arabidopsis thaliana* centromeres via loss of H3K9me2 and non-CG DNA methylation. *Genome Research*, *28*(4), 519–531. https://doi.org/10.1101/gr.227116.117.

Vincenten, N., Kuhl, L. M., Lam, I., Oke, A., Kerr, A. R. W., Hochwagen, A., et al. (2015). The kinetochore prevents centromere-proximal crossover recombination during meiosis. *eLife*, *4*, 2015. https://doi.org/10.7554/eLife.10850.

von Diezmann, L., & Rog, O. (2021). Let's get physical—Mechanisms of crossover interference. *Journal of Cell Science*, *134*(10). https://doi.org/10.1242/jcs.255745.

Wang, J., Jia, S. T., & Jia, S. (2016). New insights into the regulation of heterochromatin. *Trends in Genetics*, *32*(5), 284–294. https://doi.org/10.1016/j.tig.2016.02.005.

Weiler, K. S., & Wakimoto, B. T. (2003). Heterochromatin and gene expression in drosophila. *Annual Review of Genetics*, *29*, 577–605. https://doi.org/10.1146/ANNUREV.GE.29.120195.003045.

West, S. C., Blanco, M. G., Chan, Y. W., Matos, J., Sarbajna, S., & Wyatt, H. D. M. (2016). Resolution of recombination intermediates: Mechanisms and regulation. *Cold Spring Harbor Symposia on Quantitative Biology*, *80*, 103–109. https://doi.org/10.1101/sqb.2015.80.027649.

Westphal, T., & Reuter, G. (2002). Recombinogenic effects of suppressors of position-effect variegation in Drosophila. *Genetics*, *160*(2), 609. https://doi.org/10.1093/genetics/160.2.609.

Whitby, M. C. (2005). Making crossovers during meiosis. *Biochemical Society Transactions*, *33*(6), 1451–1455. https://doi.org/10.1042/BST20051451.

Yadav, V. K., & Claeys Bouuaert, C. (2021). Mechanism and control of meiotic DNA double-strand break formation in S. cerevisiae. *Frontiers in Cell and Development Biology*, *9*. https://doi.org/10.3389/fcell.2021.642737.

Yamamoto, M., & Miklos, G. L. G. (1978). Genetic studies on heterochromatin in Drosophila melanogaster and their implications for the functions of satellite DNA. *Chromosoma*, *66*(1), 71–98. https://doi.org/10.1007/BF00285817.

Yelina, N. E., Choi, K., Chelysheva, L., Macaulay, M., de Snoo, B., Wijnker, E., et al. (2012). Epigenetic remodeling of meiotic crossover frequency in Arabidopsis thaliana DNA methyltransferase mutants. *PLoS Genetics*, *8*(8). https://doi.org/10.1371/JOURNAL.PGEN.1002844.

Yelina, N. E., Lambing, C., Hardcastle, T. J., Zhao, X., Santos, B., & Henderson, I. R. (2015). DNA methylation epigenetically silences crossover hot spots and controls chromosomal domains of meiotic recombination in Arabidopsis. *Genes & Development*, *29*(20), 2183–2202. https://doi.org/10.1101/GAD.270876.115.

Yu, Z., Kim, Y., & Dernburg, A. F. (2016). Meiotic recombination and the crossover assurance checkpoint in Caenorhabditis elegans. *Seminars in Cell & Developmental Biology*, *54*, 106–116. https://doi.org/10.1016/J.SEMCDB.2016.03.014.

Zelazowski, M. J., Sandoval, M., Paniker, L., Hamilton, H. M., Han, J., Gribbell, M. A., et al. (2017). Age-dependent alterations in meiotic recombination cause chromosome segregation errors in spermatocytes. *Cell*, *171*(3), 601–614.e13. https://doi.org/10.1016/J.CELL.2017.08.042.

CHAPTER EIGHT

# Unwinding during stressful times: Mechanisms of helicases in meiotic recombination

**Magdalena Firlej and John R. Weir***
Structural Biochemistry of Meiosis Group, Friedrich Miescher Laboratory of the Max Planck Society, Tuebingen, Germany
*Corresponding author: e-mail address: john.weir@tuebingen.mpg.de

## Contents



## Abstract

Successful meiosis I requires that homologous chromosomes be correctly linked before they are segregated. In most organisms this physical linkage is achieved through the generation of crossovers between the homologs. Meiotic recombination co-opts and modifies the canonical homologous recombination pathway to successfully generate crossovers One of the central components of this pathway are a number of conserved DNA helicases. Helicases couple nucleic acid binding to nucleotide hydrolysis and use this activity to modify DNA or protein-DNA substrates. During meiosis I it is necessary for the cell to modulate the canonical DNA repair pathways in order to facilitate the generation of interhomolog crossovers. Many of these meiotic modulations take place in pathways involving DNA helicases, or with a meiosis specific helicase. This short review explores what is currently understood about these helicases, their interaction partners, and the role of regulatory modifications during meiosis I. We focus in particular on the molecular structure and mechanisms of these helicases.

*Current Topics in Developmental Biology*, Volume 151
ISSN 0070-2153
https://doi.org/10.1016/bs.ctdb.2022.06.004

# 1. Introduction

*"If a man insisted always on being serious, and never allowed himself a bit of fun or relaxation, he would go mad or become unstable without knowing it."*

***Herodotus***

The connection between relaxation and stability holds true not just for our general wellbeing, but specifically for our genomes. Helicases, proteinaceous machines that connect nucleic acid binding, translocation, and/or unwinding to ATP hydrolysis, play a myriad of roles in genome stability. Indeed, several helicases have key roles as anti-recombination factors, thus maintaining genome integrity. During meiosis I, however, most species *require* recombination in order to link homologous chromosomes and thus ensure their faithful segregation during anaphase I, ultimately producing viable gametes. As such meiosis I is an intricate dance between pro- and anti-recombination factors. Here, we make a whistle stop tour of eight helicases (Fig. 1) that have specific roles during meiotic recombination, examine them from a structural biochemical perspective, and briefly discuss their meiosis specific interactions and modifications, and, where possible, the implications for human health and disease.

## 1.1 Architecture and mechanism of nucleic acid helicases

Helicases all make use of a number of RecA like domains that couple nucleic acid binding to ATP hydrolysis. These RecA-like domains may be part of an intramolecular arrangement (usually two RecA-like domains) or an intermolecular arrangement (frequently six RecA-like domains arranged as a hexameric ring) (Singleton, Dillingham, & Wigley, 2007). All of the helicases we focus on through this review consist of two RecA-like domains (RecA1 and RecA2) decorated with a number of accessory domains and belong to either Superfamily 1 (SF1) or Superfamily 2 (SF2 helicases). In this review not all the "helicases" here unwind dsDNA. Many are nucleic acid translocases that do not have unwinding capability. While the RecA-like domains are highly similar, the diverse accessory domains alter the substrate binding and interaction network of each helicase. Here, we mostly focus on the factors from the budding yeast *S. cerevisiae*, but also integrate data from other organisms to try to present a more comprehensive picture.

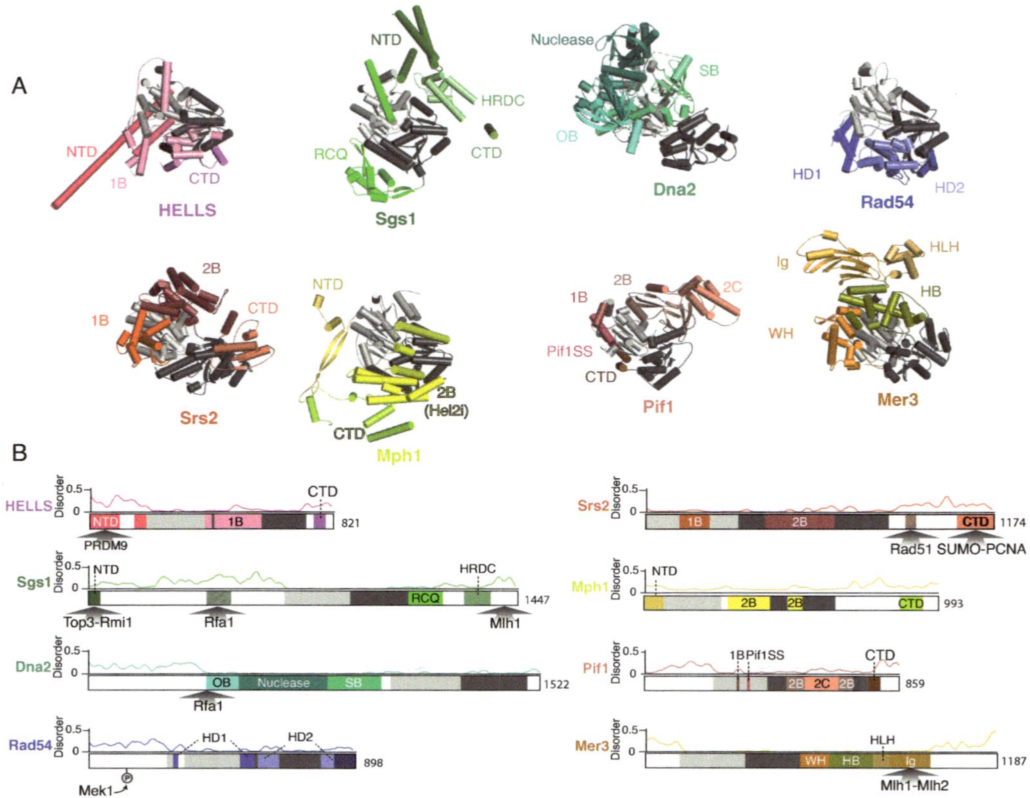


**Fig. 1** See figure legend on next page.

## 1.2 Meiotic recombination

Our understanding of meiotic recombination was described elsewhere (Hunter, 2015). Briefly, meiotic recombination is initiated through the introduction of programmed double-strand DNA breaks (DSBs). Break formation is catalysed by the meiosis specific topoisomerase Spo11 (Keeney, 2008; Keeney, Giroux, & Kleckner, 1997). Break sites are initially resected by the nuclease activity of Mre11, with further resection by Exo1 (Szankasi & Smith, 1995) and Dna2 (Budd, Choe, & Campbell, 2000) (see below). Single-stranded DNA (ssDNA) is then coated by the general recombinase Rad51, and also by the meiosis specific recombinase Dmc1 to form presynaptic filaments nucleated by Rad51, but with filaments of Dmc1 on the 3′ end, thus ensuring the dominance of Dmc1 in meiosis (Brown & Bishop, 2014). Once a presynaptic filament successfully invades a homologous sequence, the complementary strand is displaced forming a "D-loop," the first and most unstable form of joint molecule (JM) (Schwacha & Kleckner, 1994). A number of factors either act to extend, stabilize, or dissolve these nascent D-loops which give rise to more stable JMs such as single end invasions (SEIs) and double Holliday junctions (dHJs) (Hunter & Kleckner, 2001; Schwacha & Kleckner, 1995). Biased endonucleases then resolve these Holliday junctions into crossovers or non-crossovers (Allers & Lichten, 2001; Cannavo et al., 2020; Hunter & Kleckner, 2001; Kulkarni et al., 2020).

**Fig. 1** Structural overview of helicases involved in meiotic recombination. (A) Predicted structures of eight helicases central to meiotic recombination. Predicted structures of each helicase were downloaded from the publicly available AlphaFold database (https://www.alphafold.ebi.ac.uk/). Regions where the pLDDT score was $<37$ were hidden and the structures superimposed on one another using SMM and the RecA1 domain as a common reference. While crystal structures exist for many, or parts, of the helicases shown here, we have used AlphaFold predictions for a fair comparison. RecA1 shown in light grey, and RecA2 in dark grey. Domain annotations are according to previous publications. (Hu et al., 2021) (Pif1 (Lu et al., 2018), Mph1; based on RIG-1 (Luo et al., 2011). (B) Domain cartoons of helicases, coloured as in A). Disordered plot is based on flDPnn server (http://biomine.cs.vcu.edu/servers/flDPnn/), and is shown above the domain cartoon. RecA1 shown in light grey, and RecA2 in dark grey. Unstructured regions, or regions with low pLDDT score are shown in white. Interactions and modifications are shown below the domain cartoon. Note that while the Rfa1 interaction region for Sgs1 is known (Hegnauer et al., 2012), this region of Sgs1 is predicted to be unstructured.

# 2. Helicases and meiotic recombination

## 2.1 Formation of meiotic DSB breaks

The formation of Spo11 mediated double-strand DNA breaks is an event that is tightly regulated. One element of spatial regulation of DSBs is connected to the enrichment of $H3K4^{me3}$ marks (Acquaviva et al., 2013; Sommermeyer, Béneut, Chaplais, Serrentino, & Borde, 2013). In many organisms, including budding yeast, $H3K4^{me3}$ marks transcriptional start sites. In other organisms, including humans, an additional protein, PRDM9, deposits $H3K4^{me3}$ and $H3K27^{me3}$ at sequence specific regions of the genome (Baudat et al., 2010). PRDM9 consists of N-terminal methyltransferase domains, and a C-terminal array of zinc fingers (reviewed in Baudat, Imai, and de Massy (2013)). The function of PRDM9 is regulated by additional cofactors, for example, the helicase HELLS (Imai et al., 2020; Spruce et al., 2020).

### *2.1.1 HELLS helicase*

HELLS (also known as Lsh or PASG) is a SF2 family helicase (Flaus, Martin, Barton, & Owen-Hughes, 2006), and was originally identified as being involved in regulating DNA methylation at repeat elements throughout the mouse genome (Dennis, Fan, Geiman, Yan, & Muegge, 2001). While $HELLS^{-/-}$ mice are non-viable, analysis of fetal ovarian tissue revealed that while germ cells are present, there is a failure of chromosomes to undergo synapsis during meiosis I (De La Fuente et al., 2006).

Recently, HELLS was shown to interact directly with PRDM9 forming a complex both *in vivo* (Spruce et al., 2020), and *in vitro* (from Y2H experiments, (Imai et al., 2020)). The N-terminus of HELLS contains a coiled-coil domain, not found in other helicases (Fig. 1), and this domain is required for interaction with PRDM9 (Imai et al., 2020). Taken together a model has been proposed whereby the PRDM9-HELLS "pioneer" complex moves nucleosomes away from PRDM9-binding sites, thus facilitating PRDM9 DNA binding and subsequent site-specific DSB formation by the Spo11 complex (Imai et al., 2020; Spruce et al., 2020) (Fig. 2A).

Interestingly, purifications of recombinant HELLS out of bacteria showed low ATPase and chromatin remodeling activity *in vitro* (Burrage et al., 2012). Similarly to PRDM9, CDCA7 is a $Zn^{2+}$-finger protein that binds to HELLS and *in vitro* promotes robust chromatin remodeling by

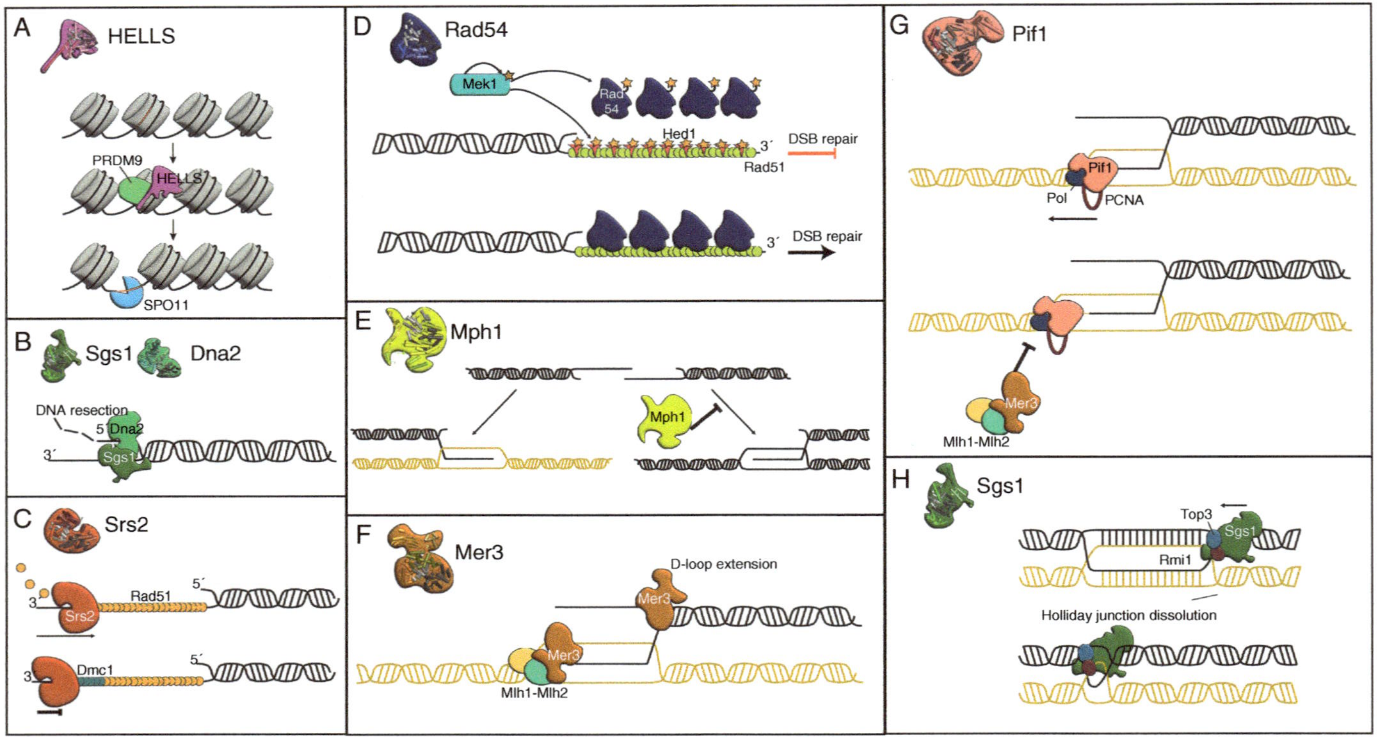


**Fig. 2** See figure legend on opposite page.

HELLS (Jenness et al., 2018). Currently, it is not clear whether CDCA7 and PRDM9 bind to the same site, or whether PRDM9 stimulates HELLS activity.

In male mice, PRDM9 binding sites are extensively methylated (Brick et al., 2018). The recruitment of HELLS and PRDM9 is required for the formation of these methylation marks (Imai et al., 2020). The role of both 5-hydroxymethylcytosine (5hmC) and 5-methylcytosine (5mC) in meiosis is unclear, however, HELLS could also play a role in the generation of these marks, since Tet proteins (that generate 5hmC (Ito et al., 2011)) have been shown to associate with HELLS (Jia et al., 2017).

## 2.2 Break resection

Following the formation of Spo11 mediated double-strand DNA breaks, the breaks are resected to generate ssDNA (reviewed in Cejka and Symington (2021)). At nascent breaks both the *endo*- and *exo*-nuclease activity of Mre11 (in complex with Rad50 and $Xrs2^{NBS1}$ as the $MRX^{MRN}$ complex) removes both Spo11 from the 5′ end and generates initial 3′ ssDNA. Subsequently, longer 3′ ssDNA is generated by the exonuclease activity

**Fig. 2** Mechanisms of nucleic acid helicases in meiosis. (A) Mechanism of HELLS helicase. PRDM9 (green) binds to specific DNA sequences. HELLS (purple) is brought to these sites *via* direct binding to PRDM9 and moves nucleosomes away from the PRDM9 DNA sequence (red), enabling cleavage by SPO11 (blue). (B) Sgs1 and Dna2 in DSB resection. The nuclease region of Dna2 (cyan) is a complementary pathway in break resection to Exo1. The nuclease activity of Dna2 is stimulated by Sgs1 (green). (C) Srs2-mediated disruption of presynaptic filaments. Srs2 (red) down regulates HR by removing Rad51 (yellow) from presynaptic filaments. Dmc1 (blue) on presynaptic filaments is refractory to Srs2 activity, thus stimulating HR in meiosis. (D) Rad54 on presynaptic filaments. During meiosis Rad54 (dark blue) can be phosphorylated by Mek1 kinase (light blue), which also targets the Rad54 cofactor Hed1 (red). The combination of these phosphorylations prevents Rad54 from catalysing the invasion of DNA duplexes by the presynaptic filament, and therefore downregulates DNA repair. (E) Mph1 and interhomolog bias. In yeast, Mph1 antagonises intersister recombination (right) thereby favouring recombination from the homologue (left). (F) Mer3 acting on nascent D-loops. Mer3 regulates D-loop extension and facilitates the formation of SEIs. Mer3 also interacts with Mlh1-Mlh2 which may form a block on the extension of D-loops, limiting the size of gene conversion tracts. (G) Pif1, Polδ and PCNA. The helicase Pif1 stimulates the polymerase activity of Polδ together with PCNA. In the context of recombination intermediates this activity is antagonised by the Mer3/Mlh1/Mlh2 complex, which acts to limit the size of gene conversion tracts. (H) Sgs1, Top3, Rmi1 form a complex (STR) which dissolves a variety of HR intermediates. Here the helicase activity of Sgs1 is being shown working on a double Holliday junction. (For interpretation of the references to colour in this figure legend, the reader is referred to the web version of this article.)

of Exo1 (Szankasi & Smith, 1995) and Dna2 (Budd et al., 2000). ssDNA is initially bound by the RPA complex (in yeast containing the proteins Rfa1, Rfa2 and Rfa3) which protects it from further degradation and formation of secondary structures (Chen, Lisby, & Symington, 2013; Deng, Yin, Petes, & Symington, 2015), and helps to "funnel" ssDNA to the downstream pathways (Fanning, Klimovich, & Nager, 2006).

#### 2.2.1 *Sgs1$^{BLM}$ and Dna2 in break resection*

Dna2 contains both a nuclease domain (Budd et al., 2000), and a C-terminal helicase domain (Bae & Seo, 2000), an organisation which is conserved in evolution (Bae et al., 1998). Our current understanding is that the helicase activity of Dna2 is required for Okazaki fragment processing and recombination (Zheng, Meng, Campbell, & Shen, 2020). Sgs1$^{BLM}$ is involved in dissolving both D-loops and dHJs (reviewed in Gupta and Schmidt (2020)) but in addition the direct binding of Sgs1 to Dna2 facilitates the nuclease activity in the latter providing a parallel resection pathway to Exo1 (Fig. 2B) (Gravel, Chapman, Magill, & Jackson, 2008; Mimitou & Symington, 2008; Sturzenegger et al., 2014; Zhu, Chung, Shim, Lee, & Ira, 2008). Likewise, Dna2 has been shown to stimulate Sgs1 activity, independent of Dna2 helicase and nuclease activities (Kasaciunaite et al., 2019; Xue et al., 2019) whereas the CtIP (Sae2 in yeast) stimulates the motor activity of Dna2 reconstituted end resection almost an order of magnitude more efficient (Ceppi et al., 2020). In budding yeast Exo1 is apparently the dominant end resection pathway (Mimitou, Yamada, & Keeney, 2017; Zakharyevich et al., 2010), whereas DNA2 in mammalian meiotic DSB resection might be particularly important in, since *Exo1*$^{-/-}$ mice do not show resection defects in spermatocytes (Paiano et al., 2020). Mutations in human *DNA2* gene have been associated with the primordial dwarfism disorder Seckel syndrome (specifically Seckel Syndrome 8 OMIM 615807) in which fertility defects, specifically repeated miscarriages, were observed (Shaheen et al., 2014).

### 2.3 Presynaptic filament formation

#### 2.3.1 *Srs2*

Srs2 is an ATP dependent 3′ to 5′ helicase that binds to Rad51 *via* amino acids 875–910 (Colavito et al., 2009). The C-terminus of Srs2 serves as an interaction region for SUMOylated PCNA as well as the target of post-translational modification (such as SUMOylation) (Armstrong,

Mohideen, & Lima, 2012; Pfander, Moldovan, Sacher, Hoege, & Jentsch, 2005) (Fig. 1B). The SIM (SUMO-interacting motif) has been shown to be involved in interaction with SUMOylated PCNA (Armstrong et al., 2012; Kolesar, Sarangi, Altmannova, Zhao, & Krejci, 2012; Papouli et al., 2005; Pfander et al., 2005). The SIM in Srs2 may be of particular importance in meiosis given the recently described pervasive role of SUMOylation during meiotic recombination (Bhagwat et al., 2021) (Table 1). The ortholog of Srs2 in mammals is unclear. A potential functional analog of Srs2 is RTEL1, a member of the XPD helicase family

**Table 1** General characteristics of helicases involved in meiotic recombination.

| Name | Functional orthologs from human (% similarity) | Molecular mass (residues) | Directionality | Direct binding partners | Meiosis specific post translational modifications |
|---|---|---|---|---|---|
| HELLS (mouse) | – | 95.1 kDa (821 a.a.) | | PRDM9 | – |
| Dna2 | DNA2 (37%) | 171.7 kDa (1522 a.a.) | | Rfa1 | – |
| Sgs1 | BLM (40%) | 163.8 kDa (1447 a.a.) | $3' \rightarrow 5'$ | Top3, Rmi1, Rfa1 | SUMO (6 sites)[a] CDK and Cdk5 phosphorylation (Grigaitis et al., 2020) |
| Rad54 | RAD54 (53%) | 101.8 kDa (898 a.a.) | Bidirectional translocase on dsDNA | Rad51 | pT132 (Niu et al., 2009) |
| Srs2 | RTEL1 (20%) | 134.3 kDa (1174 a.a.) | $3' \rightarrow 5'$ | PCNA, Rad51, Sgs1 | SUMO (4 sites)[a] |
| Mph1 | FANCM (23.5%) | 114.0 kDa (993 a.a.) | $3' \rightarrow 5'$ | – | – |
| Pif1 | PIF1 (38%) | 97.7 kDa (859 a.a.) | $5' \rightarrow 3'$ | PCNA, Polδ | SUMO (4 sites)[a] |
| Mer3 | HFM1 (37.5%) | 135 kDa (1187 a.a.) | $3' \rightarrow 5'$ | Mlh1, Mlh2, Rfc1, Pif1 | SUMO (11 sites)[a] |

[a]Bhagwat et al. (2021).

(Barber et al., 2008). Mutations in RTEL1 have been associated with several diseases such as Hoyeraal-Hreidarsson syndrome, a severe form of the bone-marrow failure and cancer predisposition disorder, and dyskeratosis congenita (Vannier, Sarek, & Boulton, 2014), and RTEL1 was shown to dissolve D-loops *in vitro* and limit crossover formation *in vivo* (Youds et al., 2010).

The formation of presynaptic filaments is antagonized by the helicase Srs2$^{RTEL1}$ through disrupting Rad51 filaments (Krejci et al., 2003; Veaute et al., 2003). The mechanism of Rad51 removal by Srs2 is dependent on the translocase activity of Srs2 (Kaniecki et al., 2017; Krejci et al., 2003). There is also direct interaction between Srs2 and Rad51, which stimulates Rad51 ATPase activity and causes filaments to disassemble (Antony et al., 2009), but this is thought to play a less prominent role as shown both *in vitro* (Kaniecki et al., 2017) and *in vivo*, by ectopic recombination assays, crossover to non-crossover ratio assessment and BIR assays (Elango et al., 2017; Jenkins et al., 2019). During meiosis this activity of Srs2 is antagonized, since Dmc1 filaments are resistant to Srs2 (Crickard, Kaniecki, Kwon, Sung, & Greene, 2018) (Fig. 2C). Given the similarity between Dmc1 and Rad51 it will be intriguing to understand how Srs2 distinguishes between Dmc1 and Rad51, and whether additional modifications play a role in these interactions in meiosis.

### 2.3.2 Rad54 (I)

Rad54 and its homolog Rdh54$^{RAD54B}$ are members of Snf2/Swi2 family of SF2 helicases. Rad54 is a highly processive translocase that forms multimers on DNA, possibly in the form of rings (Amitani, Baskin, & Kowalczykowski, 2006). Rad54 interacts directly with Rad51 *via* its N-terminus (Clever et al., 1997; Raschle, Van Komen, Chi, Ellenberger, & Sung, 2004). Budding yeast rad54Δ rdh54Δ double mutants are completely defective in meiosis, showing 1.6% spore viability, compared to 53% viability for rad54Δ and 82% for rdh54Δ single mutants (Shinohara et al., 1997). One possibility is that Rdh54$^{RAD54B}$ preferentially associates with the meiosis specific Dmc1 (Dresser et al., 1997) (whereas Rad54 interacts with Rad51). Mammalian RAD54B, however, while involved in DNA repair, does not seem to function as Rdh54 in meiosis (Wesoly et al., 2006). *In vitro* Rad54 enhances the strand invasion activity of Rad51 and it can also remodel chromatinized DNA, which is stimulated by Rad51 presynaptic filaments (Petukhova, Stratton, & Sung, 1998). It is assumed that the homology search by Rad51 is also enhanced by interaction with Rad54 (Crickard, Moevus, Kwon, Sung, & Greene, 2020; Kowalczykowski, 2015).

During meiosis Hed1 associates with Rad51, and blocks the interaction with Rad54 (Busygina et al., 2008). Indeed, the association of Hed1 with Rad51 on presynaptic filaments *in vitro* is very stable, and only unbinds when the filament is disassembled (Crickard, Kaniecki, Kwon, Sung, Lisby, et al., 2018). In contrast Hed1 does not bind to the meiosis specific Rad51 homolog Dmc1. Rad51 is required for meiosis, but its strand-exchange activity is not, thus leading to the conclusion that Rad51 is a structural accessory factor during meiosis (Cloud, Chan, Grubb, Budke, & Bishop, 2012). Taken together, Hed1 presumably functions to promote the specific functional association of Rad54 with Dmc1 during meiosis.

## 2.4 Interhomolog bias

HR during meiosis could either use the sister chromatid or the homologous chromosome as a template. Indeed, the sister chromatid is proximal to the break site and identical in sequence making it a perfect template. Nevertheless, meiotic recombination requires that the homologous chromosome is used as a template (reviewed in Lao and Hunter (2010)) This process, known as interhomolog bias, is apparently regulated at a number of levels (reviewed in Humphryes and Hochwagen (2014)). Here, we will look at the roles of two helicases, Rad54 and Mph1 in facilitating interhomolog bias.

### *2.4.1 Rad54 (II)*

Mek1 is a S/K kinase that is activated through the ATM/ATR-dependent phosphorylation of the axial protein Hop1 (Carballo, Johnson, Sedgwick, & Cha, 2008; Niu et al., 2005; Penedos et al., 2015). Mek1 phosphorylates both Rad54 (Niu et al., 2009) and Hed1 (Callender et al., 2016). Phosphorylation of Rad54 T132 inhibits its activity, which impairs DNA repair (Niu et al., 2009) (Fig. 2D). Hed1, a meiotic cofactor, binds specifically to Rad51 and prevents its association with Rad54 (Busygina et al., 2008). Phosphorylation of Hed1 by Mek1 leads to the increased protein stability of Hed1 thus further suppressing Rad51 activity (Callender et al., 2016). Taken together the model suggests DNA repair proximal to the break site, i.e., using the sister chromatid as a template, is inhibited, and thus repair is directed to the distal homolog. However, Hed1 has no effect on the

Dmc1-Rad54 interaction (Busygina et al., 2008), so there is likely a more complex network of interactions (Lao et al., 2013). Nevertheless in yeast the principal "signaller" of IH bias is Mek1 kinase, since specific inhibition of *mek1-as* gives rise to a WT ratio of interhomolog and intersister crossovers (Terentyev et al., 2010).

### 2.4.2 *Mph1$^{FANCM}$*

Mph1 is a SF2 type helicase that belongs to the FANCM family of helicases. FANCM has been implicated in Fanconi anaemia disorder, a syndrome characterised by cancer predisposition, developmental disorder, and bone marrow failure (Whitby, 2010). Most of the sequence similarity between Mph1 and FANCM is within the N-terminal helicase domain (Fig. 1B). Budding yeast Mph1 lacks the C-terminal ERCC4 nuclease domain found in other FANCM family members, (except *S. pombe* Fml1). Likewise, most FANCM family members have a double helix-hairpin-helix ($(HhH)_2$) domain, C-terminal to the ERCC4 domain, and was previously thought to be absent from Mph1. The AlphaFold2 prediction of Mph1 shows that there is very likely to be a $(HhH)_2$ like-domain present (Fig. 1A, yellow). In other FANCM family members the $(HhH)_2$ domain facilitates protein-protein interactions.

In meiosis Mph1$^{FANCM}$ has two important functions. One is to limit crossover formation, shown in both fission yeast and plants (Crismani et al., 2012; Lorenz et al., 2012). In fission yeast, it was shown that Fml1 (Mph1 homolog) dissociates the D-loops, the loss of which leading to a decrease in non-crossovers arising from SDSA (Lorenz et al., 2012). In plants, FANCM mutants show a threefold higher number of crossovers *via* a pathway that is independent of the ZMM pathway (see below for a description of ZMM proteins) (Crismani et al., 2012). In both plants and fission yeast these results are consistent with the observation in somatic cells that one function of Mph1$^{FANCM}$ is to dissolve D-loops (Prakash et al., 2009).

The second, recently reported function of Mph1, is in facilitating interhomolog bias (Sandhu et al., 2020). Mph1 has been reported to limit inter-sister joint molecule formation, thereby favoring repair from the homolog (Fig. 2E). Consistent with this idea is that *mph1* mutants become more sensitive to reduced DSB number, presumably because too few DNA repair intermediates are being funneled to the interhomolog pathway, and are thus repaired from the sister (Sandhu et al., 2020). Consistent with the meiotic functions of Mph1$^{FANCM}$ described here, FANCM mutants in humans cause both spermatogenic and premature ovarian failure (Fouquet et al., 2017; Kasak et al., 2018).

## 2.5 D-loop formation

### *2.5.1 Mer3$^{HFM1}$*

Mer3$^{HFM1}$ is a SF2 DExH box helicase, with an extended C-terminal domain (Fig. 1B). Mer3$^{HFM1}$ is the only meiosis specific helicase. Orthologs of Mer3 have been clearly identified in plants and mammals (Chen, Zhang, Timofejeva, Gerardin, & Ma, 2005; Tanaka, Miyamoto, Shouguchi-Miyata, & Ikeda, 2006). In humans, heterozygotic mutations in HFM1 lead to primary ovarian insufficiency (Wang, Zhang, Jiang, Wu, & Collaboration, 2014). *In vitro* studies on Mer3 showed that it is an active ATPase with strand separation activity working a 3′ to 5′ direction (Nakagawa, Flores-Rozas, & Kolodner, 2001) and that it might preferentially recognise Holliday junctions (Nakagawa & Kolodner, 2002b), though it is unclear how this might promote crossovers.

Further *in vitro* work demonstrated that Mer3 facilitated a heteroduplex extension by Rad51, that is, it enlarged and stabilized D-loops, stabilizing this crossover intermediate (Fig. 2F) (Mazina, Mazin, Nakagawa, Kolodner, & Kowalczykowski, 2004). In a back-to-back paper (Börner, Kleckner, & Hunter, 2004) identified Mer3 as being one of the "ZMM" proteins that facilitate type I crossover formation (together with Zip1, Zip2, Zip3, Msh4 and Msh5). Taken together this, and previous work, identifies Mer3 as an active helicase that is still part of a larger protein interaction network.

Interestingly, ATPase deficient Mer3 gave rather mild spore viability defects when compared with *mer3Δ* (Nakagawa & Kolodner, 2002a). While the effect of some residual helicase activity remains a possibility, this observation hinted at the idea that the critical role of Mer3 might lie in mediating protein: protein interactions. Beyond the concept of the "ZMM" group of proteins, the first indications of the intermolecular interactions have been recently determined. An IP-MS study on Mlh1 identified Mer3 as a potential interactor, *via* the Ig-like domain of Mer3 (Duroc et al., 2017). Deletion of Mlh1 or non-Mlh1 binding Mer3 mutant leads to an increase in the length of gene conversion tracts, both in crossovers (COs) and in non-crossovers (NCOs) (Fig. 2F). The interactions between Mlh1/ Mlh2 and Pif1 (see below) don't fully explain the strong meiotic crossover effect of *mer3Δ*, since *mlh2Δ* does not have an effect on crossover formation

(Wang, Kleckner, & Hunter, 1999), and Structurally, Mer3 is remarkably similar to the spliceosome RNA helicase Brr2 (Fig. 1). Brr2 contains two helicase "cassettes" where the second cassette has lost catalytic activity, thus functioning as a dimer (Absmeier, Santos, & Wahl, 2016). Furthermore, Brr2 has multiple protein-protein interactions that change as the spliceosome matures. We are then left with some intriguing questions. Might Mer3 also function on a template that contains RNA? RNA transcripts have been shown to have an important role in homologous recombination (Keskin et al., 2014). Does Mer3 function as a dimer? Can Mer3 be regulated by additional protein–protein interactions in addition to Mlh1-Mlh2? The last question has recently been answered with the discovery of the interaction with another helicase; Pif1.

### 2.5.2 Pif1

Pif1 is a SF1b helicase that is widely conserved throughout eukaryotes. Within the helicase domain there are several insertions. In RecA-1 there is a strand-separating 1B domain, which is thought to be stabilised by the characteristic Pif1 signature sequence (Pif1SS), (Byrd & Raney, 2017). RecA-2 is interrupted with the 2B insertion, which folds into an SH3 domain, which usually mediates protein–protein interactions (Li, 2005).

One role of Pif1 is in the regulation of telomerase, removing it from telomeres and thus regulating their length. Pif1 also removes telomerase from DSB sites, preventing the aberrant addition of telomeric DNA (Schulz & Zakian, 1994). Indeed, Pif1 is phosphorylated by Rad53 in response to DNA damage on residues in the C-terminus (T763 and T766), which prevents the recruitment of telomerase at break sites (Makovets & Blackburn, 2009). Pif1 was also found to have a role in break induced replication (BIR), and that Pif1 had a highly stimulatory effect on Pol $\partial$ activity. During BIR Pif1 migrates D-loop bubbles to mediate conservative DNA synthesis by Pol $\partial$ (Chung, Zhu, Papusha, Malkova, & Ira, 2010; Wilson et al., 2013). Pif1 interacts with the replication machinery by binding through its C-terminus to the sliding clamp PCNA (Buzovetsky et al., 2017).

IP-MS on Mer3 found both Mlh1-Mlh2 (as previously described) but also the PCNA clamp loader Rfc1 (Vernekar et al., 2021). *Mlh2Δ* gives rise to longer gene conversion tracts during meiosis (Duroc et al., 2017), an effect that is rescued when using a Pif1 mutant that cannot enter the nucleus (Vernekar et al., 2021), suggesting that Mlh1-Mlh2 may prevent Pif1 from migrating D-loop bubbles, as it does in BIR (Chung et al., 2010; Wilson et al., 2013). In an elegant *in vitro* experiment, it was confirmed that the inhibitory effect on Pif1 indeed required not only Mlh1-Mlh2, but also Mer3, which thus limits the size of gene conversion tracts (Vernekar et al., 2021) (Fig. 2G). While the exact nature of the inhibitory effect on Pif1 remains to be elucidated, a possibility is that both Pif1 and Mer3/Mlh1–2 compete for the same binding site on PCNA. With Pif1 being recruited but also inhibited during meiosis, why should it be recruited at all? Vernekar et al. proposed that the helicase activity of Pif1 could be a safeguard, in case Pol$\partial$ were to encounter secondary structure DNA elements that could block DNA repair.

## 2.6 Crossover decision

### 2.6.1 Sgs1

Sgs1 carries out a number of functions in meiosis. First, Sgs1, together with Top3 and Rmi1 (known as STR complex), can dissolve nascent D-loop structures, resulting in these DNA breaks being repaired *via* SDSA. Second, Sgs1 helicase can directly dissolve dHJs, resulting in NCOs rather than COs (Fig. 2H). Finally, and somewhat counterintuitively, Sgs1 is required for the normal distribution of meiotic COs (Hatkevich & Sekelsky, 2017; Holloway, Morelli, Borst, & Cohen, 2010). This observation is likely connected to the role of Sgs1 in preventing aberrant inter-sister recombination during meiosis (Jessop, Rockmill, Roeder, & Lichten, 2006; Oh et al., 2007), thus creating an available "pool" of ssDNA ends that can be subsequently captured by the ZMM group of proteins (Hatkevich & Sekelsky, 2017).

While humans have five RecQ helicases, the functional ortholog of Sgs1 in meiosis is Bloom Helicase (BLM). BLM is named after a multifaceted disease, Bloom's Syndrome, where the common feature is loss-of-function mutations in BLM. Bloom's Syndrome patients show growth deficiency, photosensitivity, immune deficiency, insulin resistance, and an increased risk of early onset of cancer or multiple cancers (Cunniff, Bassetti, & Ellis, 2017). Bloom Syndrome like disorders are also caused by mutations in TOP3A, RMI1 and RMI2 (Hudson et al., 2016; Martin et al., 2018), reflecting the fact that these proteins work together in a complex, known as Bloom Syndrome Complex (BTRR) in mammals (Bythell-Douglas & Deans, 2021).

The interaction of Sgs1$^{BLM}$ with Top3$^{TOP3A}$ is conserved through eukaryotes (Ahmad & Stewart, 2005; Bennett, Noirot-Gros, & Wang, 2000). While the helicase activity of Sgs1$^{BLM}$ is required for most of its functions (Bernstein et al., 2009), even the helicase defective form of Sgs1$^{BLM}$

can stimulate the decatenation activity of Top3$^{TOP3A}$ (Cejka, Plank, Dombrowski, & Kowalczykowski, 2012). Both the helicase activity of Sgs1 and Top3-Rmi1 are essential for the resolution of meiotic recombination intermediates (Kaur, De Muyt, & Lichten, 2015; Tang, Wu, Zhang, & Hunter, 2015).

Rmi1 contains an OB-fold and in mammals there are two Rmi proteins (RMI1 and RMI2). The mammalian RMI1–2 complex binds to FANCM (Deans & West, 2009), it is unclear if this interaction is conserved in yeast. During meiosis ZMM proteins must presumably act to downregulate Sgs1$^{BLM}$ activity (Jessop et al., 2006; Oh et al., 2007), yet to date there have been no reports of direct interactions between Sgs1$^{BLM}$ and the ZMM proteins despite in-depth time-resolved proteomic studies (Wild et al., 2019).

In budding yeast, resolution of dHJs is temporally controlled through the polo kinase Cdc5, which initiates crossovers with pachytene exit (Allers & Lichten, 2001; Clyne et al., 2003). In a breakthrough study the Matos laboratory has provided a detailed mechanistic explanation for the temporal regulation of crossover formation through the phosphorylation of Sgs1 (Grigaitis et al., 2020). Sgs1 is phosphorylated by CDK on up to nine consensus sites, including within the Top3/Rmi1 and Rfa1 interaction regions and the helicase domain. CDK phosphorylations stimulate the helicase activity of Sgs1 by increasing the processivity and unwinding velocity. CDK phosphorylations also serve as priming phosphorylations for Cdc5$^{PLK}$, which hyperphosphorylates Sgs1$^{BLM}$, presumably leading to downregulation of the helicase activity. As such the activation of Cdc5 provides a time window in which to allow crossovers to occur. Consistent with this argument is the observation that Cdc5$^{PLK}$ phosphorylation of the structure selective endonuclease Mus81-Mms4 enhances DNA cleavage (Matos, Blanco, Maslen, Skehel, & West, 2011), although this is complicated by the fact that Sgs1 is also required for class-I (Mus81-Mms4 independent) crossovers (Zakharyevich et al., 2010).

## 3. Summary

With the exception of Mer3/HFM1 none of the helicases above are unique to meiosis. Yet through a combination of meiosis-specific interaction partners and post-translational modifications the activity of many helicases are modulated to facilitate the goal of meiosis I, namely, the successful generation of crossover recombination events between homologous chromosomes. Genetic and proteomic studies continue to generate new interaction partners and modifications for these helicases. Recent breakthroughs in protein structure prediction have given the community access to what are likely to prove high-quality models of these helicases and their binding partners. Additional hybrid structural biological approaches combined with multicomponent biochemical reconstitution experiments and separation of function mutations will provide us with both details of these interactions and their potential physiological function.

## Acknowledgments

We thank Gerben Vader, Elisabeth Weir, and members of the Weir Laboratory for input and critical reading of the manuscript. Work in the Weir Lab is funded by the Max Planck Society, and by the German Research Foundation (WE 6513/2-1). We would also like to take the opportunity to apologise to those scientists whose excellent work we were unable to include here due to space constraints.

## References

Absmeier, E., Santos, K. F., & Wahl, M. C. (2016). Functions and regulation of the Brr2 RNA helicase during splicing. *Cell Cycle*, *15*(24), 3362–3377. https://doi.org/10.1080/15384101.2016.1249549.

Acquaviva, L., Székvölgyi, L., Dichtl, B., Dichtl, B. S., de La Roche Saint André, C., Nicolas, A., et al. (2013). The COMPASS subunit Spp1 links histone methylation to initiation of meiotic recombination. *Science*, *339*(6116), 215–218. https://doi.org/10.1126/science.1225739.

Ahmad, F., & Stewart, E. (2005). The N-terminal region of the Schizosaccharomyces pombe RecQ helicase, Rqh1p, physically interacts with topoisomerase III and is required for Rqh1p function. *Molecular Genetics and Genomics*, *273*(1), 102–114. https://doi.org/10.1007/s00438-005-1111-3.

Allers, T., & Lichten, M. (2001). Differential timing and control of noncrossover and crossover recombination during meiosis. *Cell*, *106*(1), 47–57. https://doi.org/10.1016/s0092-8674(01)00416-0.

Amitani, I., Baskin, R. J., & Kowalczykowski, S. C. (2006). Visualization of Rad54, a chromatin remodeling protein, translocating on single DNA molecules. *Molecular Cell*, *23*(1), 143–148. https://doi.org/10.1016/j.molcel.2006.05.009.

Antony, E., Tomko, E. J., Xiao, Q., Krejci, L., Lohman, T. M., & Ellenberger, T. (2009). Srs2 disassembles Rad51 filaments by a protein-protein interaction triggering ATP turnover and dissociation of Rad51 from DNA. *Molecular Cell*, *35*(1), 105–115. https://doi.org/10.1016/j.molcel.2009.05.026.

Armstrong, A. A., Mohideen, F., & Lima, C. D. (2012). Recognition of SUMO-modified PCNA requires tandem receptor motifs in Srs2. *Nature*, *483*(7387), 59–63. https://doi.org/10.1038/nature10883.

Bae, S.-H., Choi, E., Lee, K.-H., Park, J. S., Lee, S.-H., & Seo, Y.-S. (1998). Dna2 of Saccharomyces cerevisiae possesses a single-stranded DNA-specific endonuclease activity that is able to act on double-stranded DNA in the presence of ATP*. *The Journal of Biological Chemistry*, *273*(41), 26880–26890. https://doi.org/10.1074/jbc.273.41.26880.

Bae, S.-H., & Seo, Y.-S. (2000). Characterization of the enzymatic properties of the yeast Dna2 helicase/endonuclease suggests a new model for Okazaki fragment processing *. *The Journal of Biological Chemistry*, *275*(48), 38022–38031. https://doi.org/10.1074/jbc.M006513200.

Barber, L. J., Youds, J. L., Ward, J. D., McIlwraith, M. J., O'Neil, N. J., Petalcorin, M. I. R., et al. (2008). RTEL1 maintains genomic stability by suppressing homologous recombination. *Cell, 135*(2), 261–271. https://doi.org/10.1016/j.cell.2008.08.016.

Baudat, F., Buard, J., Grey, C., Fledel-Alon, A., Ober, C., Przeworski, M., et al. (2010). PRDM9 is a major determinant of meiotic recombination hotspots in humans and mice. *Science, 327*(5967), 836–840. https://doi.org/10.1126/science.1183439.

Baudat, F., Imai, Y., & de Massy, B. (2013). Meiotic recombination in mammals: Localization and regulation. *Nature Reviews. Genetics, 14*(11), 794–806. https://doi.org/10.1038/nrg3573.

Bennett, R. J., Noirot-Gros, M.-F., & Wang, J. C. (2000). Interaction between yeast Sgs1 helicase and DNA topoisomerase III *. *The Journal of Biological Chemistry, 275*(35), 26898–26905. https://doi.org/10.1016/S0021-9258(19)61459-6.

Bernstein, K. A., Shor, E., Sunjevaric, I., Fumasoni, M., Burgess, R. C., Foiani, M., et al. (2009). Sgs1 function in the repair of DNA replication intermediates is separable from its role in homologous recombinational repair. *The EMBO Journal, 28*(7), 915–925. https://doi.org/10.1038/emboj.2009.28.

Bhagwat, N. R., Owens, S. N., Ito, M., Boinapalli, J. V., Poa, P., Ditzel, A., et al. (2021). SUMO is a pervasive regulator of meiosis. *eLife, 10*, e57720. https://doi.org/10.7554/eLife.57720.

Börner, G. V., Kleckner, N., & Hunter, N. (2004). Crossover/noncrossover differentiation, synaptonemal complex formation, and regulatory surveillance at the leptotene/zygotene transition of meiosis. *Cell, 117*(1), 29–45. https://doi.org/10.1016/s0092-8674(04)00292-2.

Brick, K., Thibault-Sennett, S., Smagulova, F., Lam, K.-W. G., Pu, Y., Pratto, F., et al. (2018). Extensive sex differences at the initiation of genetic recombination. *Nature, 561*(7723), 338–342. https://doi.org/10.1038/s41586-018-0492-5.

Brown, M. S., & Bishop, D. K. (2014). DNA strand exchange and RecA homologs in meiosis. *Cold Spring Harbor Perspectives in Biology*, 7(1), a016659. https://doi.org/10.1101/cshperspect.a016659.

Budd, M. E., Choe, W.-C., & Campbell, J. L. (2000). The nuclease activity of the yeast Dna2 protein, which is related to the RecB-like nucleases, is essential in vivo *. *The Journal of Biological Chemistry, 275*(22), 16518–16529. https://doi.org/10.1074/jbc.M909511199.

Burrage, J., Termanis, A., Geissner, A., Myant, K., Gordon, K., & Stancheva, I. (2012). The SNF2 family ATPase LSH promotes phosphorylation of H2AX and efficient repair of DNA double-strand breaks in mammalian cells. *Journal of Cell Science, 125*(Pt 22), 5524–5534. https://doi.org/10.1242/jcs.111252.

Busygina, V., Sehorn, M. G., Shi, I. Y., Tsubouchi, H., Roeder, G. S., & Sung, P. (2008). Hed1 regulates Rad51-mediated recombination via a novel mechanism. *Genes & Development, 22*(6), 786–795. https://doi.org/10.1101/gad.1638708.

Buzovetsky, O., Kwon, Y., Pham, N. T., Kim, C., Ira, G., Sung, P., et al. (2017). Role of the Pif1-PCNA complex in pol δ-dependent Strand displacement DNA synthesis and break-induced replication. *Cell Reports, 21*(7), 1707–1714. https://doi.org/10.1016/j.celrep.2017.10.079.

Byrd, A. K., & Raney, K. D. (2017). Structure and function of Pif1 helicase. *Biochemical Society Transactions, 45*(5), 1159–1171. https://doi.org/10.1042/BST20170096.

Bythell-Douglas, R., & Deans, A. J. (2021). A structural guide to the bloom syndrome complex. *Structure, 29*(2), 99–113. https://doi.org/10.1016/j.str.2020.11.020.

Callender, T. L., Laureau, R., Wan, L., Chen, X., Sandhu, R., Laljee, S., et al. (2016). Mek1 down regulates Rad51 activity during yeast meiosis by phosphorylation of Hed1. *PLoS Genetics, 12*(8), e1006226. https://doi.org/10.1371/journal.pgen.1006226.

Cannavo, E., Sanchez, A., Anand, R., Ranjha, L., Hugener, J., Adam, C., et al. (2020). Regulation of the MLH1-MLH3 endonuclease in meiosis. *Nature, 586*(7830), 618–622. https://doi.org/10.1038/s41586-020-2592-2.

Carballo, J. A., Johnson, A. L., Sedgwick, S. G., & Cha, R. S. (2008). Phosphorylation of the axial element protein Hop1 by Mec1/Tel1 ensures meiotic interhomolog recombination. *Cell, 132*(5), 758–770. https://doi.org/10.1016/j.cell.2008.01.035.

Cejka, P., Plank, J. L., Dombrowski, C. C., & Kowalczykowski, S. C. (2012). Decatenation of DNA by the S. cerevisiae Sgs1-Top3-Rmi1 and RPA complex: A mechanism for disentangling chromosomes. *Molecular Cell, 47*(6), 886–896. https://doi.org/10.1016/j.molcel.2012.06.032.

Cejka, P., & Symington, L. S. (2021). DNA end resection: Mechanism and control. *Annual Review of Genetics, 55*, 285–307. https://doi.org/10.1146/annurev-genet-071719-020312.

Ceppi, I., Howard, S. M., Kasaciunaite, K., Pinto, C., Anand, R., Seidel, R., et al. (2020). CtIP promotes the motor activity of DNA2 to accelerate long-range DNA end resection. *Proceedings of the National Academy of Sciences of the United States of America, 117*(16), 8859–8869. https://doi.org/10.1073/pnas.2001165117.

Chen, H., Lisby, M., & Symington, L. S. (2013). RPA coordinates DNA end resection and prevents formation of DNA hairpins. *Molecular Cell, 50*(4), 589–600. https://doi.org/10.1016/j.molcel.2013.04.032.

Chen, C., Zhang, W., Timofejeva, L., Gerardin, Y., & Ma, H. (2005). The Arabidopsis ROCK-N-ROLLERS gene encodes a homolog of the yeast ATP-dependent DNA helicase MER3 and is required for normal meiotic crossover formation. *The Plant Journal: For Cell and Molecular Biology, 43*(3), 321–334. https://doi.org/10.1111/j.1365-313X.2005.02461.x.

Chung, W.-H., Zhu, Z., Papusha, A., Malkova, A., & Ira, G. (2010). Defective resection at DNA double-strand breaks leads to de novo telomere formation and enhances gene targeting. *PLoS Genetics, 6*(5), e1000948. https://doi.org/10.1371/journal.pgen.1000948.

Clever, B., Interthal, H., Schmuckli-Maurer, J., King, J., Sigrist, M., & Heyer, W. D. (1997). Recombinational repair in yeast: Functional interactions between Rad51 and Rad54 proteins. *The EMBO Journal, 16*(9), 2535–2544. https://doi.org/10.1093/emboj/16.9.2535.

Cloud, V., Chan, Y.-L., Grubb, J., Budke, B., & Bishop, D. K. (2012). Rad51 is an accessory factor for Dmc1-mediated joint molecule formation during meiosis. *Science, 337*(6099), 1222–1225. https://doi.org/10.1126/science.1219379.

Clyne, R. K., Katis, V. L., Jessop, L., Benjamin, K. R., Herskowitz, I., Lichten, M., et al. (2003). Polo-like kinase Cdc5 promotes chiasmata formation and cosegregation of sister centromeres at meiosis I. *Nature Cell Biology, 5*(5), 480–485. https://doi.org/10.1038/ncb977.

Colavito, S., Macris-Kiss, M., Seong, C., Gleeson, O., Greene, E. C., Klein, H. L., et al. (2009). Functional significance of the Rad51-Srs2 complex in Rad51 presynaptic filament disruption. *Nucleic Acids Research, 37*(20), 6754–6764. https://doi.org/10.1093/nar/gkp748.

Crickard, J. B., Kaniecki, K., Kwon, Y., Sung, P., & Greene, E. C. (2018). Meiosis-specific recombinase Dmc1 is a potent inhibitor of the Srs2 antirecombinase. *Proceedings of the National Academy of Sciences of the United States of America, 115*(43), E10041–E10048. https://doi.org/10.1073/pnas.1810457115.

Crickard, J. B., Kaniecki, K., Kwon, Y., Sung, P., Lisby, M., & Greene, E. C. (2018). Regulation of Hed1 and Rad54 binding during maturation of the meiosis-specific presynaptic complex. *The EMBO Journal, 37*(7), e98728. https://doi.org/10.15252/embj.201798728.

Crickard, J. B., Moevus, C. J., Kwon, Y., Sung, P., & Greene, E. C. (2020). Rad54 drives ATP hydrolysis-dependent DNA sequence alignment during homologous recombination. *Cell, 181*(6), 1380–1394. https://doi.org/10.1016/j.cell.2020.04.056.

Crismani, W., Girard, C., Froger, N., Pradillo, M., Santos, J. L., Chelysheva, L., et al. (2012). FANCM limits meiotic crossovers. *Science, 336*(6088), 1588–1590. https://doi.org/10.1126/science.1220381.

Cunniff, C., Bassetti, J. A., & Ellis, N. A. (2017). Bloom's syndrome: Clinical Spectrum, molecular pathogenesis, and cancer predisposition. *Molecular Syndromology, 8*(1), 4–23. https://doi.org/10.1159/000452082.

De La Fuente, R., Baumann, C., Fan, T., Schmidtmann, A., Dobrinski, I., & Muegge, K. (2006). Lsh is required for meiotic chromosome synapsis and retrotransposon silencing in female germ cells. *Nature Cell Biology, 8*(12), 1448–1454. https://doi.org/10.1038/ncb1513.

Deans, A. J., & West, S. C. (2009). FANCM connects the genome instability disorders Bloom's syndrome and Fanconi anemia. *Molecular Cell, 36*(6), 943–953. https://doi.org/10.1016/j.molcel.2009.12.006.

Deng, S. K., Yin, Y., Petes, T. D., & Symington, L. S. (2015). Mre11-Sae2 and RPA collaborate to prevent palindromic gene amplification. *Molecular Cell, 60*(3), 500–508. https://doi.org/10.1016/j.molcel.2015.09.027.

Dennis, K., Fan, T., Geiman, T., Yan, Q., & Muegge, K. (2001). Lsh, a member of the SNF2 family, is required for genome-wide methylation. *Genes & Development, 15*(22), 2940–2944. https://doi.org/10.1101/gad.929101.

Dresser, M. E., Ewing, D. J., Conrad, M. N., Dominguez, A. M., Barstead, R., Jiang, H., et al. (1997). DMC1 functions in a Saccharomyces cerevisiae meiotic pathway that is largely independent of the RAD51 pathway. *Genetics, 147*(2), 533–544. https://doi.org/10.1093/genetics/147.2.533.

Duroc, Y., Kumar, R., Ranjha, L., Adam, C., Guérois, R., Md Muntaz, K., et al. (2017). Concerted action of the MutLβ heterodimer and Mer3 helicase regulates the global extent of meiotic gene conversion. *eLife, 6*. https://doi.org/10.7554/eLife.21900.

Elango, R., Sheng, Z., Jackson, J., DeCata, J., Ibrahim, Y., Pham, N. T., et al. (2017). Break-induced replication promotes formation of lethal joint molecules dissolved by Srs2. *Nature Communications, 8*(1), 1790. https://doi.org/10.1038/s41467-017-01987-2.

Fanning, E., Klimovich, V., & Nager, A. R. (2006). A dynamic model for replication protein A (RPA) function in DNA processing pathways. *Nucleic Acids Research, 34*(15), 4126–4137. https://doi.org/10.1093/nar/gkl550.

Flaus, A., Martin, D. M. A., Barton, G. J., & Owen-Hughes, T. (2006). Identification of multiple distinct Snf2 subfamilies with conserved structural motifs. *Nucleic Acids Research, 34*(10), 2887–2905. https://doi.org/10.1093/nar/gkl295.

Fouquet, B., Pawlikowska, P., Caburet, S., Guigon, C., Mäkinen, M., Tanner, L., et al. (2017). A homozygous FANCM mutation underlies a familial case of non-syndromic primary ovarian insufficiency. *eLife, 6*. https://doi.org/10.7554/eLife.30490.

Gravel, S., Chapman, J. R., Magill, C., & Jackson, S. P. (2008). DNA helicases Sgs1 and BLM promote DNA double-strand break resection. *Genes & Development, 22*(20), 2767–2772. https://doi.org/10.1101/gad.503108.

Grigaitis, R., Ranjha, L., Wild, P., Kasaciunaite, K., Ceppi, I., Kissling, V., et al. (2020). Phosphorylation of the RecQ helicase Sgs1/BLM controls its DNA unwinding activity during meiosis and mitosis. *Developmental Cell, 53*(6), 706–723. https://doi.org/10.1016/j.devcel.2020.05.016.

Gupta, S. V., & Schmidt, K. H. (2020). Maintenance of yeast genome integrity by RecQ family DNA helicases. *Genes, 11*(2). https://doi.org/10.3390/genes11020205.

Hatkevich, T., & Sekelsky, J. (2017). Bloom syndrome helicase in meiosis: Pro-crossover functions of an anti-crossover protein. *BioEssays: News and Reviews in Molecular, Cellular and Developmental Biology, 39*(9). https://doi.org/10.1002/bies.201700073.

Hegnauer, A. M., Hustedt, N., Shimada, K., Pike, B. L., Vogel, M., Amsler, P., et al. (2012). An N-terminal acidic region of Sgs1 interacts with Rpa70 and recruits Rad53 kinase to stalled forks. *The EMBO Journal*, *31*(18), 3768–3783. https://doi.org/10.1038/emboj.2012.195.

Holloway, J. K., Morelli, M. A., Borst, P. L., & Cohen, P. E. (2010). Mammalian BLM helicase is critical for integrating multiple pathways of meiotic recombination. *The Journal of Cell Biology*, *188*(6), 779–789. https://doi.org/10.1083/jcb.200909048.

Hu, G., Katuwawala, A., Wang, K., Wu, Z., Ghadermarzi, S., Gao, J., et al. (2021). flDPnn: Accurate intrinsic disorder prediction with putative propensities of disorder functions. *Nature Communications*, *12*(1), 4438. https://doi.org/10.1038/s41467-021-24773-7.

Hudson, D. F., Amor, D. J., Boys, A., Butler, K., Williams, L., Zhang, T., et al. (2016). Loss of RMI2 increases genome instability and causes a bloom-like syndrome. *PLoS Genetics*, *12*(12), e1006483. https://doi.org/10.1371/journal.pgen.1006483.

Humphryes, N., & Hochwagen, A. (2014). A non-sister act: Recombination template choice during meiosis. *Experimental Cell Research*, *329*(1), 53–60. https://doi.org/10.1016/j.yexcr.2014.08.024.

Hunter, N. (2015). Meiotic recombination: The essence of heredity. *Cold Spring Harbor Perspectives in Biology*, 7(12). https://doi.org/10.1101/cshperspect.a016618.

Hunter, N., & Kleckner, N. (2001). The single-end invasion: An asymmetric intermediate at the double-strand break to double-Holliday junction transition of meiotic recombination. *Cell*, *106*(1), 59–70. https://doi.org/10.1016/s0092-8674(01)00430-5.

Imai, Y., Biot, M., Clément, J. A., Teragaki, M., Urbach, S., Robert, T., et al. (2020). PRDM9 activity depends on HELLS and promotes local 5-hydroxymethylcytosine enrichment. *eLife*, *9*. https://doi.org/10.7554/eLife.57117.

Ito, S., Shen, L., Dai, Q., Wu, S. C., Collins, L. B., Swenberg, J. A., et al. (2011). Tet proteins can convert 5-methylcytosine to 5-formylcytosine and 5-carboxylcytosine. *Science*, *333*(6047), 1300–1303. https://doi.org/10.1126/science.1210597.

Jenkins, S. S., Gore, S., Guo, X., Liu, J., Ede, C., Veaute, X., et al. (2019). Role of the Srs2-Rad51 interaction domain in crossover control in Saccharomyces cerevisiae. *Genetics*, *212*(4), 1133–1145. https://doi.org/10.1534/genetics.119.302337.

Jenness, C., Giunta, S., Müller, M. M., Kimura, H., Muir, T. W., & Funabiki, H. (2018). HELLS and CDCA7 comprise a bipartite nucleosome remodeling complex defective in ICF syndrome. *Proceedings of the National Academy of Sciences of the United States of America*, *115*(5), E876–E885. https://doi.org/10.1073/pnas.1717509115.

Jessop, L., Rockmill, B., Roeder, G. S., & Lichten, M. (2006). Meiotic chromosome synapsis-promoting proteins antagonize the anti-crossover activity of sgs1. *PLoS Genetics*, *2*(9), e155. https://doi.org/10.1371/journal.pgen.0020155.

Jia, J., Shi, Y., Chen, L., Lai, W., Yan, B., Jiang, Y., et al. (2017). Decrease in lymphoid specific helicase and 5-hydroxymethylcytosine is associated with metastasis and genome instability. *Theranostics*, 7(16), 3920–3932. https://doi.org/10.7150/thno.21389.

Kaniecki, K., De Tullio, L., Gibb, B., Kwon, Y., Sung, P., & Greene, E. C. (2017). Dissociation of Rad51 presynaptic complexes and heteroduplex DNA joints by tandem assemblies of Srs2. *Cell Reports*, *21*(11), 3166–3177. https://doi.org/10.1016/j.celrep.2017.11.047.

Kasaciunaite, K., Fettes, F., Levikova, M., Daldrop, P., Anand, R., Cejka, P., et al. (2019). Competing interaction partners modulate the activity of Sgs1 helicase during DNA end resection. *The EMBO Journal*, *38*(13), e101516. https://doi.org/10.15252/embj.2019101516.

Kasak, L., Punab, M., Nagirnaja, L., Grigorova, M., Minajeva, A., Lopes, A. M., et al. (2018). Bi-allelic recessive loss-of-function variants in FANCM cause non-obstructive azoospermia. *American Journal of Human Genetics*, *103*(2), 200–212. https://doi.org/10.1016/j.ajhg.2018.07.005.

Kaur, H., De Muyt, A., & Lichten, M. (2015). Top3-Rmi1 DNA single-Strand Decatenase is integral to the formation and resolution of meiotic recombination intermediates. *Molecular Cell, 57*(4). https://doi.org/10.1016/j.molcel.2015.01.020.

Keeney, S. (2008). Spo11 and the formation of DNA double-Strand breaks in meiosis. *Genome Dynamics and Stability, 2*, 81–123. https://doi.org/10.1007/7050_2007_026.

Keeney, S., Giroux, C. N., & Kleckner, N. (1997). Meiosis-specific DNA double-strand breaks are catalyzed by Spo11, a member of a widely conserved protein family. *Cell, 88*(3), 375–384. https://doi.org/10.1016/s0092-8674(00)81876-0.

Keskin, H., Shen, Y., Huang, F., Patel, M., Yang, T., Ashley, K., et al. (2014). Transcript-RNA-templated DNA recombination and repair. *Nature, 515*(7527), 436–439. https://doi.org/10.1038/nature13682.

Kolesar, P., Sarangi, P., Altmannova, V., Zhao, X., & Krejci, L. (2012). Dual roles of the SUMO-interacting motif in the regulation of Srs2 sumoylation. *Nucleic Acids Research, 40*(16), 7831–7843. https://doi.org/10.1093/nar/gks484.

Kowalczykowski, S. C. (2015). An overview of the molecular mechanisms of recombinational DNA repair. *Cold Spring Harbor Perspectives in Biology*, 7(11). https://doi.org/10.1101/cshperspect.a016410.

Krejci, L., Van Komen, S., Li, Y., Villemain, J., Reddy, M. S., Klein, H., et al. (2003). DNA helicase Srs2 disrupts the Rad51 presynaptic filament. *Nature, 423*(6937), 305–309. https://doi.org/10.1038/nature01577.

Kulkarni, D. S., Owens, S. N., Honda, M., Ito, M., Yang, Y., Corrigan, M. W., et al. (2020). PCNA activates the MutLγ endonuclease to promote meiotic crossing over. *Nature, 586*(7830), 623–627. https://doi.org/10.1038/s41586-020-2645-6.

Lao, J. P., Cloud, V., Huang, C.-C., Grubb, J., Thacker, D., Lee, C.-Y., et al. (2013). Meiotic crossover control by concerted action of Rad51-Dmc1 in homolog template bias and robust homeostatic regulation. *PLoS Genetics, 9*(12), e1003978. https://doi.org/10.1371/journal.pgen.1003978.

Lao, J. P., & Hunter, N. (2010). Trying to avoid your sister. *PLoS Biology, 8*(10), e1000519. https://doi.org/10.1371/journal.pbio.1000519.

Li, S. S.-C. (2005). Specificity and versatility of SH3 and other proline-recognition domains: Structural basis and implications for cellular signal transduction. *Biochemical Journal, 390*(Pt 3), 641–653. https://doi.org/10.1042/BJ20050411.

Lorenz, A., Osman, F., Sun, W., Nandi, S., Steinacher, R., & Whitby, M. C. (2012). The fission yeast FANCM ortholog directs non-crossover recombination during meiosis. *Science, 336*(6088), 1585–1588. https://doi.org/10.1126/science.1220111.

Lu, K.-Y., Chen, W.-F., Rety, S., Liu, N.-N., Wu, W.-Q., Dai, Y.-X., et al. (2018). Insights into the structural and mechanistic basis of multifunctional S. cerevisiae Pif1p helicase. *Nucleic Acids Research, 46*(3), 1486–1500. https://doi.org/10.1093/nar/gkx1217.

Luo, D., Ding, S. C., Vela, A., Kohlway, A., Lindenbach, B. D., & Pyle, A. M. (2011). Structural insights into RNA recognition by RIG-I. *Cell, 147*(2), 409–422. https://doi.org/10.1016/j.cell.2011.09.023.

Makovets, S., & Blackburn, E. H. (2009). DNA damage signalling prevents deleterious telomere addition at DNA breaks. *Nature Cell Biology, 11*(11), 1383–1386. https://doi.org/10.1038/ncb1985.

Martin, C.-A., Sarlós, K., Logan, C. V., Thakur, R. S., Parry, D. A., Bizard, A. H., et al. (2018). Mutations in TOP3A cause a bloom syndrome-like disorder. *American Journal of Human Genetics, 103*(2), 221–231. https://doi.org/10.1016/j.ajhg.2018.07.001.

Matos, J., Blanco, M. G., Maslen, S., Skehel, J. M., & West, S. C. (2011). Regulatory control of the resolution of DNA recombination intermediates during meiosis and mitosis. *Cell, 147*(1), 158–172. https://doi.org/10.1016/j.cell.2011.08.032.

Mazina, O. M., Mazin, A. V., Nakagawa, T., Kolodner, R. D., & Kowalczykowski, S. C. (2004). Saccharomyces cerevisiae Mer3 helicase stimulates 3′–5′ heteroduplex extension by Rad51. *Cell, 117*(1), 47–56. https://doi.org/10.1016/s0092-8674(04)00294-6.

Mimitou, E. P., & Symington, L. S. (2008). Sae2, Exo1 and Sgs1 collaborate in DNA double-strand break processing. *Nature*, *455*(7214), 770–774. https://doi.org/10.1038/nature07312.

Mimitou, E. P., Yamada, S., & Keeney, S. (2017). A global view of meiotic double-strand break end resection. *Science*, *355*(6320), 40–45. https://doi.org/10.1126/science.aak9704.

Nakagawa, T., Flores-Rozas, H., & Kolodner, R. D. (2001). The MER3 helicase involved in meiotic crossing over is stimulated by single-stranded DNA-binding proteins and unwinds DNA in the 3′ to 5′ direction. *The Journal of Biological Chemistry*, *276*(34), 31487–31493. https://doi.org/10.1074/jbc.M104003200.

Nakagawa, T., & Kolodner, R. D. (2002a). Saccharomyces cerevisiae Mer3 is a DNA helicase involved in meiotic crossing over. *Molecular and Cellular Biology*, *22*(10), 3281–3291. https://doi.org/10.1128/mcb.22.10.3281-3291.2002.

Nakagawa, T., & Kolodner, R. D. (2002b). The MER3 DNA helicase catalyzes the unwinding of Holliday junctions. *The Journal of Biological Chemistry*, *277*(31), 28019–28024. https://doi.org/10.1074/jbc.M204165200.

Niu, H., Wan, L., Baumgartner, B., Schaefer, D., Loidl, J., & Hollingsworth, N. M. (2005). Partner choice during meiosis is regulated by Hop1-promoted dimerization of Mek1. *Molecular Biology of the Cell*, *16*(12), 5804–5818. https://www.molbiolcell.org/doi/abs/10.1091/mbc.E05-05-0465.

Niu, H., Wan, L., Busygina, V., Kwon, Y., Allen, J. A., Li, X., et al. (2009). Regulation of meiotic recombination via Mek1-mediated Rad54 phosphorylation. *Molecular Cell*, *36*(3), 393–404. https://doi.org/10.1016/j.molcel.2009.09.029.

Oh, S. D., Lao, J. P., Hwang, P. Y.-H., Taylor, A. F., Smith, G. R., & Hunter, N. (2007). BLM ortholog, Sgs1, prevents aberrant crossing-over by suppressing formation of multichromatid joint molecules. *Cell*, *130*(2), 259–272. https://doi.org/10.1016/j.cell.2007.05.035.

Paiano, J., Wu, W., Yamada, S., Sciascia, N., Callen, E., Paola Cotrim, A., et al. (2020). ATM and PRDM9 regulate SPO11-bound recombination intermediates during meiosis. *Nature Communications*, *11*(1), 857. https://doi.org/10.1038/s41467-020-14654-w.

Papouli, E., Chen, S., Davies, A. A., Huttner, D., Krejci, L., Sung, P., et al. (2005). Crosstalk between SUMO and ubiquitin on PCNA is mediated by recruitment of the helicase Srs2p. *Molecular Cell*, *19*(1), 123–133. https://doi.org/10.1016/j.molcel.2005.06.001.

Penedos, A., Johnson, A. L., Strong, E., Goldman, A. S., Carballo, J. A., & Cha, R. S. (2015). Essential and checkpoint functions of budding yeast ATM and ATR during meiotic prophase are facilitated by differential phosphorylation of a meiotic adaptor protein, Hop1. *PLoS One*, *10*(7), e0134297. https://doi.org/10.1371/journal.pone.0134297.

Petukhova, G., Stratton, S., & Sung, P. (1998). Catalysis of homologous DNA pairing by yeast Rad51 and Rad54 proteins. *Nature*, *393*(6680), 91–94. https://doi.org/10.1038/30037.

Pfander, B., Moldovan, G.-L., Sacher, M., Hoege, C., & Jentsch, S. (2005). SUMO-modified PCNA recruits Srs2 to prevent recombination during S phase. *Nature*, *436*(7049), 428–433. https://doi.org/10.1038/nature03665.

Prakash, R., Satory, D., Dray, E., Papusha, A., Scheller, J., Kramer, W., et al. (2009). Yeast Mph1 helicase dissociates Rad51-made D-loops: Implications for crossover control in mitotic recombination. *Genes & Development*, *23*(1), 67–79. https://doi.org/10.1101/gad.1737809.

Raschle, M., Van Komen, S., Chi, P., Ellenberger, T., & Sung, P. (2004). Multiple interactions with the Rad51 recombinase govern the homologous recombination function of Rad54. *The Journal of Biological Chemistry*, *279*(50), 51973–51980. https://doi.org/10.1074/jbc.M410101200.

Sandhu, R., Monge Neria, F., Monge Neria, J., Chen, X., Hollingsworth, N. M., & Börner, G. V. (2020). DNA helicase Mph1FANCM ensures meiotic recombination between parental chromosomes by dissociating precocious displacement loops. *Developmental Cell, 53*(4), 458–472. https://doi.org/10.1016/j.devcel.2020.04.010.

Schulz, V. P., & Zakian, V. A. (1994). The saccharomyces PIF1 DNA helicase inhibits telomere elongation and de novo telomere formation. *Cell, 76*(1), 145–155. https://doi.org/10.1016/0092-8674(94)90179-1.

Schwacha, A., & Kleckner, N. (1994). Identification of joint molecules that form frequently between homologs but rarely between sister chromatids during yeast meiosis. *Cell, 76*(1), 51–63. https://doi.org/10.1016/0092-8674(94)90172-4.

Schwacha, A., & Kleckner, N. (1995). Identification of double Holliday junctions as intermediates in meiotic recombination. *Cell, 83*(5), 783–791. https://doi.org/10.1016/0092-8674(95)90191-4.

Shaheen, R., Faqeih, E., Ansari, S., Abdel-Salam, G., Al-Hassnan, Z. N., Al-Shidi, T., et al. (2014). Genomic analysis of primordial dwarfism reveals novel disease genes. *Genome Research, 24*(2), 291–299. https://doi.org/10.1101/gr.160572.113.

Shinohara, M., Shita-Yamaguchi, E., Buerstedde, J. M., Shinagawa, H., Ogawa, H., & Shinohara, A. (1997). Characterization of the roles of the Saccharomyces cerevisiae RAD54 gene and a homologue of RAD54, RDH54/TID1, in mitosis and meiosis. *Genetics, 147*(4), 1545–1556. https://doi.org/10.1093/genetics/147.4.1545.

Singleton, M. R., Dillingham, M. S., & Wigley, D. B. (2007). Structure and mechanism of helicases and nucleic acid translocases. *Annual Review of Biochemistry, 76*, 23–50. https://doi.org/10.1146/annurev.biochem.76.052305.115300.

Sommermeyer, V., Béneut, C., Chaplais, E., Serrentino, M. E., & Borde, V. (2013). Spp1, a member of the Set1 complex, promotes meiotic DSB formation in promoters by tethering histone H3K4 methylation sites to chromosome axes. *Molecular Cell, 49*(1), 43–54. https://doi.org/10.1016/j.molcel.2012.11.008.

Spruce, C., Dlamini, S., Ananda, G., Bronkema, N., Tian, H., Paigen, K., et al. (2020). HELLS and PRDM9 form a pioneer complex to open chromatin at meiotic recombination hot spots. *Genes & Development, 34*(5–6), 398–412. https://doi.org/10.1101/gad.333542.119.

Sturzenegger, A., Burdova, K., Kanagaraj, R., Levikova, M., Pinto, C., Cejka, P., et al. (2014). DNA2 cooperates with the WRN and BLM RecQ helicases to mediate long-range DNA end resection in human cells. *The Journal of Biological Chemistry, 289*(39), 27314–27326. https://doi.org/10.1074/jbc.M114.578823.

Szankasi, P., & Smith, G. R. (1995). A role for exonuclease I from S. pombe in mutation avoidance and mismatch correction. *Science, 267*(5201), 1166–1169. https://doi.org/10.1126/science.7855597.

Tanaka, K., Miyamoto, N., Shouguchi-Miyata, J., & Ikeda, J.-E. (2006). HFM1, the human homologue of yeast Mer3, encodes a putative DNA helicase expressed specifically in germ-line cells. *DNA Sequence: The Journal of DNA Sequencing and Mapping, 17*(3), 242–246. https://doi.org/10.1080/10425170600805433.

Tang, S., Wu, M. K. Y., Zhang, R., & Hunter, N. (2015). Pervasive and essential roles of the Top3-Rmi1 Decatenase orchestrate recombination and facilitate chromosome segregation in meiosis. *Molecular Cell, 57*(4). https://doi.org/10.1016/j.molcel.2015.01.021.

Terentyev, Y., Johnson, R., Neale, M. J., Khisroon, M., Bishop-Bailey, A., & Goldman, A. S. H. (2010). Evidence that MEK1 positively promotes interhomologue double-strand break repair. *Nucleic Acids Research, 38*(13), 4349–4360. https://doi.org/10.1093/nar/gkq137.

Vannier, J.-B., Sarek, G., & Boulton, S. J. (2014). RTEL1: Functions of a disease-associated helicase. *Trends in Cell Biology, 24*(7), 416–425. https://doi.org/10.1016/j.tcb.2014.01.004.

Veaute, X., Jeusset, J., Soustelle, C., Kowalczykowski, S. C., Le Cam, E., & Fabre, F. (2003). The Srs2 helicase prevents recombination by disrupting Rad51 nucleoprotein filaments. *Nature*, *423*(6937), 309–312. https://doi.org/10.1038/nature01585.

Vernekar, D. V., Reginato, G., Adam, C., Ranjha, L., Dingli, F., Marsolier, M.-C., et al. (2021). The Pif1 helicase is actively inhibited during meiotic recombination which restrains gene conversion tract length. *Nucleic Acids Research*, *49*(8), 4522–4533. https://doi.org/10.1093/nar/gkab232.

Wang, T. F., Kleckner, N., & Hunter, N. (1999). Functional specificity of MutL homologs in yeast: Evidence for three Mlh1-based heterocomplexes with distinct roles during meiosis in recombination and mismatch correction. *Proceedings of the National Academy of Sciences of the United States of America*, *96*(24), 13914–13919. https://doi.org/10.1073/pnas.96.24.13914.

Wang, J., Zhang, W., Jiang, H., Wu, B.-L., & Collaboration, P. O. I. (2014). Mutations in HFM1 in recessive primary ovarian insufficiency. *The New England Journal of Medicine*, *370*(10), 972–974. https://doi.org/10.1056/NEJMc1310150.

Wesoly, J., Agarwal, S., Sigurdsson, S., Bussen, W., Van Komen, S., Qin, J., et al. (2006). Differential contributions of mammalian Rad54 paralogs to recombination, DNA damage repair, and meiosis. *Molecular and Cellular Biology*, *26*(3), 976–989. https://doi.org/10.1128/MCB.26.3.976-989.2006.

Whitby, M. C. (2010). The FANCM family of DNA helicases/translocases. *DNA Repair*, *9*(3), 224–236. https://doi.org/10.1016/j.dnarep.2009.12.012.

Wild, P., Susperregui, A., Piazza, I., Dörig, C., Oke, A., Arter, M., et al. (2019). Network rewiring of homologous recombination enzymes during mitotic proliferation and meiosis. *Molecular Cell*, *75*(4), 859–874. https://doi.org/10.1016/j.molcel.2019.06.022.

Wilson, M. A., Kwon, Y., Xu, Y., Chung, W.-H., Chi, P., Niu, H., et al. (2013). Pif1 helicase and Polδ promote recombination-coupled DNA synthesis via bubble migration. *Nature*, *502*(7471), 393–396. https://doi.org/10.1038/nature12585.

Xue, C., Wang, W., Crickard, J. B., Moevus, C. J., Kwon, Y., Sung, P., et al. (2019). Regulatory control of Sgs1 and Dna2 during eukaryotic DNA end resection. *Proceedings of the National Academy of Sciences of the United States of America*, *116*(13), 6091–6100. https://doi.org/10.1073/pnas.1819276116.

Youds, J. L., Mets, D. G., McIlwraith, M. J., Martin, J. S., Ward, J. D., & ONeil, N. J., Rose, A. M., West, S. C., Meyer, B. J., & Boulton, S. J. (2010). RTEL-1 enforces meiotic crossover interference and homeostasis. *Science*, *327*(5970), 1254–1258. https://doi.org/10.1126/science.1183112.

Zakharyevich, K., Ma, Y., Tang, S., Hwang, P. Y.-H., Boiteux, S., & Hunter, N. (2010). Temporally and biochemically distinct activities of Exo1 during meiosis: Double-strand break resection and resolution of double Holliday junctions. *Molecular Cell*, *40*(6), 1001–1015. https://doi.org/10.1016/j.molcel.2010.11.032.

Zheng, L., Meng, Y., Campbell, J. L., & Shen, B. (2020). Multiple roles of DNA2 nuclease/helicase in DNA metabolism, genome stability and human diseases. *Nucleic Acids Research*, *48*(1), 16–35. https://doi.org/10.1093/nar/gkz1101.

Zhu, Z., Chung, W.-H., Shim, E. Y., Lee, S. E., & Ira, G. (2008). Sgs1 helicase and two nucleases Dna2 and Exo1 resect DNA double-strand break ends. *Cell*, *134*(6), 981–994. https://doi.org/10.1016/j.cell.2008.08.037.

CHAPTER NINE

# Meiotic crossover interference: Methods of analysis and mechanisms of action

**Yu-Chien Chuang and Gerald R. Smith***
Fred Hutchinson Cancer Center, Seattle, WA, United States
*Corresponding author: e-mail address: gsmith@fhcrc.org

## Contents



## Abstract

Segregation of chromosomes during meiosis, to form haploid gametes from diploid precursor cells, requires in most species formation of crossovers physically connecting homologous chromosomes. Along with sister chromatid cohesion, crossovers allow tension to be generated when chromosomes begin to segregate; tension signals that chromosome movement is proceeding properly. But crossovers too close to each other might result in less sister chromatid cohesion and tension and thus failed meiosis. Interference describes the non-random distribution of crossovers, which occur farther apart than expected from independence. We discuss both genetic and cytological methods of assaying crossover interference and models for interference, whose molecular mechanism remains to be elucidated. We note marked differences among species.

*Current Topics in Developmental Biology*, Volume 151
ISSN 0070-2153
https://doi.org/10.1016/bs.ctdb.2022.04.006

## 1. Introduction

Eukaryotes are typically generated by fusion of two special cells, one coming from each parent. Examples of these cells, called gametes, are eggs and sperm in mammals, eggs and pollen in flowering plants, and spores (or cells resulting from spore germination and growth) in fungi such as yeasts and molds. Gametes contain one copy of each chromosome and are thus haploids. Fusion of two gametes regenerates a diploid organism. The generation of haploid gametes from diploid precursors, called meiosis, requires the accurate segregation of the chromosomes, which in turn requires the pairing of the two copies of each chromosome (called homologs) and their segregation into separate cells. This process occurs with two sequential cell divisions but only one replication of the chromosomes, so that during meiosis diploids give rise to haploids (Fig. 1).

Accurate chromosome segregation at the first meiotic division (MI) requires that the two homologs, one from each of the previous parents, find each other and move to opposite sides (poles) of the cell (Fig. 1). First, each chromosome is replicated, forming sister chromatids. Then, the replicated chromosomes pair, forming a bivalent (four copies in all). In many species this pairing occurs by homologous recombination between the two copies; there is extensive DNA sequence homology only between homologs, ensuring that each chromosome pairs with its proper partner. Recombination by reciprocal exchange generates a crossover (a cytological chiasma), which connects the two chromosome pairs. When the centromeres of one pair are pulled away from the centromeres of the other pair, tension is generated if there is at least one crossover connecting them and if the sister chromatids are held together by the cohesin complex. Tension appears to be the signal that homologs are properly moving to opposite poles (Nicklas, 1997). Thus, crossovers and cohesion are critical for proper segregation and gamete formation.

Crossovers can occur at most places along a chromosome, but some regions have crossovers much more frequently than others (reviewed in Nambiar, Chuang, & Smith, 2019). For example, crossovers are rare in or near centromeres, presumably because a crossover there interrupts proper attachment of the fibers (the spindle) to the centromere (*via* the kinetochore) or the proper pulling of the centromeres to opposite poles (*e.g.*, Nambiar & Smith, 2018). In addition, two crossovers too close together may leave too little cohesin between the crossovers to firmly hold the bivalent together to generate tension upon centromere movement.

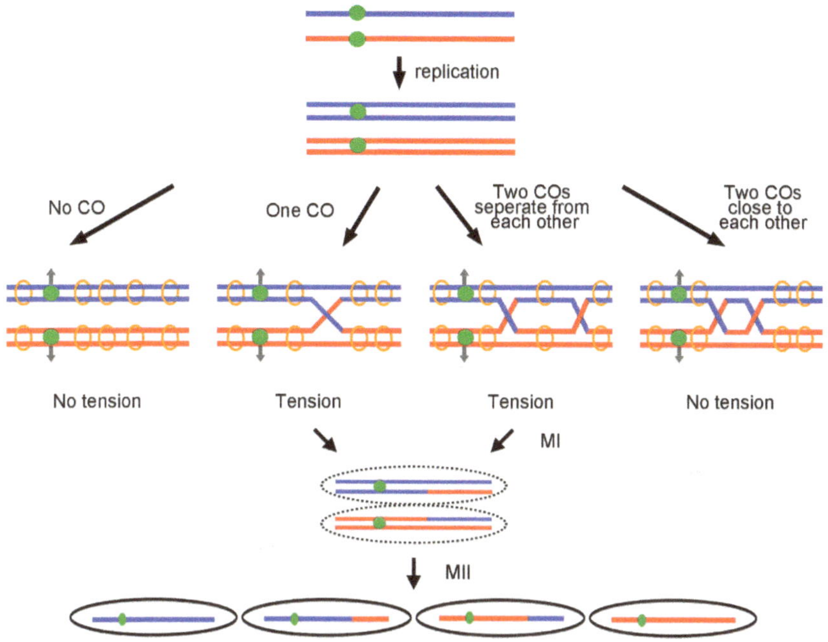


**Fig. 1** Crossovers are needed for proper chromosome segregation in meiosis. Red and blue lines indicate replicated homologs from each parent. Each line indicates a double-stranded DNA of one chromatid. Green dots represent centromeres. After replication, sister chromatids are held together by cohesins (orange circles). Gray arrows indicate the direction of homologous centromere separation at the first meiotic division (MI). (Left panel) No tension is generated between homologs when no CO is formed. (Middle two panels) Tension is formed between homologs when one CO or two well-separated COs arise. (Right panel) Two COs too close to each other may have too little sister chromatid cohesion between the crossovers to produce tension between homologs to aid segregation. The tension ensures homologous centromeres segregate properly in MI. Crossovers also generate recombinant haploid gametes after the second meiotic division (MII).

This last feature—the placement of crossovers not too close together—is called crossover interference. This chapter discusses the occurrence of interference, ways it can be measured, and models for its action. Interference was first observed in the fruit fly *Drosophila melanogaster* (Muller, 1916; Sturtevant, 1915) and has been found in nearly all species examined (Anderson, Hooker, & Stack, 2001; Barnes, Kohara, Coulson, & Hekimi, 1995; Baudat & de Massy, 2007; Copenhaver, Housworth, & Stahl, 2002; de Boer, Stam, Dietrich, Pastink, & Heyting, 2006; Fowler, Hyppa, Cromie, & Smith, 2018; Meneely, Farago, & Kauffman, 2002; Mortimer & Fogel, 1974; Perkins, 1962). The molecular mechanism of interference is still a matter of speculation, although many gene products required for interference have been identified. We first describe methods for assaying interference and discuss complications in interpreting the data, especially given differences among species. We then present proteins required for interference and finally discuss various models for interference.

## 2. Methods for measuring crossover interference

### 2.1 Genetic analysis

Two genetic methods to measure crossover interference use either random gametes (spores or next generation progeny) or complete meiotic tetrads. The former has been used with essentially all species investigated, and tetrad analysis is common with fungi. Recently, tetrad analysis has been done in other species, including *Arabidopsis thaliana* (see below) (Francis, Lam, & Copenhaver, 2006), *Zea maize* (Li, Li, & Yan, 2015), and *Mus musculus* (Cole et al., 2014). It is generally easier to get more data from random gametes than from tetrads, but in tetrads chromatid relations can be assessed and crossovers can be distinguished from gene conversions (transfer of a genetic marker from one homolog to the other with or without formation of an associated crossover). Tetrads with fewer than four viable spores are often excluded from the analyses; this bias may overlook important events associated with crossovers that, when disrupted, lead to spore death.

Using random gametes, the coefficient of coincidence (CoC) compares the observed frequency of double crossovers with that expected from independent crossovers. First used in *D. melanogaster* (Muller, 1916; Sturtevant, 1915) and subsequently in many species, it assumes recombinant chromosomes reflect crossovers and not gene conversions and it assumes no chromatid interference (*i.e.*, that 2-, 3-, and 4-strand double crossovers occur 1:2:1; Fig. 2A). CoC using three linked markers is given by the formula

$$\mathrm{CoC}\ (S_3) = \frac{\text{observed double crossover frequency}}{\text{expected double crossover frequency}} = \frac{R_{AB\ \mathrm{and}\ BC}}{R_{AB} \times R_{BC}}$$

where R values are the frequencies of single ($R_{AB}$ and $R_{BC}$) and double ($R_{AB\ \mathrm{and}\ BC}$) crossovers from a cross of this form:

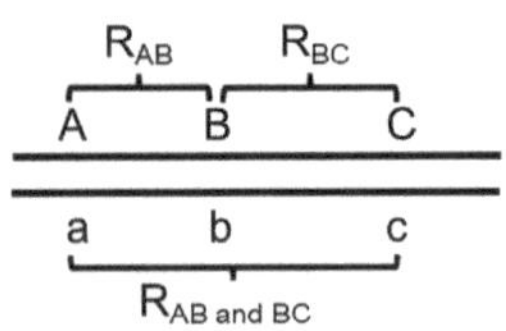

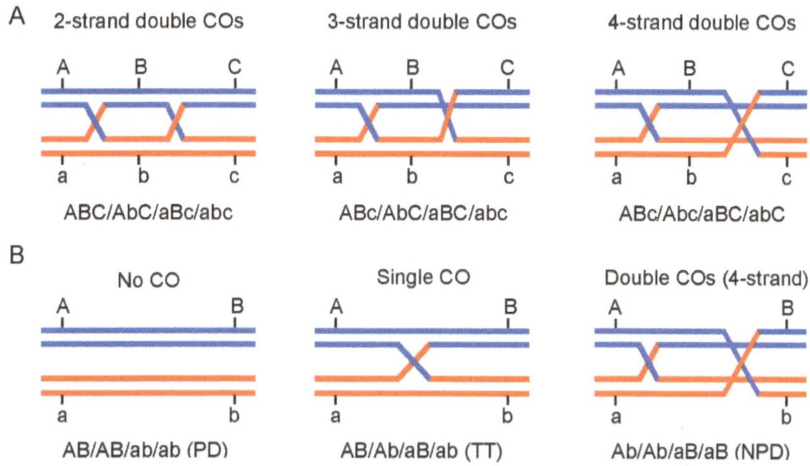


**Fig. 2** Characterizing crossovers in meiotic tetrads with genetic markers. (A) Three-marker tetrad analysis can identify two-, three-, or four-strand (chromatid) double crossovers. (B) Three types of tetrads in two-marker tetrad analysis. PD, parental di-type; TT, tetra-type; NPD, non-parental di-type. The genotypes of the four gametes are indicated below each panel.

A variation uses a fourth marker to allow comparisons with the two intervals assayed for crossovers (A–B and C–D) to be kept nearly constant in size but moved farther and farther apart.

$$\text{CoC }(S_4) = \frac{\text{observed double crossover frequency}}{\text{expected double crossover frequency}} = \frac{R_{AB\text{ and }CD}}{R_{AB} \times R_{CD}}$$

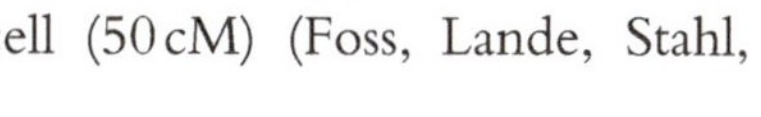


In both cases, interference (I) = 1 − CoC. In *D. melanogaster* and *Neurospora crassa* $S_4$ is lowest at short genetic distances and increases to near unity at ~40–50 centiMorgans (cM); in other words, in these species interference is strong at short distances and disappears at about the distance of one crossover per meiotic cell (50 cM) (Foss, Lande, Stahl, & Steinberg, 1993) (see Fig. 3).

Fungi and some non-vascular plants allow convenient tetrad analysis, as does the flowering plant *A. thaliana* mutant *qrt1*, which retains the four pollen from one meiosis in a single cell (Francis et al., 2006; Rhee & Somerville, 1998). CoC analysis, as above, could be used with tetrad data, but a simpler alternative uses only two markers denoting the three tetrad classes (Fig. 2B). The observed frequency of non-parental ditype (NPD) tetrads reflects 4-strand double crossovers (Fig. 2A). Assuming no chromatid interference, as is the case where tested (Emerson & Beadle, 1933; Miller et al., 2016; Mortimer & Fogel, 1974; Munz, 1994; Stahl & Housworth, 2009; Zhao, McPeek, & Speed, 1995), and no crossover interference, the expected frequency (*f*) of NPD, tetratypes (T; single crossover) and parental ditypes (PD; no crossover) can be calculated from the Haldane relation of genetic distance (in Morgans) and the observed recombinant frequency R.

$$\text{Recombinant frequency (R)} = f\text{NPD}_{\text{observed}} + \frac{1}{2} f\text{T}_{\text{observed}}$$

$$\text{genetic distance (X)} = -\frac{1}{2}\ln(1 - 2\text{R})$$

The NPD ratio $= \text{NPD}_{\text{observed}}/\text{NPD}_{\text{expected}}$, and $\text{I} = 1 - \text{NPD ratio}$.

Papazian (1952) used Haldane's formula to calculate the expected NPD frequency from the observed T frequency.

$$f\text{NPD}_{\text{expected}} = \frac{1}{2}\left[1 - f\text{T}_{\text{observed}} - \left(1 - \frac{3}{2} f\text{T}_{\text{observed}}\right)^{2/3}\right]$$

Stahl (2008) extended this analysis to include the expected frequencies of each of the three classes based on the observed recombinant frequency and Haldane's equation.

$$f\text{T}_{\text{expected}} = \frac{2}{3}\left(1 - e^{-3\text{X}}\right)$$

$$f\text{NPD}_{\text{expected}} = \frac{1}{2}\left[\left(1 - e^{-2\text{X}}\right) - f\text{T}_{\text{expected}}\right]$$

$$f\text{P}_{\text{expected}} = 1 - f\text{NPD}_{\text{expected}} - f\text{T}_{\text{expected}}$$

Papazian's analysis is not valid for T close to 2/3, but Stahl's "better way" is valid for all values. In addition, Stahl's analysis, based on all three tetrad classes, gives more reliable statistical analysis of the data than Papazian's (Stahl, 2008).

Two studies in yeasts (Munz, 1994; Shinohara et al., 2003) found similar degrees of interference using both random spore and tetrad (NPD; Papazian method) analyses, as expected.

## 2.2 Cytological analysis

A physical (cytological) method determines the position of fluorescent markers, seen as foci in a microscope, along a chromosome. Typically, the chromosomal axis is labeled uniformly, to allow distance measurements along the chromosome, and a chromosomal end is labeled uniquely, to identify an individual chromosome. A chromosome is divided into many equal-length intervals, and the frequencies of foci in each interval and the frequencies of two foci in each pair of intervals are determined. Interference is seen as foci being farther apart than expected from random positioning. Fung, Rockmill, Odell, and Roeder (2004) found that synapsis initiation complex proteins (*e.g.*, *Saccharomyces cerevisiae* Zip2 and Zip3) showed interference. Similar markers used are those thought to indicate "crossover designation," for example *Caenorhabditis elegans* COSA-1, mouse MLH1, and *A thaliana* MLH1 (Anderson, Reeves, Webb, & Ashley, 1999; Chelysheva et al., 2010; Yokoo et al., 2012). *S. cerevisiae* Zip3 promotes CO formation, is required for CO interference (see Table 2), and has been proposed to bind to DSB sites after they are "designated" by an unknown factor for CO formation (Agarwal & Roeder, 2000; Serrentino, Chaplais, Sommermeyer, & Borde, 2013). *C. elegans* COSA-1 and *A. thaliana* MLH1 are located between bivalents at or near a chiasma (Chelysheva et al., 2010; Yokoo et al., 2012). COSA-1 is also needed for CO formation (Yokoo et al., 2012). The chromosomal distribution of mouse and *A. thaliana* MLH1 foci reflect the number and distribution of crossovers (Anderson et al., 1999; Chelysheva et al., 2010). These observations support these foci being a valid proxy for crossovers.

Interference is measured as in CoC analysis: $I = 1 - CoC$, where

$$CoC = \frac{\text{observed double focus frequency}}{\text{expected double focus frequency}} = \frac{F_{AB}}{F_A \times F_B.}$$

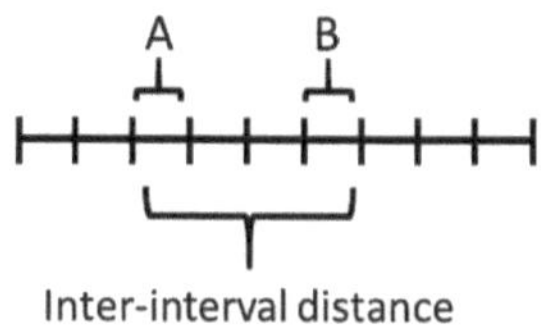


This cytological method does not require genetic markers and can be used in mutants that do not complete meiosis. But it requires spreading nuclei and high-resolution microscopy, which may be difficult for large numbers of cells. In addition, distance is measured as physical distance along the chromosomal axis, which may not be proportional to genetic map or DNA length. In *S. cerevisiae*, however, cytological interference extends over ~0.45 μm of axis, which corresponds to ~44 cM (Zhang, Liang, Hutchinson, & Kleckner, 2014; Zhang, Wang, et al., 2014), indicating equivalence of cytological and genetic interference. In *A. thaliana* and humans, the data are ~25 μm (~61 cM) and ~5 μm (~50–60 cM), respectively (Capilla-Pérez et al., 2021; Gruhn, Rubio, Broman, Hunt, & Hassold, 2013; White, Wang, Zhang, & Kleckner, 2017). [See Table 1 for relations between DNA length (Mb), genetic distance (cM), and axis length (microns) in various species.] It should be noted that several *S. cerevisiae* mutants (*e.g.*, *zip1Δ*, *sgs1Δ*, and *msh4Δ*) retain cytological interference but not genetic interference (Fung et al., 2004; Novak et al., 2001; Oh et al., 2007; Sym & Roeder, 1994). Simple explanations are that (a) *S. cerevisiae* strains from different genetic backgrounds were used to test the two methods and (b) these proteins may be required at a stage of crossover formation after crossover "designation" (*e.g.*, Zip3 binding) (De Muyt et al., 2012; Zakharyevich, Tang, Ma, & Hunter, 2012; Zhang, Wang, et al., 2014). Nevertheless, these mutants indicate some difference in cytological vs genetic interference.

## 3. Complications in interpreting interference data

While it is often said that this or that species has crossover interference, the reported strengths clearly differ among species and often across a given genome (Fig. 3). Some, such as *C. elegans*, appear to have absolute interference – two crossovers on one chromosome do not appear (Barnes et al., 1995; Meneely et al., 2002). Extensive data show strong interference for short intervals (<10 cM) in *D. melanogaster* (~0.95) and in *N. crassa* (~0.8), gradually decreasing for longer intervals, and becoming negligible at ~40–50 cM (Foss et al., 1993) (Fig. 3). *S. cerevisiae* interference ranges from near 0 up to ~0.8, but among data analyzed in Fig. 3 there is a less distinct trend with distance: intervals of similar genetic size can have different degrees of interference. According to the only report of which we are aware (Strickland, 1958), *Aspergillus nidulans* has no detectable crossover interference. *Schizosaccharomyces pombe* has statistically highly significant but weak crossover interference ($I = 0.26 \pm 0.051$) in the one pair of adjacent intervals assayed extensively (Fowler et al., 2018); other intervals may have no interference, but the data for most individual intervals reported are limited (Munz, 1994).

The ratio of genetic distance (cM) to DNA length (Mb) differs by a factor of >500 among species (Table 1), indicating that a similar level of interference in two species may reflect vastly different DNA distances. In addition, each micron of cytological axis length corresponds to different lengths of DNA. A simple interpretation is that DNA loops of increasing

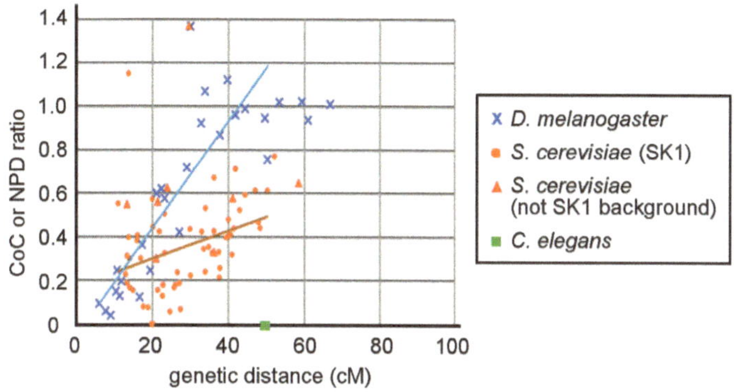


**Fig. 3** Correlation between crossover interference and genetic distance in *D. melanogaster*, *S. cerevisiae* and *C. elegans*. *D. melanogaster* CoC ($S_4$) data (Foss et al., 1993; Morgan, Bridges, & Schultz, 1935), *S. cerevisiae* NPD ratio data (Argueso et al., 2003; Argueso, Wanat, Gemici, & Alani, 2004; Chua and Roeder, 1997; He et al., 2020; Novak, Ross-Macdonald, & Roeder, 2001; Oh et al., 2007; Shinohara, Sakai, Shinohara, & Bishop, 2003; Stahl et al., 2004; Sym & Roeder, 1994; Tsubouchi, Zhao, & Roeder, 2006; Zanders & Alani, 2009) and the genetic distance of each interval are plotted. *C. elegans* makes only one crossover per chromosome (Barnes et al., 1995; Meneely et al., 2002), indicating it has complete interference up to 50cM, the genetic length of each chromosome. Most *S. cerevisiae* data were analyzed in the SK1 background (red circles), and eight data points were analyzed in other strain backgrounds (red triangles). *N. crassa* data are similar to those of *D. melanogaster* (Foss et al., 1993). *D. melanogaster* data with <5 double recombinants and *S. cerevisiae* data with <5 expected and <5 observed NPD tetrads were not included. *S. cerevisiae* data were re-analyzed using the "better way" of Stahl (2008). The blue and orange lines, respectively, show the linear regression analysis of data from *D. melanogaster* ($y = 0.025x - 0.053$; $r^2 = 0.72$) and *S. cerevisiae* ($y = 0.0064x + 0.17$; $r^2 = 0.17$); for this analysis, data with >50cM and *S. cerevisiae* data with NPD ratio >1 were omitted.

size extend from the chromosomal axis as the parameter (micron of axis length/Mb) decreases. Consequently, different genetic distances correspond to different axis lengths among species. In multicellular species the parameter (cM/μm of axis) ranges from about 2 to 10; in *S. cerevisiae*, the value is about 100 (Table 1).

Interpreting these data is complicated by additional considerations (see also Berchowitz & Copenhaver, 2010). Gene conversions, whose formation is closely related to that of crossovers, are often stated not to interfere with each other. Mortimer and Fogel (1974) and Foss and Stahl (1995) report CoC = 1.0 (95% confidence interval, or CI, = 0.75–1.45 or wider) for gene conversions at the closely linked *ARG4* and *THR1* genes of *S. cerevisiae*. Because the number of double gene conversions is small

**Table 1** Ratio of genetic distance to physical distance varies >500-fold among species.

| Species | Genetic map length (cM) | Physical map length (Mb) | Axis length (μm)[a] | Genetic length/ physical length (cM/Mb) | Axis length/ physical length (μm/Mb) | Genetic length/axis length (cM/μm) |
|---|---|---|---|---|---|---|
| *S. cerevisiae* | 4206[b] | 12[b] | 43[c] | 350 | 3.6 | 98 |
| *S. pombe* | 2100[d] | 13.8[e] | – | 150 | – | – |
| *N. crassa* | 1075[f] | 43[g] | – | 25 | – | – |
| *A. thaliana* | 597[h] | 135[i] | 240[j] | 4.4 | 1.8 | 2.5 |
| *D. melanogaster* | 287[k] | 180[l] | – | 1.6 | – | – |
| *D. melanogaster* Chr X | 65.3[k] | 24[l] | 15[m] | 2.8 | 0.63 | 4.4 |
| *C. elegans* | 300 | 103[n] | 29[o] | 2.9 | 0.28 | 10 |
| Human | 3630[p] | 3019[p] | – | 1.2 | – | – |
| Human autosomes | 3451[p] | 2740[p] | 260–320[q] | 1.3 | 0.09–0.12 | 11–13 |
| *M. musculus* | 1373[p] | 2577[p] | – | 0.53 | – | – |
| *M. musculus* autosomes | 1316[p] | 2320[p] | 150–162[r] | 0.56 | 0.06–0.07 | 8.1–8.8 |

[a]Axis length was measured by light microscopy of formaldehyde-fixed samples.
[b]Cherry et al. (1997); Mortimer, Contopoulou, and King (1992).
[c]Song et al. (2021).
[d]Egel (2004).
[e]Wood, Gwilliam, Rajandream, et al. (2002).
[f]Perkins and Barry (1977).
[g]Galagan et al. (2003).
[h]Lister and Dean (1993),
[i]The Arabidopsis Genome Initiative (2000).
[j]Morgan et al. (2021).
[k]Comeron, Ratnappan, and Bailin (2012).
[l]Adams, Celniker, Holt, et al. (2000).
[m]Page and Hawley (2001).
[n]C. elegans Sequencing Consortium (1998).
[o]Lascarez, Lagunas and Colaiácovo, unpublished data.
[p]Jensen-Seaman et al. (2004).
[q]Codina-Pascual et al. (2006).
[r]Vranis, Van der Heijden, Malki, and Bortvin (2010).

(13 and 28, respectively), the data do not rule out weak interference, either positive or negative. Sequencing 980 chromatid arms from progeny of *D. melanogaster* hybrids, Miller et al. (2016) found 291 chromatid arms with one or more gene conversions and 33 with two or more gene

conversions. 86 doubles are expected from independence; thus, CoC = 0.38 (CI = 0.28–0.49), indicating highly significant positive interference of gene conversions. Using a different method of analysis, the authors, however, concluded that gene conversions do not show interference. Miller et al. (2016) found 541 crossovers and 52 double crossovers; CoC = 0.17 (CI = 0.13–0.22), indicating strong crossover interference, as concluded by Miller et al. (2016) and others cited above for *D. melanogaster*. Similarly, for crossovers and non-crossovers (128 such doubles observed), CoC = 0.80 (CI = 0.66–0.94), indicating weak but significant interference of these events. Some but not other classes of gene conversion (6:2, 5:3, or aberrant 4:4) with an associated crossover also interfere with a nearby crossover in another interval in some species but not in others (Getz et al., 2008; Kitani, 1978; Mancera, Bourgon, Brozzi, Huber, & Steinmetz, 2008; Stahl & Foss, 2008). These results may reflect different relative activities of two pathways of recombination, one with and one without interference, as discussed in the next paragraph. Although the original definition of interference referred to genetic crossovers, more recently emphasis has been put on spatial interference of cytological markers along chromosomes, as described above. These foci are inferred to mark "designated" crossover sites, which may be closely related to, but not guaranteed to become, crossovers, as noted above. Thus, it is important to specify what is measured to indicate interference.

An added complication is the possibility of two or more types of meiotic recombination occurring in the same species. In *S. cerevisiae*, *A. thaliana*, and mice two pathways of crossover formation have been shown—one with and one without interference (*e.g.*, Argueso et al., 2004; Copenhaver et al., 2002; de los Santos et al., 2003; Higgins, Armstrong, Franklin, & Jones, 2004; Ross-Macdonald & Roeder, 1994; Stahl et al., 2004). Direct support for this view comes from studies of *S. cerevisiae zip1* and *msh5* mutants, in which the residual crossovers (20–40% of wt frequency) show little if any interference (de los Santos et al., 2003; Sym & Roeder, 1994). In an *mms4* mutant, defective in another postulated pathway, interference is not significantly different from that in wt (de los Santos et al., 2003). Some authors have proposed that the non-interfering pathway acts early, to establish chromosome pairing, and the interfering pathway acts late, to facilitate chromosome disjunction (de los Santos et al., 2003; de los Santos, Loidl, Larkin, & Hollingsworth, 2001; Getz et al., 2008). In some species, crossing over aids both events, but the degree to which the pathways differ is unclear. They could diverge at the beginning of the recombination pathway (before DSB formation) or at the end (resolution of Holliday junctions). For example, nicked Holliday junctions could arise in the "early" pathway and be resolved by Mus81-Mms4, while non-nicked (ligated) Holliday junctions could arise in the "late" pathway and be resolved by Msh4-Msh5 and Mlh1 and its partners (*e.g.*, Berchowitz & Copenhaver, 2010; Getz et al., 2008). These pathways could act differentially in different genomic intervals, further complicating the interpretation of interference strength in a given species.

Important to keep in mind is that species differ. For example, Mus81-Eme1 is essential for Holliday junction resolution and >95% of crossover formation in *S. pombe* (Boddy et al., 2001; Cromie et al., 2006; Smith, Boddy, Shanahan, & Russell, 2003), but its homolog Mus81-Mms4 is required for only ~20% of crossovers in *S. cerevisiae* (Argueso et al., 2004; de los Santos et al., 2003), although it is required for abundant viable spore-formation in both species. Some proteins essential for recombination in some species, such as Dmc1 DNA strand-exchange protein, are not apparent in other species. [Among tested species, those with Dmc1 require recombination for chromosome synapsis, but those without it do not (Stahl et al., 2004).] We know of no species in different genera in which the same set of orthologous proteins is required for meiotic recombination, emphasizing that species differ. Thus, it is risky to use data from one species in interpreting data from another species. Rather, one should state the species under discussion and not dismiss a view proposed for one species because of counter data in another species. Observations and models of interference in any species are nevertheless important for thinking about the mechanism of interference in other species.

## 4. Proteins required for crossover interference

Mutants with altered crossover interference have been identified in a variety of species. Table 2 groups the corresponding proteins according to their known functions in meiotic recombination and DNA metabolism. Homologs of some of these proteins are found in widely divergent species, suggesting some underlying similarity in interference. The molecular mechanisms by which these proteins impart interference remain largely unknown.

**Table 2** Proteins with diverse functions are required for crossover interference.

| Protein | Species | Function | Interference assay | Reference |
|---|---|---|---|---|
| **Synaptonemal complex (SC)** | | | | |
| Zip1 | *S. cerevisiae* | Transverse filament of the SC | NPD ratio | Sym and Roeder (1994) |
| SYP-1 | *C. elegans* | | Cytology | Libuda, Uzawa, Meyer, and Villeneuve (2013) |
| Zyp1 | *A. thaliana* | | Cytology | Capilla-Pérez, Durand, Hurel, et al. (2021) |
| Asy1 | *A. thaliana* | | Cytology | Lambing, Kuo, Tock, Topp, and Henderson (2020) |
| Zep1 | *O. sativa* | | Cytology | Wang et al. (2010) |
| Sycp1 | *M. musculus* | | Cytology | de Boer et al. (2006) |
| DPY-28 | *C. elegans* | Condensin I subunit; chromosomal axis formation | Cytology | Tsai et al. (2008) |
| **Regulating CO formation** | | | | |
| Msh4-Msh5 | *S. cerevisiae* | Binds Holliday junctions; facilitates CO formation | NPD ratio | Novak et al. (2001); Argueso et al. (2004); Getz et al. (2008) |
| | | | NPD ratio and CoC | Nishant, Chen, Shinohara, Shinohara, and Alani (2010) |
| | | | CoC | Anderson, Oke, Yam, Zhuge, and Fung (2015) |
| Zip3 | *S. cerevisiae* | SUMO E3 ligase; required for SC formation and CO formation | CoC | Anderson et al. (2015) |
| Mlh1-Mlh3 | *S. cerevisiae* | Endonuclease; binds Holliday junctions and may aid their resolution | NPD ratio | Argueso et al. (2003) |
| Slx4 | *S. cerevisiae* | Endonuclease; may aid resolution of Holliday junctions | NPD ratio | Higashide and Shinohara (2016) |

*Continued*

**Table 2** Proteins with diverse functions are required for crossover interference.—cont'd

| Protein | Species | Function | Interference assay | Reference |
|---|---|---|---|---|
| Sgs1 | *S. cerevisiae* | DNA helicase; chromosome synapsis; meiotic joint molecule and CO formation | NPD ratio | Oh et al. (2007) |
| | | | CoC | Anderson et al. (2015) |
| Blm | *D. melanogaster* | | CoC | Hatkevich et al. (2017) |
| Mer3 | *S. cerevisiae* | DNA helicase; promotes CO formation | NPD ratio | Nakagawa and Ogawa (1999) |
| Spo16 | *S. cerevisiae* | Facilitates SC and CO formation | NPD ratio | Shinohara, Oh, Hunter, and Shinohara (2008) |
| Spo22 (Zip4) | *S. cerevisiae* | TPR-like repeat protein; facilitates SC formation and CO formation | NPD ratio | Tsubouchi et al. (2006); Shinohara et al. (2003) |
| Pch2 | *S. cerevisiae* | Regulates chromosome synapsis and CO formation | NPD ratio | Joshi, Barot, Jamison, and Borner (2009); Zanders and Alani (2009) |
| Pch2 | *A. thaliana* | | CoC | Lambing et al. (2015) |
| RTEL-1 | *C. elegans* | Regulates CO formation | Cytology | Youds et al. (2010) |
| **DNA damage check point** | | | | |
| Tel1 | *S. cerevisiae* | Protein kinase regulating | CoC | Anderson et al. (2015) |
| Tel1 | *S. pombe* | DNA double-strand-break response | CoC | Fowler et al. (2018) |
| ATM | *M. musculus* | | Cytology | Barchi et al. (2008) |
| **Topoisomerase II and SUMOylation** | | | | |
| Top2 | *S. cerevisiae* | DNA topoisomerase; binds axial cores | Cytology | Zhang, Wang, et al. (2014) |
| Ndj1 | *S. cerevisiae* | Regulates meiotic SPB cohesion and telomere clustering | NPD ratio | Chua and Roeder (1997) |
| Slx5/8 | *S. cerevisiae* | SUMO-targeted ubiquitin ligase (STUbL) | Cytology | Zhang, Wang, et al. (2014) |
| Sir2 | *S. cerevisiae* | Activates STUbL activity of Slx5/8 | Cytology | Zhang, Wang, et al. (2014) |

## 5. Models of crossover interference

Numerous models for crossover interference have been proposed. Many have led to mathematical models that account for existing data, such as the relation of interference and genetic or cytological distance, but few have specified a molecular mechanism, including roles for proteins required for interference (Table 2). Some models can be considered genetic, in the sense of having properties related to genetic distance, or physical, in the sense of measuring kb of DNA or microns of chromosome axis in fully condensed chromosomes. In both cases, however, the features of the models often account for interference allowing exactly one crossover in the genetic distance (50 cM) required by the definition of cM and over which interference acts, at least in *D. melanogaster* and *N. crassa* (Fig. 3). (50 cM, or 0.5 Morgan, is the genetic separation generated by one crossover in a bivalent, in which half the chromosomes are recombinant in the defined interval; Fig. 2B.) Here, we discuss models in several classes. Words or phrases in quotation marks below are from the references cited.

### 5.1 Trigger model

Fox (1973) postulated that a "chiasma determination mechanism passes along the chromosome at a constant rate" and is "triggered" to form a chiasma. It continues to move along the chromosome and "requires a fixed time to become recharged" to form another chiasma. The trigger was proposed to be "some feature of the secondary structure of the chromosome," which is rare in chromosomal regions with few chiasmata, such as the heterochromatic centromeres. Although further descriptions and predictions of the mechanism were not provided, this may be the first model postulating an entity moving along the chromosome to promote a limited number of crossovers in a given region.

### 5.2 Polymerization models

Egel (1978) proposed that "nodes" (required for the "establishment of an exchange possibility") are formed only before chromosome synapsis and "serve as initiation centers at which synapsis is started." One node, presumably activated at random, starts polymerization of the synaptonemal complex (SC). As the SC grows by polymerization, it inactivates additional nodes, thereby preventing further crossovers in the region over which there was SC polymerization.

King and Mortimer (1990) also proposed a model involving polymerization, but one in which the polymer (separate from the SC) both promotes and inhibits rather than simply inhibits crossover formation. In their model, "early structures" bind randomly along the already established SC and are held within it, thereby preventing interaction with another chromosome. Each structure initiates bidirectional polymerization and "give[s] rise to the late recombination nodules," which initiate crossover formation. The growing polymer then prevents other early structures from binding or ejects those already bound but not yet polymerizing. The ejected structures can act on other SC regions without a structure-induced polymer. This feature assures that all chromosomal regions have the opportunity for crossovers, provided there are enough structures present. The nature of the early structures, the ensuing polymer, and nodules was not specified.

## 5.3 Reaction-diffusion models

In this class of models, a factor, such as a protein, diffuses in one dimension along a chromosome or bivalent for a limited distance and affects crossover formation, either positively or negatively. This factor may be restricted by the SC from moving to another bivalent. Its effect is thus limited to the chromosome or bivalent it is on, as expected for imposition of interference along but not between chromosomes.

Holliday (1977) proposed that an activating factor, limited in amount, converts an unstable crossover into a stable chiasma. The factor is proposed to be a protein that binds DNA cooperatively and forms a complex with the crossover to stabilize it. It does not interact with a non-crossover structure. Diffusion of the protein along the chromosome to the crossover site depletes the surrounding region and prevents formation of a second stable chiasma in that region. Although not stated in the model, presumably the factor is initially randomly distributed along the chromosome. Organisms that lack interference, such as *A. nidulans* (Strickland, 1958), are proposed to have an unusually high level of the activating protein. To account for interference not extending across centromeres in some species, Holliday proposed that the centromere is a barrier to diffusion of the protein.

Fujitani, Mori, and Kobayashi (2002) proposed that "random walkers" are randomly distributed along chromosomes; each diffuses randomly in one dimension and with a set rate "becomes immobilized and matures into a crossover point." Collision of two random walkers inactivates both; collision of a moving walker with an immobilized walker inactivates the moving one. This model, like those above, generates exactly one crossover in an interval of 50 cM. Mathematical models built on two parameters (the initial density of walkers and the rate at which a walker becomes immobilized) account for data in *D. melanogaster* and *N. crassa*. As in other models above, the nature of the walkers and the means by which they mature into crossovers were not stated.

Rog, Kohler, and Dernburg (2017) found that proteins of the *C. elegans* SC behave as though they are in liquid crystals—they are sensitive to 1,6-hexanediol, which disrupts such structures, and quickly diffuse from the surrounding area into a particular region (as observed after localized irradiation of a photoconvertible derivative of the SC protein SYP-3). The latter property of liquid crystals is similar to that of lipid membranes, in which proteins can readily diffuse within, but not out of, the membrane. This property would allow a protein to diffuse along the chromosome and interact with sites on that chromosome but not with those on another chromosome, as demanded for crossover interference. Thus, Zhang, Kohler, Rillo-Bohn, and Dernburg (2018) propose that SC proteins, including ZHP-3, diffuse in one dimension and a combination of positive and negative effects allows only limited numbers of crossovers (one for *C. elegans*) to form along a chromosome. Additional proteins required for interference include the SC protein SYP-1, the condensin subunit DPY-28, and the protein kinase RTEL-1 (Table 2). Homologs of SYP-1 and RTEL-1 are also required for interference in other species, suggesting that the *C. elegans* mechanism may be widespread.

Morgan et al. (2021) showed that the HEI10 protein of *A. thaliana* behaves as predicted by a reaction-diffusion model for an activator of crossover formation. HEI10 is initially nearly uniformly distributed along each bivalent and then condenses into several small foci and eventually into one to three foci, approximating the number of genetic crossovers per bivalent in other experiments. How HEI10, a putative ubiquitin E3 ligase and Zip3 homolog, promotes crossovers is unclear.

## 5.4 Counting models

Foss et al. (1993), like previous authors, pointed out that interference in *D. melanogaster* and *N. crassa* is strong for short genetic intervals but diminishes to nearly zero for intervals ∼40–50 cM or greater (Fig. 3). Note that 50 cM corresponds to about 20 Mb in *D. melanogaster* but only 2–3 Mb in *N. crassa* (Table 1). Foss et al. (1993) therefore sought an interference mechanism that measures genetic, not physical, distance. In their model "recombination-initiation events" are randomly placed along chromosomes. By unstated mechanisms, one is matured into a crossover, the next $m$ events in one direction along the chromosome are matured into non-crossovers (gene conversions or sister chromatid interaction), and the next event is matured into a crossover. The authors developed mathematical equations with the variable $m$ and found that data for *D. melanogaster* and *N. crassa* were fit well with $m = 4$ and 2, respectively. Mortimer and Fogel (1974) proposed that "recombinogenic events are distributed along the chromosome in an independent fashion" but with alternating crossovers and non-crossovers; this model is equivalent to the counting model of Foss et al. (1993) with $m = 1$.

The counting model predicts that between two neighboring crossovers there should be $m$ non-crossovers. Short intervals between two crossovers should be enriched for non-crossovers relative to the same intervals on chromosomes without flanking crossovers. Foss and Stahl (1995) tested this prediction in *S. cerevisiae* and found it did not hold. Rather, there were *fewer* non-crossovers (gene conversions) in the short interval in tetrads with flanking crossovers compared to those without flanking crossovers, suggesting interference between crossovers and non-crossovers, as noted above.

## 5.5 Mechanical models

Interference might result from a purely mechanical, rather than biochemical, property of a chromosome. Muller (1916) proposed that a chromosome is too stiff to bend near an existing crossover and thus allow formation of another crossover nearby. Kleckner et al. (2004) compared a chromosome to a stiff beam coated with a brittle ceramic film. Bending the beam would put increasing stress on the film until it cracked and relieved the surrounding stress in the film. By analogy, they proposed that formation (or designation) of a crossover requires stress (the opposite of Muller's proposal) resulting from chromosome expansion and that stress in turn is relieved by crossover formation. Thus, a crossover would not form until stress built up, but once a crossover formed, there would be no stress nearby to allow formation of a second crossover. Stress would monotonically increase with distance from the first crossover, thus allowing, with increasing probability, another crossover some distance away. Support for this model comes from the partial reduction of interference, measured as inter-focus distance between Zip3 foci, in a *top2* mutant deficient in DNA topoisomerase II, which relaxes supercoiled DNA (Zhang, Wang, et al., 2014).

In an expansion of the beam-film model (Zhang, Liang, et al., 2014), certain "precursors" are "designate[d] to eventually mature as a crossover." Thus, the act of designation imposes interference. "The precursors for CO patterning are generally assumed to be the total array of double strand break (DSB)-initiated interactions between homologs," and "yeast DSBs are evenly spaced." DSBs become "designated" to become crossovers by an unstated mechanism but are reflected by the distribution of Zip3 foci (see Cytological analysis above). The probability (frequency) curve resembles that of CoC vs cM in *D. melanogaster* (Fig. 3). In *S. cerevisiae*, the distance at which adjacent Zip3 foci occur at the frequency expected for independence is roughly 100 kb or about 50 cM. Kleckner et al. (2004) emphasize, however, that crossover designation (interference) measures axis length, not kb of DNA or cM of genetic map length.

The beam-film model has been used to formulate mathematical models for interference (Zhang, Liang, et al., 2014). These models use multiple parameters reflecting such features as the average number of precursors per chromosome, their distribution along chromosomes (random or even or intermediate), and the "designation driving force." Data from various species fit the model when appropriate values for these parameters are chosen. With nine adjustable parameters, the model fits the data from various species (Zhang, Liang, et al., 2014).

Because few molecular (biochemical) features of the beam-film model are specified, it is hard to test this model genetically. The presumed even spacing, by an unstated mechanism, of precursors (DSBs) in yeast would seem to be interference itself, perhaps one of several factors leading to crossover interference. In addition, it seems counterintuitive that crossovers rather than DSBs would relieve stress in a chromosome. The model does, however, offer a different way of thinking about the problem of crossover interference.

## 5.6 Chromosome oscillatory movement model

Hulten (2011) viewed chromosomes as long flexible entities that move with wave-like motions. Chromosome ends (telomeres, attached to the nuclear membrane) and centromeres (attached to the kinetochore and spindle apparatus) move vigorously as the homologs pair. These movements create waves along each chromosome (univalent). When the "nodes" (crests) of two waves on homologs meet, they engage and form a crossover. Because the adjacent chromosome regions are not juxtaposed, no crossovers occur there. This model accounts for chromosomal translocations and other large rearrangements, such as deletions and inversions, blocking recombination and interference in the flanking regions.

## 5.7 Clustering models

Most of the models above do not specify the proteins involved in interference. The clustering model proposed by Fowler et al. (2018) specifies particular proteins involved in several steps. This model is based on their finding that *S. pombe* linear element proteins (LinE proteins; Rec25, Rec27, and Mug20), in conjunction with another linear element protein Rec10, bind to DSB hotspots with high specificity and are required for DSB formation at most hotspots (Fowler, Gutiérrez-Velasco, Martín-Castellanos, & Smith, 2013). Rec10 is more uniformly distributed and is required for all DSB formation. Loading of LinE proteins onto chromosomes is promoted by cohesins containing meiosis-specific subunits Rec8 and Rec11. These six proteins are required for full levels of recombination. A type of Hi-C analysis showed that Rec27-bound DSB hotspots form 3D clusters over ~200 kb (Fowler et al., 2018). A DSB at one hotspot interferes with DSB formation at surrounding hotspots; this DSB interference, like hotspot clustering, is strongest for nearby hotspots and decreases to an undetectable level at ~200 kb, corresponding to ~35 cM (Fowler et al., 2018) (Table 1). DSB interference also occurs in *S. cerevisiae*, extending ~20–50 kb, corresponding to ~7–20 cM (Garcia, Gray, Allison, Cooper, & Neale, 2015). Fowler et al. (2018) proposed that DSB interference gives rise to crossover interference, just as DSBs give rise to crossovers.

In the model of Fowler et al. (2018) a protein complex, such as condensin or cohesin or both, binds to a site on a chromosome and, while remaining bound to that site, moves unidirectionally along the chromosome and forms an ever-growing loop (Hyppa, Cho, Nambiar, & Smith, 2021). Upon encountering a potential DSB hotspot in the chromosome (or pair of sister chromatids or pair of homologs) being moved along, the protein complex loads a LinE complex onto the hotspot(s), holds onto that site, and continues moving. Upon encountering the next potential hotspot, it loads another LinE complex with a probability proportional to the strength of that hotspot. After it has loaded a set number of LinE complexes at hotspots, perhaps only two, it ceases loading. A DSB formed at random in the cluster of hotspot-LinE complexes activates the Tel1 protein kinase, which blocks further DSB formation in that cluster by phosphorylating a member of the DSB-forming complex (Rec12 and its half dozen essential partners). The DSB is repaired to a crossover. Because additional DSB formation in the surrounding area (the chromosomal region in the loop) is blocked, so is crossover formation. Formation of clusters encompassing only one homolog (paired chromatids) would allow independent DSBs on the other homolog, and interference would be limited to 0.5, but clusters encompassing both homologs would allow complete interference (I = 1). Formation of clusters over different regions in individual cells would make a gradient, from complete interference to none, in the population of cells, as typically used for assessing crossover interference genetically.

In mitotic *S. pombe* cells, condensin forms topologically associated domains (TADs) over chromosomal regions of ~300 kb; cohesin forms TADs over ~80 kb (Kim, Tanizawa, Iwasaki, & Noma, 2016). These distances are about the distance over which DSB interference is observed (~200 kb, as noted above). An *S. pombe* meiotic cell forms about 60 DSBs over its 13 Mb genome, or 1 DSB per ~200 kb (Fowler, Sasaki, Milman, Keeney, & Smith, 2014). Thus, these numbers are consistent with the model. In a *tel1Δ* mutant, both DSB interference and crossover interference become negative (Fowler et al., 2018), indicating that Tel1 is required for interference. Negative interference may arise from two or more DSBs being formed in a cluster when Tel1 is not present to inactivate the DSB-forming complex. The result is coordinated action of the DSB-forming complex within a cluster and coordinated crossover formation (negative interference). This model predicts interference between gene conversions, which to our knowledge has not been reported in *S. pombe*. As noted above, gene conversions do show weak but significant interference in *D. melanogaster* and perhaps in *S. cerevisiae*.

Stahl et al. (2004) proposed that "neighboring intermediates (-to-be) [also called "attempts"] are gathered into clusters of more or less fixed size" and that exactly one member of this cluster is converted into a crossover. These clusters were proposed to be foci of Zip2-Zip3 or "late nodules" observed by electron microscopy on the SC in *D. melanogaster* (Carpenter, 2003). The mode of clustering and the molecular mechanism limiting crossovers to one per cluster were not specified.

## 6. Conclusions and future research

Crossover interference has been well documented in many species and to require a variety of proteins. The molecular mechanism remains to be determined in any species. Given the differences in interference strength (Fig. 3) and protein requirements (Table 2) in various species, it seems likely that there are multiple mechanisms. Interference strength is clearly distance-dependent in *D. melanogaster* but less so in *S. cerevisiae* and *C. elegans* (Fig. 3), suggesting that the mechanisms of interference may be quite different. Determining the mechanisms is a challenge requiring a combination of methods, including genetic and biochemical as well as cytological analyses.

## Acknowledgments

We are grateful to Sue Amundsen, Randy Hyppa, and two anonymous reviewers for helpful comments on the manuscript and Monica Colaiácovo for unpublished data. Our research is supported by research grant R35 GM118120 from the National Institutes of Health of the United States of America.

## References

Adams, M. D., Celniker, S. E., Holt, R. A., et al. (2000). The genome sequence of *Drosophila melanogaster*. *Science*, *287*, 2185–2195.

Agarwal, S., & Roeder, G. S. (2000). Zip3 provides a link between recombination enzymes and synaptonemal complex proteins. *Cell*, *102*, 245–255.

Anderson, L. K., Hooker, K. D., & Stack, S. M. (2001). The distribution of early recombination nodules on zygotene bivalents from plants. *Genetics*, *159*, 1259–1269.

Anderson, C. M., Oke, A., Yam, P., Zhuge, T., & Fung, J. C. (2015). Reduced crossover interference and increased ZMM-independent recombination in the absence of Tel1/ATM. *PLoS Genetics*, *11*, e1005478.

Anderson, L. K., Reeves, A., Webb, L. M., & Ashley, T. (1999). Distribution of crossing over on mouse synaptonemal complexes using immunofluorescent localization of MLH1 protein. *Genetics*, *151*, 1569–1579.

Argueso, J. L., Kijas, A. W., Sarin, S., Heck, J., Waase, M., & Alani, E. (2003). Systematic mutagenesis of the *Saccharomyces cerevisiae* MLH1 gene reveals distinct roles for Mlh1p in meiotic crossing over and in vegetative and meiotic mismatch repair. *Molecular and Cellular Biology*, *23*, 873–886.

Argueso, J. L., Wanat, J., Gemici, Z., & Alani, E. (2004). Competing crossover pathways act during meiosis in *Saccharomyces cerevisiae*. *Genetics*, *168*, 1805–1816.

Barchi, M., Roig, I., Di Giacomo, M., de Rooij, D. G., Keeney, S., & Jasin, M. (2008). ATM promotes the obligate XY crossover and both crossover control and chromosome axis integrity on autosomes. *PLoS Genetics*, *4*, e1000076.

Barnes, T. M., Kohara, Y., Coulson, A., & Hekimi, S. (1995). Meiotic recombination, non-coding DNA and genomic organization in *Caenorhabditis elegans*. *Genetics*, *141*, 159–179.

Baudat, F., & de Massy, B. (2007). Regulating double-stranded DNA break repair towards crossover or non-crossover during mammalian meiosis. *Chromosome Research*, *15*, 565–577.

Berchowitz, L. E., & Copenhaver, G. P. (2010). Genetic interference: Don't stand so close to me. *Current Genomics*, *11*, 91–102.

Boddy, M. N., Gaillard, P.-H. L., McDonald, W. H., Shanahan, P., Yates, J. R., & Russell, P. (2001). Mus81-Eme1 are essential components of a Holliday junction resolvase. *Cell*, *107*, 537–548.

C. elegans Sequencing Consortium. (1998). Genome sequence of the nematode *C. elegans*: A platform for investigating biology. *Science, 282*, 2012–2018.

Capilla-Pérez, L., Durand, S., Hurel, A., et al. (2021). The synaptonemal complex imposes crossover interference and heterochiasmy in *Arabidopsis*. *Proceedings of the National Academy of Sciences of the United States of America, 118*, e2023613.

Carpenter, A. T. (2003). Normal synaptonemal complex and abnormal recombination nodules in two alleles of the *Drosophila* meiotic mutant *mei-W68*. *Genetics, 163*, 1337–1356.

Chelysheva, L., Grandont, L., Vrielynck, N., le Guin, S., Mercier, R., & Grelon, M. (2010). An easy protocol for studying chromatin and recombination protein dynamics during *Arabidopsis thaliana* meiosis: Immunodetection of cohesins, histones and MLH1. *Cytogenetic and Genome Research, 129*, 143–153.

Cherry, J. M., Ball, C., Weng, S., Juvik, G., Schmidt, R., Adler, C., et al. (1997). Genetic and physical maps of *Saccharomyces cerevisiae*. *Nature, 387*, 67–73.

Chua, P. R., & Roeder, G. S. (1997). Tam1, a telomere-associated meiotic protein, functions in chromosome synapsis and crossover interference. *Genes & Development, 11*, 1786–1800.

Codina-Pascual, M., Campillo, M., Kraus, J., Speicher, M. R., Egozcue, J., Navarro, J., et al. (2006). Crossover frequency and synaptonemal complex length: Their variability and effects on human male meiosis. *Molecular Human Reproduction, 12*, 123–133.

Cole, F., Baudat, F., Grey, C., Keeney, S., de Massy, B., & Jasin, M. (2014). Mouse tetrad analysis provides insights into recombination mechanisms and hotspot evolutionary dynamics. *Nature Genetics, 46*, 1072–1080.

Comeron, J. M., Ratnappan, R., & Bailin, S. (2012). The many landscapes of recombination in *Drosophila melanogaster*. *PLoS Genetics, 8*, e1002905.

Copenhaver, G. P., Housworth, E. A., & Stahl, F. W. (2002). Crossover interference in *Arabidopsis*. *Genetics, 160*, 1631–1639.

Cromie, G. A., Hyppa, R. W., Taylor, A. F., Zakharyevich, K., Hunter, N., & Smith, G. R. (2006). Single Holliday junctions are intermediates of meiotic recombination. *Cell, 127*, 1167–1178.

de Boer, E., Stam, P., Dietrich, A. J., Pastink, A., & Heyting, C. (2006). Two levels of interference in mouse meiotic recombination. *Proceedings of the National Academy of Sciences of the United States of America, 103*, 9607–9612.

de los Santos, T., Hunter, N., Lee, C., Larkin, B., Loidl, J., & Hollingsworth, N. M. (2003). The Mus81/Mms4 endonuclease acts independently of double-Holliday junction resolution to promote a distinct subset of crossovers during meiosis in budding yeast. *Genetics, 164*, 81–94.

de los Santos, T., Loidl, J., Larkin, B., & Hollingsworth, N. M. (2001). A role for MMS4 in the processing of recombination intermediates during meiosis in *Sacchromyces cerevisiae*. *Genetics, 159*, 1511–1525.

De Muyt, A., Jessop, L., Kolar, E., Sourirajan, A., Chen, J., Dayani, Y., et al. (2012). BLM helicase ortholog Sgs1 is a central regulator of meiotic recombination intermediate metabolism. *Molecular Cell, 46*, 43–53.

Egel, R. (1978). Synaptonemal complex and crossing-over: Structural support or interference? *Heredity, 41*, 233–237.

Egel, R. (2004). The molecular biology of *Schizosacharomyces pombe*. In R. Egel (Ed.), *Genetics, genomics and beyond*. Berlin: Springer.

Emerson, S., & Beadle, G. W. (1933). Crossing-over near the spindle fiber in attached-X chromosomes of *Drosophila melanogaster*. *Zeitschrift für induktive Abstammungs- und Vererbungslehre, 65*, 129–140.

Foss, E., Lande, R., Stahl, F. W., & Steinberg, C. M. (1993). Chiasma interference as a function of genetic distance. *Genetics, 133*, 681–691.

Foss, E. J., & Stahl, F. W. (1995). A test of a counting model for chiasma interference. *Genetics*, *139*, 1201–1209.

Fowler, K. R., Gutiérrez-Velasco, S., Martín-Castellanos, C., & Smith, G. R. (2013). Protein determinants of meiotic DNA break hotspots. *Molecular Cell*, *49*, 983–996.

Fowler, K. R., Hyppa, R. W., Cromie, G. A., & Smith, G. R. (2018). Physical basis for long-distance communication along meiotic chromosomes. *Proceedings of the National Academy of Sciences of the United States of America*, *115*, e9333–e9342.

Fowler, K. R., Sasaki, M., Milman, N., Keeney, S., & Smith, G. R. (2014). Evolutionarily diverse determinants of meiotic DNA break and recombination landscapes across the genome. *Genome Research*, *24*, 1650–1664.

Fox, D. P. (1973). The control of chiasma distribution in the locust, *Schistocerca gregaria* (Forskål). *Chromosoma*, *43*, 289–328.

Francis, K. E., Lam, S. Y., & Copenhaver, G. P. (2006). Separation of *Arabidopsis* pollen tetrads is regulated by QUARTET1, a pectin methylesterase gene. *Plant Physiology*, *142*, 1004–1013.

Fujitani, Y., Mori, S., & Kobayashi, I. (2002). A reaction-diffusion model for interference in meiotic crossing over. *Genetics*, *161*, 365–372.

Fung, J. C., Rockmill, B., Odell, M., & Roeder, G. S. (2004). Imposition of crossover interference through the nonrandom distribution of synapsis initiation complexes. *Cell*, *116*, 795–802.

Galagan, J. E., Calvo, S. E., Borkovich, K. A., Selker, E. U., Read, N. D., Jaffe, D., et al. (2003). The genome sequence of the filamentous fungus *Neurospora crassa*. *Nature*, *422*, 859–868.

Garcia, V., Gray, S., Allison, R. M., Cooper, T. J., & Neale, M. J. (2015). Tel1(ATM)-mediated interference suppresses clustered meiotic double-strand-break formation. *Nature*, *520*, 114–118.

Getz, T. J., Banse, S. A., Young, L. S., Banse, A. V., Swanson, J., Wang, G. M., et al. (2008). Reduced mismatch repair of heteroduplexes reveals "non"-interfering crossing over in wild-type *Saccharomyces cerevisiae*. *Genetics*, *178*, 1251–1269.

Gruhn, J. R., Rubio, C., Broman, K. W., Hunt, P. A., & Hassold, T. (2013). Cytological studies of human meiosis: Sex-specific differences in recombination originate at, or prior to, establishment of double-strand breaks. *PLoS One*, *8*, e85075.

Hatkevich, T., Kohl, K. P., McMahan, S., Hartmann, M. A., Williams, A. M., & Sekelsky, J. (2017). Bloom syndrome helicase promotes meiotic crossover patterning and homolog disjunction. *Current Biology*, *27*, 96–102.

He, W., Rao, H., Tang, S., Bhagwat, N., Kulkarni, D. S., Ma, Y., et al. (2020). Regulated proteolysis of MutSγ controls meiotic crossing over. *Molecular Cell*, *78*, 168–183.e165.

Higashide, M., & Shinohara, M. (2016). Budding yeast SLX4 contributes to the appropriate distribution of crossovers and meiotic double-strand break formation on bivalents during meiosis. *G3 (Bethesda, Md.)*, *6*, 2033–2042.

Higgins, J. D., Armstrong, S. J., Franklin, F. C., & Jones, G. H. (2004). The *Arabidopsis* MutS homolog AtMSH4 functions at an early step in recombination: Evidence for two classes of recombination in *Arabidopsis*. *Genes & Development*, *18*, 2557–2570.

Holliday, R. (1977). Recombination and meiosis. *Philosophical Transactions of the Royal Society of London. Series B, Biological Sciences*, *277*, 359–370.

Hulten, M. A. (2011). On the origin of crossover interference: A chromosome oscillatory movement (COM) model. *Molecular Cytogenetics*, *4*, 10.

Hyppa, R. W., Cho, J. D., Nambiar, M., & Smith, G. R. (2021). Redirecting meiotic DNA break hotspot determinant proteins alters localized spatial control of DNA break formation and repair. *Nucleic Acids Research*, *50*, 899–914.

Jensen-Seaman, M. I., Furey, T. S., Payseur, B. A., Lu, Y., Roskin, K. M., Chen, C. F., et al. (2004). Comparative recombination rates in the rat, mouse, and human genomes. *Genome Research*, *14*, 528–538.

Joshi, N., Barot, A., Jamison, C., & Borner, G. V. (2009). Pch2 links chromosome axis remodeling at future crossover sites and crossover distribution during yeast meiosis. *PLoS Genetics*, *5*, e1000557.

Kim, K. D., Tanizawa, H., Iwasaki, O., & Noma, K. (2016). Transcription factors mediate condensin recruitment and global chromosomal organization in fission yeast. *Nature Genetics*, *48*, 1242–1252.

King, J. S., & Mortimer, R. K. (1990). A polymerization model of chiasma interference and corresponding computer simulation. *Genetics*, *126*, 1127–1138.

Kitani, Y. (1978). Absence of interference in association with gene conversion in *Sordaria fimicola*, and presence of interference in association with ordinary recombination. *Genetics*, *89*, 467–497.

Kleckner, N., Zickler, D., Jones, G. H., Dekker, J., Padmore, R., Henle, J., et al. (2004). A mechanical basis for chromosome function. *Proceedings of the National Academy of Sciences of the United States of America*, *101*, 12592–12597.

Lambing, C., Kuo, P. C., Tock, A. J., Topp, S. D., & Henderson, I. R. (2020). ASY1 acts as a dosage-dependent antagonist of telomere-led recombination and mediates crossover interference in *Arabidopsis*. *Proceedings of the National Academy of Sciences of the United States of America*, *117*, 13647–13658.

Lambing, C., Osman, K., Nuntasoontorn, K., West, A., Higgins, J. D., Copenhaver, G. P., et al. (2015). *Arabidopsis* PCH2 mediates meiotic chromosome remodeling and maturation of crossovers. *PLoS Genetics*, *11*, e1005372.

Li, X., Li, L., & Yan, J. (2015). Dissecting meiotic recombination based on tetrad analysis by single-microspore sequencing in maize. *Nature Communications*, *6*, 6648.

Libuda, D. E., Uzawa, S., Meyer, B. J., & Villeneuve, A. M. (2013). Meiotic chromosome structures constrain and respond to designation of crossover sites. *Nature*, *502*, 703–706.

Lister, C., & Dean, C. (1993). Recombinant inbred lines for mapping RFLP and phenotypic markers in *Arabidopsis thaliana*. *The Plant Journal*, *4*, 745–750.

Mancera, E., Bourgon, R., Brozzi, A., Huber, W., & Steinmetz, L. M. (2008). High-resolution mapping of meiotic crossovers and non-crossovers in yeast. *Nature*, *454*, 479–485.

Meneely, P. M., Farago, A. F., & Kauffman, T. M. (2002). Crossover distribution and high interference for both the X chromosome and an autosome during oogenesis and spermatogenesis in *Caenorhabditis elegans*. *Genetics*, *162*, 1169–1177.

Miller, D. E., Smith, C. B., Kazemi, N. Y., Cockrell, A. J., Arvanitakas, A. V., Blumenstiel, J. P., et al. (2016). Whole-genome analysis of individual meiotic events in *Drosophila melanogaster* reveals that noncrossover gene conversions are insensitive to interference and the centromere effect. *Genetics*, *203*, 159–171.

Morgan, T. H., Bridges, C. B., & Schultz, J. (1935). Constitution of the germinal material in relation to heredity. In *Vol. 34. Carnegie Institute Year Book* (p. 287). Carnegie Institute.

Morgan, C., Fozard, J. A., Hartley, M., Henderson, I. R., Bomblies, K., & Howard, M. (2021). Diffusion-mediated HEI10 coarsening can explain meiotic crossover positioning in *Arabidopsis*. *Nature Communications*, *12*, 4674.

Mortimer, R. K., Contopoulou, C. R., & King, J. S. (1992). Genetic and physical maps of *Saccharomyces cerevisiae*, edition 11. *Yeast*, *8*, 817–902.

Mortimer, R. K., & Fogel, S. (1974). Genetical interference and gene conversion. In R. F. Grell (Ed.), *Mechanisms in recombination* (pp. 263–275). New York and London: Plenum Press.

Muller, H. J. (1916). The mechanism of crossing-over. *American Naturalist*, *50*, 193–221. 284-305; 350-366; 421-434.

Munz, P. (1994). An analysis of interference in the fission yeast *Schizosaccharomyces pombe*. *Genetics*, *137*, 701–707.

Nakagawa, T., & Ogawa, H. (1999). The *Saccharomyces cerevisiae* MER3 gene, encoding a novel helicase-like protein, is required for crossover control in meiosis. *The EMBO Journal*, *18*, 5714–5723.

Nambiar, M., Chuang, Y. C., & Smith, G. R. (2019). Distributing meiotic crossovers for optimal fertility and evolution. *DNA Repair (Amst)*, *81*, 102648.

Nambiar, M., & Smith, G. R. (2018). Pericentromere-specific cohesin complex prevents meiotic pericentric DNA double-strand breaks and lethal crossovers. *Molecular Cell*, *71*, 540–553.

Nicklas, R. B. (1997). How cells get the right chromosomes. *Science*, *275*, 632–637.

Nishant, K. T., Chen, C., Shinohara, M., Shinohara, A., & Alani, E. (2010). Genetic analysis of baker's yeast Msh4-Msh5 reveals a threshold crossover level for meiotic viability. *PLoS Genetics*, *6*, e1001083.

Novak, J. E., Ross-Macdonald, P. B., & Roeder, G. S. (2001). The budding yeast Msh4 protein functions in chromosome synapsis and the regulation of crossover distribution. *Genetics*, *158*, 1013–1025.

Oh, S. D., Lao, J. P., Hwang, P. Y.-H., Taylor, A. F., Smith, G. R., & Hunter, N. (2007). BLM ortholog, Sgs1, prevents aberrant crossing-over by suppressing the formation of multi-chromatid joint molecules. *Cell*, *130*, 259–272.

Page, S. L., & Hawley, R. S. (2001). c(3)G encodes a *Drosophila* synaptonemal complex protein. *Genes & Development*, *15*, 3130–3143.

Papazian, H. P. (1952). The analysis of tetrad data. *Genetics*, *37*, 175–188.

Perkins, D. D. (1962). Crossing-over and interference in a multiply marked chromosome arm of *Neurospora*. *Genetics*, *47*, 1253–1274.

Perkins, D. D., & Barry, E. G. (1977). The cytogenetics of *Neurospora*. *Advances in Genetics*, *19*, 133–285.

Rhee, S. Y., & Somerville, C. R. (1998). Tetrad pollen formation in quartet mutants of *Arabidopsis thaliana* is associated with persistence of pectic polysaccharides of the pollen mother cell wall. *The Plant Journal*, *15*, 79–88.

Rog, O., Kohler, S., & Dernburg, A. F. (2017). The synaptonemal complex has liquid crystalline properties and spatially regulates meiotic recombination factors. *eLife*, *6*, e21455.

Ross-Macdonald, P., & Roeder, G. S. (1994). Mutation of a meiosis-specific MutS homolog decreases crossing over but not mismatch correction. *Cell*, *79*, 1069–1080.

Serrentino, M. E., Chaplais, E., Sommermeyer, V., & Borde, V. (2013). Differential association of the conserved SUMO ligase Zip3 with meiotic double-strand break sites reveals regional variations in the outcome of meiotic recombination. *PLoS Genetics*, *9*, e1003416.

Shinohara, M., Oh, S. D., Hunter, N., & Shinohara, A. (2008). Crossover assurance and crossover interference are distinctly regulated by the ZMM proteins during yeast meiosis. *Nature Genetics*, *40*, 299–309.

Shinohara, M., Sakai, K., Shinohara, A., & Bishop, D. K. (2003). Crossover interference in *Saccharomyces cerevisiae* requires a TID1/RDH54- and DMC1-dependent pathway. *Genetics*, *163*, 1273–1286.

Smith, G. R., Boddy, M. N., Shanahan, P., & Russell, P. (2003). Fission yeast Mus81•Eme1 Holliday junction resolvase is required for meiotic crossing over but not for gene conversion. *Genetics*, *165*, 2289–2293.

Song, M., Zhai, B., Yang, X., Tan, T., Wang, Y., Yang, X., et al. (2021). Interplay between Pds5 and Rec8 in regulating chromosome axis length and crossover frequency. *Science Advances*, *7*, eabe7920.

Stahl, F. W. (2008). On the "NPD ratio" as a test for crossover interference. *Genetics*, *179*, 701–704.
Stahl, F. W., & Foss, H. M. (2008). But see KITANI (1978). *Genetics*, *178*, 1141–1145.
Stahl, F. W., Foss, H. M., Young, L. S., Borts, R. H., Abdullah, M. F., & Copenhaver, G. P. (2004). Does crossover interference count in *Saccharomyces cerevisiae*? *Genetics*, *168*, 35–48.
Stahl, F. W., & Housworth, E. A. (2009). Methods for analysis of crossover interference in *Saccharomyces cerevisiae*. *Methods in Molecular Biology*, *557*, 35–53.
Strickland, W. N. (1958). An analysis of interference in *Aspergillus nidulans*. *Proceedings of the Royal Society of London. Series B, Biological Sciences*, *149*, 82–101.
Sturtevant, A. H. (1915). The behavior of the chromosomes as studied through linkage. *Zeitschrift für Induktive Abstammungs- und Vererbungslehre*, *13*, 234–287.
Sym, M., & Roeder, G. S. (1994). Crossover interference is abolished in the absence of a synaptonemal complex protein. *Cell*, *79*, 283–292.
The Arabidopsis Genome Initiative. (2000). Analysis of the genome sequence of the flowering plant *Arabidopsis thaliana*. *Nature*, *408*, 796–815.
Tsai, C. J., Mets, D. G., Albrecht, M. R., Nix, P., Chan, A., & Meyer, B. J. (2008). Meiotic crossover number and distribution are regulated by a dosage compensation protein that resembles a condensin subunit. *Genes & Development*, *22*, 194–211.
Tsubouchi, T., Zhao, H., & Roeder, G. S. (2006). The meiosis-specific Zip4 protein regulates crossover distribution by promoting synaptonemal complex formation together with Zip2. *Developmental Cell*, *10*, 809–819.
Vranis, N. M., Van der Heijden, G. W., Malki, S., & Bortvin, A. (2010). Synaptonemal complex length variation in wild-type male mice. *Genes*, *1*, 505–520.
Wang, M., Wang, K., Tang, D., Wei, C., Li, M., Shen, Y., et al. (2010). The central element protein ZEP1 of the synaptonemal complex regulates the number of crossovers during meiosis in rice. *Plant Cell*, *22*, 417–430.
White, M. A., Wang, S., Zhang, L., & Kleckner, N. (2017). Quantitative modeling and automated analysis of meiotic recombination. *Methods in molecular biology (Clifton, N.J.)*, *1471*, 305–323.
Wood, V., Gwilliam, R., Rajandream, M. A., et al. (2002). The genome sequence of *Schizosaccharomyces pombe*. *Nature*, *415*, 871–880.
Yokoo, R., Zawadzki, K. A., Nabeshima, K., Drake, M., Arur, S., & Villeneuve, A. M. (2012). COSA-1 reveals robust homeostasis and separable licensing and reinforcement steps governing meiotic crossovers. *Cell*, *149*, 75–87.
Youds, J. L., Mets, D. G., McIlwraith, M. J., Martin, J. S., Ward, J. D., ONeil, N. J., et al. (2010). RTEL-1 enforces meiotic crossover interference and homeostasis. *Science*, *327*, 1254–1258.
Zakharyevich, K., Tang, S., Ma, Y., & Hunter, N. (2012). Delineation of joint molecule resolution pathways in meiosis identifies a crossover-specific resolvase. *Cell*, *149*, 334–347.
Zanders, S., & Alani, E. (2009). The *pch2Δ* mutation in baker's yeast alters meiotic crossover levels and confers a defect in crossover interference. *PLoS Genetics*, *5*, e1000571.
Zhang, L., Kohler, S., Rillo-Bohn, R., & Dernburg, A. F. (2018). A compartmentalized signaling network mediates crossover control in meiosis. *eLife*, *7*, e30789.
Zhang, L., Liang, Z., Hutchinson, J., & Kleckner, N. (2014). Crossover patterning by the beam-film model: Analysis and implications. *PLoS Genetics*, *10*, e1004042.
Zhang, L., Wang, S., Yin, S., Hong, S., Kim, K. P., & Kleckner, N. (2014). Topoisomerase II mediates meiotic crossover interference. *Nature*, *511*, 551–556.
Zhao, H., McPeek, M. S., & Speed, T. P. (1995). Statistical analysis of chromatid interference. *Genetics*, *139*, 1057–1065.

## Further reading

Liu, H., Huang, J., Sun, X., & Li, J. (2018). Tetrad analysis in plants and fungi finds large differences in gene conversion rates but no GC bias. *Nature Ecology & Evolution, 2*, 164–173.

Malkova, A., Swanson, J., German, M., McCusker, J. H., Housworth, E. A., Stahl, F. W., et al. (2004). Gene conversion and crossing over along the 405-kb left arm of *Saccharomyces cerevisiae* chromosome VII. *Genetics, 168*, 49–63.

Stadler, D. R. (1959). The relationship of gene conversion to crossing over in *Neurospora*. *Proceedings of the National Academy of Sciences of the United States of America, 45*, 1625–1629.

CHAPTER TEN

# Small RNAs and their protein partners in animal meiosis

María de las Mercedes Carro[a,b], Andrew Grimson[b,c,*], and Paula E. Cohen[a,b,*]
[a]Department of Biomedical Sciences, Cornell University, Ithaca, NY, United States
[b]Cornell Reproductive Sciences Center (CoRe), Cornell University, Ithaca, NY, United States
[c]Department of Molecular Biology & Genetics, Cornell University, Ithaca, NY, United States
*Corresponding authors: e-mail address: agrimson@cornell.edu; paula.cohen@cornell.edu

## Contents



## Abstract

Meiosis is characterized by highly regulated transitions in gene expression that require diverse mechanisms of gene regulation. For example, in male mammals, transcription undergoes a global shut-down in early prophase I of meiosis, followed by increasing transcriptional activity into pachynema. Later, as spermiogenesis proceeds, the histones bound to DNA are replaced with transition proteins, which are themselves replaced with protamines, resulting in a highly condensed nucleus with repressed transcriptional activity. In addition, two specialized gene silencing events take place during prophase I: meiotic silencing of unsynapsed chromatin (MSUC), and the sex chromatin specific mechanism, meiotic sex chromosome inactivation (MSCI). Notably, conserved roles for the RNA binding protein (RBP) machinery that functions with small non-coding RNAs have been described as participating in these meiosis-specific mechanisms, suggesting that RNA-mediated gene regulation is critical for fertility in many species.

*Current Topics in Developmental Biology*, Volume 151
ISSN 0070-2153
https://doi.org/10.1016/bs.ctdb.2022.06.001

Here, we review roles of small RNAs and their associated RBPs in meiosis-related processes such as centromere function, silencing of unpaired chromatin and meiotic recombination. We will discuss the emerging evidence of non-canonical functions of these components in meiosis.

## Abbreviations

**MSCI** meiotic sex chromosome inactivation
**RBP** RNA binding proteins

# 1. Introduction

Meiosis is a specialized cell division that generates haploid gametes or spores in sexually reproducing organisms. Prophase I is the defining substage of meiosis, in which specific events take place to ensure accurate segregation of homologous (maternal and paternal) chromosomes and thus, reduction of the genetic content in daughter cells (Gray & Cohen, 2016). Chromatin structure is dynamic during prophase I, when homologous chromosomes must pair, synapse, and undergo meiotic recombination. Prior to meiosis, and progressing through prophase I, essential and stage-specific genes must be expressed within a tightly regulated timeline in order to co-ordinate these processes. As a consequence, meiosis is characterized by a stringently coordinated gene expression program. Moreover, transcriptional silencing shapes this expression program to be specific for each cell type and stage within the germline (Grive et al., 2019; Guo et al., 2018; Liu et al., 2021; Murat et al., 2021; Shi et al., 2020; Zhao et al., 2020). For example, in male mammals, transcriptional silencing of the X and Y chromosomes is a spatiotemporally-restricted event that contributes to the characteristic transcriptomic profile of pachytene spermatocytes (Turner, 2007, 2015). Even though many of the changes in transcriptome profiles during meiosis have been described, multiple aspects of the underlying regulatory mechanisms remain poorly understood.

A substantial fraction of the transcriptome is comprised of non-coding RNAs, a subset of which have roles in gene and genome regulation. In addition, cells possess different classes of RNA binding proteins (RBPs), some of which have important roles in gene regulation, often mediated by interactions with small RNAs (Hentze, Castello, Schwarzl, & Preiss, 2018). Certain RBPs and their small RNA cargoes exert their regulatory influence in the nucleus by modulating chromatin function to control transcriptional and co-transcriptional gene expression, whereas others function in the cytoplasm to control transcript stability and translation and other aspects of mRNA fate. Many small RNAs are expressed in the germline and, more specifically, during prophase I of meiosis. Interestingly, knockout (KO) models for some of the RBPs associated with small RNA biogenesis have shown them to be critical for regulation of at multiple stages of meiosis (Giauque & Bickel, 2016; Girard et al., 2021; Nakamura et al., 2007; Yadav, Mäkelä, Hyssälä, Cisneros-Montalvo, & Kotaja, 2020). One of the most intensely studied examples is the RNA interference (RNAi) pathway, a conserved mechanism of post-transcriptional gene silencing mediated by sequence-specific small RNA molecules (Bhattacharjee, Roche, & Martienssen, 2019). However, the involvement of RNAi components in transcriptional regulation has also been reported, motivated by pioneering studies in the yeast *S. pombe*. This review summarizes the established and emerging roles of RBPs and associated small RNAs as facilitators of the gene regulatory environment unique to meiosis. We will focus on studies of meiosis in animals, while also incorporating studies in other organisms that implicate unexpected roles for small RNA pathways.

## 2. RNA binding proteins in non-coding RNA pathways

RBPs comprise a heterogeneous group of proteins defined by their ability to bind to RNA through RNA binding domains and form ribonucleoprotein complexes (Corley, Burns, & Yeo, 2020). A broad classification of RBPs is based on the specific RNA species with which they interact, including the following families: mRNA-binding, pre-rRNA-binding, tRNA-binding, small nuclear RNA-binding, small nucleolar RNA-binding, and non-coding RNA-binding proteins (Gerstberger, Hafner, & Tuschl, 2014). Knowledge of RBP structure has also facilitated a functional classification based on the type and number of RNA binding domains associated with their cellular functions (for a review see Corley et al., 2020). RBPs that bind non-coding RNA have deserved special attention, in particular, RBPs involved in biogenesis and processing of regulatory non-coding RNAs known to be key players of gene regulation. Regulatory non-coding RNAs can be categorized according to their transcript length into long non-coding RNAs (lncRNAs), which are longer than 200 nucleotides (nt) long, and small non-coding RNAs, which are shorter than 200 nt. LncRNAs represent the most diverse group in terms of both their molecular mechanisms of action and the variety of regulatory functions they perform (Lee, Zhang, & Krause, 2019), a feature of lncRNAs that has greatly challenged our understanding of their functions. These diverse roles, across many cellular processes, have gained significant recognition over the past decade (Andric & Rougemaille, 2021) and have been reviewed extensively elsewhere (Quinn & Chang, 2016; Ransohoff, Wei, & Khavari, 2018; Taylor, Chu, Spektor, & Soloway, 2015). Small RNAs include microRNAs (miRNAs), endogenous small interfering RNAs (siRNAs) and PIWI-interacting RNAs (piRNAs), a classification based upon their biogenesis, mechanisms of action and functions. Recent advances in sequencing technologies have made classification of small RNAs more complex and they can now be divided further into multiple subclasses (Ghildiyal & Zamore, 2009). Here, we will focus on germline pathways that rely upon miRNAs and siRNAs, together with their associated RBPs (Table 1).

**Table 1** Biogenesis of miRNA and siRNA, associated processes and proteins.

| | **miRNA** | **Endo siRNA** |
|---|---|---|
| Organism | Animals, protists, plants, algae, viruses | Animals, fungi, protists, plants, algae |
| Genomic origin | – miRNA genes<br>– Non-canonical miRNA (mirtrons) | – Repetitive sequences in transposons, centromere and telomere regions.<br>– Convergent mRNA transcripts and sense-antisense pairs |
| Length | 21–23 nt | 21–22 nt |
| Nuclear processing | Microprocessor dependent for canonical miRNA | Microprocessor independent |
| Nuclear export | Exportin-5 dependent mechanism in mammals and flies. | *Blanks* dependent in *Drosophila*? |
| Cytoplasmic processing | Dicer dependent | Dicer dependent |
| Type of Argonaute | AGOs | AGOs<br>(S)AGOs |
| Targeted genes | Coding genes | Coding genes<br>Transposon and other repetitive sequences |

## 2.1 Biogenesis of miRNA

Biogenesis of small RNAs starts from precursor RNA molecules and involves defined multi-step pathways both in the nucleus and cytoplasm (Fig. 1). Typically, miRNA genes are transcribed by RNA polymerase II, producing primary miRNA transcripts (pri-miRNA) (Hombach, Kretz, Hombach, & Kretz, 2016). In the canonical pathway, the pri-miRNA is cleaved in an RNA structure and sequence-dependent manner by Drosha (Starega-Roslan, Witkos, Galka-Marciniak, & Krzyzosiak, 2015), a member of the RNase III family, to generate a precursor-miRNA (pre-miRNA). Drosha requires a double stranded RNA-binding domain protein (dsRBD) partner, Dgcr8 (DiGeorge syndrome critical region 8) in vertebrates and Pasha in invertebrates, to facilitate efficient processing (Table 2). Together, Drosha and Dgcr8/Pasha form the nuclear-localized microprocessor complex, which is thought to have emerged early in metazoan evolution (Bråte et al., 2018; Grimson et al., 2008). The product of this first nuclear biogenesis step is a pre-miRNA with a 2-nucleotide overhang at the 3′ end (Lee, Han, Yeom, Jin, & Kim, 2006). The majority of miRNAs originate from this canonical pathway; however, a number of variations exist. For example, microprocessor-independent pre-miRNAs can originate from the action of the splicing machinery (Salim, Kumar, Kulshreshtha, & Vivekanandan, 2021); such miRNAs were first discovered in *Drosophila melanogaster* and *Caenorhabditis elegans* (Ruby, Jan, & Bartel, 2007) and termed mirtrons. Finally, transport of pre-miRNAs from the nucleus to the cytoplasm requires contributions from additional RBPs. In mammals, Exportin-5 recognizes the 2 nucleotide 3′ overhang in the pre-miRNAs and forms a complex with the co-factor Ran, a GTPase that releases the pre-miRNA into the cytoplasm upon GTP hydrolysis (Kim, 2004; Yi, Qin, Macara, & Cullen, 2003). Evidence suggests that Exportin-1 mediates pre-miRNA transport to the cytoplasm in flies and worms, the latter lacking an ortholog for Exportin-5 (Büssing, Yang, Lai, & Grohans, 2010). Thus, multiple factors are required for miRNA biogenesis, and inhibition or loss of any of these broadly impacts miRNAs. Therefore, each of these components represents an intervention point for drug targeting or KO studies to elucidate the roles of miRNA in a given cell type or pathway.

## 2.2 Biogenesis of siRNA

Endogenous siRNAs originate from double stranded RNA (dsRNA) precursors in the nucleus, and their biogenesis does not require microprocessor

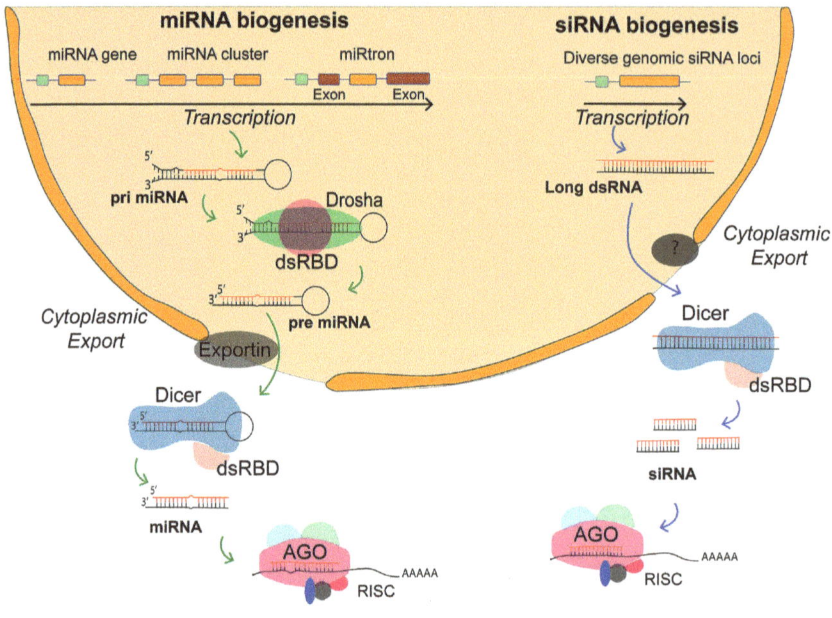


**Fig. 1** Canonical pathway for the biogenesis of miRNA and siRNA. Biogenesis of the various classes of small RNAs starts from precursor RNA molecules and involves defined multi-step pathways both in the nucleus and in the cytoplasm. miRNA genes are transcribed by RNA polymerase II as primary miRNA (pri-miRNA) or alternatively, pri-miRNAs can originate from splicing of mRNA (mirtrons). In the nucleus, the microprocessor complex, consisting of Drosha and a dsRNA binding domain protein (dsRBD) Dgcr8, or Partner of Drosha, cleaves the flanking regions of the pri-miRNA to generate precursor-miRNA (pre-miRNA) with a 2-nucleotide overhang at the 3′ end. Endogenous siRNAs start as dsRNA precursors that arise from different genomic loci in the nucleus that do not require microprocessor activity such as repetitive sequences in transposons, centromere and telomere regions, convergent mRNA transcripts and sense-antisense pairs. Pre-miRNAs are transported from the nucleus to the cytoplasm by Exportin-5 in a GTP-bound Ran dependent mechanism in mammals, while the transport of siRNA precursors to the cytoplasm is not well known. Once they reach the nucleus, both miRNAs and siRNAs precursors are processed by Dicer enzymes. Small RNA duplexes are loaded into an Argonaute protein by Dicer in collaboration with other dsRBD proteins to mediate assembly of the RNA Induced Silencing Complex (RISC). Once interacting with the Argonaute protein, the miRNA/siRNA duplex is then unwound into single stranded small RNA. Argonaute proteins can either slice the target mRNA or cause destabilization or translation repression of the mRNA via recruitment of other associated effector proteins by the RISC complex.

**Table 2** RBPs associated with miRNA and siRNA pathways in flies, worms and mice.

| | ***Drosophila*** | ***Caenorhabditis*** | **Mus** |
|---|---|---|---|
| Microprocessor | Drosha<br>Pasha | Drosha<br>Pasha | Drosha<br>DGCR8 |
| Dicer | Dcr-1 (miRNA, siRNA) and Dcr-2 (siRNA) | Dcr-1 | Dicer |
| Dicer associated dsRBPs | C3PO<br>LoQs | RDE-4 | TRBP |
| RdRP | Not present | EGO-1 (essential in the germline), Rrf-1, Rrf-2, and Rrf-3. | Not present |
| miRNA and siRNA interacting Argonautes | AGO clade (2) | AGO clade (6)<br>WAGO clade (16) | AGO clade (4) |

activity. These precursors derive from repetitive sequences found in transposons, centromeres and telomeres, convergent transcripts and other endogenous sense-antisense pairs (Carthew & Sontheimer, 2009; Golden, Gerbasi, & Sontheimer, 2008; Russo, Harrington, & Steiniger, 2016). Certain eukaryotes, such as worms and yeast (but not vertebrates or insects), possess RNA-dependent RNA polymerase (RdRP) enzymes that synthesize secondary siRNAs using primary siRNAs as templates, thus providing a system of small RNA signal amplification (Maida & Masutomi, 2011). The mechanism by which siRNA precursors are transported to the cytoplasm has not been fully characterized. Interestingly, recent work in *Drosophila* indicates that the dsRBD protein Blanks, previously shown to have a role in spermiogenesis, is involved in siRNA export (Gerbasi et al., 2011; Nitschko et al., 2020).

## 2.3 Processing of miRNA and siRNA in the cytoplasm

Within the cytoplasm, both miRNA and siRNA precursors undergo further processing by Dicer, a family of proteins conserved across eukaryotes (de Jong et al., 2009). Dicer recognizes pre-miRNAs and long dsRNAs (siRNA precursors) and cleaves them to produce miRNAs (~21–23 nt long) or siRNAs (~21–22 nt long) (Ha & Kim, 2014). Two Dicer paralogs exist in *Drosophila*, Dcr-1 and Dcr-2 (Lee et al., 2004); while Dcr-1 predominantly processes pre-miRNAs, Dcr-2 processes siRNA precursors

(Förstemann et al., 2005). However, Dcr-1 is also involved in siRNA biogenesis, as it may be required for loading of both miRNAs and certain siRNAs onto Argonaute proteins (Leuschner, Obernosterer, & Martinez, 2005). After processing by Dicer, the resulting miRNA or siRNA duplexes are loaded onto an Argonaute protein by Dicer and associated dsRBD proteins. During this process, the duplex is unwound into single stranded small RNA, discarding or degrading one strand, and resulting in the assembly of the RNA Induced Silencing Complex (RISC). The RISC is comprised of an Argonaute protein stably associated with a single stranded small RNA, referred to as the guide strand, reflecting the role of the small RNA in guiding RISC to a target RNA. The discarded strand is termed the passenger siRNA strand, or miRNA* (Ghildiyal & Zamore, 2009).

A number of dsRBD protein partners of Dicer have been described, with roles during or after initial processing. For example, in worms, Dicer interacts with RD-4, a dsRBD required for cleavage of dsRNA to siRNA, but which is not required later, during RISC assembly (Blanchard et al., 2011; Parker, Eckert, & Bass, 2006). In flies and mammals, duplex loading to Argonaute and RISC assembly by Dicer requires the protein loquacious (loQS; also known as R3D1) (Förstemann et al., 2005; Jiang et al., 2005; Saito, Ishizuka, Siomi, & Siomi, 2005) and trans-activation response RNA binding protein (TRBP) (Chendrimada et al., 2005; Wilson et al., 2015), respectively. Binding specificity and efficient passenger strand removal in *Drosophila*, is dependent on component-3 promoter of RISC (C3PO) (Li et al., 2012), which has also been found in mammalian cell lines and is critical for RISC activation (Ye et al., 2011).

## 2.4 Loading of small RNAs into Argonaute proteins and post-transcriptional gene silencing

Target recognition is based on the antisense complementarity between the guide strand of the small RNA bound by Argonaute and the target mRNA. The extend of pairing between the guide and target determine the mode of regulation and fate of the mRNA. In bilaterian animals, miRNAs are almost always partially complementary to their mRNA targets, with much of the pairing restricted to a limited and critical region of sequence at the 5′ end of the miRNA (nucleotides 2–7) called the seed region (Bartel, 2018; Lai, 2002). In contrast, siRNAs are usually fully complementary to the target mRNA. Generally, when a guide RNA is perfectly complementary to its RNA target, the Argonaute protein will catalyze endonucleolytic cleavage of the target (Liu et al., 2004; Zamore, Tuschl, Sharp, & Bartel, 2000). However, if there is imperfect complementarity between the small RNA and the target mRNA region, as is usually the case for miRNAs, the mismatched bases disrupt the catalytic interactions at the Argonaute active site, reducing or eliminating cleavage efficiency (Hutvagner & Simard, 2008; Wee, Flores-Jasso, Salomon, & Zamore, 2012). In this case, silencing occurs via recruitment of other associated effector proteins by the RISC complex that set apart the transcript from the translational machinery and induce destabilization and/or translational repression (Filipowicz, Bhattacharyya, & Sonenberg, 2008), with accelerated mRNA decay being the predominant mode of regulation in mammals (Eichhorn et al., 2014).

Argonaute protein families have evolved across the prokaryote and eukaryote domains of life (Swarts et al., 2014). In animals, they can be classified into three major groups: AGO Argonautes, which interact with mi- and siRNAs, PIWI Argonautes, which bind piRNAs, and WAGOs, worm specific argonautes that interact with secondary small RNAs produced by an RdRP (Wynant, Santos, & vanden Broeck, J., 2017). *C. elegans* is notable due to the high number of Argonaute proteins, having 27, five of which belong to the AGO clade, and due to the production of diverse types of endogenous small RNAs (Wedeles, Wu, & Claycomb, 2013; Yigit et al., 2006). *Drosophila* has five Argonaute proteins, two of them (Ago1 and Ago2) belong to the AGO clade and they bind to miRNA and siRNA (Okamura, Ishizuka, Siomi, & Siomi, 2004), with the remaining three specific for piRNAs. Mammals possess multiple AGO and PIWI argonautes, for example, humans and mice possess four AGOs and 3 or 4 (respectively) PIWIs (Thomson & Lin, 2009).

Not all Argonaute proteins have slicer activity, for example, of the four mammalian AGO proteins, AGO2 alone can cleave mRNA targets effectively (Liu et al., 2004), although AGO3 may retain a degree of slicer activity (Park et al., 2017; Park, Sim, Kehling, & Nakanishi, 2020), while AGO1 and AGO4 lack this catalytic activity. Thus, the mechanism by which a miRNA regulates its mRNA target reflects not only the extent of complementarity between the miRNA and the mRNA, but also the specific Argonaute protein onto which the small RNA is loaded (Ghildiyal & Zamore, 2009). Beyond AGO2-specific slicer activity, whether additional functional specialization exists across the four AGO paralogs common to mammals remains an unanswered question.

In the course of evolution, Argonaute proteins acquired the ability to interact with pathway-specific proteins that enabled their participation in a wide range of cellular processes (Swarts et al., 2014). In animals, GW182 proteins (proteins containing multiple glycine-tryptophan repeats) are RISC binding partners known to mediate cytoplasmic mRNA decay.

GW182 family proteins can also promote aggregation of multiple RISC complexes into ribonucleoprotein granules (Bose, Barman, Goswami, & Bhattacharyya, 2017; Eulalio, Tritschler, & Izaurralde, 2009; Patel, Barbee, & Blankenship, 2016), and facilitate interactions between multiple RISCs on a single target mRNA (Briskin, Wang, & Bartel, 2020). Concentration of argonautes and other interacting RBPs and RNAs in cytoplasmic ribonucleoprotein granules of germ cells is a conserved feature of animal species (Mukherjee & Mukherjee, 2021). These structures have a variety of names depending on the organism and origin mechanism; examples include the P-body in *Drosophila* and *C. elegans* oocytes, Balbiani body in mouse oocytes and the chromatoid body in mouse spermatocytes (Mukherjee & Mukherjee, 2021; Voronina, Seydoux, Sassone-Corsi, & Nagamori, 2011). Intriguingly, these structures appear to be dynamic during gametogenesis; they can change position within the cell, exchange material with the cytoplasm and are even able to migrate between adjacent cells through intracytoplasmic bridges (Jaglarz, Kloc, Jankowska, Szymanska, & Bilinski, 2011; Onohara, Fujiwara, Yasukochi, Himeno, & Yokota, 2010; Parker & Sheth, 2007; Ventelä, Toppari, & Parvinen, 2003). In worms, at least four distinct germ granules have been identified and each of them seems to compartmentalize specific functions associated with small RNA pathways (Sundby, Molnar, & Claycomb, 2021). It is thought that these structures could act to compartmentalize/sequester the RNAi pathway machinery, as analysis of their components showed that they are rich in non-coding RNAs and RBPs such as Argonautes, GW182, VASA homologs, among others (Meikar & Kotaja, 2014; Onohara et al., 2010; Ouyang & Seydoux, 2022). Interestingly, mRNA-RBP complexes transported into P-bodies by the GW182 family may be degraded or isolated to suppress translation in a reversible way (Eulalio et al., 2009; Hubstenberger et al., 2017; Takimoto, Wakiyama, & Yokoyama, 2009). In vertebrates, the GW182 protein TNRC6 acts as scaffold protein, tethering associated effector proteins to destabilize and translationally repress target mRNAs by inducing their decapping and deadenylation (Jonas & Izaurralde, 2015). However, the roles of germ granules are not well understood and are likely not exclusive to small RNA-associated processes.

## 3. Roles for small RNAs and associated RBPs in the germline

The germ line represents a unique system in which specific small RNA pathways contribute to the differentiation and maintenance of germ cells. Studies in animals using KO models point to diverse functions for both, miRNA and siRNA pathways during meiosis. For example, in *Drosophila* females, lack of Dcr-1, Pasha, Drosha or Ago1 causes impaired ability to sustain the meiotic cycle and loss of oocytes in the cyst (Azzam, Smibert, Lai, & Liu, 2012), indicating key roles for miRNAs in oogenesis. Furthermore, deletion of the siRNA-specific components Dcr-2 or Ago2 in *Drosophila simulans* males causes impaired meiosis and spermiogenesis; while KO males are unfertile, females exhibit only a mild fertility phenotype (Lin et al., 2018). Differences in phenotype between oogenesis and spermatogenesis in this species can be attributed to the role of siRNAs produced by Sex ratio suppressor genes involved in genomic conflict between the sex chromosomes. In this system, selfish genetic elements termed meiotic drivers that distort their transmission into gametes coexist with a suppressor element, in this case, small RNAs produced from an autosomal locus that interact with an X-linked driver (Lin et al., 2018; Muirhead & Presgraves, 2021; Vedanayagam, Lin, & Lai, 2021). Similar sexual dimorphic phenotypes are observed in the mouse in various mutants for genes encoding small RNA biogenesis associated RBPs. Loss of DGCR8 in female mice has no impact on oocyte development or fertility (Suh et al., 2010), while in testes, deletion of *Dgcr8* reduces the number of pachytene spermatocytes due to defective prophase I progression (Zimmermann et al., 2014). These, and other studies, have established that siRNAs, but not miRNAs are essential for meiosis in oocytes, with loss of Dicer associated siRNAs causing defects in spindle assembly during oocyte maturation and ovulation (Liu, Tang, He, & Rosenwaks, 2010; Ma et al., 2010; Mattiske, Han, & Mann, 2009; Murchison et al., 2007). However, miRNAs are expressed during mouse spermatogenesis (reviewed in Kotaja, 2014; Wang & Xu, 2015) where *Dicer* knockout causes loss of miRNAs and defective progression of Meiosis I (Greenlee et al., 2012; Lee et al., 2003; Modzelewski et al., 2015; Romero et al., 2011). Interestingly, Dicer KO in oocytes causes defects in maturation and ovulation, after prophase I is completed, while in spermatocytes, the meiotic defective phenotype seems to arise during prophase I, indicating differential small RNA requirements in meiosis I and II.

The detailed mechanisms giving rise to meiotic phenotypes in components of small RNA pathway mutant models is not fully understood. Given the basic cellular function of miRNAs and siRNAs in post-transcriptional gene regulation of all cell types, plus the fact that many genes encoding components of the meiosis machinery can be targeted by small RNAs, this type of gene silencing seems to be critical for meiosis. However, growing evidence suggests that small RNA gene regulation in meiosis not only occurs via the canonical post-transcriptional silencing, but also through direct involvement of components of these pathways in chromatin reorganization. In the following section, we will discuss previous work showing how RBPs involved in small RNA biogenesis and RISC assembly localize to the meiotic nucleus and relate to key prophase I processes.

## 4. Meiotic silencing of unpaired chromatin

Pairing and synapsis between homologous chromosomes are highly regulated processes that define early prophase of meiosis I. Prophase I begins in pre-leptonema with the search for homology between maternal and paternal chromosomes, aided by the sequential formation of the synaptonemal complex (SC, in most organisms, exceptions include *S. pombe*). When an appropriate match is found, in zygonema, the physical tethering between homologous chromosomes is completed by assembly of the final component of the tripartite SC. By pachynema, homologous chromosomes are tethered along their entire length by the SC, with the notable exception of the sex chromosomes, as described below. The presence of any persistent unpaired DNA triggers an evolutionarily conserved response, first studied in *Neurospora crassa*, termed Meiotic Silencing of Unpaired DNA (MSUD) (Fig. 2) (Aramayo & Metzenberg, 1996). MSUD in *N. crassa* requires Dicer (DCL-1) and Argonaute family proteins and their associated small RNAs (for a thorough review see Hammond, 2017). During the pairing of homologous chromosomes, any unpaired regions are identified by an unknown molecular mechanism that involves the SAD-6 (suppressor of ascus dominance-6) homology search protein (Samarajeewa et al., 2014). Presence of unpaired DNA stimulates bidirectional generation of single-stranded aberrant RNAs (aRNAs) from the unpaired DNA, which are then exported to a meiotic silencing complex located in the perinuclear region (Decker et al., 2015). Here, the aRNAs are a template used by the RdRP, SAD-1, to generate dsRNA, which is a substrate for DCL-1, resulting in the production of MSUD-associated siRNAs (masiRNAs). The Argonaute SMS-2, loads the masiRNAs and targets any transcript from the unpaired locus for degradation, with targeting specified by pairing between the masiRNAs and their cognate loci. The mechanism of repression appears to be post-transcriptional and within the

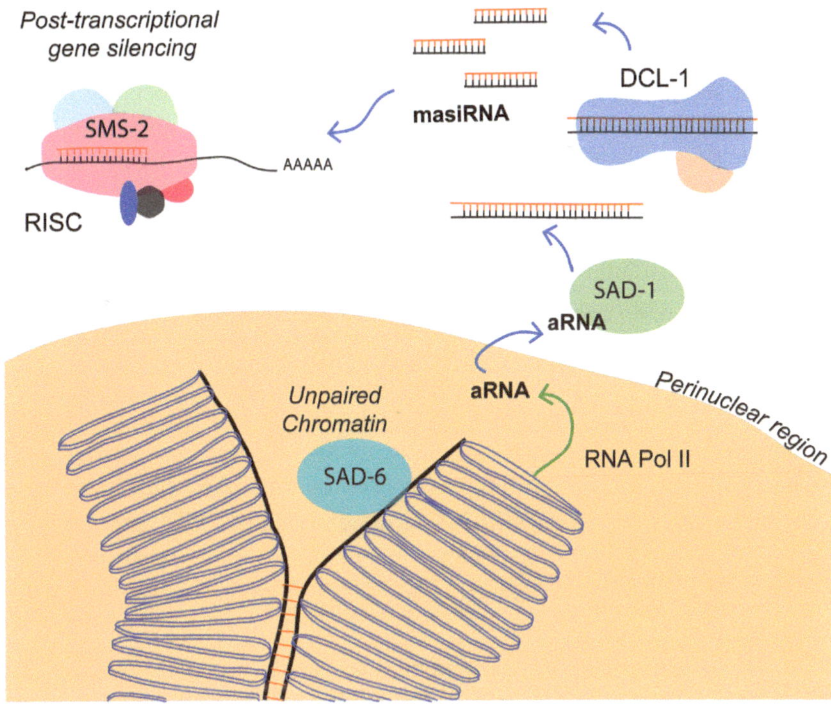


**Fig. 2** Meiotic silencing of unpaired DNA (MSUD) in the meiotic nucleus of *Neurospora crassa*. During the pairing of homologous chromosomes, regions that are unpaired are identified by an unknown molecular mechanism that involves the homology search protein SAD-6. This signal stimulates bidirectional generation of single-stranded aberrant RNAs (aRNAs) from the unpaired DNA by RNA polymerase II that are exported to the perinuclear meiotic silencing complex. Here, the aRNAs are used as a template by the RdRP SAD-1 to generate dsRNA. Subsequently, a Dicer enzyme DCL-1 and other associated factors produce siRNAs from the dsRNAs termed MSUC-associated siRNAs (masiRNAs). The Argonaute SMS-2 loads the masiRNAs and targets any transcript from the unpaired locus for degradation.

cytoplasm, since SMS-2 has been shown to be exclusively perinuclear and cytoplasmic (Xiao et al., 2021). These results establish a fundamental role for small RNAs during meiosis in *Neurospora*.

Meiotic silencing likely evolved as a surveillance system to protect against aneuploidy, transposable elements or excessive gene dosage resulting from deleterious chromosomal rearrangements (Schimenti, 2005; Turner, 2015). In contrast to MSUD, a post-transcriptional process, the equivalent pathway in worms seems to occur via transcriptional silencing. In *C. elegans*, chromatin regions that fail to pair and synapse accumulate the repressive histone modification H3K9me (Histone 3, trimethylated at lysine 9), as a result of a pathway called meiotic silencing of unpaired chromatin (Maine et al., 2005). Notably, in *C. elegans*, this meiotic silencing is dependent on the RdRP EGO1 and the Argonaute CSR-1 (Maine, 2010). Although the mechanistic details of this pathway are yet to be detailed, it has been proposed that EGO-1 and associated small RNAs are involved in recruitment of histone modifiers to the unpaired regions (Maine et al., 2005). Thus, although aspects of this pathway in *C. elegans* diverge from the *Neurospora* pathway, the fundamental role played by small RNAs is consistent.

In mammals, meiotic silencing is triggered by failure of homologous chromosomes to synapse, a process termed meiotic silencing of unsynapsed chromatin (MSUC). In mice, as autosomes become synapsed during pachynema, unsynapsed regions are marked by BRCA1, the kinase ATR and phosphorylated histone H2AX (phosphorylated by ATR at serine 139, termed γH2AX), which collectively trigger signaling cascades that result in transcriptional silencing (Baarends et al., 2005; Turner, Aprelikova, et al., 2004; Turner, Mahadevaiah, et al., 2004). Although mechanistic studies in mammals are limited, there is evidence for participation of small RNA pathways in MSUC. In mouse spermatocytes, AGO4 and AGO3 localize to autosomal regions that fail to pair, coincident with localization of BRCA1, ATR and γH2AX (Modzelewski, Holmes, Hilz, Grimson, & Cohen, 2012). Moreover, during pachynema, AGO4 may be involved in the localization of trimethylated H3K9me3, which contributes to the formation of transcriptionally silent heterochromatin. Loss of AGO4 leads to downregulation of multiple miRNA families in spermatocytes, however, the mechanism connecting these specific miRNAs and the establishment of silencing marks in unsynapsed chromosomes is unknown. Indeed, it is also possible that any role for AGO4 during MSUC involves a novel class of siRNAs, rather than miRNAs or does not require small RNAs, as Argonautes can have non-canonical functions in the nucleus of human cell lines and *Drosophila* independent of small RNAs (Acuña et al., 2020; Zaytseva et al., 2020). Nonetheless, the presence of AGO proteins at unpaired chromosomal regions during mammalian meiosis suggests additional links between MSUC and MSUD, raising the possibility that mammalian AGO proteins, likely in concert with unknown small RNAs, may be contributing to transcriptional (or co-transcriptional) silencing during MSUC, analogous to roles described for MSUD in *N. crassa*. More details on small RNA participation on meiotic silencing pathways in animals come from the study of sex chromosome silencing, a naturally occurring process in many species, which we discuss below.

## 5. Meiotic silencing of sex chromatin

Components of small RNA pathways are also implicated in the silencing of sex chromosomes during meiosis, a specialized mechanism of unpaired homolog silencing. In species with morphologically distinct sex chromosomes that contain limited homology between maternal and paternal sex chromosomes, synapsis and recombination occurs only at regions that exhibit homology, referred to as pseudo-autosomal regions (PAR). During meiosis in male mice, the presence of unpaired sex chromatin triggers a feedforward mechanism that mediates the formation of a γH2AX domain on the sex chromosomes. This domain spreads to chromatin loops along with the accumulation of characteristic repressive chromatin marks, establishing a silencing domain known as the sex body (Abe et al., 2020) (Fig. 3). This specialized form of MSUC is termed meiotic sex chromosome inactivation (MSCI) (McKee & Handel, 1993; Turner, 2007), and occurs in pachynema, at which time transcriptional silencing of the XY pair stands in marked contrast to the active transcriptional state of the autosomes (Namekawa et al., 2006; Turner, 2015). Interestingly, the machinery that leads to both MSCI and MSUC appears to have common components, including proteins of the DNA Damage Response (DDR) pathway, whose accumulation seems to trigger silencing (Ellnati et al., 2017). As in MSUC, BRCA1 accumulates on the unsynapsed sex chromosome axes and recruits the kinase ATR to phosphorylate H2AX (Turner, Aprelikova, et al., 2004). The mechanism by which γH2AX signal is spread along the chromatin loops is centered on MDC1, which binds γH2AX and enables the progressive activation of the ATR kinase through the activator protein TOPBP1 (Abe et al., 2020; Ichijima et al., 2011). TOPBP1-ATR then mediates adjacent γH2AX formation on chromatin loops to which MDC1 binds once again, and the mechanism continues progressively forwards (Ichijima et al., 2011).

Studies in mouse have shown that spermatocytes that lack *Dgcr8* or *Dicer*, thus with repression of miRNA biogenesis, or repression of both miRNA and siRNA biogenesis, respectively, show aberrant sex body establishment

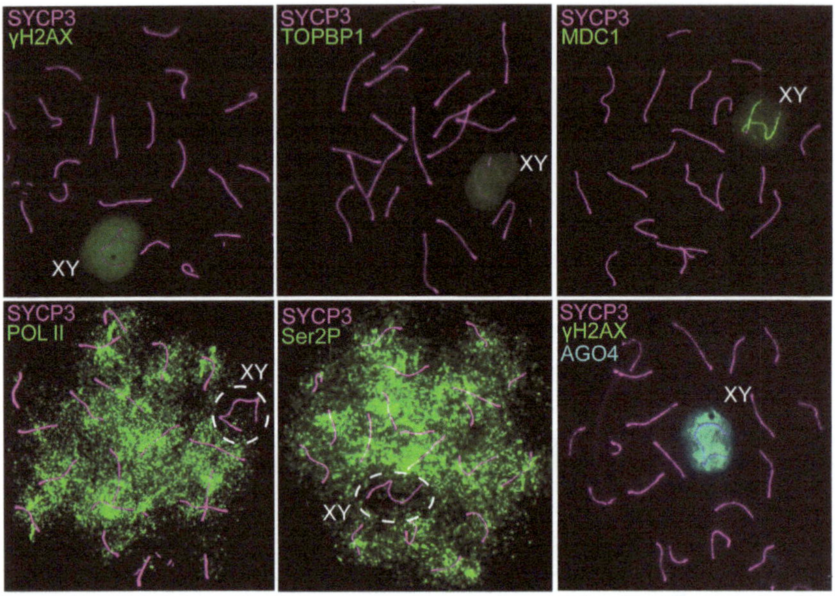


**Fig. 3** Meiotic silencing of sex chromatin in male mouse. During pachynema, the sex body becomes apparent in spermatocytes, as shown by the markers γH2AX, TOPBP1 and MDC1 (upper panels), whose signal spreads from the Synaptonemal Complex (SC) axes to chromatin loops. Both total RNA Pol II (POL II) and phosphor-Pol II (phosphorylated at Ser2; Ser2P), indicative of active transcriptional elongation, are excluded from this silencing domain. AGO4 colocalizes to the sex body. MDC1, Ser2P and AGO4 images were kindly provided by Drs. Andrew Modzelewski, Tegan Horan and Adriana Alexander.

and mis-localization of DDR factors at pachynema (Modzelewski et al., 2015). In addition, loss of DICER or AGO4 in male mice causes over-expression of X chromosome associated genes that are known to be targeted for MSCI in pachynema (Greenlee et al., 2012; Modzelewski et al., 2012). Studies have shown that AGO3, AGO4 and associated small RNAs, localize to the sex body of pachytene spermatocytes and *Ago4*-deficient germ cells show deficient silencing of X and Y chromosomes and influx of RNA pol II to the sex body (Modzelewski et al., 2012). Moreover, AGO4 might be involved in γH2AX signal establishment, since deletion of AGO4 causes aberrant morphology of the sex body and mis localization of the DDR factors, γH2AX and ATR and TOPBP1 (Modzelewski et al., 2012). This study also showed that the X-linked miRNAs localize to the sex body along with AGO4 in wild type spermatocytes, suggesting that the mechanism behind sex body establishment might involve miRNA binding. More studies are needed to establish the details of these silencing mechanisms, such as identifying the specific small RNAs that localize to the sex body and the possible target mRNAs or chromatin sites.

The existence of MSCI in *Drosophila* is controversial, and there seems to be no consensus on whether male flies exhibit sex chromosome transcriptional silencing; even less clear is whether small RNAs participate (Meiklejohn, Landeen, Cook, Kingan, & Presgraves, 2011; Vibranovski, 2014). MSCI in *C. elegans* is well documented, as X chromosomes in both males (X0) and hermaphrodites (XX) manifest reduced transcriptional activity up to diplotene (Kelly et al., 2002; Tzur et al., 2018). In hemizygous males, the single X chromosome is not able to pair during meiosis and is condensed, and interestingly, X chromosomes in hermaphrodites that are able to synapse are also silenced. Contrary to what is observed in mammals, similarities between silencing of X chromosomes during meiosis in *C. elegans* and MSUC is unclear, since both seem to respond to different pathways (Checchi & Engebrecht, 2011; Lamelza & Bhalla, 2012). Moreover, lack of synapsis might not be the trigger of X chromosome inactivation in *C. elegans* (Kelly et al., 2002; Rappaport et al., 2021). Nonetheless, loss of CSR-1 pathway factors results in mis-localization of the histone silencing mark H3K9me2 in both autosomes and X chromosomes, indicating that small RNA pathways are a common factor in sex chromosome silencing between worms and mammals (Maine et al., 2005; She, Xu, Fedotov, Kelly, & Maine, 2009).

## 6. Pericentromeric heterochromatin remodeling during meiosis

The meiotic nucleus undergoes progressive reorganization of heterochromatin as meiosis proceeds. Heterochromatin mediates proper segregation of chromosomes and facilitates long-range chromatin interactions between distant chromosomal regions. Increasing evidence suggests that certain histone modifications, which regulate the specialized architecture of meiotic chromosomes, play a fundamental role in centromere formation, pairing, synapsis, and recombination of homologs (Lam, Brick, Cheng, Pratto, & Camerini-Otero, 2019; Shi et al., 2021; Wang, Xu, Khawar, Liu, & Li, 2017). In most eukaryotes, centromeres are surrounded by densely packed pericentromeric heterochromatin (PCH) and this characteristic organization is essential to build the kinetochore (Talbert & Henikoff, 2020). For this reason, PCH and centromeres have an important role as coordinators of sister kinetochore cohesion and homologous chromosome orientation at the Metaphase I spindle, thereby ensuring accurate chromosome segregation (Brar & Amon, 2008). A general mechanism of heterochromatin formation is deacetylation of histone H3 followed by methylation at lysine 9 (H3K9me3) (Weirich, Khella, & Jeltsch, 2021), which is catalyzed by a group of histone methyltransferases that contain a SET domain and a chromodomain, such as SUV39 in animals and Clr4 in yeast. Histone methylation provides a docking site for proteins that silence transcription, recruit chromatin modifiers, and maintain heterochromatin ultrastructure. Intriguingly, many reports suggest that small RNA pathways could contribute to the accumulation of proteins associated with heterochromatin formation at the PCH (Kanellopoulou et al., 2005; Miyata et al., 2021; Pal-Bhadra et al., 2004; Park et al., 2018; Chapman, van Nues, Theron, & Bayne, 2020).

*S. pombe* was the first system in which an RNA-induced transcriptional gene silencing complex (RITS) was shown to mediate PCH formation (Verdel et al., 2004). In this process, an RdRP produces dsRNAs from bidirectional transcription of repetitive sequences at the centromeres and serves as the initial source for siRNAs generated by Dcr1 (Sugiyama, Cam, Verdel, Moazed, & Grewal, 2005). These primary siRNAs are targeted back to the repetitive sequences by the RITS complex to recruit histone modifiers such as Clr4. In yeast Clr4 mediates H3K9 methylation, which is bound by the heterochromatin-associated factor Swi6. In *S. pombe*, deletion of Ago1, Dicer or RdRP results in impaired heterochromatin formation at the centromeres and at the mating type region during meiosis and mitosis (Hall, Noma, & Grewal, n.d.; Volpe et al., 2003). Thus, it is clear that the siRNA pathway plays a major role in establishment of the higher chromatin structure of chromosomes during meiosis in *S. pombe*.

Instead of having one centromere, as in *S. pombe*, *C. elegans* possess dispersed centromeres along the length of the chromosomes. Chromatin organization of these holocentric chromosomes in the germline is also maintained by a small RNA pathway. A specific germline nuclear WAGO, HRDE-1, was shown to regulate nuclear compaction by recruitment of siRNAs to nascent transcripts in the nucleus, resulting in co-transcriptional repression (Fields & Kennedy, 2019; Perales et al., 2018). Another member of the WAGO clade, CSR-1 was shown to localize to the diakinetic chromosomes in *C. elegans* oocytes. In contrast to *S. pombe*, CSR-1 also binds to euchromatic regions in the nucleus of germ cells, where it targets nascent RNA for cleavage and also recruits chromatin-modifying factors to establish pericentromeric chromatin domains (Claycomb et al., 2009). This mechanism is dependent on small RNAs, as CSR-1 does not interact with genomic loci when small RNAs are depleted. Loss of CSR-1 and DRH-3 (Dicer related helicase) causes defects in chromosome segregation and aberrant accumulation of the repressive histone modification, H3K9me2 during meiosis (Claycomb et al., 2009; Davis et al., 2018; Duchaine et al., 2006; Nakamura et al., 2007; Yigit et al., 2006). Regardless of differences in genomic organization, recruitment of histone modifiers for chromatin compaction and PCH formation in *C. elegans* involves small RNA components acting in a non-canonical pathway reminiscent of that described in *S. pombe*.

Mouse meiotic spermatocytes and somatic cells also show PCH establishment pathway reminiscent of that in *S. pombe*, where RNA transcripts referred to as major satellite repeats are produced from the PCH and involved in heterochromatin formation (Fioriniello, Marano, Fiorillo, D'esposito, & Ragione, 2020; Fukagawa et al., 2004; Huo et al., 2020; Kanellopoulou et al., 2005; Khalil & Driscoll, 2010; Yadav et al., 2020). Using IF and *in situ* hybridization, DICER was shown to localize to heterochromatin areas of the spermatocyte nucleus, and colocalizing with major satellite transcripts by IF and RNA FISH (Yadav et al., 2020). The same authors showed that DICER interacts with major satellite repeat transcripts to produce mature small RNAs. Notably, *Dicer* and *Ago1*-null embryonic stem cells show upregulation of repetitive centromeric transcripts and defects in methylation of centromeric DNA and histone H3 (Kanellopoulou et al., 2005; Müller et al., 2022). In mouse, *Dicer* KO spermatocytes have reduced recruitment of SUV39H2 and H3K9me3 to the pericentric heterochromatin, along with a meiotic chromosome mis-segregation phenotype, suggesting a role for these small RNAs in promoting recruitment of histone modifiers to the PCH during meiosis (Yadav et al., 2020). Human somatic SUV39H1 localization to the PCH has also been shown to be regulated by small RNA binding (Johnson et al., 2017). In this regard, no evidence for an equivalent RITS complex has been detected in mammalian cells. It is important to note that *Dicer* mutants do not exhibit complete absence of heterochromatin formation, thus there are other mechanisms involved in this process. Moreover, mammals lack an RdRP to amplify small RNAs from their precursor transcripts, and therefore, the formation, spreading and maintenance of heterochromatin at the PCH likely occurs through mechanisms somewhat distinct from those in *S. pombe*. Nevertheless, key factors for heterochromatin assembly such as involvement of Dicer are conserved in all eukaryotes (Talbert & Henikoff, 2020) suggesting participation of small RNA pathways in the mechanisms involved in PCH formation. Perhaps PCH formation in mammals occurs through modeling of the chromatin environment through direct interactions between Dicer and Argonautes, along with their small RNA cargoes, with the genome, reminiscent of the *C. elegans* system.

## 7. Regulation of meiotic recombination and double strand break repair

Meiotic recombination between homologs occurs during Pachynema generates a physical link between them, known as chiasmata, which is critical for accurate segregation during meiosis I (Gray & Cohen, 2016). This process starts early in leptonema with the formation of hundreds of double stranded DNA breaks (DSB) mediated by Spo11, which resolve as either crossovers (CO) or non-crossovers, depending on the proteins recruited to this DSB hot spots and the consequent pathways regulated by the corresponding protein complexes. Several reports using mammalian cell lines have shown that small RNAs are present at DSB sites, although the biogenesis mechanism for these small RNAs has not been elucidated (Burger et al., 2017; Francia et al., 2012; Lu et al., 2018; Ohle et al., 2016; Swahari et al., 2016). In these cells, the general model proposes that DSB serve as substrates for production of small RNA termed DSB induced RNAs (diRNAS), which can be used by RBPs such as Argonaute proteins to target these same regions for either histone modifications or direct recruitment of repair factors. Small RNAs from the sequences flanking the DSB can mediate recruitment of DDR factors such as TOPBP1 (Fragkos et al., 2019) and MDC1 (Burger et al., 2017) back to the DSB. Argonaute proteins may be involved in loading of DDR factors to the DSB, since knockdown of *AGO2* impairs homologous recombination efficiency in mammalian cell lines (Wei et al., 2012). In support of this possibility, the AGO2/diRNA complex interacts with RAD51, a DDR factor, facilitating its recruitment to DNA damage sites (Gao et al., 2014) and this interaction relies on AGO2 phosphorylation by a yet to be identified kinase (Hu et al., 2021). The mentioned studies were performed using *in vitro* somatic cell cultures and this mechanism has not been shown during DSB repair in meiosis. However, TOPBP1, MDC1 and RAD51 have also been implicated in various repair processes in prophase I, and particularly in the events that result in MSCI as previously mentioned. *In vivo*, analysis of *Dicer* and *Dgcr8* KO models suggest that small RNA pathways could participate in DSB resolution and/or meiotic recombination in mouse, since KO spermatocytes show high levels of chromosomal instability (Modzelewski et al., 2015). Nonetheless, post transcriptional gene regulation through canonical miRNA pathways could be involved as well. For example, evidence shows that key crossover genes in the testes such as Separase-REC8, *MLH1* and *RAD1* are regulated by specific miRNAs (202-3p, 188-3p and 10a respectively) (Chen et al., 2022; Gao et al., 2014; Hilz, Modzelewski, Cohen, & Grimson, 2016; Song et al., 2017). Much research is needed to dissect a specific mechanism by which small RNA pathways could mediate recruitment of DDR factors and thus, influence crossover regulation and/or direct regulation of crossover machinery.

As described elsewhere within this volume, the multi-step process of crossover establishment must be tightly regulated, both temporally and spatially, to ensure correct chromosome segregation. It is well established, for example, that placement of crossovers in the vicinity of centromeres negatively influences meiotic chromosome segregation (Hassold & Hunt, 2001; Koehler, Hawley, Sherman, & Hassold, 1996). Interestingly, *S. pombe* KO mutants for Dcr1, Ago1 or the histone methyltransferase Clr4, show increased centromeric recombination (Ellermeier et al., 2010), suggesting participation of a small RNA pathway in crossover establishment through chromatin remodeling. New studies in the filamentous Ascomycete, *Sordaria macrospora*, have found that loss of Argonaute or Dicer results in increased chromosome length and altered chromatin loop/axis ratios along with increased crossover foci (Girard et al., 2021), also supporting a role of small RNAs in crossover formation through regulation of chromatin ultrastructure. Moreover, analysis of *Sordaria* KO model for Sms2 shows that proper localization of the E3 ligase Hei10 to crossovers sites relies upon presence of this Argonaute. Hei10 has been postulated as a candidate for signal integration by sensing state of SC structure formation and specific crossover events associated with recruitment of crossover factors (de Muyt et al., 2014). However, whether Hei10 interacts with Sms2 and both participate in SC formation, or recruitment of pro-crossover factors is not known. More studies in this model organism could provide mechanistic insights on whether small RNAs participate in meiotic recombination in *Sordaria*, or whether Sms2 behaves in a non-canonical fashion by directly interacting with chromatin. Nonetheless, this leading study suggests that meiotic recombination, one of the hallmarks of meiosis might also be regulated by small RNA pathways.

## 8. Concluding remarks

Gene regulation through small RNA pathways has been implicated in a diverse repertoire of meiosis associated processes. The study of the structure, localization and mechanism of action of common RBP protein effectors such as Dicer and Argonaute can shed light on the roles for these pathways in animal meiosis. Molecular insights coming from model organisms such as yeast, have made tremendous contributions to the area. The finding of a RITS complex able to directly influence transcription during the assembly of heterochromatin in *S. pombe* expanded possible roles for small RNA pathways as chromatin remodelers during meiosis in multicellular organisms. It is clear from studies in *C. elegans*, where RNAi was discovered more than two decades ago, and with a demonstrated nuclear extensive small RNA regulatory network, that chromatin dynamics are regulated by small RNA pathways during meiosis. However, the study of these processes in mammalian meiosis is, largely, lacking. Studies coming from *in vitro* somatic cell culture systems are promising, since they allowed for mechanistic studies of the roles of small RNA pathways in shared processes between meiosis and mitosis such as DNA damage response, DSB and chromatin remodeling. The lack of a reproducible *in vitro* meiosis system limits our studies of meiosis associated processes to *in vivo* approaches. Nonetheless, work discussed here provides important clues such as DICER and AGO localization to the nucleus of mammalian meiocytes, participation of these proteins in key meiotic processes such as MSUC, MSCI, PCH formation and DSB resolution that are key starting points to unravel the non-canonical roles for small RNA pathways in mammal meiosis.

It is interesting to note that presence of unpaired chromatin during chromosome synapsis is detrimental for meiosis across multiple organisms, resulting in the evolution of silencing mechanisms that appear to involve common pathways. Chromatin remodeling during the establishment of MSUC in species with dimorphic sex chromosomes might respond to several different mechanisms; moreover, it is important to acknowledge that MSUC itself is a poorly understood process. It is known that MSUC and MSCI share common effectors, such as DDR proteins, and thus it seems plausible that roles for small RNA pathways could be similar between them.

Many studies using mouse KO models for *Dicer, Dgrc8 and Ago4* have shown that these genes are key to the localization of histone repressive marks to unsynapsed chromatin, indicating the small RNA pathways are a key mechanism involved in this meiotic silencing. In order to further our understanding of this process, we need to elucidate the molecular interactions between small RNA pathway components and chromatin repressive marks. This leads to several questions: Is there a RITS or RITS-like complex in mammals that recruits chromatin modifiers to the unsynapsed chromosomal regions? Or, are Argonaute proteins mediating chromatin silencing through recruitment or spreading of DDR signals to the unsynapsed chromatin? (Fig. 4). Since MSCI is a specialized form of MSUC and most of the, albeit limited, evidence comes from the study of sex chromosome silencing in mouse pachytene spermatocytes, a detailed proteomic and RNA sequencing of the components of the sex body would contribute to solving some of these questions. The study of Argonaute proteins during meiosis is also a useful tool that could provide a list of protein and small RNA binding partners and thus, start to reveal their role in chromatin condensation.

A current challenge in the field will be differentiate between classical post-transcriptional silencing involvement of small RNA pathways in meiosis from a potential role in transcriptional regulation, and to describe the molecular components behind such a mechanism. As gene regulation in animals is ensured by an integrated network of pathways, perturbations induced by gene mutations can often be compensated through alternative pathways. As a consequence of this complexity, KO models of RBPs or a particular small RNAs do not necessarily exhibit a phenotype that reflects their true function in this context. Thus, finding the components of the pathways is likely to be challenging. Sequencing efforts using model organisms allow to identify a subset of transcripts associated with a given small RNA component. However, identifying which of the differentially expressed transcripts are direct targets of small RNAs and which of them are being differentially expressed because of other disrupted cellular processes is very challenging. In addition, annotation of small RNAs is a growing process, making identification arduous. Future studies will take advantage of our growing ability to sequence and identify small RNAs, in particular those physically associated with specific Argonaute proteins, along with their subcellular localization.

The germ cell represents a fascinating model to study small RNA pathways that will not only contribute to our understanding of silencing in the

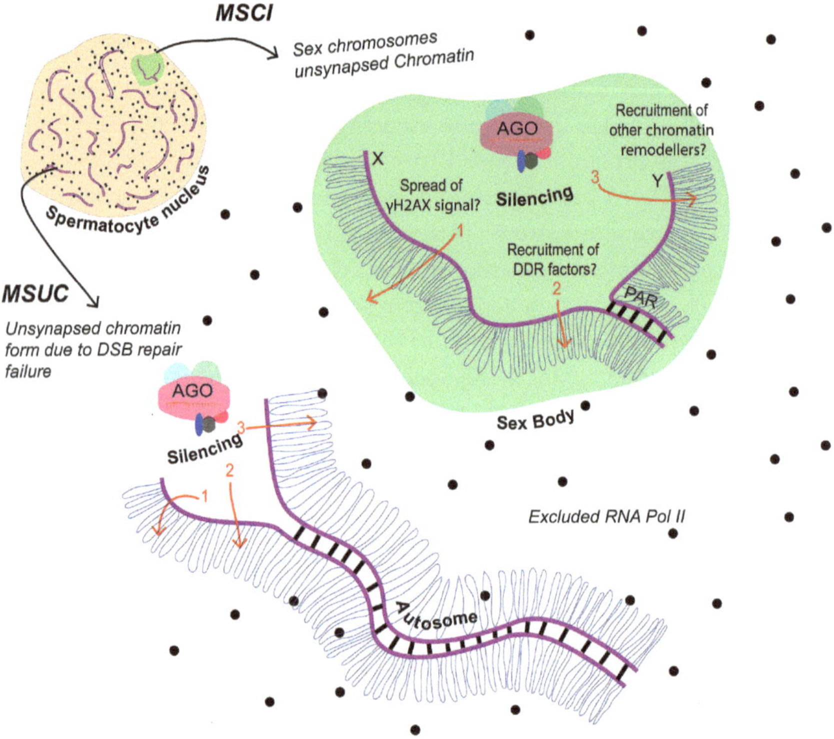


**Fig. 4** Proposed model for a role of small RNAs and Argonaute proteins in Meiotic silencing of unsynapsed chromatin (MSUC) in male mammal meiosis. AGO proteins could mediate silencing of unsynapsed chromatin regions in autosomes formed due to double strand break (DSB) failure, and in the sex body during Pachynema by regulating the localization and spreading of γH2AX (1) or recruitment of DNA damage repair factors (DDR, 2) to the chromatin. On the other hand, it is possible that small RNA-AGO complexes interact directly with chromatin to regulate the accumulation of repressive histone marks (3) in a similar mechanism to PCH formation in *S. pombe*, where an RNA-induced transcriptional gene silencing complex (RITS) mediates recruitment of histone modifiers. MSCI: Meiotic sex chromosome inactivation. PAR: pseudo autosomal region

complex processes that underlie prophase I, but also, will sum to our knowledge on involvement of small RNA pathways as transcriptional regulators, besides their well-known role in posttranscriptional gene silencing.

## Acknowledgments

The authors are grateful for funding from the Eunice Kennedy Shriver National Institute of Child Health and Human Development (P50HD076210, P50HD104454) to P.E.C. (P.I. and Project Leader) and A.G. (Project Leader), and from the Lalor Foundation (Project number 92252) to M.C.

## References

Abe, H., Alavattam, K. G., Hu, Y. C., Pang, Q., Andreassen, P. R., Hegde, R. S., et al. (2020). The initiation of meiotic sex chromosome inactivation sequesters DNA damage signaling from autosomes in mouse spermatogenesis. *Current Biology, 30*(3), 408–420. e5. https://doi.org/10.1016/j.cub.2019.11.064.

Acuña, L. I. G., Nazer, E., Rodríguez-Seguí, S. A., Pozzi, B., Buggiano, V., Marasco, L. E., et al. (2020). Nuclear role for human Argonaute-1 as an estrogen-dependent transcription coactivator. *The Journal of Cell Biology, 219*(9). https://doi.org/10.1083/JCB.201908097.

Andric, V., & Rougemaille, M. (2021). Long non-coding RNAs in the control of gametogenesis: Lessons from fission yeast. *Noncoding RNA*, 7(2), 34. MDPI AG. https://doi.org/10.3390/ncrna7020034.

Aramayo, R., & Metzenberg, R. L. (1996). Meiotic transvection in fungi. *Cell, 86*(1), 103–113. https://doi.org/10.1016/S0092-8674(00)80081-1.

Azzam, G., Smibert, P., Lai, E. C., & Liu, J. L. (2012). Drosophila Argonaute 1 and its miRNA biogenesis partners are required for oocyte formation and germline cell division. *Developmental Biology, 365*(2), 384–394. https://doi.org/10.1016/J.YDBIO.2012.03.005.

Baarends, W. M., Wassenaar, E., van der Laan, R., Hoogerbrugge, J., Sleddens-Linkels, E., Hoeijmakers, J. H. J., et al. (2005). Silencing of unpaired chromatin and histone H2A ubiquitination in mammalian meiosis. *Molecular and Cellular Biology, 25*(3), 1041–1053. https://doi.org/10.1128/MCB.25.3.1041-1053.2005/FORMAT/EPUB.

Bartel, D. P. (2018). Metazoan MicroRNAs. *Cell, 173*(1), 20–51. https://doi.org/10.1016/J.CELL.2018.03.006/ATTACHMENT/18013F94-01BD-4864-BD8E-C35285C19442/MMC1.XLSX.

Bhattacharjee, S., Roche, B., & Martienssen, R. A. (2019). RNA-induced initiation of transcriptional silencing (RITS) complex structure and function. *RNA Biology, 16*(9), 1133–1146. Taylor and Francis Inc. https://doi.org/10.1080/15476286.2019.1621624.

Blanchard, D., Parameswaran, P., Lopez-Molina, J., Gent, J., Saynuk, J. F., & Fire, A. (2011). On the nature of in vivo requirements for rde-4 in RNAi and developmental pathways in C. elegans. *RNA Biology, 8*(3), 458. https://doi.org/10.4161/RNA.8.3.14657.

Bose, M., Barman, B., Goswami, A., & Bhattacharyya, S. N. (2017). Spatiotemporal uncoupling of MicroRNA-mediated translational repression and target RNA degradation controls MicroRNP recycling in mammalian cells. *Molecular and Cellular Biology, 37*(4). https://doi.org/10.1128/MCB.00464-16.

Brar, G. A., & Amon, A. (2008). Emerging roles for centromeres in meiosis I chromosome segregation. *Nature Reviews. Genetics, 9*(12), 899–910. https://doi.org/10.1038/nrg2454.

Bråte, J., Neumann, R. S., Fromm, B., Haraldsen, A. A. B., Tarver, J. E., Suga, H., et al. (2018). Unicellular origin of the animal MicroRNA machinery. *Current Biology, 28*(20). https://doi.org/10.1016/J.CUB.2018.08.018. 3288-3295.e5.

Briskin, D., Wang, P. Y., & Bartel, D. P. (2020). The biochemical basis for the cooperative action of microRNAs. *Proceedings of the National Academy of Sciences of the United States of America, 117*(30), 17764–17774. https://doi.org/10.1073/PNAS.1920404117/SUPPL_FILE/PNAS.1920404117.SAPP.PDF.

Burger, K., Schlackow, M., Potts, M., Hester, S., Mohammed, S., & Gullerova, M. (2017). Nuclear phosphorylated Dicer processes double-stranded RNA in response to DNA damage. *The Journal of Cell Biology, 216*(8), 2373–2389. https://doi.org/10.1083/JCB.201612131.

Büssing, I., Yang, J. S., Lai, E. C., & Grohans, H. (2010). The nuclear export receptor XPO-1 supports primary miRNA processing in C. elegans and Drosophila. *The EMBO Journal, 29*(11), 1830–1839. https://doi.org/10.1038/EMBOJ.2010.82.

Carthew, R. W., & Sontheimer, E. J. (2009). *Origins and mechanisms of miRNAs and siRNAs*. https://doi.org/10.1016/j.cell.2009.01.035.

Chapman, E., van Nues, R., Theron, E., & Bayne, E. H. (2020). Mkt1 is required for RNAi-mediated silencing and establishment of heterochromatin in fission yeast. *Nucleic Acids Research*, *48*(3), 1239–1253. https://doi.org/10.1093/NAR/GKZ1157.

Checchi, P. M., & Engebrecht, J. A. (2011). Caenorhabditis elegans histone methyltransferase MET-2 shields the male X chromosome from checkpoint machinery and mediates meiotic sex chromosome inactivation. *PLoS Genetics*, 7(9), e1002267. https://doi.org/10.1371/JOURNAL.PGEN.1002267.

Chen, J., Gao, C., Luo, M., Zheng, C., Lin, X, Nig, Y., et al. (2022). MicroRNA-202 safeguards meiotic progression by preventing premature SEPARASE-mediated REC8 cleavage. *EMBO Reports*, e54298. https://doi.org/10.15252/embr.202154298. Epub ahead of print.

Chendrimada, T. P., Gregory, R. I., Kumaraswamy, E., Norman, J., Cooch, N., Nishikura, K., et al. (2005). TRBP recruits the Dicer complex to Ago2 for microRNA processing and gene silencing. *Nature*, *436*(7051), 740–744. https://doi.org/10.1038/NATURE03868.

Claycomb, J. M., Batista, P. J., Pang, K. M., Gu, W., Vasale, J. J., van Wolfswinkel, J. C., et al. (2009). The Argonaute CSR-1 and its 22G-RNA cofactors are required for holocentric chromosome segregation. *Cell*, *139*(1), 123–134. https://doi.org/10.1016/J.CELL.2009.09.014.

Corley, M., Burns, M. C., & Yeo, G. W. (2020). How RNA-binding proteins interact with RNA: Molecules and mechanisms. *Molecular Cell*, *78*(1), 9–29. Cell Press. https://doi.org/10.1016/j.molcel.2020.03.011.

Davis, G. M., Tu, S., Anderson, J. W. T., Colson, R. N., Gunzburg, M. J., Francisco, M. A., et al. (2018). The TRIM-NHL protein NHL-2 is a co-factor in the nuclear and somatic RNAi pathways in C. elegans. *eLife*, 7, e35478. https://doi.org/10.7554/ELIFE.35478.

de Jong, D., Eitel, M., Jakob, W., Osigus, H. J., Hadrys, H., DeSalle, R., et al. (2009). Multiple dicer genes in the early-diverging metazoa. *Molecular Biology and Evolution*, *26*(6), 1333–1340. https://doi.org/10.1093/MOLBEV/MSP042.

de Muyt, A., Zhang, L., Piolot, T., Kleckner, N., Espagne, E., & Zickler, D. (2014). E3 ligase Hei10: A multifaceted structure-based signaling molecule with roles within and beyond meiosis. *Genes & Development*, *28*(10), 1111–1123. https://doi.org/10.1101/GAD.240408.114.

Decker, L. M., Boone, E. C., Xiao, H., Shanker, B. S., Boone, S. F., Kingston, S. L., et al. (2015). Complex formation of RNA silencing proteins in the perinuclear region of Neurospora crassa. *Genetics*, *199*(4), 1017–1021. https://doi.org/10.1534/genetics.115.174623.

Duchaine, T. F., Wohlschlegel, J. A., Kennedy, S., Bei, Y., Conte, D., Pang, K. M., et al. (2006). Functional proteomics reveals the biochemical niche of C. elegans DCR-1 in multiple small-RNA-mediated pathways. *Cell*, *124*(2), 343–354. https://doi.org/10.1016/J.CELL.2005.11.036.

Eichhorn, S. W., Guo, H., McGeary, S. E., Rodriguez-Mias, R. A., Shin, C., Baek, D., et al. (2014). mRNA destabilization is the dominant effect of mammalian MicroRNAs by the time substantial repression ensues. *Molecular Cell*, *56*(1), 104–115. https://doi.org/10.1016/J.MOLCEL.2014.08.028.

Ellnati, E., Russell, H. R., Ojarikre, O. A., Sangrithi, M., Hirota, T., de Rooij, D. G., et al. (2017). DNA damage response protein TOPBP1 regulates X chromosome silencing in the mammalian germ line. *Proceedings of the National Academy of Sciences of the United States of America*, *114*(47), 12536–12541. https://doi.org/10.1073/pnas.1712530114.

Ellermeier, C., Higuchi, E. C., Phadnis, N., Holm, L., Geelhood, J. L., Thon, G., et al. (2010). RNAi and heterochromatin repress centromeric meiotic recombination. *Proceedings of the National Academy of Sciences of the United States of America*, *107*(19), 8701–8705. https://doi.org/10.1073/pnas.0914160107.

Eulalio, A., Tritschler, F., & Izaurralde, E. (2009). The GW182 protein family in animal cells: new insights into domains required for miRNA-mediated gene silencing. *RNA*, *15*(8), 1433. https://doi.org/10.1261/RNA.1703809.

Fields, B. D., & Kennedy, S. (2019). Chromatin compaction by small RNAs and the nuclear RNAi machinery in C. elegans. *Scientific Reports*, *9*(1), 1–9. https://doi.org/10.1038/s41598-019-45052-y.

Filipowicz, W., Bhattacharyya, S. N., & Sonenberg, N. (2008). Mechanisms of post-transcriptional regulation by microRNAs: Are the answers in sight? *Nature Reviews. Genetics*, *9*(2), 102–114. https://doi.org/10.1038/NRG2290.

Fioriniello, S., Marano, D., Fiorillo, F., D'esposito, M., & Ragione, F. D. (2020). Epigenetic factors that control pericentric heterochromatin organization in mammals. *Gene*, *11*(6). MDPI AG. https://doi.org/10.3390/genes11060595.

Förstemann, K., Tomari, Y., Du, T., Vagin, V. V., Denli, A. M., Bratu, D. P., et al. (2005). Normal microRNA maturation and germ-line stem cell maintenance requires Loquacious, a double-stranded RNA-binding domain protein. *PLoS Biology*, *3*(7), 1187–1201. https://doi.org/10.1371/JOURNAL.PBIO.0030236.

Fragkos, M., Barra, V., Egger, T., Bordignon, B., Lemacon, D., Naim, V., et al. (2019). Dicer prevents genome instability in response to replication stress. *Oncotarget*, *10*(43), 4407–4423. https://doi.org/10.18632/ONCOTARGET.27034.

Francia, S., Michelini, F., Saxena, A., Tang, D., de Hoon, M., Anelli, V., et al. (2012). Site-specific DICER and DROSHA RNA products control the DNA-damage response. *Nature*, *488*(7410), 231–235. https://doi.org/10.1038/nature11179.

Fukagawa, T., Nogami, M., Yoshikawa, M., Ikeno, M., Okazaki, T., Takami, Y., et al. (2004). Dicer is essential for formation of the heterochromatin structure in vertebrate cells. *Nature Cell Biology*, *6*(8), 784–791. https://doi.org/10.1038/NCB1155.

Gao, M., Wei, W., Li, M. M., Wu, Y. S., Ba, Z., Jin, K. X., et al. (2014). Ago2 facilitates Rad51 recruitment and DNA double-strand break repair by homologous recombination. *Cell Research*, *24*(5), 532–541. https://doi.org/10.1038/cr.2014.36.

Gerbasi, V. R., Preall, J. B., Golden, D. E., Powell, D. W., Cummins, T. D., & Sontheimer, E. J. (2011). Blanks, a nuclear siRNA/dsRNA-binding complex component, is required for Drosophila spermiogenesis. *Proceedings of the National Academy of Sciences of the United States of America*, *108*(8), 3204–3209. https://doi.org/10.1073/PNAS.1009781108.

Gerstberger, S., Hafner, M., & Tuschl, T. (2014). A census of human RNA-binding proteins. *Nature Reviews. Genetics*, *15*(12), 829–845. https://doi.org/10.1038/nrg3813.

Ghildiyal, M., & Zamore, P. D. (2009). Small silencing RNAs: An expanding universe. *Nature Reviews. Genetics*, *10*(2), 94–108. https://doi.org/10.1038/nrg2504.

Giauque, C. C., & Bickel, S. E. (2016). Heterochromatin-associated proteins hp1a and piwi collaborate to maintain the association of achiasmate homologs in drosophila oocytes. *Genetics*, *203*(1), 173–189. https://doi.org/10.1534/GENETICS.115.186460.

Girard, C., Budin, K., Boisnard, S., Zhang, L., Debuchy, R., Zickler, D., et al. (2021). RNAi-related dicer and argonaute proteins play critical roles for meiocyte formation, chromosome-axes lengths and crossover patterning in the fungus Sordaria macrospora. *Frontiers in Cell and Development Biology*, *28*(9), 684108. https://doi.org/10.3389/fcell.2021.684108.

Golden, D. E., Gerbasi, V. R., & Sontheimer, E. J. (2008). An inside job for siRNAs. *Molecular Cell*, *31*(3), 309. https://doi.org/10.1016/J.MOLCEL.2008.07.008.

Gray, S., & Cohen, P. E. (2016). Control of meiotic crossovers: From double-strand break formation to designation. *Annual Review of Genetics*, *50*, 175–210. Annual Reviews Inc. https://doi.org/10.1146/annurev-genet-120215-035111.

Greenlee, A. R., Shiao, M. S., Snyder, E., Buaas, F. W., Gu, T., Stearns, T. M., et al. (2012). Deregulated sex chromosome gene expression with male germ cell-specific loss of Dicer1. *PLoS One*, 7(10), e46359. https://doi.org/10.1371/journal.pone.0046359.

Grimson, A., Srivastava, M., Fahey, B., Woodcroft, B. I., Chiang, R. H., King, N., et al. (2008). Early origins and evolution of microRNAs and Piwi-interacting RNAs in animals. *Nature*, *455*(7217), 1193–1197. https://doi.org/10.1038/nature07415.

Grive, K. J., Hu, Y., Shu, E., Grimson, A., Elemento, O., Grenier, J. K., et al. (2019). Dynamic transcriptome profiles within spermatogonial and spermatocyte populations during postnatal testis maturation revealed by single-cell sequencing. *PLoS Genetics*, *15*(3), e1007810. https://doi.org/10.1371/journal.pgen.1007810.

Guo, J., Grow, E. J., Mlcochova, H., Maher, G. J., Lindskog, C., Nie, X., et al. (2018). The adult human testis transcriptional cell atlas. *Cell Research*, *28*(12), 1141–1157. https://doi.org/10.1038/S41422-018-0099-2.

Ha, M., & Kim, V. N. (2014). Regulation of microRNA biogenesis. *Nature Reviews. Molecular Cell Biology*, *15*(8), 509–524. Nature Publishing Group. https://doi.org/10.1038/nrm3838.

Hall, I. M., Noma, K.-I., & Grewal, S. I. S. (n.d.). RNA interference machinery regulates chromosome dynamics during mitosis and meiosis in fission yeast. https://doi.org/10.1073/pnas.232688099.

Hammond, T. M. (2017). Sixteen years of meiotic silencing by unpaired DNA. *Advances in Genetics*, *97*, 1–42. https://doi.org/10.1016/bs.adgen.2016.11.001.

Hassold, T., & Hunt, P. (2001). To err (meiotically) is human: The genesis of human aneuploidy. *Nature Reviews. Genetics*, *2*(4), 280–291. https://doi.org/10.1038/35066065.

Hentze, M. W., Castello, A., Schwarzl, T., & Preiss, T. (2018). A brave new world of RNA-binding proteins. *Nature Reviews. Molecular Cell Biology*, *19*(5), 327–341. Nature Publishing Group. https://doi.org/10.1038/nrm.2017.130.

Hilz, S., Modzelewski, A. J., Cohen, P. E., & Grimson, A. (2016). The roles of microRNAs and siRNAs in mammalian spermatogenesis. *Development*, *143*(17), 3061–3073. https://doi.org/10.1242/DEV.136721.

Hombach, S., Kretz, M., Hombach, S., & Kretz, M. (2016). Non-coding RNAs: Classification, biology and functioning. *Advances in Experimental Medicine and Biology*, *937*, 3–17. https://doi.org/10.1007/978-3-319-42059-2_1.

Hu, X., Li, Y., Zhang, T., Li, L., Chen, S., Wu, X., et al. (2021). Phosphorylation of Ago2 is required for its role in DNA double-strand break repair. *Journal of Genetics and Genomics*, *48*(4), 333–340. https://doi.org/10.1016/J.JGG.2021.03.011.

Hubstenberger, A., Courel, M., Bénard, M., Souquere, S., Ernoult-Lange, M., Chouaib, R., et al. (2017). P-body purification reveals the condensation of repressed mRNA regulons. *Molecular Cell*, *68*(1). https://doi.org/10.1016/J.MOLCEL.2017.09.003. 144-157.e5.

Huo, X., Ji, L., Zhang, Y., Lv, P., Cao, X., Wang, Q., et al. (2020). The nuclear matrix protein SAFB cooperates with major satellite RNAs to stabilize heterochromatin architecture partially through phase separation. *Molecular Cell*, 77(2). https://doi.org/10.1016/J.MOLCEL.2019.10.001. 368-383.e7.

Hutvagner, G., & Simard, M. J. (2008). Argonaute proteins: key players in RNA silencing. *Nature Reviews. Molecular Cell Biology*, *9*(1), 22–32. https://doi.org/10.1038/nrm2321.

Ichijima, Y., Ichijima, M., Lou, Z., Nussenzweig, A., Daniel Camerini-Otero, R., Chen, J., et al. (2011). MDC1 directs chromosome-wide silencing of the sex chromosomes in male germ cells. *Genes & Development*, *25*(9), 959–971. https://doi.org/10.1101/GAD.2030811.

Jaglarz, M. K., Kloc, M., Jankowska, W., Szymanska, B., & Bilinski, S. M. (2011). Nuage morphogenesis becomes more complex: Two translocation pathways and two forms of nuage coexist in Drosophila germline syncytia. *Cell and Tissue Research*, *344*(1), 169–181. https://doi.org/10.1007/S00441-011-1145-2.

Jiang, F., Ye, X., Liu, X., Fincher, L., McKearin, D., & Liu, Q. (2005). Dicer-1 and R3D1-L catalyze microRNA maturation in Drosophila. *Genes & Development*, *19*(14), 1674–1679. https://doi.org/10.1101/GAD.1334005.

Johnson, W. L., Yewdell, W. T., Bell, J. C., McNulty, S. M., Duda, Z., O'Neill, R. J., et al. (2017). RNA-dependent stabilization of SUV39H1 at constitutive heterochromatin. *eLife*, *6*. https://doi.org/10.7554/ELIFE.25299.

Jonas, S., & Izaurralde, E. (2015). Towards a molecular understanding of microRNA-mediated gene silencing. *Nature Reviews. Genetics*, *16*(7), 421–433. https://doi.org/10.1038/NRG3965.

Kanellopoulou, C., Muljo, S. A., Kung, A. L., Ganesan, S., Drapkin, R., Jenuwein, T., et al. (2005). Dicer-deficient mouse embryonic stem cells are defective in differentiation and centromeric silencing. *Genes & Development*, *19*(4), 489–501. https://doi.org/10.1101/gad.1248505.

Kelly, W. G., Schaner, C. E., Dernburg, A. F., Lee, M. H., Kim, S. K., Villeneuve, A. M., et al. (2002). X-chromosome silencing in the germline of C. elegans. *Development*, *129*(2), 479–492. https://doi.org/10.1242/dev.129.2.479.

Khalil, A. M., & Driscoll, D. J. (2010). Epigenetic regulation of pericentromeric heterochromatin during mammalian meiosis. *Cytogenetic and Genome Research*, *129*(4), 280–289. https://doi.org/10.1159/000315903.

Kim, V. N. (2004). MicroRNA precursors in motion: exportin-5 mediates their nuclear export. *Trends in Cell Biology*, *14*(4), 156–159. https://doi.org/10.1016/J.TCB.2004.02.006.

Koehler, K. E., Hawley, R. S., Sherman, S., & Hassold, T. (1996). Recombination and nondisjunction in humans and flies. *Human Molecular Genetics*, 1495–1504. 5 Spec No(REVIEW). https://doi.org/10.1093/HMG/5.SUPPLEMENT_1.1495.

Kotaja, N. (2014). MicroRNAs and spermatogenesis. *Fertility and Sterility*, *101*(6), 1552–1562. https://doi.org/10.1016/J.FERTNSTERT.2014.04.025.

Lai, E. C. (2002). Micro RNAs are complementary to 3′ UTR sequence motifs that mediate negative post-transcriptional regulation. *Nature Genetics*, *30*(4), 363–364. https://doi.org/10.1038/ng865.

Lam, K. W. G., Brick, K., Cheng, G., Pratto, F., & Camerini-Otero, R. D. (2019). Cell-type-specific genomics reveals histone modification dynamics in mammalian meiosis. *Nature Communications*, *10*(1), 1–11. https://doi.org/10.1038/s41467-019-11820-7.

Lamelza, P., & Bhalla, N. (2012). Histone methyltransferases MES-4 and MET-1 promote meiotic checkpoint activation in Caenorhabditis elegans. *PLoS Genetics*, *8*(11), e1003089. https://doi.org/10.1371/JOURNAL.PGEN.1003089.

Lee, Y., Ahn, C., Han, J., Choi, H., Kim, J., Yim, J., et al. (2003). The nuclear RNase III Drosha initiates microRNA processing. *Nature*, *425*(6956), 415–419. https://doi.org/10.1038/NATURE01957.

Lee, Y., Han, J., Yeom, K. H., Jin, H., & Kim, V. N. (2006). Drosha in primary microRNA processing. *Cold Spring Harbor Symposia on Quantitative Biology*, *71*, 51–57. https://doi.org/10.1101/SQB.2006.71.041.

Lee, Y. S., Nakahara, K., Pham, J. W., Kim, K., He, Z., Sontheimer, E. J., et al. (2004). Distinct roles for Drosophila Dicer-1 and Dicer-2 in the siRNA/miRNA silencing pathways. *Cell*, *117*(1), 69–81. https://doi.org/10.1016/S0092-8674(04)00261-2.

Lee, H., Zhang, Z., & Krause, H. M. (2019). Long noncoding RNAs and repetitive elements: Junk or intimate evolutionary partners? *Trends in Genetics*, *35*(12), 892–902. https://doi.org/10.1016/J.TIG.2019.09.006.

Leuschner, P. J. F., Obernosterer, G., & Martinez, J. (2005). MicroRNAs: Loquacious speaks out. *Current Biology*, *15*(15), R603–R605. https://doi.org/10.1016/J.CUB.2005.07.044.

Li, L., Gu, W., Liang, C., Liu, Q., Mello, C. C., & Liu, Y. (2012). The translin-TRAX complex (C3PO) is a ribonuclease in tRNA processing. *Nature Structural & Molecular Biology*, *19*(8), 824–830. https://doi.org/10.1038/nsmb.2337.

Lin, C. J., Hu, F., Dubruille, R., Vedanayagam, J., Wen, J., Smibert, P., et al. (2018). The hpRNA/RNAi pathway is essential to resolve intragenomic conflict in the Drosophila male germline. *Developmental Cell*, *46*(3). https://doi.org/10.1016/J.DEVCEL.2018.07.004. 316-326.e5.

Liu, J., Carmell, M. A., Rivas, F. V., Marsden, C. G., Thomson, J. M., Song, J. J., et al. (2004). Argonaute2 is the catalytic engine of mammalian RNAi. *Science (New York, N.Y.)*, *305*(5689), 1437–1441. https://doi.org/10.1126/SCIENCE.1102513.

Liu, X., Li, W., Yang, Y., Chen, K., Li, Y., Zhu, X., et al. (2021). Transcriptome profiling of the ovarian cells at the single-cell resolution in adult Asian seabass. *Frontiers in Cell and Development Biology*, *9*, 560. https://doi.org/10.3389/FCELL.2021.647892/BIBTEX.

Liu, H. C., Tang, Y., He, Z., & Rosenwaks, Z. (2010). Dicer is a key player in oocyte maturation. *Journal of Assisted Reproduction and Genetics*, *27*(9–10), 571–580. https://doi.org/10.1007/S10815-010-9456-X.

Lu, W. T., Hawley, B. R., Skalka, G. L., Baldock, R. A., Smith, E. M., Bader, A. S., et al. (2018). Drosha drives the formation of DNA:RNA hybrids around DNA break sites to facilitate DNA repair. *Nature Communications*, *9*(1), 1–13. https://doi.org/10.1038/s41467-018-02893-x.

Ma, J., Flemr, M., Stein, P., Berninger, P., Malik, R., Zavolan, M., et al. (2010). MicroRNA activity is suppressed in mouse oocytes. *Current Biology*, *20*(3), 265–270. https://doi.org/10.1016/J.CUB.2009.12.042.

Maida, Y., & Masutomi, K. (2011). RNA-dependent RNA polymerases in RNA silencing. *Biological Chemistry*, *392*(4), 299–304. https://doi.org/10.1515/BC.2011.035.

Maine, E. M. (2010). Meiotic silencing in Caenorhabditis elegans. *International Review of Cell and Molecular Biology*, *282*(C), 91–134. https://doi.org/10.1016/S1937-6448(10)82002-7.

Maine, E. M., Hauth, J., Ratliff, T., Vought, V. E., She, X., & Kelly, W. G. (2005). EGO-1, a putative RNA-dependent RNA polymerase, is required for heterochromatin assembly on unpaired DNA during C. elegans meiosis. *Current Biology*, *15*(21), 1972–1978. https://doi.org/10.1016/J.CUB.2005.09.049.

Mattiske, D. M., Han, L., & Mann, J. R. (2009). Meiotic maturation failure induced by DICER1 deficiency is derived from primary oocyte ooplasm. *Reproduction (Cambridge, England)*, *137*(4), 625–632. https://doi.org/10.1530/REP-08-0475.

McKee, B. D., & Handel, M. A. (1993). Sex chromosomes, recombination, and chromatin conformation. *Chromosoma*, *102*(2), 71–80. https://doi.org/10.1007/BF00356023.

Meikar, O., & Kotaja, N. (2014). Isolation of chromatoid bodies from mouse testis as a rich source of short RNAs. *Methods in Molecular Biology*, *1173*, 11–25. https://doi.org/10.1007/978-1-4939-0931-5_2.

Meiklejohn, C. D., Landeen, E. L., Cook, J. M., Kingan, S. B., & Presgraves, D. C. (2011). Sex chromosome-specific regulation in the Drosophila male germline but little evidence for chromosomal dosage compensation or meiotic inactivation. *PLoS Biology*, *9*(8), e1001126. https://doi.org/10.1371/JOURNAL.PBIO.1001126.

Miyata, K., Imai, Y., Hori, S., Nishio, M., Loo, T. M., Okada, R., et al. (2021). Pericentromeric noncoding RNA changes DNA binding of CTCF and inflammatory gene expression in senescence and cancer. *Proceedings of the National Academy of Sciences of the United States of America*, *118*(35), e2025647118. https://doi.org/10.1073/PNAS.2025647118.

Modzelewski, A. J., Hilz, S., Crate, E. A., Schweidenback, C. T. H., Fogarty, E. A., Grenier, J. K., et al. (2015). Dgcr8 and Dicer are essential for sex chromosome integrity during meiosis in males. *Journal of Cell Science*, *128*(12), 2314–2327. https://doi.org/10.1242/JCS.167148/260400/AM/DGCR8-AND-DICER-ARE-ESSENTIAL-FOR-SEX-CHROMOSOME.

Modzelewski, A. J., Holmes, R. J., Hilz, S., Grimson, A., & Cohen, P. E. (2012). AGO4 regulates entry into meiosis and influences silencing of sex chromosomes in the male mouse germline. *Developmental Cell*, *23*(2), 251–264. https://doi.org/10.1016/j.devcel.2012.07.003.

Muirhead, C. A., & Presgraves, D. C. (2021). Satellite DNA-mediated diversification of a sex-ratio meiotic drive gene family in Drosophila. *Nature Ecology & Evolution*, *5*(12), 1604–1612. https://doi.org/10.1038/s41559-021-01543-8.

Mukherjee, N., & Mukherjee, C. (2021). Germ cell ribonucleoprotein granules in different clades of life: From insects to mammals. *Wiley Interdisciplinary Reviews. RNA*, *12*(4), e1642. https://doi.org/10.1002/WRNA.1642.

Müller, M., Fäh, T., Schaefer, M., Hermes, V., Luitz, J., Stalder, P., et al. (2022). AGO1 regulates pericentromeric regions in mouse embryonic stem cells. *Life Science Alliance*, *5*(6), e202101277. https://doi.org/10.26508/LSA.202101277.

Murat, F., Mbengue, N., Winge, S. B., Trefzer, T., Leushkin, E., Sepp, M., et al. (2021). The molecular evolution of spermatogenesis across mammals. *BioRxiv*, *2021*(11), 08.467712. https://doi.org/10.1101/2021.11.08.467712.

Murchison, E. P., Stein, P., Xuan, Z., Pan, H., Zhang, M. Q., Schultz, R. M., et al. (2007). Critical roles for Dicer in the female germline. *Genes & Development*, *21*(6), 682–693. https://doi.org/10.1101/GAD.1521307.

Nakamura, M., Ando, R., Nakazawa, T., Yudazono, T., Tsutsumi, N., Hatanaka, N., et al. (2007). Dicer-related drh-3 gene functions in germ-line development by maintenance of chromosomal integrity in Caenorhabditis elegans. *Genes to Cells*, *12*(9), 997–1010. https://doi.org/10.1111/J.1365-2443.2007.01111.X.

Namekawa, S. H., Park, P. J., Zhang, L. F., Shima, J. E., McCarrey, J. R., Griswold, M. D., et al. (2006). Postmeiotic sex chromatin in the male germline of mice. *Current Biology*, *16*(7), 660–667. https://doi.org/10.1016/J.CUB.2006.01.066.

Nitschko, V., Kunzelmann, S., Fröhlich, T., Fröhlich, F., Arnold, G. J., Orstemann, K. F., et al. (2020). Trafficking of siRNA precursors by the dsRBD protein Blanks in Drosophila. *Nucleic Acids Research*, *48*(7), 3906–3921. https://doi.org/10.1093/nar/gkaa072.

Ohle, C., Tesorero, R., Schermann, G., Dobrev, N., Sinning, I., & Fischer, T. (2016). Transient RNA-DNA Hybrids are required for efficient double-strand break repair. *Cell*, *167*(4). https://doi.org/10.1016/J.CELL.2016.10.001. 1001-1013.e7.

Okamura, K., Ishizuka, A., Siomi, H., & Siomi, M. C. (2004). Distinct roles for Argonaute proteins in small RNA-directed RNA cleavage pathways. *Genes & Development*, *18*(14), 1655–1666. https://doi.org/10.1101/GAD.1210204.

Onohara, Y., Fujiwara, T., Yasukochi, T., Himeno, M., & Yokota, S. (2010). Localization of mouse vasa homolog protein in chromatoid body and related nuage structures of mammalian spermatogenic cells during spermatogenesis. *Histochemistry and Cell Biology*, *133*(6), 627–639. https://doi.org/10.1007/S00418-010-0699-5/FIGURES/8.

Ouyang, J. P. T., & Seydoux, G. (2022 Jan). Nuage condensates: Accelerators or circuit breakers for sRNA silencing pathways? *RNA (New York, N.Y.)*, *28*(1), 58–66. https://doi.org/10.1261/RNA.079003.121. rna.079003.121.

Pal-Bhadra, M., Leibovitch, B. A., Gandhi, S. G., Rao, M., Bhadra, U., Birchler, J. A., et al. (2004). Heterochromatic silencing and HP1 localization in Drosophila are dependent on the RNAi machinery. *Science (New York, N.Y.)*, *303*(5658), 669–672. https://doi.org/10.1126/SCIENCE.1092653.

Park, J., Lee, H., Han, N., Kwak, S., Lee, H. T., Kim, J. H., et al. (2018). Long non-coding RNA ChRO1 facilitates ATRX/DAXX-dependent H3.3 deposition for transcription-associated heterochromatin reorganization. *Nucleic Acids Research*, *46*(22), 11759–11775. https://doi.org/10.1093/NAR/GKY923.

Park, M. S., Phan, H. D., Busch, F., Hinckley, S. H., Brackbill, J. A., Wysocki, V. H., et al. (2017). Human Argonaute3 has slicer activity. *Nucleic Acids Research*, *45*(20), 11867–11877. https://doi.org/10.1093/nar/gkx916.

Park, M. S., Sim, G. Y., Kehling, A. C., & Nakanishi, K. (2020). Human Argonaute2 and Argonaute3 are catalytically activated by different lengths of guide RNA. *Proceedings of the National Academy of Sciences of the United States of America*, *117*(46), 28576–28578. https://doi.org/10.1073/pnas.2015026117.

Parker, G. S., Eckert, D. M., & Bass, B. L. (2006). RDE-4 preferentially binds long dsRNA and its dimerization is necessary for cleavage of dsRNA to siRNA. *RNA (New York, N.Y.)*, *12*(5), 807–818. https://doi.org/10.1261/RNA.2338706.

Parker, R., & Sheth, U. (2007). P bodies and the control of mRNA translation and degradation. *Molecular Cell*, *25*(5), 635–646. https://doi.org/10.1016/J.MOLCEL.2007.02.011.

Patel, P. H., Barbee, S. A., & Blankenship, J. T. (2016). GW-bodies and P-bodies constitute two separate pools of sequestered non-translating RNAs. *PLoS One*, *11*(3), e0150291. https://doi.org/10.1371/JOURNAL.PONE.0150291.

Perales, R., Pagano, D., Wan, G., Fields, B. D., Saltzman, A. L., & Kennedy, S. G. (2018). Transgenerational epigenetic inheritance is negatively regulated by the HERI-1 chromodomain protein. *Genetics*, *210*(4), 1287–1299. https://doi.org/10.1534/GENETICS.118.301456.

Quinn, J. J., & Chang, H. Y. (2016). Unique features of long non-coding RNA biogenesis and function. *Nature Reviews. Genetics*, *17*(1), 47–62. Nature Publishing Group. https://doi.org/10.1038/nrg.2015.10.

Ransohoff, J. D., Wei, Y., & Khavari, P. A. (2018). The functions and unique features of long intergenic non-coding RNA. *Nature Reviews. Molecular Cell Biology*, *19*(3), 143–157. Nature Publishing Group. https://doi.org/10.1038/nrm.2017.104.

Rappaport, Y., Achache, H., Falk, R., Murik, O., Ram, O., & Tzur, Y. B. (2021). Bisection of the X chromosome disrupts the initiation of chromosome silencing during meiosis in Caenorhabditis elegans. *Nature Communications*, *12*(1), 4802. https://doi.org/10.1038/s41467-021-24815-0.

Romero, Y., Meikar, O., Papaioannou, M. D., Conne, B., Grey, C., Weier, M., et al. (2011). Dicer1 Depletion in male germ cells leads to infertility due to cumulative meiotic and spermiogenic defects. *PLoS One*, *6*(10), e25241. https://doi.org/10.1371/JOURNAL.PONE.0025241.

Ruby, J. G., Jan, C. H., & Bartel, D. P. (2007). Intronic microRNA precursors that bypass Drosha processing. *Nature*, *448*(7149), 83–86. https://doi.org/10.1038/nature05983.

Russo, J., Harrington, A. W., & Steiniger, M. (2016). Antisense transcription of retrotransposons in Drosophila: an origin of endogenous small interfering RNA precursors. *Genetics*, *202*(1), 107–121. https://doi.org/10.1534/GENETICS.115.177196.

Saito, K., Ishizuka, A., Siomi, H., & Siomi, M. C. (2005). Processing of pre-microRNAs by the Dicer-1-Loquacious complex in Drosophila cells. *PLoS Biology*, *3*(7), 1202–1212. https://doi.org/10.1371/JOURNAL.PBIO.0030235.

Salim, U., Kumar, A., Kulshreshtha, R., & Vivekanandan, P. (2021). Biogenesis, characterization, and functions of mirtrons. *Wiley Interdisciplinary Reviews. RNA*, *13*(1), e1680. https://doi.org/10.1002/WRNA.1680.

Samarajeewa, D. A., Sauls, P. A., Sharp, K. J., Smith, Z. J., Xiao, H., Groskreutz, K. M., et al. (2014). Efficient detection of unpaired DNA requires a member of the rad54-like family of homologous recombination proteins. *Genetics*, *198*(3), 895–904. https://doi.org/10.1534/genetics.114.168187.

Schimenti, J. (2005). Synapsis or silence. *Nature Genetics*, *37*(1), 11–13. https://doi.org/10.1038/ng0105-11.

She, X., Xu, X., Fedotov, A., Kelly, W. G., & Maine, E. M. (2009). Regulation of heterochromatin assembly on unpaired chromosomes during Caenorhabditis elegans meiosis by components of a small RNA-mediated pathway. *PLoS Genetics*, *5*(8), e1000624. https://doi.org/10.1371/journal.pgen.1000624.

Shi, Z., Lim, C., Tran, V., Cui, K., Zhao, K., & Chen, X. (2020). Single-cyst transcriptome analysis of Drosophila male germline stem cell lineage. *Development*, *147*(8), dev184259. https://doi.org/10.1242/DEV.184259.

Shi, J., Ma, Y., Hua, H., Liu, Y., Li, W., Yu, H., et al. (2021). Dynamic histone H3 modifications regulate meiosis initiation via respiration. *Frontiers in Cell and Development Biology*, *9*, 683. https://doi.org/10.3389/FCELL.2021.646214/BIBTEX.

Song, W. Y., Meng, H., Wang, X. G., Jin, H. X., Yao, G. D., Shi, S. L., et al. (2017). Reduced microRNA-188-3p expression contributes to apoptosis of spermatogenic cells in patients with azoospermia. *Cell Proliferation*, *50*(1), e12297. https://doi.org/10.1111/CPR.12297.

Starega-Roslan, J., Witkos, T. M., Galka-Marciniak, P., & Krzyzosiak, W. J. (2015). Sequence features of Drosha and Dicer cleavage sites affect the complexity of isomiRs. *International Journal of Molecular Sciences*, *16*(4), 8110–8127. https://doi.org/10.3390/IJMS16048110.

Sugiyama, T., Cam, H., Verdel, A., Moazed, D., & Grewal, S. I. S. (2005). RNA-dependent RNA polymerase is an essential component of a self-enforcing loop coupling heterochromatin assembly to siRNA production. *Proceedings of the National Academy of Sciences of the United States of America*, *102*(1), 152–157. https://doi.org/10.1073/PNAS.0407641102.

Suh, N., Baehner, L., Moltzahn, F., Melton, C., Shenoy, A., Chen, J., et al. (2010). MicroRNA function is globally suppressed in mouse oocytes and early embryos. *Current Biology*, *20*(3), 271. https://doi.org/10.1016/J.CUB.2009.12.044.

Sundby, A. E., Molnar, R. I., & Claycomb, J. M. (2021). Connecting the dots: Linking Caenorhabditis elegans small RNA pathways and germ granules. *Trends in Cell Biology*, *31*(5), 387–401. https://doi.org/10.1016/J.TCB.2020.12.012.

Swahari, V., Nakamura, A., Baran-Gale, J., Garcia, I., Crowther, A. J., Sons, R., et al. (2016). Essential function of dicer in resolving DNA damage in the rapidly dividing cells of the developing and malignant cerebellum. *Cell Reports*, *14*(2), 216–224. https://doi.org/10.1016/J.CELREP.2015.12.037.

Swarts, D. C., Makarova, K., Wang, Y., Nakanishi, K., Ketting, R. F., Koonin, E. V., et al. (2014). The evolutionary journey of Argonaute proteins. *Nature Structural & Molecular Biology*, *21*(9), 743–753. Nature Publishing Group. https://doi.org/10.1038/nsmb.2879.

Takimoto, K., Wakiyama, M., & Yokoyama, S. (2009). Mammalian GW182 contains multiple Argonaute-binding sites and functions in microRNA-mediated translational repression. *RNA (New York, N.Y.)*, *15*(6), 1078–1089. https://doi.org/10.1261/RNA.1363109.

Talbert, P. B., & Henikoff, S. (2020). What makes a centromere? *Experimental Cell Research*, *389*(2), 111895. https://doi.org/10.1016/J.YEXCR.2020.111895.

Taylor, D. H., Chu, E. T. J., Spektor, R., & Soloway, P. D. (2015). Long non-coding RNA regulation of reproduction and development. *Molecular Reproduction and Development*, *82*(12), 932–956. https://doi.org/10.1002/mrd.22581.

Thomson, T., & Lin, H. (2009). The biogenesis and function of PIWI proteins and piRNAs: Progress and prospect. *Annual Review of Cell and Developmental Biology*, *25*, 355–376. https://doi.org/10.1146/annurev.cellbio.24.110707.175327.

Turner, J. M. A. (2007). Meiotic sex chromosome inactivation. *Development*, *134*(10), 1823–1831. https://doi.org/10.1242/dev.000018.

Turner, J. M. A. (2015). Meiotic silencing in mammals. *Annual Review of Genetics, 49*, 395–412. https://doi.org/10.1146/ANNUREV-GENET-112414-055145.

Turner, J. M. A., Aprelikova, O., Xu, X., Wang, R., Kim, S., Chandramouli, G. V. R., et al. (2004). BRCA1, histone H2AX phosphorylation, and male meiotic sex chromosome inactivation. *Current Biology, 14*(23), 2135–2142. https://doi.org/10.1016/J.CUB.2004.11.032.

Turner, J. M. A., Mahadevaiah, S. K., Fernandez-Capetillo, O., Nussenzweig, A., Xu, X., Deng, C. X., et al. (2004). Silencing of unsynapsed meiotic chromosomes in the mouse. *Nature Genetics, 37*(1), 41–47. https://doi.org/10.1038/ng1484.

Tzur, Y. B., Winter, E., Gao, J., Hashimshony, T., Yanai, I., & Colaiácovo, M. P. (2018). Spatiotemporal gene expression analysis of the Caenorhabditis elegans germline uncovers a syncytial expression switch. *Genetics, 210*(2), 587–605. https://doi.org/10.1534/GENETICS.118.301315.

Vedanayagam, J., Lin, C. J., & Lai, E. C. (2021). Rapid evolutionary dynamics of an expanding family of meiotic drive factors and their hpRNA suppressors. *Nature Ecology & Evolution, 5*(12), 1613–1623. https://doi.org/10.1038/s41559-021-01592-z.

Ventelä, S., Toppari, J., & Parvinen, M. (2003). Intercellular organelle traffic through cytoplasmic bridges in early spermatids of the rat: mechanisms of haploid gene product sharing. *Molecular Biology of the Cell, 14*(7), 2768. https://doi.org/10.1091/MBC.E02-10-0647.

Verdel, A., Jia, S., Gerber, S., Sugiyama, T., Gygi, S., Grewal, S. I. S., et al. (2004). RNAi-mediated targeting of heterochromatin by the RITS complex. *Science (New York, N.Y.), 303*(5658), 672–676. https://doi.org/10.1126/SCIENCE.1093686.

Vibranovski, M. D. (2014). Meiotic sex chromosome inactivation in Drosophila. *Journal of Genomics, 2*, 104–117. https://doi.org/10.7150/jgen.8178.

Volpe, T., Schramke, V., Hamilton, G. L., White, S. A., Teng, G., Martienssen, R. A., et al. (2003). RNA interference is required for normal centromere function in fission yeast. *Chromosome Research, 11*(2), 137–146. https://doi.org/10.1023/A:1022815931524.

Voronina, E., Seydoux, G., Sassone-Corsi, P., & Nagamori, I. (2011). RNA granules in germ cells. *Cold Spring Harbor Perspectives in Biology, 3*(12), a002774. https://doi.org/10.1101/CSHPERSPECT.A002774.

Wang, L., & Xu, C. (2015). Role of microRNAs in mammalian spermatogenesis and testicular germ cell tumors. *Reproduction, 149*(3), R127–R137. https://doi.org/10.1530/REP-14-0239.

Wang, L., Xu, Z., Khawar, M. B., Liu, C., & Li, W. (2017). The histone codes for meiosis. *Reproduction (Cambridge, England), 154*(3), R65–R79. https://doi.org/10.1530/REP-17-0153.

Wedeles, C. J., Wu, M. Z., & Claycomb, J. M. (2013). A multitasking Argonaute: Exploring the many facets of C. elegans CSR-1. *Chromosome Research, 21*(6–7), 573–586. https://doi.org/10.1007/S10577-013-9383-7/TABLES/1.

Wee, L. M., Flores-Jasso, C. F., Salomon, W. E., & Zamore, P. D. (2012). Argonaute divides its RNA guide into domains with distinct functions and RNA-binding properties. *Cell, 151*(5), 1055–1067. https://doi.org/10.1016/J.CELL.2012.10.036.

Wei, W., Ba, Z., Gao, M., Wu, Y., Ma, Y., Amiard, S., et al. (2012). A role for small RNAs in DNA double-strand break repair. *Cell, 149*(1), 101–112. https://doi.org/10.1016/j.cell.2012.03.002.

Weirich, S., Khella, M. S., & Jeltsch, A. (2021). Structure, activity and function of the suv39h1 and suv39h2 protein lysine methyltransferases. *Life, 11*(7), 703. MDPI AG. https://doi.org/10.3390/life11070703.

Wilson, R. C., Tambe, A., Kidwell, M. A., Noland, C. L., Schneider, C. P., & Doudna, J. A. (2015). Dicer-TRBP complex formation ensures accurate mammalian MicroRNA biogenesis. *Molecular Cell, 57*(3), 397–407. https://doi.org/10.1016/J.MOLCEL.2014.11.030/ATTACHMENT/08369455-A78E-402D-99D5-A15CAF4E159D/MMC2.ZIP.

Wynant, N., Santos, D., & vanden Broeck, J. (2017). The evolution of animal Argonautes: Evidence for the absence of antiviral AGO Argonautes in vertebrates. *Scientific Reports*, 7(1), 1–13. https://doi.org/10.1038/s41598-017-08043-5.

Xiao, H., Vierling, M. M., Kennedy, R. F., Boone, E. C., Decker, L. M., Sy, V. T., et al. (2021). Involvement of RNA granule proteins in meiotic silencing by unpaired DNA. *G3 (Bethesda, Md.)*, *11*(10), jkab179. https://doi.org/10.1093/G3JOURNAL/JKAB179.

Yadav, R. P., Mäkelä, J. A., Hyssälä, H., Cisneros-Montalvo, S., & Kotaja, N. (2020). DICER regulates the expression of major satellite repeat transcripts and meiotic chromosome segregation during spermatogenesis. *Nucleic Acids Research*, *48*(13), 7135–7153. https://doi.org/10.1093/nar/gkaa460.

Ye, X., Huang, N., Liu, Y., Paroo, Z., Huerta, C., Li, P., et al. (2011). Structure of C3PO and mechanism of human RISC activation. *Nature Structural & Molecular Biology*, *18*(6), 650–657. https://doi.org/10.1038/nsmb.2032.

Yi, R., Qin, Y., Macara, I. G., & Cullen, B. R. (2003). Exportin-5 mediates the nuclear export of pre-microRNAs and short hairpin RNAs. *Genes & Development*, *17*(24), 3011–3016. https://doi.org/10.1101/GAD.1158803.

Yigit, E., Batista, P. J., Bei, Y., Pang, K. M., Chen, C. C. G., Tolia, N. H., et al. (2006). Analysis of the C. elegans Argonaute family reveals that distinct Argonautes act sequentially during RNAi. *Cell*, *127*(4), 747–757. https://doi.org/10.1016/J.CELL.2006.09.033.

Zamore, P. D., Tuschl, T., Sharp, P. A., & Bartel, D. P. (2000). RNAi: Double-stranded RNA directs the ATP-dependent cleavage of mRNA at 21 to 23 nucleotide intervals. *Cell*, *101*(1), 25–33. https://doi.org/10.1016/S0092-8674(00)80620-0.

Zaytseva, O., Mitchell, N. C., Guo, L., Marshall, O. J., Parsons, L. M., Hannan, R. D., et al. (2020). Transcriptional repression of Myc underlies the tumour suppressor function of AGO1 in Drosophila. *Development*, *147*(11), dev190231. https://doi.org/10.1242/DEV.190231/225316.

Zhao, Z. H., Ma, J. Y., Meng, T. G., Wang, Z. B., Yue, W., Zhou, Q., et al. (2020). Single-cell RNA sequencing reveals the landscape of early female germ cell development. *The FASEB Journal*, *34*(9), 12634–12645. https://doi.org/10.1096/FJ.202001034RR.

Zimmermann, C., Romero, Y., Warnefors, M., Bilican, A., Borel, C., Smith, L. B., et al. (2014). Germ cell-specific targeting of DICER or DGCR8 reveals a novel role for endo-siRNAs in the progression of mammalian spermatogenesis and male fertility. *PLoS One*, *9*(9), e107023. https://doi.org/10.1371/JOURNAL.PONE.0107023.

CHAPTER ELEVEN

# Checkpoint control in meiotic prophase: Idiosyncratic demands require unique characteristics

**Vivek B. Raina[a], Maud Schoot Uiterkamp[b,c,d], and Gerben Vader[b,c,d,*]**
[a]Department of Biochemistry and Molecular Biophysics, Columbia University Medical Center, New York City, NY, United States
[b]Center for Reproductive Medicine, Reproductive Biology Laboratory, Amsterdam Reproduction and Development Research Institute, Amsterdam UMC, University of Amsterdam, Amsterdam, The Netherlands
[c]Cancer Center Amsterdam, Cancer Biology and Immunology, Amsterdam, The Netherlands
[d]Section of Oncogenetics, Department of Human Genetics, Amsterdam UMC, Vrije Universiteit, Amsterdam, The Netherlands
*Corresponding author: e-mail address: g.vader@amsterdamumc.nl

## Contents



## Abstract

Chromosomal transactions such as replication, recombination and segregation are monitored by cell cycle checkpoint cascades. These checkpoints ensure the proper execution of processes that are needed for faithful genome inheritance from one cell to the next, and across generations. In meiotic prophase, a specialized checkpoint monitors defining events of meiosis: programmed DNA break formation, followed by dedicated repair through recombination based on interhomolog (IH) crossovers. This checkpoint shares molecular characteristics with canonical DNA damage checkpoints active during somatic cell cycles. However, idiosyncratic requirements of meiotic prophase have introduced unique features in this signaling cascade. In this review, we discuss the unique features of the meiotic prophase checkpoint. While being related to canonical DNA damage checkpoint cascades, the meiotic prophase checkpoint also shows similarities with the spindle assembly checkpoint (SAC) that guards chromosome segregation. We highlight these emerging similarities in the signaling logic of the checkpoints that govern meiotic prophase and chromosome segregation, and how thinking of these similarities can help us better understand meiotic prophase control. We also discuss work showing that, when aberrantly expressed, components of the meiotic prophase

*Current Topics in Developmental Biology*, Volume 151
ISSN 0070-2153
https://doi.org/10.1016/bs.ctdb.2022.04.007

checkpoint might alter DNA repair fidelity and chromosome segregation in cancer cells. Considering checkpoint function in light of demands imposed by the special characteristics of meiotic prophase helps us understand checkpoint integration into the meiotic cell cycle machinery.

## 1. Mitotic and meiotic checkpoints

Equal assortment of chromosomes during cell division ensures inheritance of genetic information from one generation to the next. Mitosis produces genetically identical cells, whereas meiosis (which in multicellular organisms is restricted to dedicated germ cells) leads to the formation of reproductive cells (i.e., gametes—sperm and egg, or spores) that contain half the number of chromosomes of the parental cell (Petronczki, Siomos, & Nasmyth, 2003). Regulatory mechanisms—cell cycle checkpoints (Hartwell & Weinert, 1989)—monitor the execution of molecular events, and establish order during cell cycle progression. *DNA damage checkpoints* monitor genome integrity (Weinert & Hartwell, 1988). Upon detection of DNA lesions, signaling cascades lead to cell cycle arrest and the activation of DNA repair. Halting cell division when genome integrity is challenged minimizes the propagation of a damaged genome, and DNA damage checkpoint failure leads to genome destabilization and the development of cancer (Molinaro, Martoriati, & Cailliau, 2021). A distinct checkpoint monitors chromosome segregation. The *spindle assembly checkpoint* (SAC) guards the separation of chromatids to prevent genomically imbalanced daughter cells (Hoyt, Totis, & Roberts, 1991; Li & Murray, 1991). Chromosome missegregation can lead to the generation of aneuploid cells. Aneuploidy is associated with cancer, developmental defects and reproductive failure (Hassold & Hunt, 2001; Vasudevan et al., 2021). Indeed, mutations in genes required for SAC function contribute to increased cancer risk and to reproductive failure (Baker et al., 2004; Hanks et al., 2004; Limwongse, Schwartz, Bocian, & Robin, 1999; Yost et al., 2017; Zhang et al., 2020).

During meiosis, the formation of haploid products from a diploid progenitor cell requires modifications to cell cycle events, and also of checkpoint function (MacQueen & Hochwagen, 2011; Subramanian & Hochwagen, 2014). The *meiotic prophase checkpoint* is active when DNA lesions are introduced in the form of DNA double strand breaks (DSBs) and subsequently repaired into crossovers. Checkpoint dysfunction impacts meiotic chromosome assortment and gamete production (Rinaldi, Bloom, & Schimenti, 2020; Rinaldi, Bolcun-Filas, Kogo, Kurahashi, & Schimenti, 2017). We discuss how the demands of meiotic DSB formation and repair have led to characteristics of the meiotic prophase checkpoint (i.e., what it detects, how it influences cell cycle progression, and why), and how it shares molecular as well as conceptual similarities with canonical DNA damage checkpoints and the SAC.

## 2. Unique requirements of meiotic prophase

To appreciate the unique nature of the meiotic prophase checkpoint it is needed to consider the "goal" of DSB formation and repair: to enable the formation of crossover repair products between homologous chromosomes. DSBs are introduced and repaired into crossover products during meiotic prophase. Crossover formation generates linkages between homologous chromosomes. This linking is needed for faithful segregation of homologous chromosome pairs in meiosis (Petronczki et al., 2003). DSB formation occurs at defined regions within the genome (i.e., hotspots) and is driven by Spo11—a topoisomerase-like enzyme—in concert with additional factors (Bergerat et al., 1997; Keeney, Giroux, & Kleckner, 1997; Keeney, Lange, & Mohibullah, 2014; Lam & Keeney, 2014). DSB repair uses homologous recombination (HR), a repair pathway that uses sequences found elsewhere in the genome to repair DSB lesions (Hunter, 2015). In mitotically-dividing cells, HR prefers to utilize templates present on sister chromatids (Bzymek, Thayer, Oh, Kleckner, & Hunter, 2010). Meiotic HR-based repair requires usage of templates present on homologous chromosomes (i.e., interhomolog (IH)-directed repair). The switch from intersister to interhomolog-directed repair (i.e., IH bias) characterizes meiotic DNA repair (Humphryes & Hochwagen, 2014; Hunter, 2015; Schwacha & Kleckner, 1994, 1997). An IH crossover results in a reciprocal exchange of chromosome arms of homologs which, in combination with sister chromatid cohesion, generates homolog linkages (Petronczki et al., 2003). The placement of crossovers too close to each other can lead to a failure to establish homolog linkages that support efficient chromosome segregation (Hollis et al., 2020). Thus, crossovers should not be established in close vicinity of each other—a process called "crossover interference" (Berchowitz & Copenhaver, 2010; Muller, 1916; Otto & Payseur, 2019; Sturtevant, 1913, 1915). Concomitantly with the establishment of IH-interactions (via HR-based invasion events), a meiosis-specific structure—the synaptonemal complex (SC)—is formed in between homologs (Cahoon & Hawley, 2016; Hunter, 2015). The proteins that drive establishment and polymerization of the SC influence crossover generation and interference, and SC formation drives progression through the meiotic program.

Every chromosome pair needs to receive at least one IH crossover, and the placement of DSBs and crossovers relative to each other should be controlled. The meiotic prophase checkpoint creates the biochemical circumstances that allow specific DSB repair (i.e., IH crossover repair), with particular requirements (i.e., crossover interference). Crucially, key components of the meiotic prophase checkpoint are also directly responsible for steering these processes (as will be discussed below). Checkpoint function is thus intricately connected with the idiosyncratic characteristics of meiotic DSB repair.

## 3. The meiotic prophase checkpoint

How does the meiotic prophase checkpoint determine control of the Spo11-dependent DSB formation and repair? Detailed knowledge regarding checkpoint function derives mostly from studies in the budding yeast *Saccharomyces cerevisiae*, and that work is central to our review. When appropriate we will mention names of mammalian genes in parentheses. Spo11 remains covalently attached to the 5′-termini of cleaved DNA ends, upon which it is released by the endonucleolytic activity of Sae2 (CtIP) (Clerici, Mantiero, Lucchini, & Longhese, 2005; Farah, Cromie, & Smith, 2009; Keeney & Kleckner, 1995; McKee & Kleckner, 1997; Prinz, Amon, & Klein, 1997), followed by further 5′-3′ resection facilitated by Exo1, the MRX (Mre11-Rad50-Xrs2) (MRN) complex, and Sae2 (Cannavo & Cejka, 2014; Cannavo, Reginato, & Cejka, 2019; Garcia, Phelps, Gray, & Neale, 2011; Hartsuiker et al., 2009; Nairz & Klein, 1997; Rothenberg, Kohli, & Ludin, 2009; Tsubouchi & Ogawa, 1998; Zakharyevich et al., 2010). The DNA damage checkpoint kinases Tel1 (ATM) and Mec1 (ATR) are activated at Spo11-mediated DSBs (Awasthi, Foiani, & Kumar, 2015; Marechal & Zou, 2013; Paull, 2015). Immediate signaling downstream of Spo11-mediated DSB in meiosis is similar to the response in mitotic cells (e.g., Carballo & Cha, 2007; Ho & Burgess, 2011; Lydall, Nikolsky, Bishop, & Weinert, 1996; Refolio, Cavero, Marcon, Freire, & San-Segundo, 2011), with some notable

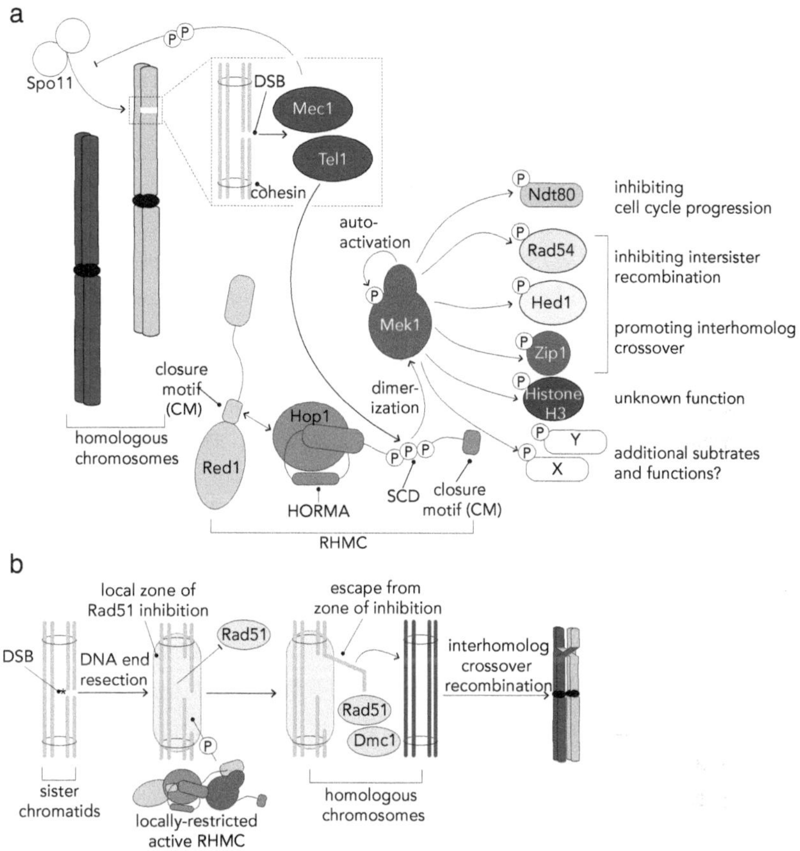


**Fig. 1** (A) Schematic of the core meiotic prophase checkpoint, with indicated interdependencies. RHMC, Red1-Hop1-Mek1 complex; DSB, DNA double strand break; SCD, SQ/TQ cluster domain. (B) The "spatial separation" model for the role of RHMC/Mek1 in control IH-based repair.

exceptions as discussed below (Fig. 1A). Tel1 is recruited to DNA ends via the MRX complex (Fukunaga, Kwon, Sung, & Sugimoto, 2011; Shiotani & Zou, 2009; Zou & Elledge, 2003). Mec1 is activated at RPA-coated 3′-single stranded DNA tails generated after 5′-3′ resection of DNA, via an interaction with Ddc2 (ATRIP) (Refolio et al., 2011; Shiotani & Zou, 2009; Zou & Elledge, 2003). Mec1/Tel1 phosphorylate a myriad of downstream targets (Chen, Albuquerque, Liang, Suhandynata, & Zhou, 2010). Mec1/Tel1 activation in canonical DNA damage responses leads to the activation of downstream checkpoint kinases, among them the Rad53 (CHK2) kinase (Awasthi et al., 2015; Marechal & Zou, 2013;

Paull, 2015). Mec1/Tel1-dependent recruitment of Rad9 (53BP1) to DSB sites triggers activation of Rad53 (Gilbert, Green, & Lowndes, 2001; Schwartz et al., 2002; Sweeney et al., 2005). This event does not occur in response to Spo11-dependent DSB formation, although Rad53 and Rad9 are expressed in meiosis, and Rad53 can be activated in response to exogenously introduced DNA damage (Cartagena-Lirola, Guerini, Manfrini, Lucchini, & Longhese, 2008). The failure of Rad53 activation in response to Spo11-DSBs likely involves a unique chromatin environment, which precludes Rad9 recruitment (Cartagena-Lirola et al., 2008; Usui & Shinohara, 2021). The meiotic checkpoint diverges from canonical DNA damage checkpoints at this point of the signaling cascade. A key meiosis-specific Mec1/Tel1 substrate is Hop1 in budding yeast (meiotic HORMADs—HORMAD1/2 in mammals) (Carballo, Johnson, Sedgwick, & Cha, 2008; Penedos et al., 2015). Hop1 is a HORMA domain protein—a structural domain also present in other factors, such as Mad2, Rev7 and p31COMET (Aravind & Koonin, 1998; Berchowitz & Copenhaver, 2010; Gu, Desai, & Corbett, 2022; Rosenberg & Corbett, 2015). HORMA domains exist in unique topological states (referred to as "open/unbuckled (O/U)" or "closed (C)") (Rosenberg & Corbett, 2015). "Closed" HORMA domains can capture binding partners via topological embrace of a peptide termed a "closure motif" or CM (Kim et al., 2014). The CM is present in binding partners of Hop1. Hop1 associates with chromosomes via a CM-mediated interaction with Red1 (SYCP2/3) (de los Santos & Hollingsworth, 1999; West et al., 2019). In addition, Hop1 itself contains a COOH-terminal CM, and these factors thus have the potential to form intramolecular HORMA-CM interactions and potentially larger multimeric assemblies (Friedman, Hollingsworth, & Byers, 1994; Kim et al., 2014; Niu et al., 2005). Red1-Hop1 (likely together with cohesin) forms a meiosis-specific chromosome organization, which guides Spo11 activity (Panizza et al., 2011). Mec1/Tel1 phosphorylate Hop1 within its SQ/TQ Cluster Domain (SCD) region (Carballo et al., 2008; Penedos et al., 2015), enabling Hop1 to act as an adaptor for the meiosis-specific kinase Mek1 (Carballo et al., 2008; Cheng, Chuang, Shen, Lin, & Wang, 2013; Penedos et al., 2015; Xie et al., 2018). Association of Mek1 with Red1-Hop1 establishes the Red1-Hop1-Mek1 complex (RHMC), a key meiotic prophase checkpoint complex (Fig. 1A). The Forkhead-associated (FHA) domain of Mek1 interacts with phosphorylated Hop1, resulting in the chromosomal recruitment and activation of Mek1 (Carballo et al., 2008; Niu et al., 2007, 2005; Penedos et al., 2015). This activation is proposed to act through controlled,

Hop1-mediated dimerization of Mek1 (Niu et al., 2005), and how this regulation is achieved structurally should be an important focus for future research. Mek1 thus takes the place of Rad53 in the canonical DNA damage checkpoint: within RHMC, Mek1 forms a signaling intermediate downstream of Mec1/Tel1 (Hunter, 2008). Hop1 is the equivalent of Rad9: Mec1/Tel1-dependent phosphorylation of Hop1 links Mec1/Tel1 to Mek1. Mek1 phosphorylates a myriad of substrates (Callender et al., 2016; Chen et al., 2018, 2015; Govin et al., 2010; Kniewel et al., 2017; Niu et al., 2007, 2005, 2009; Suhandynata, Wan, Zhou, & Hollingsworth, 2016; Sung, Shokat, Gygi, & Hollingsworth, 2009) and influences behavior downstream of DSB sensing (Fig. 1A). Without Mek1 activity, IH bias of meiotic HR-based repair is not instated and cells exit from meiotic prophase (Callender & Hollingsworth, 2010; Prugar, Burnett, Chen, & Hollingsworth, 2017; Schaefer, Loidl, & Hollingsworth, 2005), concomitant with DSBs being repaired via sister chromatid templates or in the presence of unrepaired DSBs (Carballo et al., 2008). Mek1 thus establishes meiosis-specific DNA repair and mediates the checkpoint response. The functional outcome of Mek1 with respect to IH bias is centered around Rad51. Rad51 (and its meiosis-specific homolog Dmc1 (Bishop, Park, Xu, & Kleckner, 1992)) is a RecA recombinase that facilitates strand invasion during HR (Bishop et al., 1992; Chen, Yang, & Pavletich, 2008; San Filippo, Sung, & Klein, 2008). Rad51/Dmc1-ssDNA nucleoprotein filaments can invade DNA double helices to identify identical repair templates. Mek1-dependent phosphorylation of Rad54, a binding partner of Rad51, attenuates Rad51-Rad54 complex formation (Niu et al., 2009; Sung et al., 2009). In addition, Mek1 phosphorylates Hed1, a meiosis-specific interaction partner of Rad51, whose binding interferes with Rad51-Rad54 association (Busygina et al., 2012, 2008; Callender et al., 2016; Crickard et al., 2018; Tsubouchi & Shirleen Roeder, 2006). Mek1-dependent stabilization of Hed1 negatively influences Rad51 strand exchange activity (Busygina et al., 2008; Callender et al., 2016). Hence, Mek1 activity inhibits Rad51-mediated HR. It should be noted that the way by which Rad51 promotes HR in meiosis is highly specialized. The strand exchange activity of Rad51 (which is essential to allow mitotic HR) is dispensable for meiotic HR, and, in meiosis, Rad51 acts as an accessory factor for Dmc1 exchange activity (Cloud, Chan, Grubb, Budke, & Bishop, 2012). The interplay between Rad51, Dmc1 and Mek1-dependent regulation during establishment of interhomolog bias is an important topic for future work (Lao et al., 2013).

How can the activity of Mek1 toward Rad51 be reconciled with the promotion of IH-based repair, and checkpoint activity? Mek1 activity is proposed to be restricted by the spatial organization of chromatin around Spo11-dependent DSB sites, conceivably via spatially-restricted Mec1/Tel1 activity and localized Mek1 binding, driven by Red1/Hop1 (Goldfarb & Lichten, 2010; Kniewel et al., 2017; Subramanian et al., 2016; Suhandynata et al., 2016) (Fig. 1B). With sister chromatid cohesion ensuring proximity of the unbroken sister chromatid to the broken DNA molecule, this would lead to a situation in which the sister template is present within a zone of inhibition. This would disfavor "local" IS-based repair which, combined with the stabilization of Dmc1-coated nucleofilaments by the Hop2/Mnd1 complex (Chen et al., 2004; Henry et al., 2006; Petukhova et al., 2005; Tsubouchi & Shirleen Roeder, 2002), causes a push toward IH-based repair. Additional steps in the establishment of IH bias are likely controlled by Mek1: it might counteract cohesin establishment, potentially increasing DSB end mobility to facilitate distant template searches (Hong et al., 2013; Kim et al., 2010). Via phosphorylation of the SC-component Zip1, Mek1 also promotes the establishment of repair intermediates that can mature into crossovers (Chen et al., 2015; Subramanian et al., 2016).

Mek1 also controls meiotic progression, which becomes apparent when Spo11-generated DSBs cannot be repaired (for example in cells lacking Dmc1): the meiotic prophase arrest of *dmc1*Δ cells depends on Mek1 activity (Carballo et al., 2008; Xu, Weiner, & Kleckner, 1997). This effect of Mek1 is likely a combination between a direct effect on the cell cycle machinery that drives prophase progression, and an indirect effect on the maintenance of DNA templates that are "checkpoint-proficient." A direct role for budding yeast Mek1 is found in the inhibition of Ndt80 activity, likely via a direct interaction and phosphorylation of Ndt80 (Chen et al., 2018; Prugar et al., 2017) (Fig. 1A). Ndt80 is a transcription factor, whose activity is essential for meiotic prophase exit through the activation of a set of genes that drive progression past prophase (Chu & Herskowitz, 1998). Disrupting the Ndt80-Mek1 interaction leads to a partial checkpoint defect, pointing to additional roles/substrates for Mek1 (Chen et al., 2018; Prugar et al., 2017). Mek1-dependent inhibition of intersister repair could also indirectly affect checkpoint proficiency: as long as cells cannot repair DSBs via IH-based repair, Mek1 ensures the presence of unrepaired DSBs through its local inhibitory effect on Rad51. Unrepaired breaks (and associated ssDNA tracts) sustain Mec1/Tel1-dependent checkpoint signaling. Under repair-defective conditions, Mek1 would thus enforce Mec1/Tel1 activity, triggering persistent checkpoint activation. Conversely, upon loss of Mek1 activity, DSBs can rapidly be repaired using sister chromatid templates, precipitating a decline of checkpoint-catalyzing DNA intermediates (Niu et al., 2005). The model posits that Mek1 activity toward Rad51 is restricted to a defined zone of activity, limited to the sister chromatid (Fig. 1A). On the other hand, the effect of Mek1 toward Ndt80 might argue for a (rapidly diffusible) nucleoplasmic pool of active Mek1, or for the presence of diffusible Ndt80, as is indeed suggested by a study on the regulation of Ndt80 localization by checkpoint signaling (Wang, Chang, Wu, & Tung, 2011). We need to reconcile these two potentially contradictory models (i.e., one where Mek1 activity needs to be locally restricted, and one where the checkpoint/Mek1 induces a pan-nuclear effect). What is the (dynamic) distribution of active Mek1 kinase? If a diffusible, active pool of Mek1 exists, what is its biochemical composition? Alternatively, is Ndt80 recruited to DSB sites, to be modified by Mek1?

Meiotic chromosomes undergo condensation and synapse, via SC polymerization (Cahoon & Hawley, 2016; Hunter, 2015). In budding yeast and mice, synapsis is dependent on DSB formation (Baudat, Manova, Yuen, Jasin, & Keeney, 2000; Giroux, Dresser, & Tiano, 1989; Romanienko & Camerini-Otero, 2000). In organisms such as worms and fruit flies, it is independent of DSBs (Dernburg et al., 1998) (MacQueen, Colaiácovo, McDonald, & Villeneuve, 2002; McKim et al., 1998; Vazquez, Belmont, & Sedat, 2002). SC assembly defects impact crossover formation, chromosome segregation and fertility (Gladstone, Obeso, Chuong, & Dawson, 2009; Judis, Chan, Schwartz, Seftel, & Hassold, 2004; Miyamoto et al., 2003). Synapsis defects lead to prophase delays, demonstrating that this condition leads to checkpoint activation. Factors required to halt cell cycle progression in response to SC defects have been identified. The AAA+ ATPase Pch2 (TRIP13) is required for the meiotic delay caused by SC defects, whereas the absence of Pch2 is not (or less) relevant for checkpoint arrest in *dmc1Δ* cells (Ho & Burgess, 2011; Hochwagen, Tham, Brar, & Amon, 2005; San-Segundo & Roeder, 1999; Wu & Burgess, 2006). This suggested that different checkpoints exist: one that responds to recombination defects and one that responds to synapsis defects (Wu & Burgess, 2006). Work in *C. elegans* points indeed to a direct connection between synapsis defects, checkpoint activation and Pch2 (Bhalla & Dernburg, 2005), whereas in *D. melanogaster* Pch2 also responds to DSB repair defects (Joyce & McKim, 2009). Subsequent observations, however, argue that, at least in budding yeast, these observations can also be explained by the presence of a single checkpoint cascade. First, SC formation is connected to DSB formation and crossover repair, cell cycle responses to synapsis defects rely on DSB formation, and are relayed by Hop1 and Mek1 (Chuang, Cheng, & Wang, 2012; Subramanian et al., 2016; Wu, Ho, & Burgess, 2010). Synapsis defects lead to delays in crossover formation, and continuous Spo11-dependent DSB formation (Subramanian et al., 2016; Thacker, Mohibullah, Zhu, & Keeney, 2014). Since DNA repair intermediates are key sites of SC nucleation (Agarwal & Roeder, 2000; Fung, Rockmill, Odell, & Roeder, 2004), this observation likely represents feedback regulation to ensure CO establishment. In budding yeast, synapsis defects likely trigger a cell cycle delay "indirectly," i.e., via signaling created at DSB repair sites. Under certain conditions, Pch2/TRIP13 is also required to instate a cell cycle arrest in response to unrepairable DSBs (Raina & Vader, 2020) (see below). Hence, these findings point to a shared checkpoint response pathway that originates at DSB sites, and of which Pch2 is a mediator.

## 4. The logic of the spindle assembly checkpoint

To rationalize meiotic checkpoint behavior, we point to analogies with the checkpoint that monitors chromosome segregation, i.e., the SAC. Our discussion of the SAC is centered around analogies with the meiotic prophase checkpoint, and we thus refer the reader to more complete SAC reviews for more detailed and complex information (e.g., Corbett, 2017; Lara-Gonzalez, Pines, & Desai, 2021; Musacchio, 2015). We also note that our discussion of the SAC is centered around its mitotic role in mediating faithful sister chromatid segregation, and is based on work done in a myriad of model systems. In this section, we will mostly use mammalian/human genes. Furthermore, since the role of Pch2 in the SAC is mostly based on work on TRIP13 in mammalian systems, we will refer to this factor as Pch2/TRIP13 here. The SAC prevents anaphase entry and cohesin cleavage until chromosomes are properly attached to microtubules of the spindle. Chromatids—upon severing of cohesin complexes—need to be moved toward opposite sides of the cell, driven by forces generated by microtubules of the spindle apparatus. Productive chromosome-spindle interactions, in which sister chromatids are bi-oriented on the mitotic spindle, result in correct segregation. Spindle-chromosomes interactions are established at kinetochores, large assemblies that bridge chromatin and microtubules. Kinetochores also generate the SAC signal:

assembly of the mitotic checkpoint complex (MCC) is catalyzed at unattached kinetochores (Faesen et al., 2017; Kulukian, Han, & Cleveland, 2009; Lara-Gonzalez, Kim, Oegema, Corbett, & Desai, 2021; Lara-Gonzalez, Pines, et al., 2021; Musacchio, 2015; Piano et al., 2021; Rieder, Cole, Khodjakov, & Sluder, 1995) (Fig. 2A). The SAC is active during every (unperturbed) mitosis, and can be activated by single unattached kinetochores (Collin, Nashchekina, Walker, & Pines, 2013; Dick & Gerlich, 2013; Lara-Gonzalez, Pines, et al., 2021; Musacchio, 2015; Musacchio & Salmon, 2007; Rieder et al., 1995). The MCC is composed of Mad2, Mad3(BubR1), Bub3 and Cdc20 (Kulukian et al., 2009; Sudakin, Chan, & Yen, 2001). Mad2 is a HORMA protein, BubR1 is a pseudokinase (Suijkerbuijk et al., 2012), and Bub3 is a 7-bladed β-propeller WD40 domain protein (Larsen & Harrison, 2004). Cdc20 is an essential activator of an E3-ubiquitin ligase called the Anaphase promoting complex/cyclosome (APC/C) (Dawson, Roth, & Artavanis-Tsakonas, 1995; Visintin, Prinz, & Amon, 1997). Upon formation, the MCC diffuses into the cytoplasm to capture and inhibit the APC/C, as such preventing anaphase onset (Lara-Gonzalez, Pines, et al., 2021; Musacchio, 2015). After bi-orientation, MCC disassembly allows APC/C-Cdc20 to become active toward two of its substrates, cyclin B and securin (Cohen-Fix, Peters, Kirschner, & Koshland, 1996; Miniowitz-Shemtov, Teichner, Sitry-Shevah, & Hershko, 2010; Murray, Solomon, & Kirschner, 1989; Primorac & Musacchio, 2013; Shirayama, Tóth, Gálová, & Nasmyth, 1999). APC/C ubiquitinates cyclin B, triggering proteasomal degradation and resulting in the inactivation of Cdk1, and anaphase onset (Dawson et al., 1995; Irniger, Piatti, Michaelis, & Nasmyth, 1995; King et al., 1995; Sudakin et al., 1995). APC/C-mediated ubiquitination likewise drives degradation of securin (Ciosk et al., 1998; Cohen-Fix et al., 1996; Funabiki et al., 1996; Primorac & Musacchio, 2013), liberating a protease called separase. Separase cleaves the kleisin subunit of cohesin, resulting in loss of sister chromatid cohesion (Nasmyth, Peters, & Uhlmann, 2001; Uhlmann, Lottspeich, & Nasmyth, 1999; Uhlmann, Wernic, Poupart, Koonin, & Nasmyth, 2000). Thus, the SAC prevents the loss of sister chromatid cohesion and inhibits the premature entry of cells into anaphase under conditions where not all chromosomes are properly attached to microtubules.

A central component of the SAC is Mad2, which can switch between two topological states (i.e., O-Mad2 and C-Mad2) (Alfieri, Chang, & Barford, 2018; Mapelli & Musacchio, 2007; Musacchio, 2015) (Fig. 2A). To capture Cdc20 within the MCC, this topological transition is crucial:

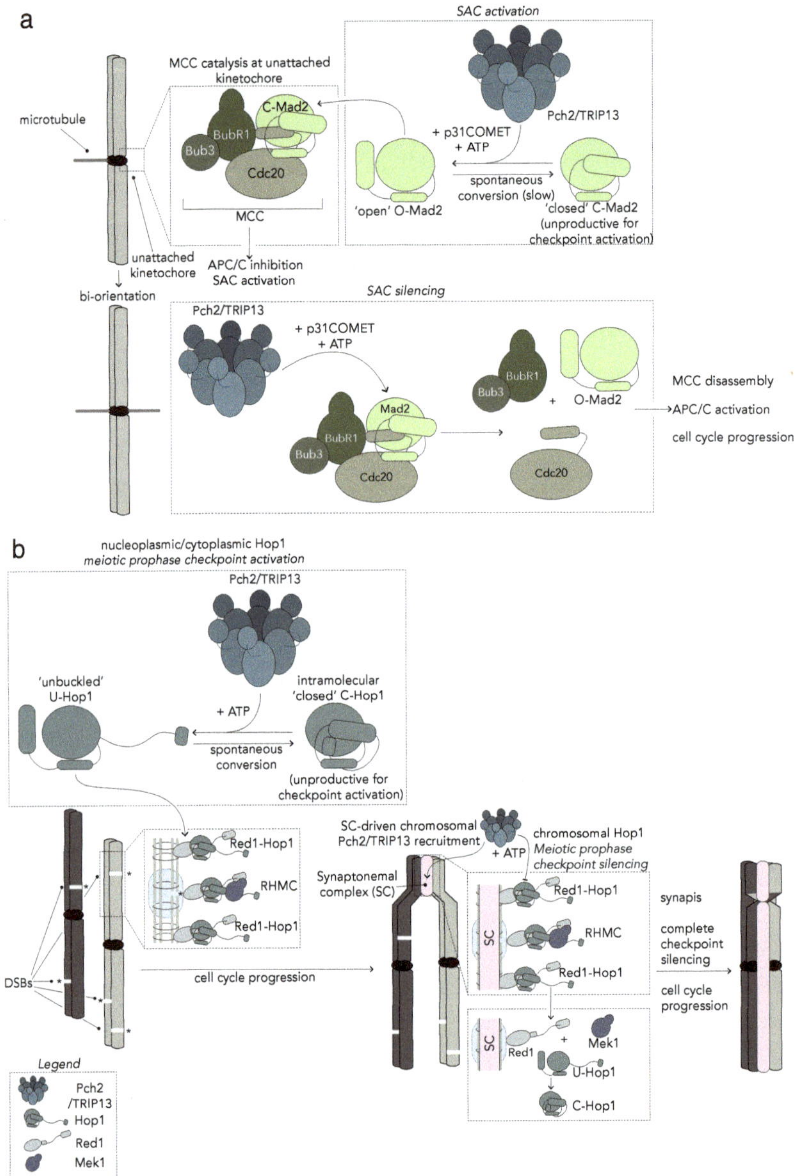


**Fig. 2** (A) Schematic depicting the dual roles played by Pch2/TRIP13 in both the activation and silencing of the SAC during chromosome segregation. MCC, mitotic checkpoint complex. (B) Schematic depicting the dual role of Pch2/TRIP13 during the meiotic prophase checkpoint. Activation and silencing functions are set by the localization of Pch2/TRIP13, where the formation of the synaptonemal complex determines toggling between activating and silencing activities of Pch2/TRIP13. SC, synaptonemal complex.

Cdc20 can be bound via a topological embrace by C-Mad2 (Luo, Tang, Rizo, & Yu, 2002; Mapelli, Massimiliano, Santaguida, & Musacchio, 2007; Sironi et al., 2002). At unattached kinetochores, the catalysis of O-Mad2 to C-Mad2 is a central step in MCC formation and SAC function (Musacchio, 2015). In order to keep the MCC catalysis cycle dynamic and productive, free, unbound O-Mad2 should be available for incorporation into the MCC (Kim et al., 2018). Once the SAC signal needs to be silenced (i.e., when all chromosomes have reached proper bi-orientation), Cdc20 should be released from the MCC (reviewed in Gu et al., 2022). Both these reactions depend on Pch2/TRIP13—the same enzyme that is a central regulator of the meiotic prophase checkpoint (Eytan et al., 2014; Kim et al., 2018; Ma & Poon, 2016, 2018; Marks et al., 2017; Tipton et al., 2012; Wang et al., 2014; Ye et al., 2015). In budding yeast, Pch2/TRIP13 is a meiosis-specific factor and as such likely does not play a role in the biochemical wiring of the mitotic SAC. It is currently unclear how budding yeast SAC signaling operates without input from Pch2/TRIP13-driven activity. Pch2/TRIP13 acts on HORMA domains by leveraging energy provided by ATP hydrolysis (Alfieri et al., 2018; Brulotte et al., 2017; Chen, Jomaa, Ortega, & Alani, 2014; Ye et al., 2017, 2015). C-Mad2 can exist on its own (i.e., as "monomeric" C-Mad2, which is not proficient in MCC incorporation (Ma & Poon, 2018) or within a protein complex (i.e., the MCC)). Pch2/TRIP13 loss leads to a defect in SAC functioning, which can be explained by the role of Pch2/TRIP13 in the (re)generation of monomeric O-Mad2: SAC function requires that sufficient O-Mad2 is available for MCC incorporation. Without Pch2/TRIP13, the pool of "monomeric" O-Mad2 quickly declines (Kim et al., 2018), preventing efficient MCC generation, thus leading to SAC deficiency. Is there also an effect of Pch2/TRIP13 on MCC-bound C-Mad2? Such an effect is expected to cause a SAC silencing defect, and this is exactly what is seen upon Pch2/TRIP13 depletion in cells that are at the metaphase-to-anaphase condition (Eytan et al., 2014; Kim et al., 2018; Wang et al., 2014). The dynamic assembly and disassembly of MCC, via interplay between the AAA+ ATPase Pch2/TRIP13 and Mad2, is thus critical for SAC signaling (Fig. 2A).

In addition to bi-orientation, several non-productive kinetochore-microtubule interactions occur. For example, sister kinetochores can be associated with microtubules that emanate from the same pole of the spindle. How are such interactions sensed and corrected? The Chromosomal Passenger Complex (CPC), with Aurora-B kinase as the enzymatic core, is key for this: it distinguishes whether kinetochore-microtubule interactions are bi-oriented or not (Hindriksen, Lens, & Hadders, 2017; Krenn & Musacchio, 2015; van der Waal, Hengeveld, van der Horst, & Lens, 2012). The model by which it does so is referred to as the "dog-on-a-leash" model, whereby only bi-orientation results in spatial separation of CPC pool from kinetochore substrates (i.e., substrates become positioned outside the zone of Aurora-B activity) (Hindriksen et al., 2017; Krenn & Musacchio, 2015; Liu, Vader, Vromans, Lampson, & Lens, 2009; Tanaka et al., 2002; Welburn et al., 2010). Aurora-B phosphorylates microtubule-interacting factors that reside on the kinetochore, such as Ndc80 (Hindriksen et al., 2017; Krenn & Musacchio, 2015; Liu et al., 2009; Tanaka et al., 2002; Welburn et al., 2010; Wimbish & DeLuca, 2020). Phosphorylation promotes microtubule detachment, and thus formation of unattached kinetochores. In addition to creating unattached, SAC-proficient kinetochores, this enables a new round of attachments. As such the CPC is central to the process of "error-correction." The CPC also influences the SAC directly (Hindriksen et al., 2017; Krenn & Musacchio, 2015; Roy, Han, Fontan, Jema, & Joglekar, 2021; Santaguida, Vernieri, Villa, Ciliberto, & Musacchio, 2011; Vader et al., 2007). In addition to Aurora-B, other components of the SAC contribute to the process of error-correction (Cairo & Lacefield, 2020). As such, SAC signaling is intricately connected to the promotion of a desired attachment state (i.e., bi-orientation).

## 5. The meiotic prophase checkpoint: Logic and open questions in light of the SAC

Analogies between the wiring of the meiotic prophase checkpoint and the SAC are abundant. The SAC and the meiotic prophase checkpoint are integrated into unperturbed cell cycle progression. The sites where signal generation occurs (i.e., unattached kinetochores or DSBs) are intermediates of chromosome segregation or meiotic HR. Both checkpoint cascades are integrated into the regulation of the actual process the checkpoint networks are monitoring. For the SAC, several central players in this network control "error correction." In meiotic prophase, components of the checkpoint network influence DSB formation and repair. Mec1/Tel1 activation requires DSB formation, but these kinases also control DSB activity (Cooper, Wardell, Garcia, & Neale, 2014; Gray, Allison, Garcia, Goldman, & Neale, 2013; Mohibullah & Keeney, 2017) (Fig. 1A). This is needed to prevent DSB formation too close to other DSBs (Garcia, Gray, Allison, Cooper, & Neale, 2015; Johnson et al., 2021; Lukaszewicz, Lange, Keeney, & Jasin, 2021; Prieler et al., 2021), events that increase the risk for (micro)deletions or amplifications. The RHMC is also involved in the outcome of meiotic prophase. Active Mek1 couples DSB detection with promoting "productive" DNA repair outcomes, by discriminating productive (i.e., IH-based) from non-productive (i.e., IS-based) repair. This dual role is conceptually analogous to Aurora-B/CPC-mediated error correction. Both systems rely on specific kinase activation, driven by the assembly of multi-protein complexes (i.e., the RHMC and the CPC, respectively). The CPC activates and localizes Aurora-B, and a similar picture is emerging for the RHMC: Red1 forms a connection between Hop1-Mek1 and chromatin, whereas Hop1 integrates Mek1 with upstream signaling. Mek1 is able to promote homolog-based repair in a conceptually similar manner to way by which the CPC promotes bi-orientation: both systems are proposed to promote a metabolic outcome (i.e., IH repair or bi-orientation) by inhibiting non-desired configurations. The CPC distinguishes whether a microtubule interaction with kinetochores is correct or not via a spatial separation (Krenn & Musacchio, 2015). A "spatial separation" model has been proposed for Mek1: by being locally-restricted—dictated by RHMC and local Mec1/Tel1 signaling (Awasthi et al., 2015; Marechal & Zou, 2013; Paull, 2015)—it "promotes" distant IH-based repair by disfavoring IS-based repair (Goldfarb & Lichten, 2010; Subramanian et al., 2016). Only when a repair template/intermediate "escapes" the local zone of inhibition (an event analogous to spatial separation of kinetochore substrates from Aurora-B/CPC) can HR inhibition be overcome. The role of Aurora-B in SAC function is a combination between an "indirect" effect and a direct effect on SAC function. This again is reminiscent of the dual role played by Mek1. The functional similarities between these two systems highlight outstanding questions regarding RHMC/Mek1 function. Understanding where (and how) the CPC is localized has proven essential to understand error correction. To understand how Mek1 establishes a spatial separation-like mechanism, it will be key to build a comprehensive model where Mek1 localizes and how DSB ends can overcome local inhibition.

The central complexes of the SAC and the meiotic checkpoint (i.e., the MCC and RHMC), exhibit fascinating similarities (Fig. 2). Both complexes contain a HORMA domain-containing factor: Mad2 in the MCC and Hop1 in the RHMC (Rosenberg & Corbett, 2015). The capture of

CM-containing factors by Mad2 or Hop1 (Cdc20 and Red1, respectively) establishes these checkpoint complexes. Both the MCC and RHMC contain a protein harboring a kinase domain (BubR1 and Mek1, respectively), although BubR1 is a pseudokinase and does not enzymatically contribute to signaling (Elowe, 2021; Suijkerbuijk et al., 2012). Related to RHMC/MCC function is the central role of Pch2/TRIP13 in both checkpoints (Rosenberg & Corbett, 2015). In the SAC, Pch2/TRIP13 forms an enzyme/client module with Mad2, whereas in meiotic prophase, Pch2/TRIP13 collaborates with meiotic HORMADs (e.g., Hop1). Pch2/TRIP13 can fuel SAC activation, as well as silencing, as discussed earlier. Based on the roles for Pch2/TRIP13 toward the topological state of HORMA domains (as in the SAC), a crucial requirement for meiotic checkpoint signaling appears to be the generation of signaling-competent Hop1 (Fig. 2B). Analogous to the SAC, in which Pch2/TRIP13 fuels the generation of O-Mad2, non-chromosomal Pch2 fuels checkpoint function by generating such a pool of Hop1, to be incorporated into chromosome-activated RHMC. Meiotic HORMA proteins are unique because they encode a CM within their C-terminal domains (Kim et al., 2014; Rosenberg & Corbett, 2015; West, Komives, & Corbett, 2018). Intramolecular CM-HORMA interactions would exclude Hop1 molecules from participating in RHMC generation (West et al., 2018, 2019). Pch2/TRIP13 likely disengages Hop1's HORMA domain from its own CM (in intra-C-Hop1) to generate U-Hop1 (Chen et al., 2014; West et al., 2018; Ye et al., 2017, 2015). Subsequent Red1-driven chromosomal incorporation makes Hop1 available for phosphorylation by DSB-dependent Mec1/Tel1, driving activation of RHMC. The control of a key chromosomal event by a non-chromosomal enzymatic activity is analogous to kinetochore-based MCC generation relying on cytoplasmic Pch2/TRIP13 (Kim et al., 2018; Ma & Poon, 2016, 2018; Marks et al., 2017; Nelson, Hwang, Chen, & Bhalla, 2015). The meiotic prophase arrest of *dmc1Δ* cells does not require Pch2 (Hochwagen et al., 2005; Wu & Burgess, 2006), unless Hop1 levels are interfered with (Raina & Vader, 2020; Wu & Burgess, 2006). Thus, in *dmc1Δ* cells, sufficient checkpoint strength can be generated, even without Pch2 available to generate high levels of U-Hop1. In cells that lack Pch2, lower concentrations of U-Hop1 will be present—levels are set by the total levels of Hop1, and the equilibrium between intramolecular U-Hop1 and C-Hop1 (Herruzo et al., 2021; Raina & Vader, 2020; West et al., 2018; Yang, Hu, Portheine, Chuenban, & Schnittger, 2020; Ye et al., 2015). In *dmc1Δ* cells, extensive ssDNA resection takes place, leading to strong activation of

Mec1. Reduced availability of U-Hop1 is thus sufficient to fuel sustained RHMC generation under conditions where Mec1/Tel1 signaling is chronically high (as in *dmc1Δ* cells). Indeed, Pch2 becomes indispensable under conditions of reduced Mec1/Tel1 activity, such as when DSB resection is defective (Wu & Burgess, 2006). This behavior has implications for the observed roles of Pch2 in cell cycle arrest in response to synapsis defects. Although DSBs in *zip1Δ* cells generally disappear with slower kinetics to those in unperturbed cells (Börner, Kleckner, & Hunter, 2004; Thacker et al., 2014), the delay/arrest observed in *zip1Δ* cells depends on Spo11-dependent DSBs (Börner et al., 2004; González-Arranz et al., 2018). Under less "strong" Mec1/Tel1-activating conditions, Pch2 thus becomes required for efficient RHMC formation. This action is connected to the total levels of Hop1: reducing Hop1 levels further impairs checkpoint responses to synapsis defects (Raina & Vader, 2020), while increasing Hop1 levels alleviates the need for Pch2 function (Herruzo et al., 2016). Increasing Hop1 levels can thus overcome the need for Pch2-dependent topological catalysis of C-Hop1 into U-Hop1. This is similar to what is seen during SAC function: increasing Mad2 levels renders the SAC function of TRIP13 non-essential (Kim et al., 2018; Marks et al., 2017).

Pch2/TRIP13 also acts to silence SAC function (Eytan et al., 2014; Marks et al., 2017 ; Raina & Vader, 2020), and in meiotic prophase, Pch2/TRIP13 is equally responsible for checkpoint silencing (Marks et al., 2017; Raina & Vader, 2020) (Fig. 2B). Pch2/TRIP13 activates the prophase checkpoint through a non-chromosomal effect on Hop1, likely driven by a cytoplasmic pool of Pch2 (Herruzo et al., 2021; Raina & Vader, 2020; West et al., 2018; Yang et al., 2020; Ye et al., 2015). However, Pch2 is recruited to specific regions of meiotic chromosomes, in a manner that is influenced by the SC (Börner, Barot, & Kleckner, 2008; Cardoso da Silva & Vader, 2021; Deshong, Ye, Lamelza, & Bhalla, 2014; Ji et al., 2016; San-Segundo & Roeder, 1999; Subramanian et al., 2016; Wojtasz et al., 2009). Chromosomal Pch2/TRIP13 removes Hop1—by releasing it from Red1—thus triggering RHMC disassembly (Börner et al., 2008; Gu et al., 2022; Rosenberg & Corbett, 2015; San-Segundo & Roeder, 1999; Subramanian et al., 2016; Wojtasz et al., 2009) (Fig. 2B). Hop1 is needed for efficient Spo11-dependent DSB formation (Carballo et al., 2008; Mao-Draayer, Galbraith, Pittman, Cool, & Malone, 1996), and Pch2-dependent removal of Hop1 from chromosomes can ostensibly be envisioned to reduce DSB activity (Rousová et al., 2021; San-Segundo & Roeder, 1999; Subramanian et al., 2016; Vader et al., 2011). In total, this effect is thus expected to extinguish checkpoint signaling. The observation that deletion of *PCH2* in otherwise unperturbed cells leads to a delay in prophase progression (Hochwagen et al., 2005; Raina & Vader, 2020; San-Segundo & Roeder, 1999) is in support of a checkpoint silencing role. When conditions are generated under which DNA repair cannot take place (i.e., in cells lacking both Dmc1 and Rad51), this leads to a failure to arrest in meiotic prophase (Raina & Vader, 2020; Shinohara, Gasior, Ogawa, Kleckner, & Bishop, 1997). This checkpoint "override" is associated with premature SC formation, and requires Pch2 function (Raina & Vader, 2020). Thus, when SC is present on chromosomes, Pch2 can silence the checkpoint, even when DSBs cannot be repaired. Hence, SC formation (normally coupled to the establishment of IH crossover-proficient repair conditions) recruits Pch2 to chromosomes, leading to dismantlement of RHMC and checkpoint silencing. Therefore, Pch2/TRIP13 activity can, depending on the molecular context of its meiotic substrate, promote or silence checkpoint function, similar to what is seen in the SAC. In addition, the SC is a signaling conduit that dictates Pch2 function. Since SC polymerization is initiated at defined DNA repair sites that are destined to mature into IH crossovers (Chelysheva et al., 2012; Giroux et al., 1989; Lynn, Soucek, & Börner, 2007), the SC couples repair outcomes with checkpoint control. SC polymerization is thus conceptually similar to the role of microtubule-kinetochore in dictating SAC functionality (i.e., determining the "on/off" state of checkpoint catalysis, on an individual chromosome basis). The Pch2/TRIP13-HORMA module is emerging as a central component of meiotic prophase checkpoint and SAC signaling. It will be key to understand the molecular basis of SC-dependent Pch2/TRIP13 recruitment (Cardoso da Silva & Vader, 2021). Is additional regulation (e.g., of Pch2/TRIP13 enzymatic activity or substrate preference) enabling toggling between activation and silencing? Both checkpoints rely on kinase activation, and roles for phosphatases in promoting checkpoint silencing seem likely (Falk, Chan, Hoffmann, & Hochwagen, 2010; Lara-Gonzalez, Pines, et al., 2021; Saurin, 2018).

Finally, we note that in certain cases, the connections between meiotic prophase control and SAC signaling can be reversed: in *C. elegans*, SAC components regulate the meiotic prophase checkpoint (Bohr, Nelson, Giacopazzi, Lamelza, & Bhalla, 2018; Bohr, Nelson, Klee, & Bhalla, 2015; Devigne & Bhalla, 2021). Are these observations unique to this organism, or do they indicate a deeper intertwined relationship between these two signaling cascades? In any case, a conceptual framework, inspired by analogies with other checkpoint cascades and founded on genetic dependencies and biochemical rules, should enable leaps in our understanding of the unique network that guards chromosome interactions in meiotic prophase.

## 6. Meiotic checkpoint factors in cancer?

Genes involved in the meiotic prophase checkpoint are mostly germ line expressed. However, cancer cells often show aberrant expression of meiotic factors, including checkpoint factors. This has led to the idea that these genes endow cancer cells with altered DNA repair and checkpoint dynamics. We will discuss pathological effects of expression of meiotic checkpoint factors in cancer cells.

Imbalanced levels of HORMA proteins influence checkpoint function and reliance on Pch2/TRIP13 (Herruzo et al., 2016; Ma & Poon, 2016; Marks et al., 2017; Raina & Vader, 2020). Aberrant expression of meiotic HORMADs (HORMAD1/2 in humans) (Gao, Kardos, et al., 2018; Li et al., 2021; Liu et al., 2012, 2020; Nichols et al., 2018; Viet et al., 2021; Wang et al., 2018; Watkins et al., 2015; Zong et al., 2021) might interfere with Pch2/TRIP13/Mad2-mediated SAC signaling. HORMAD1 expression can indeed lead to increased micronuclei formation and chromosomal instability (Watkins et al., 2015). Micronuclei are indicative of mitotic errors, and it should be established whether HORMAD1-mediated modulation of SAC function potentially underlies these defects. In most organisms—budding yeast appears to be an exception—Pch2/TRIP13 collaborates with p31COMET, to functionally act on Mad2 and meiotic HORMADs (Alfieri et al., 2018; Brulotte et al., 2017; Eytan et al., 2014; Giacopazzi, Vong, Devigne, & Bhalla, 2020; Habu, Kim, Weinstein, & Matsumoto, 2002; Ji et al., 2016; Mapelli et al., 2006; Marks et al., 2017; Puchades, Sandate, & Lander, 2020; Raina & Vader, 2020; Varetti, Guida, Santaguida, Chiroli, & Musacchio, 2011; Xia et al., 2004; Yang et al., 2007; Ye et al., 2017). p31COMET forms a dimer with Mad2 (and likely meiotic HORMADs). These dimers are recognized by Pch2/TRIP13: p31COMET thus acts by bridging AAA+ ATPase to client (Alfieri et al., 2018; Balboni, Yang, Komaki, Brun, & Schnittger, 2020; Brulotte et al., 2017; Giacopazzi et al., 2020; Puchades et al., 2020; Ye et al., 2017). Aberrant expression of meiotic HORMADs could potentially act by titrating away p31COMET, crippling Pch2/TRIP13-mediated function. Rev7 is classically known to control translesion synthesis (de Krijger, Boersma, & Jacobs, 2021), and was recently recognized to also control the choice between HR and non-homologous end-joining-based DNA repair, as part of the Shieldin complex (Boersma et al., 2015; Clairmont et al., 2020; Dev et al., 2018; Findlay et al., 2018; Gao, Feng, et al., 2018; Ghezraoui et al., 2018; Gupta et al., 2018; Noordermeer et al., 2018; Sarangi, Clairmont, Galli, Moreau, & D'Andrea, 2020; Setiaputra & Durocher, 2019; Tomida et al., 2018). Pch2/TRIP13 also influences these Rev7-mediated processes, although the dependence of these reactions on p31COMET is currently still a matter of debate (Clairmont et al., 2020; de Krijger, Föhr, et al., 2021; Noordermeer et al., 2018; Sarangi, Clairmont, & D'Andrea, 2020; Sarangi, Clairmont, Galli, et al., 2020; Setiaputra & Durocher, 2019; Xie et al., 2021). Aberrant HORMAD1/2 presence can conceivably thus also impact Rev7-dependent dynamics.

In meiosis, RHMC is a determinant of checkpoint function as well as IH-based repair. Indeed, the re-expression of HORMAD1/2 and the human Red1-like genes SYCP2/3 (Chung et al., 2013; Costa, Boroni, & Soares, 2018; Hosoya & Miyagawa, 2021; Hosoya et al., 2011; Masterson et al., 2015; Slebos et al., 2006; Wu & Tuo, 2019) is in certain cases linked to altered HR dynamics, although the exact effects appear variable. For example, a study found that HORMAD1 causes an inhibition of HR-based DNA repair (Watkins et al., 2015), reminiscent of the meiotic function of RHMC, where it is associated with an inhibition of Rad51-mediated DNA repair. HORMAD1 has also been reported to positively affect HR (Gao, Kardos, et al., 2018; Liu et al., 2020; Nichols et al., 2018), as well as influence DNA mismatch repair (Liu et al., 2020), in both cases through incompletely understood mechanisms. The reason behind these discrepancies is unclear, but in light of the established functional relationship between HORMA protein and Pch2/TRIP13 protein levels, cell-specific differences in protein levels might play a role. In any case, HR defects are often seen in hereditary as well as sporadic cancers, and caused by a myriad of genetic alterations, such as loss of the BRCA1/2 tumor suppressors (Stok, Kok, van den Tempel, & van Vugt, 2021; Zimmer, Reinhardt, & Medová, 2020). HR deficiency is a driver of malignancy, and is exploited in patient-specific cancer therapies (Toh & Ngeow, 2021). Aberrant expression of meiotic factors might constitute an alternative mechanism by which cancer cells can acquire altered DNA repair characteristics. In this light, the finding that SYCP3 (a functional homolog of Red1) can directly associate with BRCA2 (Hosoya et al., 2011), causing an inhibition of HR, is noteworthy. When examining aberrant re-expression of Hop1/Red1 homologs as potential drivers of tumorigenesis the third component of the budding yeast RHMC, Mek1, should also be considered. Mek1 is a meiosis-specific Rad53/Chk2-like kinase. It is currently not understood whether an equally specialized human kinase exists, and the human equivalent of the RHMC complex could instead employ the canonical DNA damage checkpoint kinase Chk2 (Abdu, Brodsky, & Schüpbach, 2002; Rinaldi et al., 2020, 2017). It will be interesting to investigate whether, and if so how, meiotic HORMADs and/or SYCP2/3 might interfere with canonical checkpoint roles of Chk2 in safeguarding genome stability in somatic cells (Smith, Southgate, Tweddle, & Curtin, 2020). Understanding how meiotic factors influence somatic DNA repair could lead to novel therapeutic avenues in combatting cancer (Ashworth & Lord, 2018), while it will also continue to shed light on inner workings of the checkpoint machinery that safeguards recombination during sexual reproduction.

## Acknowledgments

Work in the Vader laboratory is financially supported by an Amsterdam UMC Research fellowship to G.V. We thank Valentina Piano (Max Planck Institute of Molecular Physiology, Dortmund, Germany) for comments on the manuscript. We apologize to those whose work could not be cited due to space constraints.

## References

Abdu, U., Brodsky, M., & Schüpbach, T. (2002). Activation of a meiotic checkpoint during *Drosophila* oogenesis regulates the translation of *Gurken* through *Chk2/Mnk*. *Current Biology*, *12*, 1645–1651.

Agarwal, S., & Roeder, G. S. (2000). Zip3 provides a link between recombination enzymes and synaptonemal complex proteins. *Cell*, *102*, 245–255.

Alfieri, C., Chang, L., & Barford, D. (2018). Mechanism for remodelling of the cell cycle checkpoint protein MAD2 by the ATPase TRIP13. *Nature*, *559*, 274–278.

Aravind, L., & Koonin, E. V. (1998). The HORMA domain: A common structural denominator in mitotic checkpoints, chromosome synapsis and DNA repair. *Trends in Biochemical Sciences*, *23*, 284–286.

Ashworth, A., & Lord, C. J. (2018). Synthetic lethal therapies for cancer: What's next after PARP inhibitors? *Nature Reviews. Clinical Oncology*, *15*, 564–576.

Awasthi, P., Foiani, M., & Kumar, A. (2015). ATM and ATR signaling at a glance. *Journal of Cell Science*, *128*, 4255–4262.

Baker, D. J., Jeganathan, K. B., Cameron, J. D., Thompson, M., Juneja, S., Kopecka, A., et al. (2004). BubR1 insufficiency causes early onset of aging-associated phenotypes and infertility in mice. *Nature Genetics*, *36*, 744–749.

Balboni, M., Yang, C., Komaki, S., Brun, J., & Schnittger, A. (2020). *COMET* functions as a *PCH2* cofactor in regulating the HORMA domain protein *ASY1*. *Current Biology*, *30*, 4113–4127.e6.

Baudat, F., Manova, K., Yuen, J. P., Jasin, M., & Keeney, S. (2000). Chromosome synapsis defects and sexually dimorphic meiotic progression in mice lacking Spo11. *Molecular Cell, 6*, 989–998.

Berchowitz, L. E., & Copenhaver, G. P. (2010). Genetic interference: Don't stand so close to me. *Current Genomics, 11*, 91–102.

Bergerat, A., de Massy, B., Gadelle, D., Varoutas, P. C., Nicolas, A., & Forterre, P. (1997). An atypical topoisomerase II from *Archaea* with implications for meiotic recombination. *Nature, 386*, 414–417.

Bhalla, N., & Dernburg, A. F. (2005). A conserved checkpoint monitors meiotic chromosome synapsis in *Caenorhabditis elegans*. *Science, 310*, 1683–1686.

Bishop, D. K., Park, D., Xu, L., & Kleckner, N. (1992). *DMC1*: A meiosis-specific yeast homolog of *E. coli recA* required for recombination, synaptonemal complex formation, and cell cycle progression. *Cell, 69*, 439–456.

Boersma, V., Moatti, N., Segura-Bayona, S., Peuscher, M. H., van der Torre, J., Wevers, B. A., et al. (2015). MAD2L2 controls DNA repair at telomeres and DNA breaks by inhibiting 5' end resection. *Nature, 521*, 537–540.

Bohr, T., Nelson, C. R., Giacopazzi, S., Lamelza, P., & Bhalla, N. (2018). Shugoshin is essential for meiotic prophase checkpoints in *C. elegans*. *Current Biology, 28*, 3199–3211.e3.

Bohr, T., Nelson, C. R., Klee, E., & Bhalla, N. (2015). Spindle assembly checkpoint proteins regulate and monitor meiotic synapsis in C. elegans. *The Journal of Cell Biology, 211*, 233–242.

Börner, G. V., Barot, A., & Kleckner, N. (2008). Yeast Pch2 promotes domainal axis organization, timely recombination progression, and arrest of defective recombinosomes during meiosis. *Proceedings of the National Academy of Sciences of the United States of America, 105*, 3327–3332.

Börner, G. V., Kleckner, N., & Hunter, N. (2004). Crossover/noncrossover differentiation, synaptonemal complex formation, and regulatory surveillance at the leptotene/zygotene transition of meiosis. *Cell, 117*, 29–45.

Brulotte, M. L., Jeong, B.-C., Li, F., Li, B., Yu, E. B., Wu, Q., et al. (2017). Mechanistic insight into TRIP13-catalyzed Mad2 structural transition and spindle checkpoint silencing. *Nature Communications, 8*, 1956.

Busygina, V., Saro, D., Williams, G., Leung, W.-K., Say, A. F., Sehorn, M. G., et al. (2012). Novel attributes of Hed1 affect dynamics and activity of the Rad51 presynaptic filament during meiotic recombination. *The Journal of Biological Chemistry*. https://doi.org/10.1074/jbc.m111.297309.

Busygina, V., Sehorn, M. G., Shi, I. Y., Tsubouchi, H., Shirleen Roeder, G., & Sung, P. (2008). Hed1 regulates Rad51-mediated recombination via a novel mechanism. *Genes & Development*. https://doi.org/10.1101/gad.1638708.

Bzymek, M., Thayer, N. H., Oh, S. D., Kleckner, N., & Hunter, N. (2010). Double Holliday junctions are intermediates of DNA break repair. *Nature, 464*, 937–941.

Cahoon, C. K., & Hawley, R. S. (2016). Regulating the construction and demolition of the synaptonemal complex. *Nature Structural & Molecular Biology, 23*, 369–377.

Cairo, G., & Lacefield, S. (2020). Establishing correct kinetochore-microtubule attachments in mitosis and meiosis. *Essays in Biochemistry, 64*, 277–287.

Callender, T. L., & Hollingsworth, N. M. (2010). Mek1 suppression of meiotic double-strand break repair is specific to sister chromatids, chromosome autonomous and independent of Rec8 cohesin complexes. *Genetics, 185*, 771–782.

Callender, T. L., Laureau, R., Wan, L., Chen, X., Sandhu, R., Laljee, S., et al. (2016). Mek1 down regulates Rad51 activity during yeast meiosis by phosphorylation of Hed1. *PLoS Genetics, 12*, e1006226.

Cannavo, E., & Cejka, P. (2014). Sae2 promotes dsDNA endonuclease activity within Mre11-Rad50-Xrs2 to resect DNA breaks. *Nature*, *514*, 122–125.

Cannavo, E., Reginato, G., & Cejka, P. (2019). Stepwise 5' DNA end-specific resection of DNA breaks by the Mre11-Rad50-Xrs2 and Sae2 nuclease ensemble. *Proceedings of the National Academy of Sciences of the United States of America*, *116*, 5505–5513.

Carballo, J. A., & Cha, R. S. (2007). Meiotic roles of Mec1, a budding yeast homolog of mammalian ATR/ATM. *Chromosome Research*, *15*, 539–550.

Carballo, J. A., Johnson, A. L., Sedgwick, S. G., & Cha, R. S. (2008). Phosphorylation of the axial element protein Hop1 by Mec1/Tel1 ensures meiotic interhomolog recombination. *Cell*, *132*, 758–770.

Cardoso da Silva, R., & Vader, G. (2021). Getting there: Understanding the chromosomal recruitment of the AAA+ ATPase Pch2/TRIP13 during meiosis. *Current Genetics*, *67*, 553–565.

Cartagena-Lirola, H., Guerini, I., Manfrini, N., Lucchini, G., & Longhese, M. P. (2008). Role of the *Saccharomyces cerevisiae* Rad53 checkpoint kinase in signaling double-strand breaks during the meiotic cell cycle. *Molecular and Cellular Biology*, *28*, 4480–4493.

Chelysheva, L., Vezon, D., Chambon, A., Gendrot, G., Pereira, L., Lemhemdi, A., et al. (2012). The Arabidopsis *HEI10* is a new ZMM protein related to Zip3. *PLoS Genetics*, *8*, e1002799.

Chen, S.-H., Albuquerque, C. P., Liang, J., Suhandynata, R. T., & Zhou, H. (2010). A proteome-wide analysis of kinase-substrate network in the DNA damage response. *The Journal of Biological Chemistry*. https://doi.org/10.1074/jbc.m110.106989.

Chen, X., Gaglione, R., Leong, T., Bednor, L., de Los Santos, T., Luk, E., et al. (2018). Mek1 coordinates meiotic progression with DNA break repair by directly phosphorylating and inhibiting the yeast pachytene exit regulator Ndt80. *PLoS Genetics*, *14*, e1007832.

Chen, C., Jomaa, A., Ortega, J., & Alani, E. E. (2014). Pch2 is a hexameric ring ATPase that remodels the chromosome axis protein Hop1. *Proceedings of the National Academy of Sciences of the United States of America*, *111*, E44–E53.

Chen, Y.-K., Leng, C.-H., Olivares, H., Lee, M.-H., Chang, Y.-C., Kung, W.-M., et al. (2004). Heterodimeric complexes of Hop2 and Mnd1 function with Dmc1 to promote meiotic homolog juxtaposition and strand assimilation. *Proceedings of the National Academy of Sciences of the United States of America*, *101*, 10572–10577. https://doi.org/10.1073/pnas.0404195101.

Chen, X., Suhandynata, R. T., Sandhu, R., Rockmill, B., Mohibullah, N., Niu, H., et al. (2015). Phosphorylation of the synaptonemal complex protein Zip1 regulates the crossover/noncrossover decision during yeast meiosis. *PLoS Biology*, *13*, e1002329.

Chen, Z., Yang, H., & Pavletich, N. P. (2008). Mechanism of homologous recombination from the RecA-ssDNA/dsDNA structures. *Nature*, *453*, 489–484.

Cheng, Y.-H., Chuang, C.-N., Shen, H.-J., Lin, F.-M., & Wang, T.-F. (2013). Three distinct modes of Mec1/ATR and Tel1/ATM activation illustrate differential checkpoint targeting during budding yeast early meiosis. *Molecular and Cellular Biology*, *33*, 3365–3376.

Chu, S., & Herskowitz, I. (1998). Gametogenesis in yeast is regulated by a transcriptional cascade dependent on Ndt80. *Molecular Cell*, *1*, 685–696.

Chuang, C.-N., Cheng, Y.-H., & Wang, T.-F. (2012). Mek1 stabilizes Hop1-Thr318 phosphorylation to promote interhomolog recombination and checkpoint responses during yeast meiosis. *Nucleic Acids Research*, *40*, 11416–11427.

Chung, J.-Y., Kitano, H., Takikita, M., Cho, H., Noh, K. H., Kim, T. W., et al. (2013). Synaptonemal complex protein 3 as a novel prognostic marker in early stage non-small cell lung cancer. *Human Pathology*, *44*, 472–479.

Ciosk, R., Zachariae, W., Michaelis, C., Shevchenko, A., Mann, M., & Nasmyth, K. (1998). An ESP1/PDS1 complex regulates loss of sister chromatid cohesion at the metaphase to anaphase transition in yeast. *Cell*, *93*, 1067–1076.

Clairmont, C. S., Sarangi, P., Ponnienselvan, K., Galli, L. D., Csete, I., Moreau, L., et al. (2020). TRIP13 regulates DNA repair pathway choice through REV7 conformational change. *Nature Cell Biology*, *22*, 87–96.

Clerici, M., Mantiero, D., Lucchini, G., & Longhese, M. P. (2005). The *Saccharomyces cerevisiae* Sae2 protein promotes resection and bridging of double strand break ends. *The Journal of Biological Chemistry*, *280*, 38631–38638.

Cloud, V., Chan, Y.-L., Grubb, J., Budke, B., & Bishop, D. K. (2012). Rad51 is an accessory factor for Dmc1-mediated joint molecule formation during meiosis. *Science*, *337*, 1222–1225.

Cohen-Fix, O., Peters, J. M., Kirschner, M. W., & Koshland, D. (1996). Anaphase initiation in *Saccharomyces cerevisiae* is controlled by the APC-dependent degradation of the anaphase inhibitor Pds1p. *Genes & Development*, *10*, 3081–3093.

Collin, P., Nashchekina, O., Walker, R., & Pines, J. (2013). The spindle assembly checkpoint works like a rheostat rather than a toggle switch. *Nature Cell Biology*, *15*, 1378–1385.

Cooper, T. J., Wardell, K., Garcia, V., & Neale, M. J. (2014). Homeostatic regulation of meiotic DSB formation by ATM/ATR. *Experimental Cell Research*, *329*, 124–131.

Corbett, K. D. (2017). Molecular mechanisms of spindle assembly checkpoint activation and silencing. *Progress in Molecular and Subcellular Biology*, *56*, 429–455.

Costa, R. L., Boroni, M., & Soares, M. A. (2018). Distinct co-expression networks using multi-omic data reveal novel interventional targets in HPV-positive and negative head-and-neck squamous cell cancer. *Scientific Reports*, *8*, 15254.

Crickard, J. B., Brooks Crickard, J., Kaniecki, K., Kwon, Y., Sung, P., Lisby, M., et al. (2018). Regulation of Hed1 and Rad54 binding during maturation of the meiosis-specific presynaptic complex. *The EMBO Journal*. https://doi.org/10.15252/embj.201798728.

Dawson, I. A., Roth, S., & Artavanis-Tsakonas, S. (1995). The Drosophila cell cycle gene *fizzy* is required for normal degradation of cyclins A and B during mitosis and has homology to the *CDC20* gene of *Saccharomyces cerevisiae*. *The Journal of Cell Biology*, *129*, 725–737.

de Krijger, I., Boersma, V., & Jacobs, J. J. L. (2021). REV7: Jack of many trades. *Trends in Cell Biology*, *31*, 686–701.

de Krijger, I., Föhr, B., Pérez, S. H., Vincendeau, E., Serrat, J., Thouin, A. M., et al. (2021). MAD2L2 dimerization and TRIP13 control shieldin activity in DNA repair. *Nature Communications*, *12*, 5421.

de los Santos, T., & Hollingsworth, N. M. (1999). Red1p, a *MEK1*-dependent phosphoprotein that physically interacts with Hop1p during meiosis in yeast. *The Journal of Biological Chemistry*, *274*, 1783–1790.

Dernburg, A. F., McDonald, K., Moulder, G., Barstead, R., Dresser, M., & Villeneuve, A. M. (1998). Meiotic recombination in *C. elegans* initiates by a conserved mechanism and is dispensable for homologous chromosome synapsis. *Cell*, *94*, 387–398.

Deshong, A. J., Ye, A. L., Lamelza, P., & Bhalla, N. (2014). A quality control mechanism coordinates meiotic prophase events to promote crossover assurance. *PLoS Genetics*, *10*, e1004291.

Dev, H., Chiang, T.-W. W., Lescale, C., de Krijger, I., Martin, A. G., Pilger, D., et al. (2018). Shieldin complex promotes DNA end-joining and counters homologous recombination in BRCA1-null cells. *Nature Cell Biology*, *20*, 954–965.

Devigne, A., & Bhalla, N. (2021). Mad1's ability to interact with Mad2 is essential to regulate and monitor meiotic synapsis in C. elegans. *PLoS Genetics*, *17*, e1009598.

Dick, A. E., & Gerlich, D. W. (2013). Kinetic framework of spindle assembly checkpoint signalling. *Nature Cell Biology*, *15*, 1370–1377.

Elowe, S. (2021). Adventures of the undead at kinetochores. *Molecular & Cellular Oncology*, *8*, 1876511.

Eytan, E., Wang, K., Miniowitz-Shemtov, S., Sitry-Shevah, D., Kaisari, S., Yen, T. J., et al. (2014). Disassembly of mitotic checkpoint complexes by the joint action of the AAA-ATPase TRIP13 and p31(comet). *Proceedings of the National Academy of Sciences of the United States of America*, *111*, 12019–12024.

Faesen, A. C., Thanasoula, M., Maffini, S., Breit, C., Müller, F., van Gerwen, S., et al. (2017). Basis of catalytic assembly of the mitotic checkpoint complex. *Nature*, *542*, 498–502.

Falk, J. E., Chan, A. C.-H., Hoffmann, E., & Hochwagen, A. (2010). A Mec1- and PP4-dependent checkpoint couples centromere pairing to meiotic recombination. *Developmental Cell*, *19*, 599–611.

Farah, J. A., Cromie, G. A., & Smith, G. R. (2009). Ctp1 and Exonuclease 1, alternative nucleases regulated by the MRN complex, are required for efficient meiotic recombination. *Proceedings of the National Academy of Sciences of the United States of America*, *106*, 9356–9361.

Findlay, S., Heath, J., Luo, V. M., Malina, A., Morin, T., Coulombe, Y., et al. (2018). SHLD2/FAM35A co-operates with REV7 to coordinate DNA double-strand break repair pathway choice. *The EMBO Journal*, *37*, e100158. https://doi.org/10.15252/embj.2018100158.

Friedman, D. B., Hollingsworth, N. M., & Byers, B. (1994). Insertional mutations in the yeast *HOP1* gene: Evidence for multimeric assembly in meiosis. *Genetics*, *136*, 449–464.

Fukunaga, K., Kwon, Y., Sung, P., & Sugimoto, K. (2011). Activation of protein kinase Tel1 through recognition of protein-bound DNA ends. *Molecular and Cellular Biology*, *31*, 1959–1971.

Funabiki, H., Yamano, H., Kumada, K., Nagao, K., Hunt, T., & Yanagida, M. (1996). Cut2 proteolysis required for sister-chromatid seperation in fission yeast. *Nature*, *381*, 438–441.

Fung, J. C., Rockmill, B., Odell, M., & Roeder, G. S. (2004). Imposition of crossover interference through the nonrandom distribution of synapsis initiation complexes. *Cell*, *116*, 795–802.

Gao, S., Feng, S., Ning, S., Liu, J., Zhao, H., Xu, Y., et al. (2018). An OB-fold complex controls the repair pathways for DNA double-strand breaks. *Nature Communications*, *9*, 3925. https://doi.org/10.1038/s41467-018-06407-7.

Gao, Y., Kardos, J., Yang, Y., Tamir, T. Y., Mutter-Rottmayer, E., Weissman, B., et al. (2018). The cancer/testes (CT) antigen HORMAD1 promotes homologous recombinational DNA repair and radioresistance in lung adenocarcinoma cells. *Scientific Reports*, *8*, 15304.

Garcia, V., Gray, S., Allison, R. M., Cooper, T. J., & Neale, M. J. (2015). Tel1(ATM)-mediated interference suppresses clustered meiotic double-strand-break formation. *Nature*, *520*, 114–118.

Garcia, V., Phelps, S. E. L., Gray, S., & Neale, M. J. (2011). Bidirectional resection of DNA double-strand breaks by Mre11 and Exo1. *Nature*, *479*, 241–244.

Ghezraoui, H., Oliveira, C., Becker, J. R., Bilham, K., Moralli, D., Anzilotti, C., et al. (2018). 53BP1 cooperation with the REV7-shieldin complex underpins DNA structure-specific NHEJ. *Nature*, *560*, 122–127.

Giacopazzi, S., Vong, D., Devigne, A., & Bhalla, N. (2020). PCH-2 collaborates with CMT-1 to proofread meiotic homolog interactions. *PLoS Genetics*, *16*, e1008904.

Gilbert, C. S., Green, C. M., & Lowndes, N. F. (2001). Budding yeast Rad9 is an ATP-dependent Rad53 activating machine. *Molecular Cell*, *8*, 129–136.

Giroux, C. N., Dresser, M. E., & Tiano, H. F. (1989). Genetic control of chromosome synapsis in yeast meiosis. *Genome*, *31*, 88–94.

Gladstone, M. N., Obeso, D., Chuong, H., & Dawson, D. S. (2009). The synaptonemal complex protein Zip1 promotes bi-orientation of centromeres at meiosis I. *PLoS Genetics*, *5*, e1000771.

Goldfarb, T., & Lichten, M. (2010). Frequent and efficient use of the sister chromatid for DNA double-strand break repair during budding yeast meiosis. *PLoS Biology*, *8*(10), e1000520. https://doi.org/10.1371/journal.pbio.1000520.

González-Arranz, S., Cavero, S., Morillo-Huesca, M., Andújar, E., Pérez-Alegre, M., Prado, F., et al. (2018). Functional impact of the H2A.Z histone variant during meiosis in *Saccharomyces cerevisiae*. *Genetics*, *209*, 997–1015.

Govin, J., Dorsey, J., Gaucher, J., Rousseaux, S., Khochbin, S., & Berger, S. L. (2010). Systematic screen reveals new functional dynamics of histones H3 and H4 during gametogenesis. *Genes & Development*, *24*, 1772–1786.

Gray, S., Allison, R. M., Garcia, V., Goldman, A. S. H., & Neale, M. J. (2013). Positive regulation of meiotic DNA double-strand break formation by activation of the DNA damage checkpoint kinase Mec1(ATR). *Open Biology*, *3*, 130019.

Gu, Y., Desai, A., & Corbett, K. D. (2022). Evolutionary dynamics and molecular mechanisms of HORMA domain protein signaling. *Annual Review of Biochemistry*. https://doi.org/10.1146/annurev-biochem-090920-103246.

Gupta, R., Somyajit, K., Narita, T., Maskey, E., Stanlie, A., Kremer, M., et al. (2018). DNA repair network analysis reveals shieldin as a key regulator of NHEJ and PARP inhibitor sensitivity. *Cell*, *173*, 972–988.e23.

Habu, T., Kim, S. H., Weinstein, J., & Matsumoto, T. (2002). Identification of a MAD2-binding protein, CMT2, and its role in mitosis. *The EMBO Journal*, *21*, 6419–6428.

Hanks, S., Coleman, K., Reid, S., Plaja, A., Firth, H., Fitzpatrick, D., et al. (2004). Constitutional aneuploidy and cancer predisposition caused by biallelic mutations in BUB1B. *Nature Genetics*, *36*, 1159–1161.

Hartsuiker, E., Mizuno, K., Molnar, M., Kohli, J., Ohta, K., & Carr, A. M. (2009). Ctp1CtIP and Rad32Mre11 nuclease activity are required for Rec12Spo11 removal, but Rec12Spo11 removal is dispensable for other MRN-dependent meiotic functions. *Molecular and Cellular Biology*, *29*, 1671–1681.

Hartwell, L. H., & Weinert, T. A. (1989). Checkpoints: Controls that ensure the order of cell cycle events. *Science*, *246*, 629–634. https://doi.org/10.1126/science.2683079.

Hassold, T., & Hunt, P. (2001). To err (meiotically) is human: The genesis of human aneuploidy. *Nature Reviews. Genetics*, *2*, 280–291.

Henry, J. M., Camahort, R., Rice, D. A., Florens, L., Swanson, S. K., Washburn, M. P., et al. (2006). Mnd1/Hop2 facilitates Dmc1-dependent interhomolog crossover formation in meiosis of budding yeast. *Molecular and Cellular Biology*, *26*, 2913–2923.

Herruzo, E., Lago-Maciel, A., Baztán, S., Santos, B., Carballo, J. A., & San-Segundo, P. A. (2021). Pch2 orchestrates the meiotic recombination checkpoint from the cytoplasm. *PLoS Genetics*, *17*, e1009560.

Herruzo, E., Ontoso, D., González-Arranz, S., Cavero, S., Lechuga, A., & San-Segundo, P. A. (2016). The Pch2 AAA+ ATPase promotes phosphorylation of the Hop1 meiotic checkpoint adaptor in response to synaptonemal complex defects. *Nucleic Acids Research*, *44*, 7722–7741.

Hindriksen, S., Lens, S. M. A., & Hadders, M. A. (2017). The ins and outs of Aurora B inner centromere localization. *Frontiers in Cell and Development Biology*, *5*, 112.

Ho, H.-C., & Burgess, S. M. (2011). Pch2 acts through Xrs2 and Tel1/ATM to modulate interhomolog bias and checkpoint function during meiosis. *PLoS Genetics*, 7, e1002351.

Hochwagen, A., Tham, W.-H., Brar, G. A., & Amon, A. (2005). The FK506 binding protein Fpr3 counteracts protein phosphatase 1 to maintain meiotic recombination checkpoint activity. *Cell*, *122*, 861–873.

Hollis, J. A., Glover, M. L., Schlientz, A. J., Cahoon, C. K., Bowerman, B., Wignall, S. M., et al. (2020). Excess crossovers impede faithful meiotic chromosome segregation in *C. elegans*. *PLoS Genetics*, *16*, e1009001.

Hong, S., Sung, Y., Yu, M., Lee, M., Kleckner, N., & Kim, K. P. (2013). The logic and mechanism of homologous recombination partner choice. *Molecular Cell*, *51*, 440–453.

Hosoya, N., & Miyagawa, K. (2021). Synaptonemal complex proteins modulate the level of genome integrity in cancers. *Cancer Science*, *112*, 989–996.

Hosoya, N., Okajima, M., Kinomura, A., Fujii, Y., Hiyama, T., Sun, J., et al. (2011). Synaptonemal complex protein SYCP3 impairs mitotic recombination by interfering with BRCA2. *EMBO Reports*, *13*, 44–51.

Hoyt, M. A., Totis, L., & Roberts, B. T. (1991). *S. cerevisiae* genes required for cell cycle arrest in response to loss of microtubule function. *Cell*, *66*, 507–517.

Humphryes, N., & Hochwagen, A. (2014). A non-sister act: Recombination template choice during meiosis. *Experimental Cell Research*, *329*, 53–60.

Hunter, N. (2008). Hop1 and the meiotic DNA-damage response. *Cell*, *132*, 731–732.

Hunter, N. (2015). Meiotic recombination: The essence of heredity. *Cold Spring Harbor Perspectives in Biology*, 7, a016618. https://doi.org/10.1101/cshperspect.a016618.

Irniger, S., Piatti, S., Michaelis, C., & Nasmyth, K. (1995). Genes involved in sister chromatid separation are needed for B-type cyclin proteolysis in budding yeast. *Cell*, *81*, 269–278.

Ji, J., Tang, D., Shen, Y., Xue, Z., Wang, H., Shi, W., et al. (2016). P31comet, a member of the synaptonemal complex, participates in meiotic DSB formation in rice. *Proceedings of the National Academy of Sciences of the United States of America*, *113*, 10577–10582.

Johnson, D., Crawford, M., Cooper, T., Claeys Bouuaert, C., Keeney, S., Llorente, B., et al. (2021). Concerted cutting by Spo11 illuminates meiotic DNA break mechanics. *Nature*, *594*, 572–576.

Joyce, E. F., & McKim, K. S. (2009). Drosophila *PCH2* is required for a pachytene checkpoint that monitors double-strand-break-independent events leading to meiotic crossover formation. *Genetics*, *181*, 39–51.

Judis, L., Chan, E. R., Schwartz, S., Seftel, A., & Hassold, T. (2004). Meiosis I arrest and azoospermia in an infertile male explained by failure of formation of a component of the synaptonemal complex. *Fertility and Sterility*, *81*, 205–209.

Keeney, S., Giroux, C. N., & Kleckner, N. (1997). Meiosis-specific DNA double-strand breaks are catalyzed by Spo11, a member of a widely conserved protein family. *Cell*, *88*, 375–384.

Keeney, S., & Kleckner, N. (1995). Covalent protein-DNA complexes at the 5' strand termini of meiosis-specific double-strand breaks in yeast. *Proceedings of the National Academy of Sciences*, *92*, 11274–11278. https://doi.org/10.1073/pnas.92.24.11274.

Keeney, S., Lange, J., & Mohibullah, N. (2014). Self-organization of meiotic recombination initiation: General principles and molecular pathways. *Annual Review of Genetics*, *48*, 187–214.

Kim, D. H., Han, J. S., Ly, P., Ye, Q., McMahon, M. A., Myung, K., et al. (2018). TRIP13 and APC15 drive mitotic exit by turnover of interphase- and unattached kinetochore-produced MCC. *Nature Communications*, *9*, 4354.

Kim, Y., Rosenberg, S. C., Kugel, C. L., Kostow, N., Rog, O., Davydov, V., et al. (2014). The chromosome axis controls meiotic events through a hierarchical assembly of HORMA domain proteins. *Developmental Cell*, *31*, 487–502.

Kim, K. P., Weiner, B. M., Zhang, L., Jordan, A., Dekker, J., & Kleckner, N. (2010). Sister cohesion and structural axis components mediate homolog bias of meiotic recombination. *Cell*, *143*, 924–937.

King, R. W., Peters, J. M., Tugendreich, S., Rolfe, M., Hieter, P., & Kirschner, M. W. (1995). A 20S complex containing CDC27 and CDC16 catalyzes the mitosis-specific conjugation of ubiquitin to cyclin B. *Cell*, *81*, 279–288.

Kniewel, R., Murakami, H., Liu, Y., Ito, M., Ohta, K., Hollingsworth, N. M., et al. (2017). Histone H3 threonine 11 phosphorylation is catalyzed directly by the meiosis-specific kinase Mek1 and provides a molecular readout of Mek1 activity in vivo. *Genetics*, *207*, 1313–1333.

Krenn, V., & Musacchio, A. (2015). The Aurora B kinase in chromosome bi-orientation and spindle checkpoint signaling. *Frontiers in Oncology*, *5*, 225. https://doi.org/10.3389/fonc.2015.00225.

Kulukian, A., Han, J. S., & Cleveland, D. W. (2009). Unattached kinetochores catalyze production of an anaphase inhibitor that requires a Mad2 template to prime Cdc20 for BubR1 binding. *Developmental Cell*, *16*, 105–117.

Lam, I., & Keeney, S. (2014). Mechanism and regulation of meiotic recombination initiation. *Cold Spring Harbor Perspectives in Biology*, 7, a016634.

Lao, J. P., Cloud, V., Huang, C.-C., Grubb, J., Thacker, D., Lee, C.-Y., et al. (2013). Meiotic crossover control by concerted action of Rad51-Dmc1 in homolog template bias and robust homeostatic regulation. *PLoS Genetics*, *9*, e1003978.

Lara-Gonzalez, P., Kim, T., Oegema, K., Corbett, K., & Desai, A. (2021). A tripartite mechanism catalyzes Mad2-Cdc20 assembly at unattached kinetochores. *Science*, *371*, 64–67.

Lara-Gonzalez, P., Pines, J., & Desai, A. (2021). Spindle assembly checkpoint activation and silencing at kinetochores. *Seminars in Cell & Developmental Biology*, *117*, 86–98.

Larsen, N. A., & Harrison, S. C. (2004). Crystal structure of the spindle assembly checkpoint protein Bub3. *Journal of Molecular Biology*, *344*, 885–892.

Li, J., Huan, J., Yang, F., Chen, H., Wang, M., & Heng, X. (2021). Identification and validation of a seizure-free-related gene signature for predicting poor prognosis in lower-grade gliomas. *International Journal of General Medicine*, *14*, 7399–7410.

Li, R., & Murray, A. W. (1991). Feedback control of mitosis in budding yeast. *Cell*, *66*, 519–531.

Limwongse, C., Schwartz, S., Bocian, M., & Robin, N. H. (1999). Child with mosaic variegated aneuploidy and embryonal rhabdomyosarcoma. *American Journal of Medical Genetics*, *82*, 20–24.

Liu, M., Chen, J., Hu, L., Shi, X., Zhou, Z., Hu, Z., et al. (2012). *HORMAD2/CT46.2*, a novel cancer/testis gene, is ectopically expressed in lung cancer tissues. *Molecular Human Reproduction*, *18*, 599–604. https://doi.org/10.1093/molehr/gas033.

Liu, D., Vader, G., Vromans, M. J. M., Lampson, M. A., & Lens, S. M. A. (2009). Sensing chromosome bi-orientation by spatial separation of Aurora B kinase from kinetochore substrates. *Science*, *323*, 1350–1353. https://doi.org/10.1126/science.1167000.

Liu, K., Wang, Y., Zhu, Q., Li, P., Chen, J., Tang, Z., et al. (2020). Aberrantly expressed HORMAD1 disrupts nuclear localization of MCM8-MCM9 complex and compromises DNA mismatch repair in cancer cells. *Cell Death & Disease*, *11*, 519.

Lukaszewicz, A., Lange, J., Keeney, S., & Jasin, M. (2021). De novo deletions and duplications at recombination hotspots in mouse germlines. *Cell*, *184*, 5970–5984.e18.

Luo, X., Tang, Z., Rizo, J., & Yu, H. (2002). The Mad2 spindle checkpoint protein undergoes similar major conformational changes upon binding to either Mad1 or Cdc20. *Molecular Cell*, *9*, 59–71.

Lydall, D., Nikolsky, Y., Bishop, D. K., & Weinert, T. (1996). A meiotic recombination checkpoint controlled by mitotic checkpoint genes. *Nature*, *383*, 840–843.

Lynn, A., Soucek, R., & Börner, G. V. (2007). ZMM proteins during meiosis: Crossover artists at work. *Chromosome Research*, *15*, 591–605.

Ma, H. T., & Poon, R. Y. C. (2016). TRIP13 regulates both the activation and inactivation of the spindle-assembly checkpoint. *Cell Reports*, *14*, 1086–1099.

Ma, H. T., & Poon, R. Y. C. (2018). TRIP13 functions in the establishment of the spindle assembly checkpoint by replenishing O-MAD2. *Cell Reports*, *22*, 1439–1450. https://doi.org/10.1016/j.celrep.2018.01.027.

MacQueen, A. J., Colaiácovo, M. P., McDonald, K., & Villeneuve, A. M. (2002). Synapsis-dependent and -independent mechanisms stabilize homolog pairing during meiotic prophase in *C. elegans*. *Genes & Development*, *16*, 2428–2442.

MacQueen, A. J., & Hochwagen, A. (2011). Checkpoint mechanisms: The puppet masters of meiotic prophase. *Trends in Cell Biology*, *21*, 393–400.

Mao-Draayer, Y., Galbraith, A. M., Pittman, D. L., Cool, M., & Malone, R. E. (1996). Analysis of meiotic recombination pathways in the yeast *Saccharomyces cerevisiae*. *Genetics*, *144*, 71–86.

Mapelli, M., Filipp, F. V., Rancati, G., Massimiliano, L., Nezi, L., Stier, G., et al. (2006). Determinants of conformational dimerization of Mad2 and its inhibition by p31comet. *The EMBO Journal*, *25*, 1273–1284.

Mapelli, M., Massimiliano, L., Santaguida, S., & Musacchio, A. (2007). The Mad2 conformational dimer: Structure and implications for the spindle assembly checkpoint. *Cell*, *131*, 730–743.

Mapelli, M., & Musacchio, A. (2007). MAD contortions: Conformational dimerization boosts spindle checkpoint signaling. *Current Opinion in Structural Biology*, *17*, 716–725.

Marechal, A., & Zou, L. (2013). DNA Damage Sensing by the ATM and ATR Kinases. *Cold Spring Harbor Perspectives in Biology*, *5*, a012716. https://doi.org/10.1101/cshperspect.a012716.

Marks, D. H., Thomas, R., Chin, Y., Shah, R., Khoo, C., & Benezra, R. (2017). Mad2 overexpression uncovers a critical role for TRIP13 in mitotic exit. *Cell Reports*, *19*, 1832–1845.

Masterson, L., Sorgeloos, F., Winder, D., Lechner, M., Marker, A., Malhotra, S., et al. (2015). Deregulation of SYCP2 predicts early stage human papillomavirus-positive oropharyngeal carcinoma: A prospective whole transcriptome analysis. *Cancer Science*, *106*, 1568–1575.

McKee, A. H., & Kleckner, N. (1997). A general method for identifying recessive diploid-specific mutations in Saccharomyces cerevisiae, its application to the isolation of mutants blocked at intermediate stages of meiotic prophase and characterization of a new gene SAE2. *Genetics*, *146*, 797–816.

McKim, K. S., Green-Marroquin, B. L., Sekelsky, J. J., Chin, G., Steinberg, C., Khodosh, R., et al. (1998). Meiotic synapsis in the absence of recombination. *Science*, *279*, 876–878.

Miniowitz-Shemtov, S., Teichner, A., Sitry-Shevah, D., & Hershko, A. (2010). ATP is required for the release of the anaphase-promoting complex/cyclosome from inhibition by the mitotic checkpoint. *Proceedings of the National Academy of Sciences of the United States of America*, *107*, 5351–5356.

Miyamoto, T., Hasuike, S., Yogev, L., Maduro, M. R., Ishikawa, M., Westphal, H., et al. (2003). Azoospermia in patients heterozygous for a mutation in SYCP3. *Lancet*, *362*, 1714–1719.

Mohibullah, N., & Keeney, S. (2017). Numerical and spatial patterning of yeast meiotic DNA breaks by Tel1. *Genome Research*, *27*, 278–288.

Molinaro, C., Martoriati, A., & Cailliau, K. (2021). Proteins from the DNA damage response: Regulation, dysfunction, and anticancer strategies. *Cancers*, *13*, 3819. https://doi.org/10.3390/cancers13153819.

Muller, H. J. (1916). The mechanism of crossing-over. *The American Naturalist, 50*, 592. https://doi.org/10.1086/279534.

Murray, A. W., Solomon, M. J., & Kirschner, M. W. (1989). The role of cyclin synthesis and degradation in the control of maturation promoting factor activity. *Nature, 339*, 280–286.

Musacchio, A. (2015). The molecular biology of spindle assembly checkpoint signaling dynamics. *Current Biology, 25*, R1002–R1018.

Musacchio, A., & Salmon, E. D. (2007). The spindle-assembly checkpoint in space and time. *Nature Reviews. Molecular Cell Biology, 8*, 379–393.

Nairz, K., & Klein, F. (1997). *mre11S*—A yeast mutation that blocks double-strand-break processing and permits nonhomologous synapsis in meiosis. *Genes & Development, 11*, 2272–2290.

Nasmyth, K., Peters, J. M., & Uhlmann, F. (2001). Splitting the chromosome: Cutting the ties that bind sister chromatids. *Novartis Foundation Symposium, 237*, 113–133. discussion 133-8, 158–63.

Nelson, C. R., Hwang, T., Chen, P.-H., & Bhalla, N. (2015). *TRIP13PCH-2* promotes Mad2 localization to unattached kinetochores in the spindle checkpoint response. *The Journal of Cell Biology, 211*, 503–516.

Nichols, B. A., Oswald, N. W., McMillan, E. A., McGlynn, K., Yan, J., Kim, M. S., et al. (2018). HORMAD1 is a negative prognostic indicator in lung adenocarcinoma and specifies resistance to oxidative and genotoxic stress. *Cancer Research, 78*, 6196–6208.

Niu, H., Li, X., Job, E., Park, C., Moazed, D., Gygi, S. P., et al. (2007). Mek1 kinase is regulated to suppress double-strand break repair between sister chromatids during budding yeast meiosis. *Molecular and Cellular Biology, 27*, 5456–5467.

Niu, H., Wan, L., Baumgartner, B., Schaefer, D., Loidl, J., & Hollingsworth, N. M. (2005). Partner choice during meiosis is regulated by Hop1-promoted dimerization of Mek1. *Molecular Biology of the Cell, 16*, 5804–5818.

Niu, H., Wan, L., Busygina, V., Kwon, Y., Allen, J. A., Li, X., et al. (2009). Regulation of meiotic recombination via Mek1-mediated Rad54 phosphorylation. *Molecular Cell, 36*, 393–404.

Noordermeer, S. M., Adam, S., Setiaputra, D., Barazas, M., Pettitt, S. J., Ling, A. K., et al. (2018). The shieldin complex mediates 53BP1-dependent DNA repair. *Nature, 560*, 117–121.

Otto, S. P., & Payseur, B. A. (2019). Crossover interference: Shedding light on the evolution of recombination. *Annual Review of Genetics, 53*, 19–44.

Panizza, S., Mendoza, M. A., Berlinger, M., Huang, L., Nicolas, A., Shirahige, K., et al. (2011). Spo11-accessory proteins link double-strand break sites to the chromosome axis in early meiotic recombination. *Cell, 146*, 372–383.

Paull, T. T. (2015). Mechanisms of ATM activation. *Annual Review of Biochemistry, 84*, 711–738.

Penedos, A., Johnson, A. L., Strong, E., Goldman, A. S., Carballo, J. A., & Cha, R. S. (2015). Essential and checkpoint functions of budding yeast ATM and ATR during meiotic prophase are facilitated by differential phosphorylation of a meiotic adaptor protein, Hop1. *PLoS One, 10*, e0134297.

Petronczki, M., Siomos, M. F., & Nasmyth, K. (2003). Un ménage à quatre: The molecular biology of chromosome segregation in meiosis. *Cell, 112*, 423–440.

Petukhova, G. V., Pezza, R. J., Vanevski, F., Ploquin, M., Masson, J.-Y., & Daniel Camerini-Otero, R. (2005). The Hop2 and Mnd1 proteins act in concert with Rad51 and Dmc1 in meiotic recombination. *Nature Structural & Molecular Biology, 12*, 449–453. https://doi.org/10.1038/nsmb923.

Piano, V., Alex, A., Stege, P., Maffini, S., Stoppiello, G. A., Huis In 't Veld, P. J., et al. (2021). CDC20 assists its catalytic incorporation in the mitotic checkpoint complex. *Science, 371*, 67–71.

Prieler, S., Chen, D., Huang, L., Mayrhofer, E., Zsótér, S., Vesely, M., et al. (2021). Spo11 generates gaps through concerted cuts at sites of topological stress. *Nature*, *594*, 577–582.

Primorac, I., & Musacchio, A. (2013). Panta rhei: The APC/C at steady state. *The Journal of Cell Biology*, *201*, 177–189.

Prinz, S., Amon, A., & Klein, F. (1997). Isolation of *COM1*, a new gene required to complete meiotic double-strand break-induced recombination in *Saccharomyces cerevisiae*. *Genetics*, *146*, 781–795.

Prugar, E., Burnett, C., Chen, X., & Hollingsworth, N. M. (2017). Coordination of double strand break repair and meiotic progression in yeast by a Mek1-Ndt80 negative feedback loop. *Genetics*, *206*, 497–512.

Puchades, C., Sandate, C. R., & Lander, G. C. (2020). The molecular principles governing the activity and functional diversity of AAA+ proteins. *Nature Reviews. Molecular Cell Biology*, *21*, 43–58.

Raina, V. B., & Vader, G. (2020). Homeostatic control of meiotic prophase checkpoint function by Pch2 and Hop1. *Current Biology*, *30*, 4413–4424.e5.

Refolio, E., Cavero, S., Marcon, E., Freire, R., & San-Segundo, P. A. (2011). The Ddc2/ATRIP checkpoint protein monitors meiotic recombination intermediates. *Journal of Cell Science*, *124*, 2488–2500.

Rieder, C. L., Cole, R. W., Khodjakov, A., & Sluder, G. (1995). The checkpoint delaying anaphase in response to chromosome monoorientation is mediated by an inhibitory signal produced by unattached kinetochores. *The Journal of Cell Biology*, *130*, 941–948.

Rinaldi, V. D., Bloom, J. C., & Schimenti, J. C. (2020). Oocyte elimination through DNA damage signaling from CHK1/CHK2 to p53 and p63. *Genetics*, *215*, 373–378.

Rinaldi, V. D., Bolcun-Filas, E., Kogo, H., Kurahashi, H., & Schimenti, J. C. (2017). The DNA damage checkpoint eliminates mouse oocytes with chromosome synapsis failure. *Molecular Cell*, *67*, 1026–1036.e2.

Romanienko, P. J., & Camerini-Otero, R. D. (2000). The mouse Spo11 gene is required for meiotic chromosome synapsis. *Molecular Cell*, *6*, 975–987.

Rosenberg, S. C., & Corbett, K. D. (2015). The multifaceted roles of the HORMA domain in cellular signaling. *The Journal of Cell Biology*, *211*, 745–755.

Rothenberg, M., Kohli, J., & Ludin, K. (2009). Ctp1 and the MRN-complex are required for endonucleolytic Rec12 removal with release of a single class of oligonucleotides in fission yeast. *PLoS Genetics*, *5*, e1000722.

Rousová, D., Nivsarkar, V., Altmannova, V., Raina, V. B., Funk, S. K., Liedtke, D., et al. (2021). Novel mechanistic insights into the role of Mer2 as the keystone of meiotic DNA break formation. *eLife*, *10*. https://doi.org/10.7554/eLife.72330.

Roy, B., Han, S. J. Y., Fontan, A. N., Jema, S., & Joglekar, A. P. (2021). Aurora B phosphorylates Bub1 to promote spindle assembly checkpoint signaling. *Current Biology*, *32*, 237–247.e6. https://doi.org/10.1016/j.cub.2021.10.049.

San Filippo, J., Sung, P., & Klein, H. (2008). Mechanism of eukaryotic homologous recombination. *Annual Review of Biochemistry*, *77*, 229–257.

San-Segundo, P. A., & Roeder, G. S. (1999). Pch2 links chromatin silencing to meiotic checkpoint control. *Cell*, *97*, 313–324.

Santaguida, S., Vernieri, C., Villa, F., Ciliberto, A., & Musacchio, A. (2011). Evidence that Aurora B is implicated in spindle checkpoint signalling independently of error correction. *The EMBO Journal*, *30*, 1508–1519.

Sarangi, P., Clairmont, C. S., & D'Andrea, A. D. (2020). Disassembly of the shieldin complex by TRIP13. *Cell Cycle*, *19*, 1565–1575.

Sarangi, P., Clairmont, C. S., Galli, L. D., Moreau, L. A., & D'Andrea, A. D. (2020). p31 promotes homologous recombination by inactivating REV7 through the TRIP13 ATPase. *Proceedings of the National Academy of Sciences of the United States of America*, *117*, 26795–26803.

Saurin, A. T. (2018). Kinase and phosphatase cross-talk at the kinetochore. *Frontiers in Cell and Development Biology*, *6*, 62.
Schaefer, D., Loidl, J., & Hollingsworth, N. M. (2005). Partner choice during meiosis is regulated by Hop1-promoted dimerization of Mek1. *Molecular Biology of the Cell*, *16*, 5804–5818.
Schwacha, A., & Kleckner, N. (1994). Identification of joint molecules that form frequently between homologs but rarely between sister chromatids during yeast meiosis. *Cell*, *76*, 51–63.
Schwacha, A., & Kleckner, N. (1997). Interhomolog bias during meiotic recombination: Meiotic functions promote a highly differentiated interhomolog-only pathway. *Cell*, *90*, 1123–1135.
Schwartz, M. F., Duong, J. K., Sun, Z., Morrow, J. S., Pradhan, D., & Stern, D. F. (2002). Rad9 phosphorylation sites couple Rad53 to the *Saccharomyces cerevisiae* DNA damage checkpoint. *Molecular Cell*, *9*, 1055–1065.
Setiaputra, D., & Durocher, D. (2019). Shieldin—The protector of DNA ends. *EMBO Reports*, *20*, e47560. https://doi.org/10.15252/embr.201847560.
Shinohara, A., Gasior, S., Ogawa, T., Kleckner, N., & Bishop, D. K. (1997). *Saccharomyces cerevisiae* recA homologues *RAD51* and *DMC1* have both distinct and overlapping roles in meiotic recombination. *Genes to Cells*, *2*, 615–629.
Shiotani, B., & Zou, L. (2009). Single-stranded DNA orchestrates an ATM-to-ATR switch at DNA breaks. *Molecular Cell*, *33*, 547–558.
Shirayama, M., Tóth, A., Gálová, M., & Nasmyth, K. (1999). APC(Cdc20) promotes exit from mitosis by destroying the anaphase inhibitor Pds1 and cyclin Clb5. *Nature*, *402*, 203–207.
Sironi, L., Mapelli, M., Knapp, S., De Antoni, A., Jeang, K.-T., & Musacchio, A. (2002). Crystal structure of the tetrameric Mad1-Mad2 core complex: Implications of a "safety belt" binding mechanism for the spindle checkpoint. *The EMBO Journal*, *21*, 2496–2506.
Slebos, R. J. C., Yi, Y., Ely, K., Carter, J., Evjen, A., Zhang, X., et al. (2006). Gene expression differences associated with human papillomavirus status in head and neck squamous cell carcinoma. *Clinical Cancer Research*, *12*, 701–709.
Smith, H. L., Southgate, H., Tweddle, D. A., & Curtin, N. J. (2020). DNA damage checkpoint kinases in cancer. *Expert Reviews in Molecular Medicine*, *22*, e2.
Stok, C., Kok, Y. P., van den Tempel, N., & van Vugt, M. A. T. M. (2021). Shaping the BRCAness mutational landscape by alternative double-strand break repair, replication stress and mitotic aberrancies. *Nucleic Acids Research*, *49*, 4239–4257.
Sturtevant, A. H. (1913). The linear arrangement of six sex-linked factors in *Drosophila*, as shown by their mode of association. *Journal of Experimental Zoology*, *14*, 43–59. https://doi.org/10.1002/jez.1400140104.
Sturtevant, A. H. (1915). The behavior of the chromosomes as studied through linkage. *Zeitschrift für Induktive Abstammungs- und Vererbungslehre*, *13*, 234–287. https://doi.org/10.1007/bf01792906.
Subramanian, V. V., & Hochwagen, A. (2014). The meiotic checkpoint network: Step-by-step through meiotic prophase. *Cold Spring Harbor Perspectives in Biology*, *6*, a016675.
Subramanian, V. V., MacQueen, A. J., Vader, G., Shinohara, M., Sanchez, A., Borde, V., et al. (2016). Chromosome synapsis alleviates Mek1-dependent suppression of meiotic DNA repair. *PLoS Biology*, *14*, e1002369.
Sudakin, V., Chan, G. K., & Yen, T. J. (2001). Checkpoint inhibition of the APC/C in HeLa cells is mediated by a complex of BUBR1, BUB3, CDC20, and MAD2. *The Journal of Cell Biology*, *154*, 925–936.
Sudakin, V., Ganoth, D., Dahan, A., Heller, H., Hershko, J., Luca, F. C., et al. (1995). The cyclosome, a large complex containing cyclin-selective ubiquitin ligase activity, targets cyclins for destruction at the end of mitosis. *Molecular Biology of the Cell*, *6*, 185–197.

Suhandynata, R. T., Wan, L., Zhou, H., & Hollingsworth, N. M. (2016). Identification of putative Mek1 substrates during meiosis in *Saccharomyces cerevisiae* using quantitative phosphoproteomics. *PLoS One*, *11*, e0155931.
Suijkerbuijk, S. J. E., van Dam, T. J. P., Karagöz, G. E., von Castelmur, E., Hubner, N. C., Duarte, A. M. S., et al. (2012). The vertebrate mitotic checkpoint protein BUBR1 is an unusual pseudokinase. *Developmental Cell*, *22*, 1321–1329.
Sung, P., Shokat, K. M., Gygi, S. P., & Hollingsworth, N. M. (2009). Regulation of meiotic recombination via Mek1-mediated Rad54 phosphorylation. *Molecular Cell*, *36*(3), 393–404.
Sweeney, F. D., Yang, F., Chi, A., Shabanowitz, J., Hunt, D. F., & Durocher, D. (2005). Saccharomyces cerevisiae Rad9 acts as a Mec1 adaptor to allow Rad53 activation. *Current Biology*, *15*, 1364–1375.
Tanaka, T. U., Rachidi, N., Janke, C., Pereira, G., Galova, M., Schiebel, E., et al. (2002). Evidence that the Ipl1-Sli15 (Aurora kinase-INCENP) complex promotes chromosome bi-orientation by altering kinetochore-spindle pole connections. *Cell*, *108*, 317–329.
Thacker, D., Mohibullah, N., Zhu, X., & Keeney, S. (2014). Homologue engagement controls meiotic DNA break number and distribution. *Nature*, *510*, 241–246.
Tipton, A. R., Wang, K., Oladimeji, P., Sufi, S., Gu, Z., & Liu, S.-T. (2012). Identification of novel mitosis regulators through data mining with human centromere/kinetochore proteins as group queries. *BMC Cell Biology*, *13*, 15.
Toh, M., & Ngeow, J. (2021). Homologous recombination deficiency: Cancer predispositions and treatment implications. *The Oncologist*, *26*, e1526–e1537. https://doi.org/10.1002/onco.13829.
Tomida, J., Takata, K.-I., Bhetawal, S., Person, M. D., Chao, H.-P., Tang, D. G., et al. (2018). FAM35A associates with REV7 and modulates DNA damage responses of normal and BRCA1-defective cells. *The EMBO Journal*, *37*, e99543. https://doi.org/10.15252/embj.201899543.
Tsubouchi, H., & Ogawa, H. (1998). A novel mre11 mutation impairs processing of double-strand breaks of DNA during both mitosis and meiosis. *Molecular and Cellular Biology*, *18*, 260–268.
Tsubouchi, H., & Shirleen Roeder, G. (2002). The Mnd1 protein forms a complex with Hop2 to promote homologous chromosome pairing and meiotic double-strand break repair. *Molecular and Cellular Biology*, *22*, 3078–3088. https://doi.org/10.1128/mcb.22.9.3078-3088.2002.
Tsubouchi, H., & Shirleen Roeder, G. (2006). Budding yeast Hed1 down-regulates the mitotic recombination machinery when meiotic recombination is impaired. *Genes & Development*, *20*, 1766–1775. https://doi.org/10.1101/gad.1422506.
Uhlmann, F., Lottspeich, F., & Nasmyth, K. (1999). Sister-chromatid separation at anaphase onset is promoted by cleavage of the cohesin subunit Scc1. *Nature*, *400*, 37–42.
Uhlmann, F., Wernic, D., Poupart, M. A., Koonin, E. V., & Nasmyth, K. (2000). Cleavage of cohesin by the CD clan protease separin triggers anaphase in yeast. *Cell*, *103*, 375–386.
Usui, T., & Shinohara, A. (2021). Rad9, a 53BP1 ortholog of budding yeast, is insensitive to Spo11-induced double-strand breaks during meiosis. *Frontiers in Cell and Development Biology*, *9*, 635383.
Vader, G., Blitzblau, H. G., Tame, M. A., Falk, J. E., Curtin, L., & Hochwagen, A. (2011). Protection of repetitive DNA borders from self-induced meiotic instability. *Nature*, *477*, 115–119.
Vader, G., Cruijsen, C. W. A., van Harn, T., Vromans, M. J. M., Medema, R. H., & Lens, S. M. A. (2007). The chromosomal passenger complex controls spindle checkpoint function independent from its role in correcting microtubule kinetochore interactions. *Molecular Biology of the Cell*, *18*, 4553–4564.

van der Waal, M. S., Hengeveld, R. C. C., van der Horst, A., & Lens, S. M. A. (2012). Cell division control by the chromosomal passenger complex. *Experimental Cell Research, 318*, 1407–1420.

Varetti, G., Guida, C., Santaguida, S., Chiroli, E., & Musacchio, A. (2011). Homeostatic control of mitotic arrest. *Molecular Cell, 44*, 710–720.

Vasudevan, A., Schukken, K. M., Sausville, E. L., Girish, V., Adebambo, O. A., & Sheltzer, J. M. (2021). Aneuploidy as a promoter and suppressor of malignant growth. *Nature Reviews. Cancer, 21*, 89–103.

Vazquez, J., Belmont, A. S., & Sedat, J. W. (2002). The dynamics of homologous chromosome pairing during male Drosophila meiosis. *Current Biology, 12*, 1473–1483.

Viet, C. T., Yu, G., Asam, K., Thomas, C. M., Yoon, A. J., Wongworawat, Y. C., et al. (2021). The REASON score: An epigenetic and clinicopathologic score to predict risk of poor survival in patients with early stage oral squamous cell carcinoma. *Biomarker Research, 9*, 42.

Visintin, R., Prinz, S., & Amon, A. (1997). CDC20 and CDH1: A family of substrate-specific activators of APC-dependent proteolysis. *Science, 278*, 460–463.

Wang, Y., Chang, C.-Y., Wu, J.-F., & Tung, K.-S. (2011). Nuclear localization of the meiosis-specific transcription factor Ndt80 is regulated by the pachytene checkpoint. *Molecular Biology of the Cell, 22*, 1878–1886.

Wang, K., Sturt-Gillespie, B., Hittle, J. C., Macdonald, D., Chan, G. K., Yen, T. J., et al. (2014). Thyroid hormone receptor interacting protein 13 (TRIP13) AAA-ATPase is a novel mitotic checkpoint-silencing protein. *The Journal of Biological Chemistry, 289*, 23928–23937. https://doi.org/10.1074/jbc.m114.585315.

Wang, X., Tan, Y., Cao, X., Kim, J. A., Chen, T., Hu, Y., et al. (2018). Epigenetic activation of HORMAD1 in basal-like breast cancer: Role in Rucaparib sensitivity. *Oncotarget, 9*, 30115–30127.

Watkins, J., Weekes, D., Shah, V., Gazinska, P., Joshi, S., Sidhu, B., et al. (2015). Genomic complexity profiling reveals that HORMAD1 overexpression contributes to homologous recombination deficiency in triple-negative breast cancers. *Cancer Discovery, 5*, 488–505.

Weinert, T. A., & Hartwell, L. H. (1988). The RAD9 gene controls the cell cycle response to DNA damage in Saccharomyces cerevisiae. *Science, 241*, 317–322. https://doi.org/10.1126/science.3291120.

Welburn, J. P. I., Vleugel, M., Liu, D., Yates, J. R., 3rd, Lampson, M. A., Fukagawa, T., et al. (2010). Aurora B phosphorylates spatially distinct targets to differentially regulate the kinetochore-microtubule interface. *Molecular Cell, 38*, 383–392.

West, A. M. V., Komives, E. A., & Corbett, K. D. (2018). Conformational dynamics of the Hop1 HORMA domain reveal a common mechanism with the spindle checkpoint protein Mad2. *Nucleic Acids Research, 46*, 279–292.

West, A. M., Rosenberg, S. C., Ur, S. N., Lehmer, M. K., Ye, Q., Hagemann, G., et al. (2019). A conserved filamentous assembly underlies the structure of the meiotic chromosome axis. *eLife, 8*, e40372. https://doi.org/10.7554/eLife.40372.

Wimbish, R. T., & DeLuca, J. G. (2020). Hec1/Ndc80 tail domain function at the kinetochore-microtubule interface. *Frontiers in Cell and Development Biology, 8*, 43.

Wojtasz, L., Daniel, K., Roig, I., Bolcun-Filas, E., Xu, H., Boonsanay, V., et al. (2009). Mouse HORMAD1 and HORMAD2, two conserved meiotic chromosomal proteins, are depleted from synapsed chromosome axes with the help of TRIP13 AAA-ATPase. *PLoS Genetics, 5*, e1000702.

Wu, H.-Y., & Burgess, S. M. (2006). Two distinct surveillance mechanisms monitor meiotic chromosome metabolism in budding yeast. *Current Biology, 16*, 2473–2479.

Wu, H.-Y., Ho, H.-C., & Burgess, S. M. (2010). Mek1 kinase governs outcomes of meiotic recombination and the checkpoint response. *Current Biology, 20*, 1707–1716. https://doi.org/10.1016/j.cub.2010.09.016.

Wu, C., & Tuo, Y. (2019). SYCP2 expression is a novel prognostic biomarker in luminal A/B breast cancer. *Future Oncology*, *15*, 817–826.

Xia, G., Luo, X., Habu, T., Rizo, J., Matsumoto, T., & Yu, H. (2004). Conformation-specific binding of p31comet antagonizes the function of Mad2 in the spindle checkpoint. *The EMBO Journal*, *23*, 3133–3143. https://doi.org/10.1038/sj.emboj.7600322.

Xie, C., He, C., Jiang, Y., Yu, H., Cheng, L., Nshogoza, G., et al. (2018). Structural insights into the recognition of phosphorylated Hop1 by Mek1. *Acta Crystallographica. Section D, Structural Biology*, *74*, 1027–1038.

Xie, W., Wang, S., Wang, J., de la Cruz, M. J., Xu, G., Scaltriti, M., et al. (2021). Molecular mechanisms of assembly and TRIP13-mediated remodeling of the human Shieldin complex. *Proceedings of the National Academy of Sciences of the United States of America*, *118*, e2024512118. https://doi.org/10.1073/pnas.2024512118.

Xu, L., Weiner, B. M., & Kleckner, N. (1997). Meiotic cells monitor the status of the interhomolog recombination complex. *Genes & Development*, *11*, 106–118.

Yang, C., Hu, B., Portheine, S. M., Chuenban, P., & Schnittger, A. (2020). State changes of the HORMA protein *ASY1* are mediated by an interplay between its closure motif and *PCH2*. *Nucleic Acids Research*, *48*, 11521–11535.

Yang, M., Li, B., Tomchick, D. R., Machius, M., Rizo, J., Yu, H., et al. (2007). p31comet blocks Mad2 activation through structural mimicry. *Cell*, *131*, 744–755.

Ye, Q., Kim, D. H., Dereli, I., Rosenberg, S. C., Hagemann, G., Herzog, F., et al. (2017). The AAA+ ATPase TRIP13 remodels HORMA domains through N-terminal engagement and unfolding. *The EMBO Journal*, *36*, 2419–2434.

Ye, Q., Rosenberg, S. C., Moeller, A., Speir, J. A., Su, T. Y., & Corbett, K. D. (2015). TRIP13 is a protein-remodeling AAA+ ATPase that catalyzes MAD2 conformation switching. *eLife*, *4*, e07367. https://doi.org/10.7554/eLife.07367.

Yost, S., de Wolf, B., Hanks, S., Zachariou, A., Marcozzi, C., Clarke, M., et al. (2017). Biallelic *TRIP13* mutations predispose to Wilms tumor and chromosome missegregation. *Nature Genetics*, *49*, 1148–1151.

Zakharyevich, K., Ma, Y., Tang, S., Hwang, P. Y.-H., Boiteux, S., & Hunter, N. (2010). Temporally and biochemically distinct activities of Exo1 during meiosis: Double-strand break resection and resolution of double Holliday junctions. *Molecular Cell*, *40*, 1001–1015.

Zhang, Z., Li, B., Fu, J., Li, R., Diao, F., Li, C., et al. (2020). Bi-allelic missense pathogenic variants in *TRIP13* cause female infertility characterized by oocyte maturation arrest. *American Journal of Human Genetics*, *107*, 15–23.

Zimmer, Y., Reinhardt, C., & Medová, M. (2020). *Exploiting DNA damage response in the era of precision oncology*. Frontiers Media SA.

Zong, B., Sun, L., Peng, Y., Wang, Y., Yu, Y., Lei, J., et al. (2021). [Corrigendum] HORMAD1 promotes docetaxel resistance in triple negative breast cancer by enhancing DNA damage tolerance. *Oncology Reports*, *46*, 195. https://doi.org/10.3892/or.2021.8146.

Zou, L., & Elledge, S. J. (2003). Sensing DNA damage through ATRIP recognition of RPA-ssDNA complexes. *Science*, *300*, 1542–1548.

CHAPTER TWELVE

# PCH-2 and meiotic HORMADs: A module for evolutionary innovation in meiosis?

**Needhi Bhalla***
Department of Molecular, Cell and Developmental Biology, University of California, Santa Cruz, Santa Cruz, CA, United States
*Corresponding author: e-mail address: nbhalla@ucsc.edu

## Contents



## Abstract

Sexual reproduction and the specialized cell division it relies upon, meiosis, are biological processes that present an incredible degree of both evolutionary conservation and divergence. One clear example of this paradox is the role of the evolutionarily ancient PCH-2/HORMAD module during meiosis. On one hand, the complex, and sometimes disparate, meiotic defects observed when PCH-2 and/or the meiotic HORMADS are mutated in different model systems have prevented a straightforward characterization of their conserved functions. On the other hand, these functional variations demonstrate the impressive molecular rewiring that accompanies evolution of the meiotic processes these factors are involved in. While the defects observed in *pch-2* mutants appear to vary in different systems, in this review, I argue that PCH-2 has a conserved meiotic function: to coordinate meiotic recombination with synapsis to ensure an appropriate number and distribution of crossovers. Further, given the dramatic variation in how the events of recombination and synapsis are themselves regulated in different model systems, the mechanistic differences in PCH-2 and meiotic HORMAD function make biological sense when viewed as species-specific elaborations layered onto this fundamental, conserved role.

*Current Topics in Developmental Biology*, Volume 151
ISSN 0070-2153
https://doi.org/10.1016/bs.ctdb.2022.07.001

## 1. Introduction

Sexual reproduction relies on the specialized cell division, meiosis, to generate haploid gametes from diploid progenitors for fertilization. Because meiosis evolved from mitosis (Wilkins & Holliday, 2009), they share common mechanisms that drive cell cycle progression. However, since sexual reproduction is subject to a remarkable degree of selection pressure (Swanson & Vacquier, 2002), meiotic mechanisms often exhibit an impressive degree of evolutionary innovation. Studying meiosis in a variety of model systems reinforces this intriguing paradox, suggesting that this specialized cell division presents a unique opportunity for evolution to molecularly tinker with mechanisms that govern cell cycle control and chromosome behavior while maintaining core functions. In this way, our studies of meiosis benefit not only from identifying what is highly conserved but also what has diverged, potentially in response to species-specific requirements.

A good example of a meiotic event that exhibits both extensive conservation and divergence is meiotic recombination. In most organisms, crossover recombination is the ultimate goal of meiosis. During meiotic prophase, homologous chromosomes undergo progressively intimate interactions that ultimately culminate in crossover recombination. After homologs pair, a macromolecular complex, called the synaptonemal complex (SC), is assembled between them in a process called synapsis. Pairing and synapsis are essential for crossover recombination to generate the linkages, or chiasmata, between homologous chromosomes that direct meiotic chromosome segregation (Bhalla & Dernburg, 2008). Because of this interdependence, some organisms, such as budding yeast, mice and plants, mechanistically link pairing, synapsis and recombination in meiotic prophase (Baudat, Manova, Yuen, Jasin, & Keeney, 2000; Giroux, Dresser, & Tiano, 1989; Grelon, Vezon, Gendrot, & Pelletier, 2001; Romanienko & Camerini-Otero, 2000). However, this interdependence does not necessarily require the mechanistic coupling of these events, as *C. elegans* and *Drosophila* demonstrate (Dernburg et al., 1998; McKim et al., 1998). Other important differences include the proteins that direct programmed double strand break (DSB) formation, the proteins that promote DNA strand exchange during homologous repair and whether organisms rely on meiosis-specific crossover recombination pathways, mechanisms that are common to both meiosis or mitosis, or both, to produce crossovers (Gray & Cohen, 2016).

In addition to the fundamental role that recombination plays in ensuring accurate chromosome segregation, crossover recombination accomplishes another important function: it generates new haplotypes for natural selection to act upon to drive evolution. Thus, the distribution of crossovers may need to be as tightly regulated as their presence to assure a random assortment of alleles on a population level. Further, recombination's ability to unlink deleterious combinations of alleles must be balanced with maintaining beneficial ones. The significance of controlling both crossover number and distribution is clearly illustrated by the existence of mechanisms such as crossover assurance, in which every pair of homologous chromosomes gets at least one crossover; crossover homeostasis, in which the number of crossovers remains relatively invariant even if the number of recombination precursors change; and crossover interference, in which the presence of a crossover inhibits the formation of a crossover nearby (Gray & Cohen, 2016).

The AAA-ATPase PCH-2, also known as Pch2 in *S. cerevisiae*, PCH2 in plants and *Drosophila* and TRIP13 in mammals, is an evolutionary ancient AAA-ATPase that structurally remodels a family of proteins with HORMA domains (HORMADs) to modify their function (Rosenberg & Corbett, 2015; Vader, 2015). This PCH-2/HORMAD module has been adapted for a wide variety of molecular functions, including meiotic crossover recombination and checkpoint control, the focus of this review. However, attempts to specifically define this module's "conserved" role in meiotic crossover recombination and checkpoint control have been challenging for the field. In this review, I will argue that the conserved function of the PCH-2/HORMAD module is to coordinate meiotic recombination with synapsis to control the number and distribution of crossovers. Thus, the supposed enigmatic role of PCH-2 in meiosis could be seen as a natural product of the evolutionary variation in how meiotic recombination and/or synapsis is regulated in different model systems.

## 2. The PCH-2/HORMAD module

The specifics of the PCH-2/HORMAD module have been reviewed elsewhere (Rosenberg & Corbett, 2015; Vader, 2015). In general, this evolutionarily ancient module (Burroughs, Zhang, Schaffer, Iyer, & Aravind, 2015; Tromer, van Hooff, Kops, & Snel, 2019) is defined by the ability of HORMAD proteins to adopt at least two conformations, open and closed, and the requirement of PCH-2 to convert the closed conformation

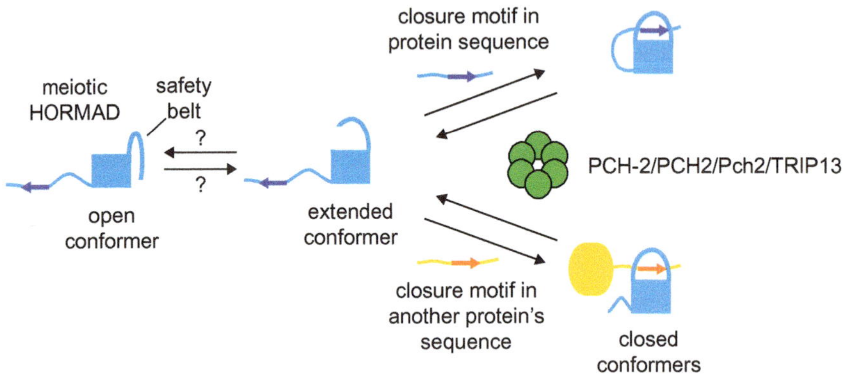


**Fig. 1** Model for PCH-2/Pch2/PCH2/TRIP13 in remodeling meiotic HORMADs. Schematic of different conformers of meiotic HORMADs that have been demonstrated to exist in vitro (extended and closed) and their remodeling by PCH-2 and its orthologs. Whether meiotic HORMADs also adopt the open conformation is unknown.

to the open one. HORMADs adopt the closed conformation when they bind a short peptide sequence either present in their own sequence or within other proteins (the closure motif) and their C-terminal "safety belt" wraps around the closure motif to stabilize this interaction (Fig. 1). This topological interaction requires PCH-2's ability to consume energy in the form of ATP to remodel and liberate HORMADs (Ye et al., 2015). When free, most HORMADs adopt an open conformation, in which this safety belt is discretely tucked against the HORMA domain (Fig. 1). Given that HORMADs typically participate in a variety of signaling mechanisms, PCH-2 function is essential to recycle HORMADs and make them available for future use.

Systems from bacteria to humans have exploited the PCH-2/HORMAD module to carry out fundamental biological processes. In most of these cases, the closed version of HORMADs is the active version of the protein. To prevent the spread of bacteriophage, infected bacteria use a HORMAD protein that adopts the closed conformation upon binding foreign phage proteins to initiate a signaling pathway that culminates in death before phage can replicate (Ye et al., 2020). In the eukaryotic spindle checkpoint, the closed version of the HORMAD, Mad2, initiates and amplifies the signal from unattached kinetochores to arrest or delay the cell cycle (Lara-Gonzalez, Westhorpe, & Taylor, 2012). During eukaryotic DNA repair, the closed version of Rev7, another HORMAD, independently regulates both translesion synthesis and DNA repair pathway choice, depending on which protein partner it binds (Clairmont et al., 2020; Malik et al., 2020). And in meiosis, the prophase events of pairing, synapsis and inter-homolog recombination are driven by the closed conformation of meiotic HORMADs assembling on meiotic chromosomes (Kim et al., 2014).

Structural and biochemical analysis of Mad2's remodeling by PCH-2 have provided important insight into its molecular mechanics: The hexameric ATPase partially unfolds the N terminus of Mad2, dissociating it from its protein partner containing the closure motif and converting it from the closed to the open state (Alfieri, Chang, & Barford, 2018; Ye et al., 2017). This interaction is likely to be relevant for PCH-2's remodeling of meiotic HORMADs (Yang, Hu, Portheine, Chuenban, & Schnittger, 2020; Ye et al., 2017). However, instead of an open conformation, meiotic HORMADs adopt an "extended" state in which the safety belt is no longer tucked against the HORMA domain (West, Komives, & Corbett, 2017) (Fig. 1). In contrast to Mad2, attempts to undertake similar structural and biochemical studies with meiotic HORMADs and PCH-2 have been less informative, resulting in major unanswered questions about the role of these proteins during meiosis. For example, what are the biological consequences of PCH-2's remodeling of meiotic HORMADs? Does this remodeling occur in the nucleoplasm, the cytoplasm, on chromosomes or some combination of the three? What is the biological relevance of the unlocked version of meiotic HORMADs? And are the answers to these questions the same in all systems?

## 3. The role of meiotic HORMADs

At the outset of meiotic prophase, a combination of proteins, including cohesins, axis core proteins and meiotic HORMADs, assemble on chromosomes and structure them in preparation for the meiosis-specific events of interhomolog pairing, synapsis and recombination (Ur & Corbett, 2021). Cohesins organize meiotic chromosomes into linear loop arrays and axis core proteins scaffold these loop arrays into an axial structure that can support pairing, synapsis and recombination (Ur & Corbett, 2021). In both fission and budding yeasts, plants and mice, axis core proteins also play an essential role in recruiting meiotic HORMADs to the axis. In these systems, axis core proteins, such as Red1 in budding yeast, Rec10/Rec27 in fission yeast, SYCP2/SYCP3 in mice and ASY3/ASY4 in plants, contain closure motifs that allow meiotic HORMADs to adopt the closed conformation on meiotic chromosomes (Ur & Corbett, 2021), driving pairing, synapsis and interhomolog recombination. In *C. elegans*, axis core proteins appear dispensable for axis formation and meiotic HORMADs are recruited on chromosomes directly by cohesins in this system (Kim et al., 2014).

Despite their presence in almost all systems studied to date, meiotic HORMADs demonstrate some functional divergence of their own. Indeed, since PCH-2 controls the function of multiple HORMADS in some systems, potentially constraining its evolutionary trajectory, evolution may specifically target meiotic HORMADs to test out new functions. Meiotic HORMADs are essential for the formation of DSBs in mice, *C. elegans* and both fission and budding yeasts (Daniel et al., 2011; Goodyer et al., 2008; Latypov et al., 2010; Mao-Draayer, Galbraith, Pittman, Cool, & Malone, 1996) but not in *Arabidopsis* (Sanchez-Moran, Santos, Jones, & Franklin, 2007). Meiotic HORMADs are depleted or redistributed from meiotic chromosomes in response to synapsis in mice, budding yeast and plants (Borner, Kleckner, & Hunter, 2004; Cuacos et al., 2021; Lambing et al., 2015; Wojtasz et al., 2009) but not in *C. elegans* (Couteau & Zetka, 2005; Goodyer et al., 2008; Martinez-Perez & Villeneuve, 2005; Zetka, Kawasaki, Strome, & Muller, 1999). They play a role in signaling meiotic progression in budding yeast, *C. elegans* and mice (Daniel et al., 2011; Kim, Kostow, & Dernburg, 2015; Kogo, Tsutsumi, Inagaki, et al., 2012; Kogo, Tsutsumi, Ohye, et al., 2012; Martinez-Perez & Villeneuve, 2005; San-Segundo & Roeder, 1999; Shin, McGuire, & Rajkovic, 2013; Silva et al., 2014; Wojtasz et al., 2012) but not in fission yeast (Brown, Jarosinska, & Lorenz, 2018). As another clear example of their divergence, the number of meiotic HORMADS varies between organisms: budding and fission yeast have one meiotic HORMAD (Hop1) (Hollingsworth, Goetsch, & Byers, 1990; Lorenz et al., 2004), plants have two (ASY1 and ASY2) (Caryl, Armstrong, Jones, & Franklin, 2000), mice have two (HORMAD 1 and 2) (Wojtasz et al., 2009), *C. elegans* have four (HTP-3, HIM-3, HTP-1 and HTP-2) (Couteau & Zetka, 2005; Goodyer et al., 2008; Martinez-Perez & Villeneuve, 2005; Zetka et al., 1999) and *Drosophila* appear to have none (van Hooff, Tromer, van Wijk, Snel, & Kops, 2017). Highlighting how meiotic HORMADs may diverge in response to species-specific requirements, evolutionary analysis in plants show that meiotic HORMADs experience positive selection in response to increases in ploidy (Hollister et al., 2012; Yant et al., 2013).

## 4. PCH-2/Pch2/PCH2/TRIP13 in meiosis

Pch2 was originally identified in *S. cerevisiae* as a pachytene checkpoint factor that enforced a meiotic arrest when there were defects in synapsis (San-Segundo & Roeder, 1999). This functional link with synapsis was reinforced when work in *C. elegans* identified PCH-2 as essential for a checkpoint response when there were defects in synapsis but not when there were defects in recombination (Bhalla & Dernburg, 2005) and with the observation that the only model system that appeared to be missing Pch2 was fission yeast, in which homologous chromosomes do not synapse (Wu & Burgess, 2006). However, this simplistic model for Pch2/PCH-2 function was complicated by observations that budding yeast *pch2* mutants themselves exhibited recombination defects, including delays in the formation of crossovers (Borner, Barot, & Kleckner, 2008; Wu & Burgess, 2006), an increase in inter-sister recombination (Zanders, Sonntag Brown, Chen, & Alani, 2011), and effects on crossover assurance (Chakraborty et al., 2017), interference and homeostasis (Joshi, Barot, Jamison, & Borner, 2009; Zanders & Alani, 2009). Subsequent experiments in both *C. elegans* and mice verified PCH-2/TRIP13's participation in recombination (Deshong, Ye, Lamelza, & Bhalla, 2014; Li & Schimenti, 2007; Roig et al., 2010), namely in crossover assurance and interference, further confounding its identification as simply a synapsis checkpoint protein. Even with the recognition that Pch2/TRIP13 depleted or redistributed meiotic HORMADs as chromosomes synapse (Borner et al., 2008; Wojtasz et al., 2009), the recombination defects observed in *pch-2* mutants were difficult to explain.

As the role of PCH-2 in recombination continued to perplex the meiosis community, a more nuanced picture of the regulation of recombination began to emerge. It had always been appreciated that DSBs typically vastly outnumber crossovers. With more careful cytological analysis of crossover formation, there was a growing realization that the process of meiotic crossover recombination is implemented progressively during meiotic prophase (Cole et al., 2012; Joshi, Brown, Bishop, & Borner, 2015; Morgan et al., 2021; Yokoo et al., 2012). Thus, decisions about which DSBs will become crossovers occur gradually over meiotic prophase, with a persistent winnowing of crossover precursors until all aspects of crossover control, including assurance, interference and homeostasis, are satisfied.

More importantly, in all systems, no matter if synapsis is dependent on or independent of recombination, this winnowing either overlaps completely (as in *C. elegans* (Yokoo et al., 2012)) or partially (as in budding yeast, plants, and mice (Capilla-Perez et al., 2021; Cole et al., 2012; Joshi et al., 2015; Morgan et al., 2021)) with a period when chromosomes are synapsed, suggesting that there may be some mechanism to coordinate crossover recombination and synapsis during this progressive implementation of crossover recombination and its control. That the variation of recombination rates correlates better with length of the synaptonemal complex than with the amount of DNA provides additional support for the existence of such a mechanism (Lynn et al., 2002; Ruiz-Herrera et al., 2017).

Altogether, this growing body of knowledge provided a new framework within which to consider PCH-2's function and its connection to both synapsis and recombination. In this review, I will discuss the wide-ranging literature on PCH-2's meiotic role and show that PCH-2's effect on recombination can be explained in some systems through the coordination of recombination with the mechanisms that control pairing and synapsis. Further, I will argue that this role is generalizable to other systems despite the fact that different organisms use different mechanisms to control pairing and synapsis. In this way, what has appeared as disparate and contradictory observations about the role of PCH-2 in different organisms can actually be integrated into a more unified model that describes PCH-2's conserved function. This role for PCH-2 may have broader implications beyond the regulation of crossover number and control, given the suggestion that variations in pachytene checkpoint stringency may play important roles in genetic incompatibility and speciation (Li, Barringer, & Barbash, 2009).

For the purposes of this review, I will generally use the *C. elegans* nomenclature for PCH-2. However, when discussing specific model systems, I will use the nomenclature for that organism (Pch2 for yeast, PCH2 for *Drosophila* and plants and TRIP13 for mammals). Further, as I discuss the details of PCH-2 function in specific contexts (pairing, synapsis, recombination, etc.), I will highlight common themes that provide insight into the core function(s) of PCH-2 across model systems, providing a more complete synthesis at the end of this review.

## 5. Localization of PCH-2

PCH-2's localization provides our first illustration of a core, conserved function with species-specific embellishments and very likely reports on where PCH-2 is remodeling meiotic HORMADs. In almost all organisms that it has been characterized in, with the exception of *Drosophila* (Joyce & McKim, 2010), PCH-2 localizes to meiotic chromosomes (Deshong et al., 2014; Joshi et al., 2009; Lambing et al., 2015; San-Segundo & Roeder, 1999). When meiotic HORMADs are assembling on chromosomes to structure them for pairing, synapsis and recombination, PCH-2 can be found as puncta on meiotic chromosomes in budding yeast, plants and *C. elegans* (Deshong et al., 2014; Joshi et al., 2009; Lambing et al., 2015; San-Segundo & Roeder, 1999). In the plant *Brassica* and budding yeast, these puncta colocalize with other proteins that mark sites where crossover precursors initiate synapsis (Lambing et al., 2015; Joshi et al., 2009), consistent with the link between recombination and synapsis. Later, after synapsis, PCH-2 localizes either as puncta or as linear stretches along the synaptonemal complex in budding yeast, plants and *C. elegans* (Deshong et al., 2014; Joshi et al., 2009; Lambing et al., 2015; San-Segundo & Roeder, 1999). In C. *elegans*, this localization to the synaptonemal complex coincides with when chromosomes are competent for crossover recombination (Deshong et al., 2014) and its ejection from the synaptonemal complex depends on the cell cycle polo-like kinase, PLK-2 (Roelens et al., 2019).

A series of elegant experiments in budding yeast have highlighted a more complicated picture of Pch2 localization, some of which may be unique to this system. Before its localization to the synaptonemal complex, Pch2 is found at the nucleolus, the region of the nucleus where repetitive ribosomal (r)DNA is localized (San-Segundo & Roeder, 1999). Pch2's enrichment at the nucleolus appears to be specifically required to limit DSB formation at this repetitive sequence in budding yeast (Vader et al., 2011) and is not functionally connected to its role regulating meiotic progression (Herruzo, Santos, Freire, Carballo, & San-Segundo, 2019). Further, targeting Pch2 to the cytoplasm supports its ability to arrest the cell cycle when there are synapsis defects (Herruzo et al., 2021), indicating that Pch2's remodeling of meiotic HORMADs and effect on meiotic progression can take place away from meiotic chromosomes. Plants exhibit a similar phenomenon, in which PCH2's ability to remodel one of its meiotic HORMADs, ASY1, is essential to shuttle it into meiotic nuclei to structure meiotic chromosomes in preparation for pairing, synapsis and recombination (Yang et al., 2020).

Altogether, these data suggest that PCH-2's primary meiotic role in most systems is to remodel meiotic HORMADs on meiotic chromosomes, both prior to and during synapsis, to regulate pairing, synapsis, recombination and meiotic progression, potentially coordinating these events. In some systems, such as in budding yeast and plants, this remodeling of meiotic HORMADs can also happen away from chromosomes with functional implications for both meiotic progression and interhomolog recombination. Meanwhile, in *Drosophila*, where PCH2 localizes to the nuclear periphery, PCH2's role in regulating recombination and meiotic progression appears to only occur away from meiotic chromosomes, potentially presenting one extreme of evolutionary innovation in PCH2's meiotic role.

## 6. Role in pairing and synapsis

As mentioned above, PCH-2's localization to meiotic chromosomes in most systems suggests a role for PCH-2 in controlling pairing and synapsis. The clearest case for this role for PCH-2 is in *C. elegans*. In this organism, both pairing and synapsis are accelerated in *pch-2* mutants (Deshong et al., 2014), indicating that PCH-2 actively inhibits these events. This inhibition involves the destabilization of incorrect homolog interactions during pairing and synapsis, since a mutant version of PCH-2 that can bind its substrates but is not able to remodel them results in non-homologous synapsis (Giacopazzi, Vong, Devigne, & Bhalla, 2020). Thus, PCH-2 proofreads homolog interactions that underlie pairing and synapsis in *C. elegans*, contributing both to their accuracy and completion.

It is currently unclear whether PCH-2's role in proofreading homolog interactions that underlie accurate homolog pairing and synapsis is conserved in other systems, particularly those that use recombination to initiate these events, but preliminary data is promising. In plants and mice, synapsis is incomplete in *pch2/Trip13* mutants (Lambing et al., 2015; Miao et al., 2013; Roig et al., 2010). However, in plants, this phenotype is likely to be an indirect result of the requirement for PCH2 to shuttle meiotic HORMADs into the nucleus during meiotic prophase (Yang et al., 2020). In mice, the defects in synapsis are limited to regions that might present a challenge for synapsis, such as pericentric regions (Roig et al., 2010), suggesting a more direct role in regulating synapsis. Further, the localization of Pch2/PCH2 in budding yeast and plants as puncta that colocalize with proteins involved in recombination and synapsis (Lambing et al., 2015; Joshi et al., 2009) supports this proposed role in regulating pairing and synapsis.

## 7. Role in recombination

Soon after Pch2's identification as a checkpoint factor, a role for Pch2 in recombination began to emerge in multiple systems but the observations did not easily lend themselves to the development of a coherent model. Two clear themes materialize from the analysis of *pch-2* mutants in different systems: (1) there is an effect on the number and distribution of crossovers; and (2) this effect is variable and non-uniform. In budding yeast *pch2* mutants, there is a general increase in the number of crossovers on mid-sized and large chromosomes (Zanders & Alani, 2009), loss of assurance on small chromosomes (Chakraborty et al., 2017), and a loss of interference in some genetic intervals but not others (Joshi et al., 2009; Zanders & Alani, 2009). In plant and *Drosophila pch2* mutants, recombination is elevated in some genetic intervals and reduced and even unaffected in others (Joyce & McKim, 2009; Lambing et al., 2015), leading to similar variable effects on interference. And in murine *Trip13* mutants, this variable effect on crossover number and distribution manifests as a reduction in crossover assurance, a reduction in the spacing between crossovers on large chromosomes and changes in the position of crossovers on shorter chromosomes (Roig et al., 2010). The variability of these mutant phenotypes, and the lack of a regular pattern, has definitely contributed to the difficulty in developing a unified, integrated model of PCH-2 function in the field.

This challenge may be addressed by experiments in *C. elegans*, which benefits from a clearer understanding of synapsis initiation and the distribution of DSB formation across chromosomes, allowing distinct patterns to appear from the array of observed recombination defects in *pch*-2 mutants. Once again, in this system, mutation of *pch-2* produces a loss of crossover assurance and non-uniform effects on recombination on two chromosomes tested, an autosome and the sex chromosome (Deshong et al., 2014). However, when analyzed in the context of pairing, synapsis and DSB formation, these recombination defects appear less mysterious. First, recombination is less severely affected in *pch-2* mutants at pairing center ends of chromosomes (Deshong et al., 2014). Pairing Centers are cis-acting sites near one end of each chromosome that promote homolog pairing and synapsis (MacQueen et al., 2005). Therefore, these regions are often the first to pair, the first to synapse and possibly, the first to initiate early events in crossover recombination. Altogether, these data suggest that PCH-2 limits this temporal bias in crossover formation, potentially to more widely distribute recombination events to regions distal to pairing centers.

Second, recombination is less severely affected in *pch-2* mutants in chromosomal regions where DSBs are enriched (Deshong et al., 2014). *C. elegans* autosomes do not experience DSB formation, and therefore crossover formation, uniformly: chromosome ends are enriched for DSB formation and crossovers, relative to the central regions of the chromosome (Nadarajan, Altendorfer, Saito, Martinez-Garcia, & Colaiacovo, 2021; Yu, Kim, & Dernburg, 2016). Therefore, in addition to its possible role limiting the temporal bias in crossover formation, PCH-2 may also be limiting the effect of DSB enrichment to more widely distribute crossovers across the genome. Since the central regions of autosomes are also gene rich, in comparison to chromosome ends, PCH-2's role in redistributing crossovers to this region may specifically be an attempt to create new haplotypes for evolution to act on, despite the paucity of DSBs. It remains an open question, however, whether these patterns are seen on the remaining four autosomes in *C. elegans*.

In addition to this hypothesized role during unperturbed meiosis, additional work in *Drosophila* and *C. elegans* recombination mutants provides further evidence for PCH2/PCH-2 controlling crossover distribution. In both of these systems, defects in recombination alters the distribution of crossovers across chromosomes and this redistribution partially relies on PCH2/PCH-2 (Deshong et al., 2014; Joyce & McKim, 2010). This effect on crossover distribution is accompanied by protracted PCH-2 activity: In *C. elegans*, PCH-2 persists on meiotic chromosomes while in *Drosophila*, PCH-2's remains at the nuclear periphery longer (Deshong et al., 2014; Joyce & McKim, 2010). In Drosophila, overexpression of PCH2 is sufficient to recapitulate these effects on crossover distribution (Joyce & McKim, 2010). Thus, both the absence and the persistence of PCH-2 activity during meiotic prophase has consequences on the distribution of crossovers, consistent with a direct role in this process.

This context now provides a useful way to frame and potentially reanalyze the puzzling phenotypes observed in budding yeast, plants and mice. In these systems, DSBs both initiate crossover recombination and act as potential sites for synapsis initiation (Baudat et al., 2000; Giroux et al., 1989; Grelon et al., 2001; Romanienko & Camerini-Otero, 2000). Similar to *C. elegans*, they are also not uniformly distributed and specific regions, known as "hot spots," tend to be more favorable for DSB formation (Gray & Cohen, 2016) and presumably, the resulting downstream events, such as the decision to become a crossover and initiation of synapsis. Therefore, given the observation that PCH-2 limits crossovers where pairing and synapsis initiate and DSBs are enriched in *C. elegans*, an analogous role for Pch2/TRIP13/PCH2 in systems such as budding yeast, mice and plants might be to similarly limit the frequency with which these "hot spots" would become crossovers to ensure a wider distribution of crossovers throughout the genome and on a population level. Since a reduction in the number of DSBs may exacerbate the bias for "hot spots" to become crossovers, this role may also explain why Pch2 is essential for crossover homeostasis (Joshi et al., 2009; Zanders & Alani, 2009).

How PCH-2 mechanistically ensures the correct number and distribution of crossovers is currently unclear but in budding yeast and mice, it involves both the regulation of DSB formation and interhomolog bias via meiotic HORMADs. A direct interaction between meiotic HORMADs and an essential member of the DSB forming machinery (Mer2 in budding yeast and IHO1 in mice) explains how PCH-2 controls DSB formation through its removal and redistribution of meiotic HORMADs (Panizza et al., 2011; Stanzione et al., 2016). PCH-2's participation in the installation of homolog bias is less clear. During crossover recombination, DNA repair is deliberately directed away from the sister chromatid, the preferred template during homologous recombination in mitosis, toward the homolog. Meiotic HORMADs are essential for homolog bias (Schwacha & Kleckner, 1994). More specifically, in budding yeast, phosphorylation of the HORMAD, Hop1, by conserved DNA damage kinases promotes homolog bias (Carballo, Johnson, Sedgwick, & Cha, 2008). Pch2 contributes to this phosphorylation event, and homolog bias, likely in part by regulating the availability of Hop1 (Herruzo et al., 2016). However, since Pch2 interacts with the N-terminus of the DNA repair protein Xrs2 in a two-hybrid assay and removal of this region in Xrs2 recapitulates phenotypes observed in *pch2* mutants (Ho & Burgess, 2011), Pch2's regulation of interhomolog bias might be more direct and occur on chromosomes. Given the knowledge that recombination is progressively implemented, Pch2/TRIP13/PCH2 may participate in destabilizing some crossover precursors involving meiotic HORMADs on chromosomes to favor the stabilization of others and promote this redistribution, both before and during synapsis. Whether Pch2 contributes to interhomolog bias indirectly, directly, or some combination of the two, this role likely explains the increase in inter-sister recombination observed in budding yeast *pch2* mutants (Zanders et al., 2011).

*C. elegans* differs from budding yeast and mice in that the regulation of DSB formation appears to be distinct from that of interhomolog recombination (Deshong et al., 2014). This partition has been clearly demonstrated in *Drosophila* (Joyce & McKim, 2009) and suggests that in both of these ecdysozoans, PCH-2/PCH2's role in controlling crossover number and distribution is primarily through control of interhomolog recombination (Deshong et al., 2014). In *C. elegans*, this control is likely through the remodeling of meiotic HORMADs on chromosomes, given their conserved role in promoting interhomolog crossover recombination (Couteau, Nabeshima, Villeneuve, & Zetka, 2004; Couteau & Zetka, 2011; Hayashi, Chin, & Villeneuve, 2007; Martinez-Perez & Villeneuve, 2005) and PCH-2's localization to chromosomes when they are competent for meiotic recombination (Deshong et al., 2014). This functional partition is not because these two systems have uncoupled DSB formation from synapsis (Dernburg et al., 1998; McKim et al., 1998) since Arabidopsis, which needs DSBs for synapsis (Grelon et al., 2001), also does not rely on meiotic HORMADs for DSB formation (Sanchez-Moran et al., 2007). Instead, this separation of function may reflect yet another set of species-specific requirements for meiotic HORMADs and their regulation by PCH-2 in meiotic progression and crossover recombination in these systems. The major question in *Drosophila* is how PCH2 accomplishes this effect on crossover distribution from the nuclear periphery (Joyce & McKim, 2010) and in the apparent absence of a meiotic HORMAD (van Hooff et al., 2017).

## 8. Role in regulating meiotic progression

Pch2 was first identified in budding yeast as a checkpoint factor that restored meiotic progression when there were defects in synapsis, providing a clear link between Pch2 function and a role in regulating meiotic progression (San-Segundo & Roeder, 1999). Unfortunately, deciphering this role was complicated by observations that *pch2* single mutants produced defects in meiotic progression (Hochwagen, Tham, Brar, & Amon, 2005; Wu & Burgess, 2006). These contradictory phenotypes can be reconciled when we recognize that Pch2 remodels Hop1 both on and away from chromosomes to promote meiotic progression (Herruzo et al., 2021; Raina & Vader, 2020). As had been suggested by dosage experiments with Hop1

(Bailis, Smith, & Roeder, 2000; Herruzo et al., 2016; Raina & Vader, 2020), Pch2 ensures there is enough Hop1 available for functional checkpoint signaling and this can occur away from chromosomes (Herruzo et al., 2021). However, once synapsis is complete, Pch2 is recruited to synapsed chromosomes to remove or redistribute meiotic HORMADs from chromosomes as an important signal for meiotic progression to continue (Raina & Vader, 2020).

Mechanistically, Pch2's depletion of Hop1 from budding yeast chromosomes in response to synapsis promotes meiotic progression by reducing the activity of Mek1 kinase, a key regulator of inter-homolog recombination (Niu et al., 2005; Subramanian et al., 2016; West et al., 2017). When Mek1 activity is down regulated, its inhibition of the transcription factor, Ndt80 is relieved and Ndt80 is capable of driving the transcription of factors essential for exit from pachytene (Chen et al., 2018; Prugar, Burnett, Chen, & Hollingsworth, 2017). By contrast, the depletion of meiotic HORMADs from murine meiotic chromosomes by TRIP13 during synapsis promotes meiotic progression by effectively silencing the DNA damage response. However, what this checkpoint response looks like is different based on sex. During spermatogenesis, when the X and Y chromosomes fail to completely synapse, components of the DNA damage response participate in a whole scale remodeling of these chromosomes to silence them in a process called meiotic sex chromosomes inactivation (MSCI) (Turner, 2007). MSCI is essential for the progression of spermatogenesis and events that disrupt it, even indirectly, lead to an early block in meiotic progression (Turner, 2007). For example, defects in recombination and/or synapsis produce persistent DNA damage on autosomes, preventing the effective accumulation of the DNA repair factors on sex chromosomes and early steps in MSCI (Mahadevaiah et al., 2008). TRIP13's removal of meiotic HORMADs during autosomal synapsis downregulates DSB formation and alleviates the block to sister repair (Daniel et al., 2011; Li & Schimenti, 2007; Wojtasz et al., 2009), reducing the number of persistent DSBs on autosomes that may compete for DNA repair factors with unsynapsed sex chromosomes. In this way, MSCI is an indirect readout for meiotic progression and quality control. Since oocytes have two X chromosomes and do not experience MSCI, meiotic progression is assessed directly through the completion of DNA repair, although it appears that there are dose-dependent responses depending on the load of DNA damage: Defects in recombination that produce many persistent repair intermediates block meiosis earlier than defects that only disrupt synapsis or produce fewer repair intermediates (Rinaldi, Bolcun-Filas, Kogo, Kurahashi, & Schimenti, 2017). Thus, despite the mechanistic differences in how meiotic progression is promoted, synapsis, and the concomitant removal/redistribution of meiotic HORMADs by Pch2/TRIP13, is an essential reporter of effective quality control during meiosis in both budding yeast and mice.

What about in systems in which synapsis precedes and does not rely on DSB formation and the initiation of recombination? The role of meiotic HORMADs in *C. elegans* in regulating meiotic progression shares some similarities and some differences with budding yeast and mice. In this system, meiotic HORMADs also assemble on meiotic chromosomes to drive pairing, synapsis and recombination but do not get removed in response to synapsis (Couteau & Zetka, 2005; Goodyer et al., 2008; Martinez-Perez & Villeneuve, 2005; Zetka et al., 1999). In contrast to what is observed in budding yeast and mice and inferred in plants, it is the binding of three of the four meiotic HORMADs, HIM-3, HTP-1 and HTP-2, on chromosomes that promotes meiotic progression (Kim et al., 2015). Some evidence suggests that a soluble version of HTP-1 that fails to assemble on meiotic chromosomes, and presumably is present as an extended conformer, delays meiotic prophase (Silva et al., 2014). This model for how meiotic HORMADs might regulate meiotic progression in *C. elegans*, in which a soluble version that fails to bind meiotic chromosomes delays meiosis, is in direct contrast to the model presented in budding yeast and mice, in which the removal of meiotic HORMADs from meiotic chromosomes by Pch2/TRIP13 drives meiotic progression and may signal that the downstream mechanisms that impact meiotic progression will be specific to *C. elegans*.

From this assortment of checkpoint mechanisms in budding yeast, mice and *C. elegans*, one coherent picture that emerges is that PCH-2 and/or meiotic HORMADs consistently coordinates the cell cycle with the events of recombination and synapsis in multiple systems but in ways that are layered on species-specific regulation of meiotic progression. Indeed, the situation in *Drosophila*, in which PCH2 delays meiosis in response to defects (Joyce & McKim, 2009, 2010) despite the apparent absence of a meiotic HORMAD in its genome (van Hooff et al., 2017), may be one extreme example that highlights this point. Given that defects in meiosis produce varying effects on meiotic progression, both within species and between species, it is entirely possible that this variation in PCH-2 dependent checkpoint responses is a defining feature in the evolution of this process.

## 9. Requirement for a cofactor

Another major difference that highlights the differences between PCH-2's role in different meiotic systems is the requirement of a cofactor. Structural and biochemical studies with Mad2 have reinforced that PCH-2/TRIP13 uses a cofactor, CMT-1/p31$^{comet}$, to mediate interaction with its HORMAD substrates (Alfieri et al., 2018; Ye et al., 2015). The expansion of the HORMAD/PCH2 module to include an adaptor is conserved in plants and *C. elegans* (Balboni, Yang, Komaki, Brun, & Schnittger, 2020; Giacopazzi et al., 2020; Ji et al., 2016). In Arabidopsis, COMET facilitates PCH2's ability to shuttle the meiotic HORMAD, ASY1, into the nucleus in early prophase and PCH2's depletion of ASY1 from meiotic chromosomes in response to synapsis (Balboni et al., 2020). In rice, mutation of *P31$^{comet}$* function produces defects in DSB formation, homolog pairing and crossover recombination (Ji et al., 2016). In *C. elegans*, unlike in Arabidopsis, mutation of *cmt-1* does not affect PCH-2's localization to meiotic chromosomes but recapitulates the phenotype of a mutant version of PCH-2 that binds substrates but cannot remodel them (Giacopazzi et al., 2020), implicating CMT-1 in PCH-2's ability to hydrolyze ATP and remodel HORMADs but not in its PCH-2's ability to bind its substrates. By contrast, other organisms, such as budding yeast, do not have an obvious CMT-1/p31comet ortholog (van Hooff et al., 2017), raising the question of what the molecular mechanics of meiotic HORMAD remodeling look like in this system.

## 10. A unified model of PCH-2's role in meiotic prophase

How to integrate this complex and disparate data into an integrated model of PCH-2 function in meiotic prophase? The theme that emerges is an important meiotic role for PCH-2 and its orthologs with common features and species-specific embellishments. For example, a common feature to all systems is that PCH-2 remodels meiotic HORMADs in solution (to make them available for loading onto meiotic chromosomes and/or shuttle them into the nucleus) (see #1 in Fig. 2) and/or on chromosomes to regulate the role of meiotic HORMADs in inter-homolog recombination, affecting the number and distribution of crossovers.

For systems such as budding yeast, plants and mice, in which crossover recombination is initiated prior to synapsis, Pch2/PCH2/TRIP13's role in regulating crossover number and distribution is central and occurs prior to

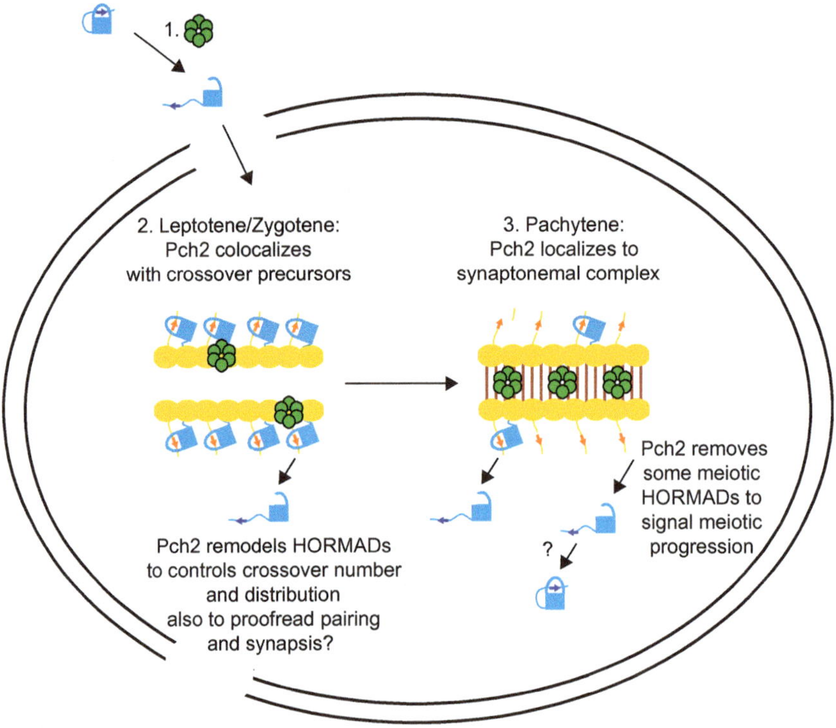


**Fig. 2** Model for Pch2/TRIP13/PCH2 function in systems in which recombination precedes synapsis, such as budding yeast, plants and mice. 1. Pch2 (green hexamer) remodels meiotic HORMADs from their closed conformation, bound to its own closure motif, to the extended conformation to enable its entry into meiotic nuclei and assembly on meiotic chromosomes. 2. During leptotene/zygotene, Pch2 forms puncta on meiotic chromosomes, colocalizing with crossover precursors, remodeling meiotic HORMADs from closed to extended to contribute to the gradual implementation of crossover number and distribution and possibly proofread homolog pairing and synapsis. Yellow ovals represent axis proteins that contain closure motifs and recruit meiotic HORMADs. 3. During pachytene, Pch2 localizes to the synaptonemal complex between synapsed homologous chromosomes, depleting meiotic HORMADs to limit meiotic recombination, drive meiotic progression and participate in the last stages of implementing crossover number and distribution through the remodeling of a reduced pool of meiotic HORMADs on meiotic chromosomes.

synapsis, possibly to ensure a wider distribution of crossovers beyond DSB "hot spots." However, since crossover precursors are essential for robust pairing and synapsis in these systems (Baudat et al., 2000; Giroux et al., 1989; Grelon et al., 2001; Romanienko & Camerini-Otero, 2000) and Pch2/PCH2 colocalizes with crossover precursors in budding yeast and plants (Lambing et al., 2015; Joshi et al., 2009), this remodeling on chromosomes is also likely relevant for pairing and synapsis (see #2 in Fig. 2), possibly proofreading them for accuracy.

Once synapsis initiates in these systems, PCH-2 localizes to SC (see #3 in Fig. 2) to deplete or redistribute meiotic HORMADS and signal successful synapsis, the progression of most crossover precursors into appropriate DNA repair pathways and that meiosis should progress. This depletion also restores the sister chromatid as a viable DNA repair template, allowing for the completion of DNA repair, and in some systems, downregulates DSB formation. Further, this depletion may be selective: work in budding yeast has shown that its meiotic HORMAD, Hop1, persists at chromosome ends, leading to an increase in DSBs on shorter chromosomes (Subramanian et al., 2019), thus guaranteeing they get crossovers. Finally, although meiotic HORMADs are depleted from chromosomes during synapsis, they are not completely removed (Cuacos et al., 2021; Lambing et al., 2015; Wojtasz et al., 2009), suggesting roles for both meiotic HORMADs and Pch2/PCH2/TRIP13 even when chromosomes are synapsed, such as contributing to the gradual implementation of crossovers and their control and distribution during synapsis.

In systems in which synapsis precedes recombination, such as *C. elegans*, the data for PCH-2's role in pairing and synapsis is clearest (Deshong et al., 2014; Giacopazzi et al., 2020) (Fig. 3A) and is likely accomplished by PCH-2 remodeling meiotic HORMADs on chromosomes directly involved in driving pairing and synapsis. Once chromosomes synapse, PCH-2 localizes to the SC and remodels meiotic HORMADs on these synapsed chromosomes to coordinate control of meiotic recombination so that it is more widely distributed than simply at the first places to pair, synapse or experience DSBs. This remodeling also helps implement crossover assurance and interference (Fig. 3).

PCH2's role is clearly different in *Drosophila*, where PCH2 appears to regulate meiotic progression and when chromosomes are competent for crossover formation from the nuclear periphery (Joyce & McKim, 2010), raising the obvious question: what is its meiotic substrate(s)? One possible candidate is Mad2 which controls cell cycle progression (Rodriguez-Bravo et al., 2014) and has been shown to regulate meiotic synapsis in *C. elegans* through its localization at the nuclear periphery (Devigne & Bhalla, 2021). Another compelling candidate is Rev7, either through its capacity in regulating recombination (Clairmont et al., 2020) or cell cycle progression (Listovsky & Sale, 2013; Pfleger, Salic, Lee, & Kirschner, 2001).

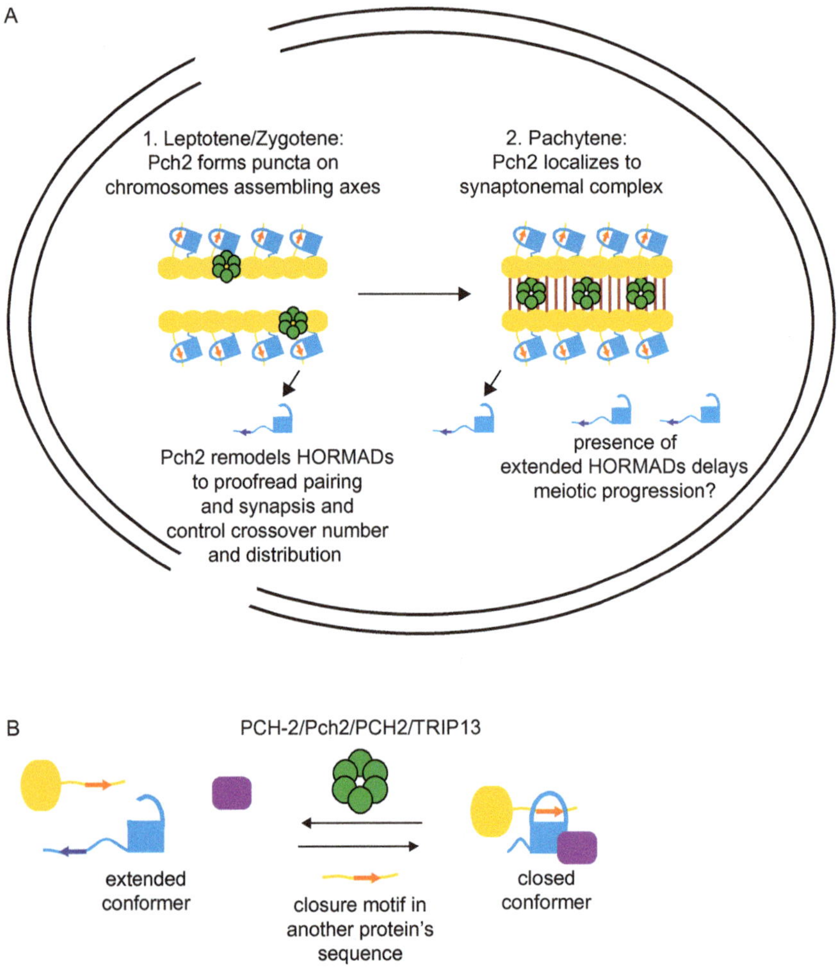


**Fig. 3** Model for PCH-2 function in systems in which synapsis precedes recombination and PCH-2 localizes to the synaptonemal complex, such as *C. elegans*. (A) 1. During leptotene/zygotene, Pch2 forms puncta on meiotic chromosomes, remodeling meiotic HORMADs from closed to extended to control to proofread homolog pairing and synapsis. Yellow ovals represent axis proteins that contain closure motifs and recruit meiotic HORMADs. In this system, the assembly of meiotic HORMADs on chromosomes drives meiotic progression. 2. During pachytene, PCH-2 localizes to the synaptonemal complex between synapsed homologous chromosomes, meiotic HORMADs remain on chromosomes and remodeling meiotic HORMADs from closed to extended to contribute to the gradual implementation of crossover number and distribution. The presence of soluble extended conformers of at least one meiotic HORMAD, HTP-1, delays meiotic progression. (B) A model for how PCH-2's remodeling of meiotic HORMADs could destabilize interhomolog intermediates involved in pairing, synapsis and/or recombination on meiotic chromosomes, contributing to their proofreading and regulation.

Pch2/PCH2's remodeling of HORMADs in the cytoplasm to make them available for their roles in meiotic nuclei is clear in both budding yeast and plants (Herruzo et al., 2021; Yang et al., 2020). What does PCH-2's remodeling of meiotic HORMADs directly on chromosomes accomplish, aside from their depletion to signal meiotic progression in some systems? Aside from their adoption of multiple conformers to accomplish different functions, another hallmark of HORMADs is a dimerization interface that enables interactions with other factors (Rosenberg & Corbett, 2015). One possibility is that PCH-2's remodeling of meiotic HORMADs on chromosomes destabilizes interactions at this interface that are required for pairing, synapsis and/or interhomolog recombination (Fig. 3B). This model could explain PCH-2's proofreading function during pairing and synapsis and potentially its role in controlling the number and distribution of crossovers. In particular, PCH-2's ability to destabilize some crossover intermediates to favor their distribution elsewhere in the genome during the progressive implementation of crossover recombination could account for its observed involvement in crossover assurance, interference and homeostasis in almost all systems to date.

A final set of species-specific embellishments are the variations in molecular mechanisms that produce the checkpoint responses that rely on HORMADs and are regulated by PCH2 in response to meiotic defects. It seems clear that, when confronted with meiotic defects, the PCH-2/HORMAD module often exploits meiotic processes specific to species, such as MSCI during mammalian spermatogenesis, to implement its effect on meiotic progression. These species-specific mechanisms could also explain the variation in stringency of pachytene checkpoint responses among different species, even among sexes in the same species. Since these differences in checkpoint stringency have been used to explain the ability of some organisms to tolerate genetic incompatibilities and changes in ploidy (Li et al., 2009), it is possible that future studies of the PCH-2/HORMAD module will expand this discussion to our understanding of speciation and hybrid sterility.

## 11. Conclusion

The role of PCH-2 in meiosis has long been enigmatic and confusing, defying an easy and accessible characterization of its conserved function among the variety of systems in which meiosis is currently studied. However, its conservation, even in systems lacking meiotic HORMADs, suggests an important role in meiotic regulation. When assessed in the context of the multiple, different ways that essential events in meiosis are themselves regulated in these systems, a conserved function for PCH-2 becomes clearer and differences can be interpreted as species-specific embellishments on this conserved function. Indeed, given that sexual reproduction is known for evolutionary innovation, these differences may provide as much insight into how organisms coordinate recombination with synapsis as conservation and similarities do.

## References

Alfieri, C., Chang, L., & Barford, D. (2018). Mechanism for remodelling of the cell cycle checkpoint protein MAD2 by the ATPase TRIP13. *Nature, 559*, 274–278.

Bailis, J. M., Smith, A. V., & Roeder, G. S. (2000). Bypass of a meiotic checkpoint by overproduction of meiotic chromosomal proteins. *Molecular and Cellular Biology, 20*, 4838–4848.

Balboni, M., Yang, C., Komaki, S., Brun, J., & Schnittger, A. (2020). COMET Functions as a PCH2 cofactor in regulating the HORMA domain protein ASY1. *Current Biology, 30*(4113–4127), e4116.

Baudat, F., Manova, K., Yuen, J. P., Jasin, M., & Keeney, S. (2000). Chromosome synapsis defects and sexually dimorphic meiotic progression in mice lacking Spo11. *Molecular Cell, 6*, 989–998.

Bhalla, N., & Dernburg, A. F. (2005). A conserved checkpoint monitors meiotic chromosome synapsis in Caenorhabditis elegans. *Science, 310*, 1683–1686.

Bhalla, N., & Dernburg, A. F. (2008). Prelude to a division. *Annual Review of Cell and Developmental Biology, 24*, 397–424.

Borner, G. V., Barot, A., & Kleckner, N. (2008). Yeast Pch2 promotes domainal axis organization, timely recombination progression, and arrest of defective recombinosomes during meiosis. *Proceedings of the National Academy of Sciences of the United States of America, 105*, 3327–3332.

Borner, G. V., Kleckner, N., & Hunter, N. (2004). Crossover/noncrossover differentiation, synaptonemal complex formation, and regulatory surveillance at the leptotene/zygotene transition of meiosis. *Cell, 117*, 29–45.

Brown, S. D., Jarosinska, O. D., & Lorenz, A. (2018). Genetic interactions between the chromosome axis-associated protein Hop1 and homologous recombination determinants in Schizosaccharomyces pombe. *Current Genetics, 64*, 1089–1104.

Burroughs, A. M., Zhang, D., Schaffer, D. E., Iyer, L. M., & Aravind, L. (2015). Comparative genomic analyses reveal a vast, novel network of nucleotide-centric systems in biological conflicts, immunity and signaling. *Nucleic Acids Research, 43*, 10633–10654.

Capilla-Perez, L., Durand, S., Hurel, A., Lian, Q., Chambon, A., Taochy, C., et al. (2021). The synaptonemal complex imposes crossover interference and heterochiasmy in Arabidopsis. *Proceedings of the National Academy of Sciences of the United States of America, 118*.

Carballo, J. A., Johnson, A. L., Sedgwick, S. G., & Cha, R. S. (2008). Phosphorylation of the axial element protein Hop1 by Mec1/Tel1 ensures meiotic interhomolog recombination. *Cell, 132*, 758–770.

Caryl, A. P., Armstrong, S. J., Jones, G. H., & Franklin, F. C. (2000). A homologue of the yeast HOP1 gene is inactivated in the Arabidopsis meiotic mutant asy1. *Chromosoma, 109*, 62–71.

Chakraborty, P., Pankajam, A. V., Lin, G., Dutta, A., Krishnaprasad, G. N., Tekkedil, M. M., et al. (2017). Modulating crossover frequency and interference for obligate crossovers in Saccharomyces cerevisiae meiosis. *G3 (Bethesda)*, 7, 1511–1524.

Chen, X., Gaglione, R., Leong, T., Bednor, L., de Los Santos, T., Luk, E., et al. (2018). Mek1 coordinates meiotic progression with DNA break repair by directly phosphorylating and inhibiting the yeast pachytene exit regulator Ndt80. *PLoS Genetics*, *14*, e1007832.

Clairmont, C. S., Sarangi, P., Ponnienselvan, K., Galli, L. D., Csete, I., Moreau, L., et al. (2020). TRIP13 regulates DNA repair pathway choice through REV7 conformational change. *Nature Cell Biology*, *22*, 87–96.

Cole, F., Kauppi, L., Lange, J., Roig, I., Wang, R., Keeney, S., et al. (2012). Homeostatic control of recombination is implemented progressively in mouse meiosis. *Nature Cell Biology*, *14*, 424–430.

Couteau, F., Nabeshima, K., Villeneuve, A., & Zetka, M. (2004). A component of C. elegans meiotic chromosome axes at the interface of homolog alignment, synapsis, nuclear reorganization, and recombination. *Current Biology*, *14*, 585–592.

Couteau, F., & Zetka, M. (2005). HTP-1 coordinates synaptonemal complex assembly with homolog alignment during meiosis in C. elegans. *Genes & Development*, *19*, 2744–2756.

Couteau, F., & Zetka, M. (2011). DNA damage during meiosis induces chromatin remodeling and synaptonemal complex disassembly. *Developmental Cell*, *20*, 353–363.

Cuacos, M., Lambing, C., Pachon-Penalba, M., Osman, K., Armstrong, S. J., Henderson, I. R., et al. (2021). Meiotic chromosome axis remodelling is critical for meiotic recombination in Brassica rapa. *Journal of Experimental Botany*, *72*, 3012–3027.

Daniel, K., Lange, J., Hached, K., Fu, J., Anastassiadis, K., Roig, I., et al. (2011). Meiotic homologue alignment and its quality surveillance are controlled by mouse HORMAD1. *Nature Cell Biology*, *13*, 599–610.

Dernburg, A. F., McDonald, K., Moulder, G., Barstead, R., Dresser, M., & Villeneuve, A. M. (1998). Meiotic recombination in C. elegans initiates by a conserved mechanism and is dispensable for homologous chromosome synapsis. *Cell*, *94*, 387–398.

Deshong, A. J., Ye, A. L., Lamelza, P., & Bhalla, N. (2014). A quality control mechanism coordinates meiotic prophase events to promote crossover assurance. *PLoS Genetics*, *10*, e1004291.

Devigne, A., & Bhalla, N. (2021). Mad1's ability to interact with Mad2 is essential to regulate and monitor meiotic synapsis in C. elegans. *PLoS Genetics*, *17*, e1009598.

Giacopazzi, S., Vong, D., Devigne, A., & Bhalla, N. (2020). PCH-2 collaborates with CMT-1 to proofread meiotic homolog interactions. *PLoS Genetics*, *16*, e1008904.

Giroux, C. N., Dresser, M. E., & Tiano, H. F. (1989). Genetic control of chromosome synapsis in yeast meiosis. *Genome*, *31*, 88–94.

Goodyer, W., Kaitna, S., Couteau, F., Ward, J. D., Boulton, S. J., & Zetka, M. (2008). HTP-3 links DSB formation with homolog pairing and crossing over during C. elegans meiosis. *Developmental Cell*, *14*, 263–274.

Gray, S., & Cohen, P. E. (2016). Control of meiotic crossovers: From double-strand break formation to designation. *Annual Review of Genetics*, *50*, 175–210.

Grelon, M., Vezon, D., Gendrot, G., & Pelletier, G. (2001). AtSPO11-1 is necessary for efficient meiotic recombination in plants. *The EMBO Journal*, *20*, 589–600.

Hayashi, M., Chin, G. M., & Villeneuve, A. M. (2007). C. elegans germ cells switch between distinct modes of double-strand break repair during meiotic prophase progression. *PLoS Genetics*, *3*, e191.

Herruzo, E., Lago-Maciel, A., Baztan, S., Santos, B., Carballo, J. A., & San-Segundo, P. A. (2021). Pch2 orchestrates the meiotic recombination checkpoint from the cytoplasm. *PLoS Genetics*, *17*, e1009560.

Herruzo, E., Ontoso, D., Gonzalez-Arranz, S., Cavero, S., Lechuga, A., & San-Segundo, P. A. (2016). The Pch2 AAA+ ATPase promotes phosphorylation of the Hop1 meiotic checkpoint adaptor in response to synaptonemal complex defects. *Nucleic Acids Research, 44*, 7722–7741.

Herruzo, E., Santos, B., Freire, R., Carballo, J. A., & San-Segundo, P. A. (2019). Characterization of Pch2 localization determinants reveals a nucleolar-independent role in the meiotic recombination checkpoint. *Chromosoma, 128*, 297–316.

Ho, H. C., & Burgess, S. M. (2011). Pch2 acts through Xrs2 and Tel1/ATM to modulate interhomolog bias and checkpoint function during meiosis. *PLoS Genetics, 7*, e1002351.

Hochwagen, A., Tham, W. H., Brar, G. A., & Amon, A. (2005). The FK506 binding protein Fpr3 counteracts protein phosphatase 1 to maintain meiotic recombination checkpoint activity. *Cell, 122*, 861–873.

Hollingsworth, N. M., Goetsch, L., & Byers, B. (1990). The HOP1 gene encodes a meiosis-specific component of yeast chromosomes. *Cell, 61*, 73–84.

Hollister, J. D., Arnold, B. J., Svedin, E., Xue, K. S., Dilkes, B. P., & Bomblies, K. (2012). Genetic adaptation associated with genome-doubling in autotetraploid Arabidopsis arenosa. *PLoS Genetics, 8*, e1003093.

Ji, J., Tang, D., Shen, Y., Xue, Z., Wang, H., Shi, W., et al. (2016). P31comet, a member of the synaptonemal complex, participates in meiotic DSB formation in rice. *Proceedings of the National Academy of Sciences of the United States of America, 113*, 10577–10582.

Joshi, N., Barot, A., Jamison, C., & Borner, G. V. (2009). Pch2 links chromosome axis remodeling at future crossover sites and crossover distribution during yeast meiosis. *PLoS Genetics, 5*, e1000557.

Joshi, N., Brown, M. S., Bishop, D. K., & Borner, G. V. (2015). Gradual implementation of the meiotic recombination program via checkpoint pathways controlled by global DSB levels. *Molecular Cell, 57*, 797–811.

Joyce, E. F., & McKim, K. S. (2009). Drosophila PCH2 is required for a pachytene checkpoint that monitors double-strand-break-independent events leading to meiotic crossover formation. *Genetics, 181*, 39–51.

Joyce, E. F., & McKim, K. S. (2010). Chromosome axis defects induce a checkpoint-mediated delay and interchromosomal effect on crossing over during Drosophila meiosis. *PLoS*. Genet 6.

Kim, Y., Kostow, N., & Dernburg, A. F. (2015). The chromosome axis mediates feedback control of CHK-2 to ensure crossover formation in C. elegans. *Developmental Cell, 35*, 247–261.

Kim, Y., Rosenberg, S. C., Kugel, C. L., Kostow, N., Rog, O., Davydov, V., et al. (2014). The chromosome axis controls meiotic events through a hierarchical assembly of HORMA domain proteins. *Developmental Cell, 31*, 487–502.

Kogo, H., Tsutsumi, M., Inagaki, H., Ohye, T., Kiyonari, H., & Kurahashi, H. (2012). HORMAD2 is essential for synapsis surveillance during meiotic prophase via the recruitment of ATR activity. *Genes to Cells, 17*, 897–912.

Kogo, H., Tsutsumi, M., Ohye, T., Inagaki, H., Abe, T., & Kurahashi, H. (2012). HORMAD1-dependent checkpoint/surveillance mechanism eliminates asynaptic oocytes. *Genes to Cells, 17*, 439–454.

Lambing, C., Osman, K., Nuntasoontorn, K., West, A., Higgins, J. D., Copenhaver, G. P., et al. (2015). Arabidopsis PCH2 mediates meiotic chromosome remodeling and maturation of crossovers. *PLoS Genetics, 11*, e1005372.

Lara-Gonzalez, P., Westhorpe, F. G., & Taylor, S. S. (2012). The spindle assembly checkpoint. *Current Biology, 22*, R966–R980.

Latypov, V., Rothenberg, M., Lorenz, A., Octobre, G., Csutak, O., Lehmann, E., et al. (2010). Roles of Hop1 and Mek1 in meiotic chromosome pairing and recombination partner choice in Schizosaccharomyces pombe. *Molecular and Cellular Biology, 30*, 1570–1581.

Li, X. C., Barringer, B. C., & Barbash, D. A. (2009). The pachytene checkpoint and its relationship to evolutionary patterns of polyploidization and hybrid sterility. *Heredity (Edinburgh)*, *102*, 24–30.

Li, X. C., & Schimenti, J. C. (2007). Mouse pachytene checkpoint 2 (trip13) is required for completing meiotic recombination but not synapsis. *PLoS Genetics*, *3*, e130.

Listovsky, T., & Sale, J. E. (2013). Sequestration of CDH1 by MAD2L2 prevents premature APC/C activation prior to anaphase onset. *The Journal of Cell Biology*, *203*, 87–100.

Lorenz, A., Wells, J. L., Pryce, D. W., Novatchkova, M., Eisenhaber, F., McFarlane, R. J., et al. (2004). S. pombe meiotic linear elements contain proteins related to synaptonemal complex components. *Journal of Cell Science*, *117*, 3343–3351.

Lynn, A., Koehler, K. E., Judis, L., Chan, E. R., Cherry, J. P., Schwartz, S., et al. (2002). Covariation of synaptonemal complex length and mammalian meiotic exchange rates. *Science*, *296*, 2222–2225.

MacQueen, A. J., Phillips, C. M., Bhalla, N., Weiser, P., Villeneuve, A. M., & Dernburg, A. F. (2005). Chromosome sites play dual roles to establish homologous synapsis during meiosis in C. elegans. *Cell*, *123*, 1037–1050.

Mahadevaiah, S. K., Bourc'his, D., de Rooij, D. G., Bestor, T. H., Turner, J. M., & Burgoyne, P. S. (2008). Extensive meiotic asynapsis in mice antagonises meiotic silencing of unsynapsed chromatin and consequently disrupts meiotic sex chromosome inactivation. *The Journal of Cell Biology*, *182*, 263–276.

Malik, R., Kopylov, M., Gomez-Llorente, Y., Jain, R., Johnson, R. E., Prakash, L., et al. (2020). Structure and mechanism of B-family DNA polymerase zeta specialized for translesion DNA synthesis. *Nature Structural & Molecular Biology*, *27*, 913–924.

Mao-Draayer, Y., Galbraith, A. M., Pittman, D. L., Cool, M., & Malone, R. E. (1996). Analysis of meiotic recombination pathways in the yeast Saccharomyces cerevisiae. *Genetics*, *144*, 71–86.

Martinez-Perez, E., & Villeneuve, A. M. (2005). HTP-1-dependent constraints coordinate homolog pairing and synapsis and promote chiasma formation during C. elegans meiosis. *Genes & Development*, *19*, 2727–2743.

McKim, K. S., Green-Marroquin, B. L., Sekelsky, J. J., Chin, G., Steinberg, C., Khodosh, R., et al. (1998). Meiotic synapsis in the absence of recombination. *Science*, *279*, 876–878.

Miao, C., Tang, D., Zhang, H., Wang, M., Li, Y., Tang, S., et al. (2013). Central region component1, a novel synaptonemal complex component, is essential for meiotic recombination initiation in rice. *Plant Cell*, *25*, 2998–3009.

Morgan, C., Fozard, J. A., Hartley, M., Henderson, I. R., Bomblies, K., & Howard, M. (2021). Diffusion-mediated HEI10 coarsening can explain meiotic crossover positioning in Arabidopsis. *Nature Communications*, *12*, 4674.

Nadarajan, S., Altendorfer, E., Saito, T. T., Martinez-Garcia, M., & Colaiacovo, M. P. (2021). HIM-17 regulates the position of recombination events and GSP-1/2 localization to establish short arm identity on bivalents in meiosis. *Proceedings of the National Academy of Sciences of the United States of America*, *118*.

Niu, H., Wan, L., Baumgartner, B., Schaefer, D., Loidl, J., & Hollingsworth, N. M. (2005). Partner choice during meiosis is regulated by Hop1-promoted dimerization of Mek1. *Molecular Biology of the Cell*, *16*, 5804–5818.

Panizza, S., Mendoza, M. A., Berlinger, M., Huang, L., Nicolas, A., Shirahige, K., et al. (2011). Spo11-accessory proteins link double-strand break sites to the chromosome axis in early meiotic recombination. *Cell*, *146*, 372–383.

Pfleger, C. M., Salic, A., Lee, E., & Kirschner, M. W. (2001). Inhibition of Cdh1-APC by the MAD2-related protein MAD2L2: a novel mechanism for regulating Cdh1. *Genes & Development*, *15*, 1759–1764.

Prugar, E., Burnett, C., Chen, X., & Hollingsworth, N. M. (2017). Coordination of double strand break repair and meiotic progression in yeast by a Mek1-Ndt80 negative feedback loop. *Genetics*, *206*, 497–512.

Raina, V. B., & Vader, G. (2020). Homeostatic control of meiotic prophase checkpoint function by Pch2 and Hop1. *Current Biology*, *30*(22), 4413–4424.e5.

Rinaldi, V. D., Bolcun-Filas, E., Kogo, H., Kurahashi, H., & Schimenti, J. C. (2017). The DNA damage checkpoint eliminates mouse oocytes with chromosome synapsis failure. *Molecular Cell*, *67*(1026–1036), e1022.

Rodriguez-Bravo, V., Maciejowski, J., Corona, J., Buch, H. K., Collin, P., Kanemaki, M. T., et al. (2014). Nuclear pores protect genome integrity by assembling a premitotic and Mad1-dependent anaphase inhibitor. *Cell*, *156*, 1017–1031.

Roelens, B., Barroso, C., Montoya, A., Cutillas, P., Zhang, W., Woglar, A., et al. (2019). Spatial regulation of polo-like kinase activity during Caenorhabditis elegans meiosis by the nucleoplasmic HAL-2/HAL-3 complex. *Genetics*, *213*, 79–96.

Roig, I., Dowdle, J. A., Toth, A., de Rooij, D. G., Jasin, M., & Keeney, S. (2010). Mouse TRIP13/PCH2 is required for recombination and normal higher-order chromosome structure during meiosis. *PLoS Genet*, *6*.

Romanienko, P. J., & Camerini-Otero, R. D. (2000). The mouse Spo11 gene is required for meiotic chromosome synapsis. *Molecular Cell*, *6*, 975–987.

Rosenberg, S. C., & Corbett, K. D. (2015). The multifaceted roles of the HORMA domain in cellular signaling. *The Journal of Cell Biology*, *211*, 745–755.

Ruiz-Herrera, A., Vozdova, M., Fernandez, J., Sebestova, H., Capilla, L., Frohlich, J., et al. (2017). Recombination correlates with synaptonemal complex length and chromatin loop size in bovids-insights into mammalian meiotic chromosomal organization. *Chromosoma*, *126*, 615–631.

Sanchez-Moran, E., Santos, J. L., Jones, G. H., & Franklin, F. C. (2007). ASY1 mediates AtDMC1-dependent interhomolog recombination during meiosis in Arabidopsis. *Genes & Development*, *21*, 2220–2233.

San-Segundo, P. A., & Roeder, G. S. (1999). Pch2 links chromatin silencing to meiotic checkpoint control. *Cell*, *97*, 313–324.

Schwacha, A., & Kleckner, N. (1994). Identification of joint molecules that form frequently between homologs but rarely between sister chromatids during yeast meiosis. *Cell*, *76*, 51–63.

Shin, Y. H., McGuire, M. M., & Rajkovic, A. (2013). Mouse HORMAD1 is a meiosis i checkpoint protein that modulates DNA double- strand break repair during female meiosis. *Biology of Reproduction*, *89*, 29.

Silva, N., Ferrandiz, N., Barroso, C., Tognetti, S., Lightfoot, J., Telecan, O., et al. (2014). The fidelity of synaptonemal complex assembly is regulated by a signaling mechanism that controls early meiotic progression. *Developmental Cell*, *31*, 503–511.

Stanzione, M., Baumann, M., Papanikos, F., Dereli, I., Lange, J., Ramlal, A., et al. (2016). Meiotic DNA break formation requires the unsynapsed chromosome axis-binding protein IHO1 (CCDC36) in mice. *Nature Cell Biology*, *18*, 1208–1220.

Subramanian, V. V., MacQueen, A. J., Vader, G., Shinohara, M., Sanchez, A., Borde, V., et al. (2016). Chromosome synapsis alleviates Mek1-dependent suppression of meiotic DNA repair. *PLoS Biology*, *14*, e1002369.

Subramanian, V. V., Zhu, X., Markowitz, T. E., Vale-Silva, L. A., San-Segundo, P. A., Hollingsworth, N. M., et al. (2019). Persistent DNA-break potential near telomeres increases initiation of meiotic recombination on short chromosomes. *Nature Communications*, *10*, 970.

Swanson, W. J., & Vacquier, V. D. (2002). The rapid evolution of reproductive proteins. *Nature Reviews. Genetics*, *3*, 137–144.

Tromer, E. C., van Hooff, J. J. E., Kops, G., & Snel, B. (2019). Mosaic origin of the eukaryotic kinetochore. *Proceedings of the National Academy of Sciences of the United States of America, 116*, 12873–12882.

Turner, J. M. (2007). Meiotic sex chromosome inactivation. *Development, 134*, 1823–1831.

Ur, S. N., & Corbett, K. D. (2021). Architecture and dynamics of meiotic chromosomes. *Annual Review of Genetics, 55*, 497–526.

Vader, G. (2015). Pch2(TRIP13): Controlling cell division through regulation of HORMA domains. *Chromosoma, 124*, 333–339.

Vader, G., Blitzblau, H. G., Tame, M. A., Falk, J. E., Curtin, L., & Hochwagen, A. (2011). Protection of repetitive DNA borders from self-induced meiotic instability. *Nature, 477*, 115–119.

van Hooff, J. J., Tromer, E., van Wijk, L. M., Snel, B., & Kops, G. J. (2017). Evolutionary dynamics of the kinetochore network in eukaryotes as revealed by comparative genomics. *EMBO Reports, 18*(9), 1559–1571.

West, A. M. V., Komives, E. A., & Corbett, K. D. (2017). Conformational dynamics of the Hop1 HORMA domain reveal a common mechanism with the spindle checkpoint protein Mad2. *Nucleic Acids Research.*

Wilkins, A. S., & Holliday, R. (2009). The evolution of meiosis from mitosis. *Genetics, 181*, 3–12.

Wojtasz, L., Cloutier, J. M., Baumann, M., Daniel, K., Varga, J., Fu, J., et al. (2012). Meiotic DNA double-strand breaks and chromosome asynapsis in mice are monitored by distinct HORMAD2-independent and -dependent mechanisms. *Genes & Development, 26*, 958–973.

Wojtasz, L., Daniel, K., Roig, I., Bolcun-Filas, E., Xu, H., Boonsanay, V., et al. (2009). Mouse HORMAD1 and HORMAD2, two conserved meiotic chromosomal proteins, are depleted from synapsed chromosome axes with the help of TRIP13 AAA-ATPase. *PLoS Genetics, 5*, e1000702.

Wu, H. Y., & Burgess, S. M. (2006). Two distinct surveillance mechanisms monitor meiotic chromosome metabolism in budding yeast. *Current Biology, 16*, 2473–2479.

Yang, C., Hu, B., Portheine, S. M., Chuenban, P., & Schnittger, A. (2020). State changes of the HORMA protein ASY1 are mediated by an interplay between its closure motif and PCH2. *Nucleic Acids Research, 48*, 11521–11535.

Yant, L., Hollister, J. D., Wright, K. M., Arnold, B. J., Higgins, J. D., Franklin, F. C. H., et al. (2013). Meiotic adaptation to genome duplication in Arabidopsis arenosa. *Current Biology, 23*, 2151–2156.

Ye, Q., Kim, D. H., Dereli, I., Rosenberg, S. C., Hagemann, G., Herzog, F., et al. (2017). The AAA+ ATPase TRIP13 remodels HORMA domains through N-terminal engagement and unfolding. *The EMBO Journal.*

Ye, Q., Lau, R. K., Mathews, I. T., Birkholz, E. A., Watrous, J. D., Azimi, C. S., et al. (2020). HORMA domain proteins and a Trip13-like ATPase regulate bacterial cGAS-like enzymes to mediate bacteriophage immunity. *Molecular Cell*, 77(709–722), e707.

Ye, Q., Rosenberg, S. C., Moeller, A., Speir, J. A., Su, T. Y., & Corbett, K. D. (2015). TRIP13 is a protein-remodeling AAA+ ATPase that catalyzes MAD2 conformation switching. *eLife, 4*.

Yokoo, R., Zawadzki, K. A., Nabeshima, K., Drake, M., Arur, S., & Villeneuve, A. M. (2012). COSA-1 reveals robust homeostasis and separable licensing and reinforcement steps governing meiotic crossovers. *Cell, 149*, 75–87.

Yu, Z., Kim, Y., & Dernburg, A. F. (2016). Meiotic recombination and the crossover assurance checkpoint in Caenorhabditis elegans. *Seminars in Cell & Developmental Biology, 54*, 106–116.

Zanders, S., & Alani, E. (2009). The pch2Delta mutation in baker's yeast alters meiotic crossover levels and confers a defect in crossover interference. *PLoS Genetics*, *5*, e1000571.

Zanders, S., Sonntag Brown, M., Chen, C., & Alani, E. (2011). Pch2 modulates chromatid partner choice during meiotic double-strand break repair in Saccharomyces cerevisiae. *Genetics*, *188*, 511–521.

Zetka, M. C., Kawasaki, I., Strome, S., & Muller, F. (1999). Synapsis and chiasma formation in Caenorhabditis elegans require HIM-3, a meiotic chromosome core component that functions in chromosome segregation. *Genes & Development*, *13*, 2258–2270.

CHAPTER THIRTEEN

# *In vitro* spermatogenesis: Why meiotic checkpoints matter

**Qijing Lei, Ans M.M. van Pelt, and Geert Hamer***
Center for Reproductive Medicine, Reproductive Biology Laboratory, Amsterdam Reproduction and Development Research Institute, Amsterdam UMC, University of Amsterdam, Amsterdam, The Netherlands
*Corresponding author: e-mail address: g.hamer@amsterdamumc.nl

## Contents



## Abstract

Successful *in vitro* spermatogenesis would generate functional haploid spermatids, and thus, form the basis for novel approaches to treat patients with impaired spermatogenesis or develop alternative strategies for male fertility preservation. Several culture strategies, including cell cultures using various stem cells and *ex vivo* cultures of testicular tissue, have been investigated to recapitulate spermatogenesis *in vitro*. Although some studies have described complete meiosis and subsequent generation of functional spermatids, key meiotic events, such as chromosome synapsis and homologous recombination required for successful meiosis and faithful *in vitro*-derived gametes, are often not reported. To guarantee the generation of *in vitro*-formed spermatids without persistent DNA double-strand breaks (DSBs) and chromosomal aberrations, criteria to evaluate whether all meiotic events are completely executed *in vitro* need to be established. *In vivo*, these meiotic events are strictly monitored by meiotic checkpoints that eliminate aberrant spermatocytes. To establish criteria to evaluate *in vitro* meiosis, we review the meiotic events and checkpoints that have been investigated by previous *in vitro* spermatogenesis studies. We found that, although major meiotic events such as initiation of DSBs and recombination, complete chromosome synapsis, and XY-body

*Current Topics in Developmental Biology*, Volume 151
ISSN 0070-2153
https://doi.org/10.1016/bs.ctdb.2022.04.009

formation can be achieved *in vitro*, crossover formation, chiasmata frequency, and checkpoint mechanisms have been mostly ignored. In addition, complete spermiogenesis, during which round spermatids differentiate into elongated spermatids, has not been achieved *in vitro* by various cell culture strategies. Finally, we discuss the implications of meiotic checkpoints for *in vitro* spermatogenesis protocols and future clinical use.

## 1. Introduction

Spermatogenesis, the development of haploid spermatids from mitotic precursor cells in the seminiferous tubules of the testis, is comprised of three main phases: self-renewal and differentiation of spermatogonial stem cells (SSCs), meiosis of spermatocytes, and maturation of haploid spermatids (Jan et al., 2012). In principle, this process supports the life-long production of approximately 1000 sperm per second. However, an estimated 10%–15% of couples suffer from subfertility, of which roughly 50% is due to a male factor (Kumar & Singh, 2015). Of these cases, a significant proportion is diagnosed with non-obstructive azoospermia (NOA), the lack of spermatozoa in semen due to impaired spermatogenesis (Van Der Steeg et al., 2011). Unfortunately, in approximately half of these cases, no underlying cause of spermatogenic failure can be identified (Krausz & Riera-Escamilla, 2018; Tüttelmann, Ruckert, & Röpke, 2018), and an estimated 10% of NOA cases may be caused by meiotic aberrations, for instance impaired recombination, that can activate a meiotic checkpoint, thereby leading to spermatocyte apoptosis (Gonsalves et al., 2004). These checkpoints normally eliminate spermatocytes with unrepaired DNA damage or incorrectly oriented chromosomes, and thus, prevent the chromosomal aberrations from being transmitted to the offspring. In patients harboring a mutation in a gene that is normally involved in meiotic DNA double-strand-break (DSB) repair or meiotic chromosome synapsis, these checkpoints induce spermatocyte apoptosis and cause meiotic arrest (Jan et al., 2018; Tournaye, Krausz, & Oates, 2017). Unfortunately, there are no treatment options for cases where no spermatids are formed.

In some cases, male infertility is caused by external factors, for instance cancer treatments, such as chemo- and radiotherapy, that initiate apoptosis of chemo- and radio-sensitive spermatogonia (Meistrich, 2013). In these cases, fertility can be preserved in adult patients by cryopreservation of retrieved spermatozoa. However, this is not an option for prepubertal boys in which spermatogenesis has not yet been initiated. A potential future treatment to preserve fertility in prepubertal boys is cryopreservation of SSCs prior to the gonadotoxic treatment, followed by SSC propagation and subsequent auto-transplantation at a reproductive age (Struijk, Mulder, Van Der Veen, Van Pelt, & Repping, 2013). In the future, this method can also be used to correct mutations causing spermatogenic arrest in isolated SSCs. In this case, auto-transplantation of corrected SSCs can potentially restore spermatogenic failure (Mulder et al., 2016). However, because the modified genotype will be transmitted to subsequent generations, the ethics and safety of genome editing methods in the germline should be addressed before clinical application. Moreover, the testes of patients suffering from meiotic arrest still contain endogenous (uncorrected) germ cells, and thus, the question of whether these endogenous germ cells can hinder homing and development of transplanted SSCs still remains to be resolved (Lei & Hamer, 2021; Morimoto et al., 2021; Wang et al., 2021).

Instead of SSC transplantation to the testis, many research groups are exploring the generation of functional spermatozoa *in vitro* for male fertility treatment and preservation. Even in cases where germ cells are completely absent, for instance in patients suffering from Sertoli cell only syndrome, functional spermatozoa can theoretically be generated *in vitro* from induced pluripotent stem cells (iPSCs) derived from the patient's somatic tissues. The current strategies to generate functional spermatozoa *in vitro* include differentiation of embryonic stem cells (ESCs) (Ishikura et al., 2021; Zhou et al., 2016), iPSCs (Hayashi, Ohta, Kurimoto, Aramaki, & Saitou, 2011; Hendriks, Dondorp, et al., 2015; Hendriks, Dancet, van Pelt, Hamer, & Repping, 2015) or SSCs (Lei et al., 2020; Sun et al., 2018), and *ex vivo* cultures of testicular fragments (Perrard et al., 2016; Portela et al., 2019; Sato et al., 2011, 2015). Although generation of round spermatid-like cells has been reported in different species using various culture strategies, the standards and methods to assess the quality of the *in vitro*-generated germ cells remain controversial, especially for meiotic cells. Handel and colleagues (Handel, Eppig, & Schimenti, 2014) have proposed rigorous criteria to evaluate whether gametogenesis has indeed been recapitulated outside of the gonads, which assess meiotic chromosome organization, synaptonemal complex formation, and recombination, as well as the capacity to generate viable euploid offspring using the *in vitro*-derived gametes. Although recombination is one of the most important meiotic events, also completion of chromosome synapsis and meiotic sex chromosome inactivation (MSCI) during the first meiotic prophase and chiasmata frequency during the first meiotic division (MI) should also be added to the list. Incomplete synapsis of homologous chromosomes, which reduces MSCI, impairs recombination and crossover formation. Because crossovers are required for proper orientation of the homologous chromosome pairs in bivalents, they are essential for correct segregation of the homologous chromosomes during MI (Hassold & Hunt, 2001). Moreover, meiotic recombination, chromosome synapsis, MSCI, and chromosome alignment during the meiotic M-phases are all strictly monitored by checkpoint mechanisms that can induce spermatocyte apoptosis (Lane & Kauppi, 2019; Subramanian & Hochwagen, 2014). However, we recently observed that in an *in vitro* spermatogenesis culture system involving mouse SSCs, despite associating with both aberrant chromosome synapsis and recombination, *in vitro*-derived spermatocyte-like cells still proceeded to the meiotic M-phase and, occasionally, even formed spermatid-like cells (Lei et al., 2020). In this culture system, meiotic checkpoints did not eliminate all aberrant spermatocytes (Lei et al., 2020). Thus, we argue that to prevent the formation of gametes with genomic damage or aneuploidies, the functionality of meiotic checkpoints should also be investigated for *in vitro* gametogenesis.

In this review, several meiotic events, including chromosome synapsis completion, MSCI, crossover formation, and chiasmata frequency, which are monitored by meiotic checkpoint mechanisms, are included in the list of benchmarks to evaluate *in vitro*-derived gametes as previously defined by Handel and colleagues (Handel et al., 2014). Subsequently, articles investigating *in vitro* spermatogenesis are included and systematically screened in relation to these events. In addition, the clinical prospects of *in vitro*-derived gametes, as well as the criteria that are applied to assess *in vitro* meiosis in these articles, are discussed.

## 2. Key meiotic events during meiosis

The generation of gametes *in vitro* without persistent DSBs or chromosomal aberrations, such as aneuploidies, can only be guaranteed when all key meiotic events are completely executed. *In vivo*, male meiosis is part of the process of spermatogenesis and initiates with the differentiation of mitotic spermatogonia into pre-leptotene spermatocytes under the influence of retinoic acid, which is secreted by the somatic Sertoli cells (Sou, Pryce, Tee, & McClurg, 2021). In the testis, meiotic initiation and progression are spatio-temporally regulated by Sertoli cells and the other surrounding germ cells (Endo, Freinkman, de Rooij, & Page, 2017). These different testicular cell types are spatially associated, and their interactions determine the timing and progression of the stages of spermatogenesis (Endo et al., 2017). *In vitro*, most aspects of the complex physiological regulation in the testis are lost and must be mimicked by adding testicular cells, hormones, or growth factors. Here, we briefly describe the key meiotic events of meiotic prophase I and metaphase stages that can be directly compared with *in vitro* spermatogenesis studies. Details on how these key meiotic events are recapitulated *in vitro* in these studies are presented in Table 1.

## 2.1 Key events of meiotic prophase I

Meiotic prophase I can be subdivided into four cytological stages based on the progression of synapsis of the homologous chromosomes facilitated by the synaptonemal complex (SC): leptonema (formation of axial elements of the SC), zygonema (initiation of synapsis by transversal and central element proteins of the SC), pachynema (full synapsis), and diplonema (disassembly of the SC) (Fig. 1A and B). Early meiotic events, including initiation of DSBs, meiotic recombination, recognition, and pairing of homologous chromosomes, start at the leptotene stage, while at the pachytene stage, chromosome synapsis and most DSB repairs are completed, meiotic crossovers are formed, and MSCI is established by localizing the X and Y chromosomes in the XY body (Fig. 1A and C) (Bolcun-Filas & Handel, 2018).

### *2.1.1 DSB formation, early DSB repair, and chromosome synapsis in vitro*

Only one study has been able to successfully differentiate mouse ESC-derived primordial germ cell-like cells (PGCLCs) into spermatogonia-like cells and leptotene spermatocytes *in vitro* (Ishikura et al., 2016). Although SCs were formed, further meiotic progression was not observed in this culture system. Nevertheless, the generated spermatogonia-like cells could be propagated *in vitro* and initiate spermatogenesis after transplantation into the seminiferous tubules of infertile mice (Ishikura et al., 2016). In addition, *in vitro* spermatogenesis involving SSCs has been widely investigated. A crucial finding has been that retinoic acid (RA) alone can initiate spermatogonial differentiation and early meiotic entry *in vitro* (Dann et al., 2008). When using juvenile Sertoli cells as feeder layer cells, in combination with RA treatment, zygotene spermatocytes can be generated, in which DSB formation and chromosome synapsis initiation can be observed (Wang et al., 2016). Interestingly, even pachytene spermatocytes with XY-body formation were occasionally observed in this culture system. Moreover, a recent study described that nutrient restriction in combination with RA

**Table 1** *In vitro* spermatogenesis using human or mouse ESCs, iPSCs and SSCs as starting point by cell culture strategies

| | | | | | Meiotic prophase | | |
|---|---|---|---|---|---|---|---|
| | | | | | Leptotene | | Zygotene |
| **Author, year** | **Species** | **Start cell types** | **Culture specifics** | **Types of feeder layers** | **Initiation of DSBs** | **Formation of axial elements** | **Initiation of synapsis** |
| Geijsen et al. (2004) | Mouse | ESCs | Embryoid body (EB) formation in combination with Retinoic Acid (RA) treatment | MEFs | Not described | Not described | Not described |
| Nayernia et al. (2006) | Mouse | ESCs | Embryoid body (EB) formation in combination with RA treatment | MEFs | Not described | Yes | Not described |
| Kee, Angeles, Flores, Nguyen, and Pera (2009) | Human | ESCs | By silencing and overexpressing of DAZL, DAZ and BOULE gene combined with stepwise differentiation protocol | Cultured in matrigel-coated plates | Yes | Yes | Yes |
| Eguizabal et al. (2011) | Human | ESCs and iPSCs | Stepwise differentiation protocol | Primary human foreskin fibroblasts (iHFFs)<br>Primary human foreskin fibroblasts (iHFFs)<br>Primary human foreskin fibroblasts (iHFFs)<br>Primary human foreskinfibroblasts (iHFFs) | Yes | Yes | Not described |

| | | | | | | | |
|---|---|---|---|---|---|---|---|
| Panula et al. (2011) | Human | iPSCs | BMP4, 7 and 8b supplement | Cultured in matrigel-coated plates | Not described | Yes | Not described |
| Easley IV et al. (2012) | Human | ESCs and iPSCs | Mouse SSC culture condition | Feeder-free<br>Feeder-free | Not described | Not described | Not described |
| Zhou et al. (2016) | Mouse | ESCs | Stepwise differentiation protocol | Neonatal testicular somatic cells | Yes | Yes | Yes |
| Zhao et al. (2018) | Human | iPSCs | Stepwise differentiation protocol | Feeder-free | Not described | Yes | Not described |
| Xu et al. (2020) | Human | ESCs | Stepwise differentiation protocol | Cultured in combinational poly-lysine and laminin-coated plates | Not described | Yes | Not described |
| Feng et al. (2002) | Mouse | A telomerase-immortalized GSC line | SCF supplement | Feeder-free | Not described | Yes | Yes |
| Dann et al. (2008) | Mouse | GSCs | Retinoic acid treatment | MEFs | Yes | Yes | Not described |
| Nolte et al. (2010) | Mouse | A multipotent adult GSC line | Embryoid body (EB) formation in combination with retinoic Acid treatment | Feeder-free | Not described | Yes | Yes |
| Wang et al. (2016) | Mouse | GSCs | Retinoic acid treatment | Pup Sertoli cells | Yes | Yes | Yes |
| Sun et al. (2018) | Human | GSCs | Using a medium containing RA, SCF, BMP4 and KSR | Cultured in matrigel-coated plates | Yes | Yes | Yes |
| Lei et al. (2020) | Mouse | GSCs | Stepwise differentiation protocol | Sertoli cell line (SK49) | Yes | Yes | Yes |

*Continued*

**Table 1** *In vitro* spermatogenesis using human or mouse ESCs, iPSCs and SSCs as starting point by cell culture strategies—cont'd

| | | | | | Meiotic prophase | | |
|---|---|---|---|---|---|---|---|
| | | | | | Leptotene | | Zygotene |
| **Author, year** | **Species** | **Start cell types** | **Culture specifics** | **Types of feeder layers** | **Initiation of DSBs** | **Formation of axial elements** | **Initiation of synapsis** |
| Lei, Zhang, van Pelt, and Hamer (2021) | Mouse | GSCs | Stepwise differentiation protocol | Sertoli cell line (SK49) | Yes | Yes | Yes |
| Zhang, Gunewardena, and Wang (2021) | Mouse | GSCs | Nutrient restriction in combination with a stepwise differentiation protocol | MEFs | Yes | Yes | Yes |

| | Meiotic prophase | | | | Metaphases | Post-meiosis | Offspring | |
|---|---|---|---|---|---|---|---|---|
| | Pachytene | | | Diplotene | Metaphase I | Round spermatids (RS) | Embryonic development | Offspring |
| **Author, year** | **Completion of synapsis** | **XY body formation** | **Crossover formation** | **Desynapsis** | **Bivalents or Univalents** | | | |
| Geijsen et al. (2004) | Not described | Not described | Not described | Not described | Not described | Yes | Blastocysts | Not described |
| Nayernia et al. (2006) | Not described | Not described | Not described | Not described | Not described | Yes | 2-cell state | Yes |
| Kee et al. (2009) | Not described | Not described | Not described | Not described | Not described | Yes | Not described | Not described |

| | | | | | | | | |
|---|---|---|---|---|---|---|---|---|
| Eguizabal et al. (2011) | Not described | Not described | Not described | Not described | Not described | only iPSCs can form RS | Not described | Not described |
| Panula et al. (2011) | Not described | Not described | Not described | Not described | Not described | Yes | Not described | Not described |
| Easley IV et al. (2012) | Not described | Not described | Not described | Not described | Not described | Yes | Not described | Not described |
| Zhou et al. (2016) | Yes | Yes | Not described | Yes | Bivalents | Yes | 2-cell stages | Yes |
| Zhao et al., 2018) | Not described | Not described | Not described | Not described | Not described | Yes | Not described | Not described |
| Xu et al. (2020) | Not described | Not described | Yes | Not described | Not described | Yes | Not described | Not described |
| Feng et al. (2002) | Not described | Not described | Not described | Early chiasmata | Not described | Yes | Not described | Not described |
| Dann et al. (2008) | Not described | Not described | Not described | Not described | Not described | Not described | Not described | Not described |
| Nolte et al. (2010) | Not described | Not described | Not described | Not described | Not described | Yes | 2-cell stages | Yes |
| Wang et al. (2016) | Yes, but occasionally | Yes, but occasionally | Not described | Not described | Not described | Not described | Not described | Not described |

*Continued*

**Table 1** *In vitro* spermatogenesis using human or mouse ESCs, iPSCs and SSCs as starting point by cell culture strategies—cont'd

| | Meiotic prophase | | | | Metaphases | Post-meiosis | Offspring | |
|---|---|---|---|---|---|---|---|---|
| | Pachytene | | | Diplotene | Metaphase I | Round spermatids (RS) | Embryonic development | Offspring |
| **Author, year** | **Completion of synapsis** | **XY body formation** | **Crossover formation** | **Desynapsis** | **Bivalents or Univalents** | | | |
| Sun et al. (2018) | Yes | Yes | Yes | Not described | Not described | Yes | 8-cell stage[a] | No |
| Lei et al. (2020) | No | No | No | No | Univalents | Yes | Not described | Not described |
| Lei et al. (2021) | Yes | Yes | Yes, but very few | No | Univalents | No | Not described | Not described |
| Zhang et al. (2021) | Yes | Not described | Not described | Not described | Not described | Not described | Not described | Not described |

[a] *In vitro*-formed human round spermatids used to fertilize mouse oocytes.

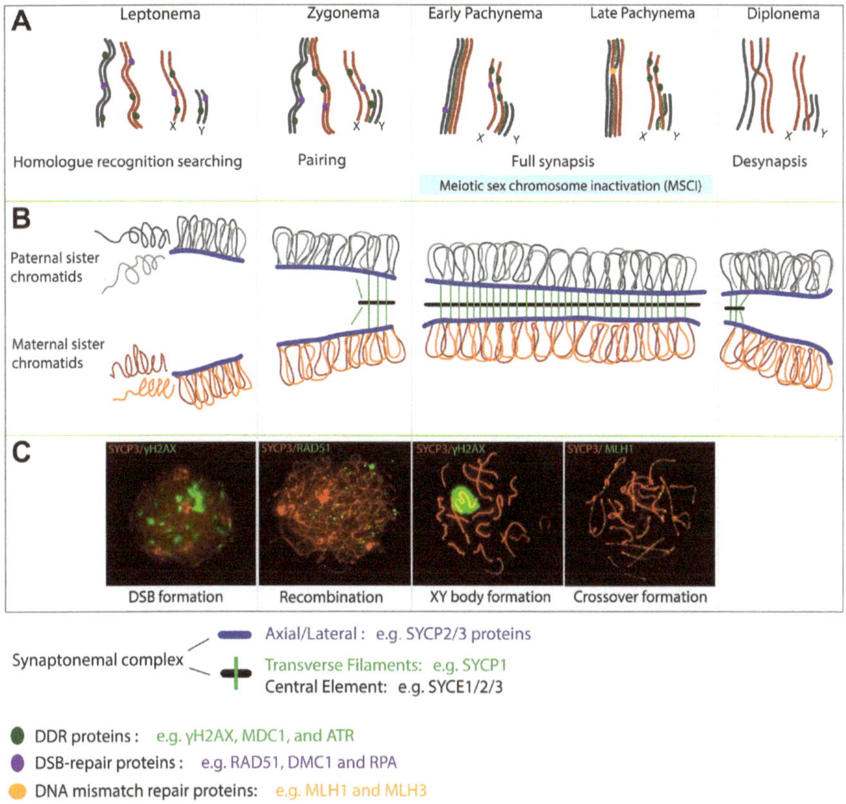


**Fig. 1** Schematic illustration of key meiotic events during meiotic prophase I. The cytological stages of the first meiotic prophase are depicted from left to right (A). Homologous chromosomes are indicated in gray (paternal) and red (maternal). Some key synaptonemal complex (SC) components and DNA repair proteins are illustrated in the upper (A) and middle (B) panels. For the immunofluorescent images (C), mouse spermatocytes stained for SYCP3 (red) together with γH2AX (green), RAD51 (green) or MLH1 (green).

treatment, followed by the use of culture medium containing melatonin, FSH, TGFβ, BPE, and dihydrotestosterone, with the support of mouse embryonic fibroblasts (MEFs), can induce the differentiation of mouse spermatogonia into pachytene-like cells in which meiotic DSB formation and chromosome synapsis can be recapitulated (Zhang et al., 2021).

### *2.1.2 Complete chromosome synapsis, establishment of MSCI, XY body formation, and crossovers* in vitro

Several studies have reported that pluripotent stem cells can be induced to undergo meiosis and form post-meiotic cells *in vitro* (Easley IV et al., 2012;

Eguizabal et al., 2011; Geijsen et al., 2004; Kee et al., 2009; Nayernia et al., 2006; Panula et al., 2011; Xu et al., 2020; Zhao et al., 2018; Zhou et al., 2016). Of these studies, however, only Zhou et al., who used mouse ESCs, described most key meiotic events *in vitro* such as DSB formation, recombination initiation, complete chromosome synapsis, and XY-body formation. Although chiasmata structures in MI-spermatocytes were described, evidence for complete meiotic DSB repair and crossover formation was not provided (Zhou et al., 2016). In addition, another study demonstrated that human ESC-derived spermatocytes can display MLH1-foci, whereas other meiotic events, such as complete chromosome synapsis and recombination, were not described (Xu et al., 2020). Therefore, the details of meiotic DSB repair and crossover formation *in vitro* remain largely unclear using pluripotent stem cells as the starting cell type.

Meanwhile, we observed many meiotic challenges when starting *in vitro* spermatogenesis with mouse primary SSCs such as incomplete homologous chromosome synapsis, lack of a discernable XY body, and incomplete meiotic crossover formation (Lei et al., 2020). Despite these issues, we observed that many *in vitro*-generated spermatocytes were not eliminated by meiotic checkpoints, which would induce spermatocyte apoptosis *in vivo*. Instead, these cells proceeded to MI, leading to the generation of univalent chromosome pairs (pairs of sister chromatids) at MI and, possibly, the formation of aneuploid spermatid-like cells (Lei et al., 2020). In an optimized culture system, in which we refreshed the required Sertoli cell feeder layer after meiotic initiation, complete chromosome synapsis and XY-body formation were achieved *in vitro*. However, meiotic recombination and crossover formation still appeared insufficient (Lei et al., 2021). In another study, using human male germline stem cells, crossover formation could be observed after the induction of meiosis in culture plates coated with Matrigel and media supplemented with RA, stem cell factor (SCF), bone morphogenetic protein 4 (BPM4), and knockout serum replacement (KSR), although chiasmata formation in MI-spermatocytes was not observed (see below) (Sun et al., 2018).

In total, several meiotic events of meiotic prophase I, such as DSB initiation, recombination, chromosome synapsis completion, and XY-body-formation, have been reported by several researchers using ESCs (Zhou et al., 2016) or SSCs (Lei et al., 2021; Sun et al., 2018; Wang et al., 2016). However, evidence for the accuracy of meiotic DSB repair and subsequent crossover formation is lacking for *in vitro* spermatogenesis.

Moreover, MSCI, an epigenetic silencing process that is regulated by, among others, protein ubiquitination and histone methylation (*e.g.*, H3K9me2) or phosphorylation (*e.g.*, H2AFY), as reviewed by Turner (2007), has not been studied *in vitro*. Proof of MSCI not only requires recognition of the unsynapsed chromosomes by SC proteins combined with DNA damage response (DDR) proteins, but also of transcriptional silencing, for instance di- and tri-methylated histone H3 lysine 9 (H3K9me2 and H3K9me3) or reduced expression of the Y-chromosomal genes *Zfy1/2* (Ichijima, Sin, & Namekawa, 2012; Royo et al., 2010; Turner, 2007). Although XY-body formation is a typical hallmark of MSCI and can be achieved *in vitro*, its detection alone does not fully prove the establishment of MSCI.

## 2.2 Key events of the meiotic metaphases

At least one genetic crossover per homologous chromosome pair is required to keep chromosomes together prior to the first meiotic M-phase. During MI, they are visible as chiasmata, which are required to prevent chromosome nondisjunction and subsequent aneuploid spermatid formation. Here, we briefly describe chiasmata formation and chromosome segregation at this stage, as well as review chiasmata frequency in studies of *in vitro* spermatogenesis.

### 2.2.1 *Chiasmata formation and chromosome segregation during meiotic metaphases* in vivo

Each pair of homologous chromosomes, or bivalents, contains at least one or two chiasmata developed from meiotic crossovers, enabling alignment to the equatorial plane and proper segregation during MI. Once all bivalents are aligned to the equatorial plane, cohesion proteins distal to the centromeres are removed so that homologous chromosomes can disjoin, while cohesion proteins proximal to the centromeres are retained until MII to prevent premature segregation of sister chromatids (Lane & Kauppi, 2019; Webster & Schuh, 2017). The homologous chromosome pairs are subsequently pulled to opposite poles by microtubule spindles. During MII, centromeric cohesion is also removed, followed by segregation of sister chromatids and formation of four haploid daughter cells.

### 2.2.2 *Chiasmata formation* in vitro

Too few chiasmata is one of the most frequent meiotic chromosomal abnormalities described in infertile men (Egozcue et al., 2005; Hulten, Eliasson, & Tillinger, 1970). The investigation of chiasmata frequency in MI-spermatocytes using cytogenetic techniques, such as fluorescence *in-situ* hybridization or meiotic chromosome spreads, can be applied to determine the cause of infertility (Egozcue et al., 2005; Hulten et al., 1970; Skakkebaek, Bryant, & Philip, 1973; Templado, Marina, & Egozcue, 1976; Uroz, Rajmil, & Templado, 2011). However, *in vitro*, as presented in Table 1, only four studies have analyzed the karyotypes of MI-spermatocytes (Feng et al., 2002; Lei et al., 2020, 2021; Zhou et al., 2016). Of these, Zhou et al. reported the presence of normal chiasmata in bivalents (Zhou et al., 2016). Using a telomerase-immortalized spermatogonial cell line, Feng et al. observed the presence of early chiasmata-like structures in diplotene spermatocytes. However, no chiasmata frequency in MI-spermatocytes nor other meiotic events, including complete chromosome synapsis, recombination, crossover formation, and XY-body formation, were investigated (Feng et al., 2002). In addition, a study using human spermatogonia reported that meiotic crossovers could be detected *in vitro*, but subsequent chiasmata formation at MI was not described (Sun et al., 2018). Our recent studies demonstrated that *in vitro*-generated MI spermatocytes contained univalents instead of bivalents, most likely due to insufficient crossover formation in pachytene spermatocytes *in vitro* (Lei et al., 2020, 2021). If these cells were to complete meiosis *in vitro*, then this would most likely lead to aneuploid spermatid-like cells (Uroz & Templado, 2012).

## 2.3 Spermiogenesis—The formation of elongated spermatids

After the completion of meiosis, round spermatids are formed that, *in vivo,* will subsequently undergo drastic cytological and morphological postmeiotic changes, including nuclear condensation and elongation, in a process called spermiogenesis (Jan et al., 2012). Similar to the (pre)meiotic transitions, these changes are coordinated by Sertoli cells that interact with spermatids and secrete RA to induce their elongation. In addition, pachytene spermatocytes that interact with the same Sertoli cells as these elongating spermatids also produce RA, which regulates timely spermatid development (Endo et al., 2017). Spermiogenesis, especially the histone to protamine transition that is required for nuclear condensation, also involves induction and repair of DSBs (Laberge & Boissonneault, 2005; Marcon & Boissonneault, 2004). Thus, when round spermatids are used to fertilize oocytes *via* round spermatid injection (ROSI), these early spermatids may still harbor DSBs. In addition, there is the widely shared view that the use of ROSI may lead to an increased risk of imprinting disorders linked to epigenetic abnormalities associating with improper methylation patterns in immature spermatids (Gross, Hanson, & Hotaling, 2020). Therefore, for future clinical application of *in vitro*-generated spermatids, elongated spermatids would be preferable. However, when the aforementioned complex physiological environment produced by Sertoli cells and interacting germ cells is lacking, proper spermatid elongation may be challenging to mimic *in vitro*. Here, we review the current studies on the generation of round and elongated spermatids *in vitro*.

### 2.3.1 Formation of round spermatid-like cells and elongated spermatid-like cells in vitro

In principle, ESCs have the potential to differentiate into any cell type (Murry & Keller, 2008). For instance, several studies have reported the differentiation of mouse ESCs to PGCLCs, SSC-like cells, or germline stem cell (GSC)-like cells *in vitro*, followed by transplantation into the testes of infertile mice, after which these cells could initiate spermatogenesis and form elongated spermatids in the seminiferous tubules (Hayashi et al., 2011; Ishikura et al., 2016; Li et al., 2019). In addition, when these ESC-derived GSC-like cells were transplanted into recipient mouse testes followed by *ex vivo* differentiation in testicular fragments, round and elongated spermatid formation was observed (Ishikura et al., 2021). Regardless, spermatogenesis still relies on the *in vivo* or *ex vivo* microenvironment provided by somatic cells within the seminiferous tubules. Several studies have also reported that round spermatid-like cells could be achieved *in vitro* using both mouse (Geijsen et al., 2004; Nayernia et al., 2006) and human (Easley IV et al., 2012) ESCs. However, patient ESCs are rarely available for clinical use, which severely limits their use in human fertility treatment or preservation. Another option could be the use of patient iPSCs, which can also differentiate to generate round spermatid-like cells (Easley IV et al., 2012; Zhao et al., 2018). However, the generation of iPSCs requires further reprogramming of somatic cells, and the safety of iPSC-generated reproductive cells has not been sufficiently investigated to warrant their further use. Fortunately, many patients with impaired spermatogenesis still possess SSCs that can be successfully isolated and cultured *in vitro* (Mulder et al., 2016). Because SSCs are already primed to ultimately differentiate into gametes, one could argue that SSCs are a more suitable starting point for *in vitro* spermatogenesis than other stem cell types. For instance, several studies have reported the generation of round spermatid-like cells using a mouse telomerase-immortalized spermatogonial cell line (Feng et al., 2002), a mouse multipotent adult germline stem cell line (Nolte et al., 2010), and mouse (Lei et al., 2020) or human SSCs (Sun et al., 2018).

In addition to using various cell culture strategies, *ex vivo* cultures of testicular tissue fragments have been widely investigated in different species. To date, *ex vivo* generation of elongated spermatids can be achieved by culturing mouse neonatal (Sato et al., 2011) and mature (Sato et al., 2015) testis tissue fragments and human fetal gonadal fragments (Yuan et al., 2020). In one study, human haploid germ cells could be generated by using prepubertal testis tissues (De Michele et al., 2018). For adult human testis tissues, morphologically mature spermatozoa could be histologically observed in one study (Perrard et al., 2016). However, other studies have reported that human testicular organotypic culture cannot generate spermatids *in vitro* (Medrano et al., 2018; Portela et al., 2019).

Overall, the generation of elongated spermatids requires the support of an *in vivo* or *ex vivo* testicular microenvironment, and the direct differentiation of ESCs, iPSCs, and SSCs *in vitro* using cell culture strategies so far ends with the formation of round spermatid-like cells.

## 3. Meiotic checkpoint mechanisms

Similar to other cell cycle checkpoints, meiotic checkpoints provide a control mechanism to arrest the cell cycle or induce apoptosis, that is, if something goes wrong, such as when DNA is damaged (Fig. 2) (Carr, 1996). Given that meiosis can be particularly error-prone, several checkpoints strictly monitor chromosome synapsis and recombination to ensure proper crossover formation during meiotic prophase I (Subramanian & Hochwagen, 2014). The prophase recombination checkpoint is tightly regulated by canonical DNA DDR proteins, namely, the conserved checkpoint sensor kinases ATR/ATM and the checkpoint mediator kinases CHK1/2, which induce apoptosis of abnormal spermatocytes (MacQueen & Hochwagen, 2011; Subramanian & Hochwagen, 2014) generally by the activation of the p53 or p63 protein in mice (Bolcun-Filas, Rinaldi, White, & Schimenti, 2014; Marcet-Ortega et al., 2017; Odorisio, Rodriguez, Evans, Clarke, & Burgoyne, 1998) and the p63 protein in humans (Jan et al., 2018). Homologous chromosome synapsis is indirectly checked upon the establishment of MSCI. When the autosomes are fully synapsed, the non-homologous regions of the X and Y chromosomes are rapidly silenced because they remain unsynapsed. However, incomplete synapsis of autosomes sequesters silencing factors away from the X and Y

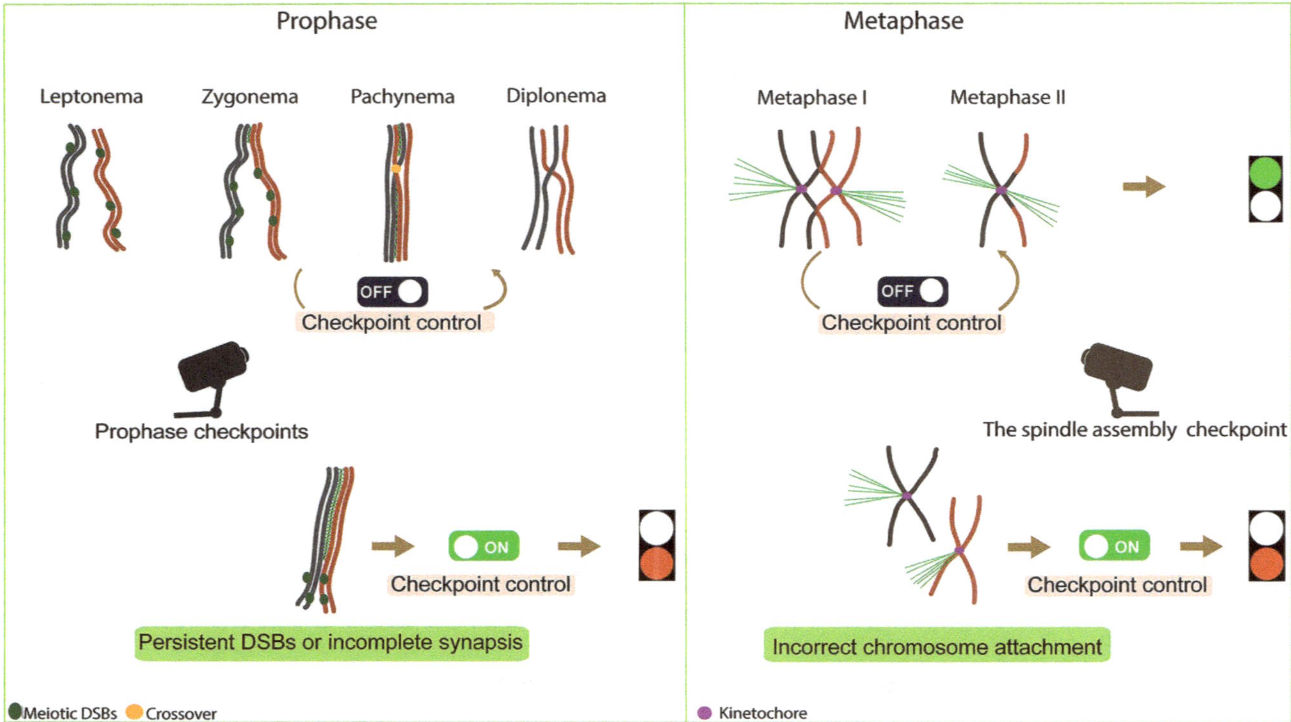


**Fig. 2** Schematic illustration of meiotic checkpoints. Meiotic checkpoints exist in the prophase and metaphase stages to monitor key meiotic events and eliminate aberrant spermatocytes. Meiotic recombination and homologous chromosome synapsis are checked by prophase checkpoints. Moreover, during metaphase stages, the spindle assembly checkpoint (SAC) is activated once chromosomes are not correctly attached on the spindle.

chromosomes to silence the unsynapsed autosomal regions. This causes incomplete transcriptional repression of the Y chromosomal *Zfy* genes (Turner, 2007), which causes apoptosis of mouse (Royo et al., 2010) and human (Jan et al., 2018) spermatocytes. Details on the regulation and function of meiotic prophase checkpoints have been reviewed extensively by Hochwagen and colleagues (Kar & Hochwagen, 2021; MacQueen & Hochwagen, 2011; Subramanian & Hochwagen, 2014), Turner (2015), and Raina and colleagues in this issue of Current Topics in Developmental Biology.

After meiotic prophase I, the spindle assembly checkpoint (SAC) ensures the proper attachment of chromosomes to the meiotic spindles at metaphase stages to prevent abnormal chromosome segregation during the two subsequent meiotic divisions. The meiotic metaphase starts with the attachment of kinetochores to microtubules that form the meiotic spindle. The SAC, which is comprised of mitotic checkpoint complex proteins such as MAD1, MAD2, BUB2, BUB3, BUBR1, and CDC20, inhibits the anaphase promoting complex, and thereby meiotic progression, until sufficient microtubules are attached, thereby providing the proper amount of tension by pulling the homologous chromosomes (first meiotic metaphase) or sister chromatids (second meiotic metaphase) to the opposite spindle poles (Lane & Kauppi, 2019). The regulation and function of the SAC in general has been reviewed in great detail by Musacchio (2015), and for meiosis specifically by Petronczki and colleagues (Petronczki, Siomos, & Nasmyth, 2003), and Lane and Kauppi (2019).

## 3.1 Investigation of meiotic checkpoints *in vitro*

Interestingly, these checkpoint control mechanisms seem not fully functional during *in vitro* meiosis. When starting with mouse spermatogonia cultured on SK49 immortalized Sertoli cells, although *in vitro* pachytene spermatocytes displayed unsynapsed chromosomes and lacked meiotic crossover formation, these cells still developed into MI-spermatocytes and occasionally formed round spermatid-like cells (Lei et al., 2020). Moreover, in a further optimized culture system, chromosome synapsis was complete, XY-body formation appeared normal, and meiotic DSBs and subsequent recombination was initiated. However, meiotic DSBs were not completely repaired and insufficient crossovers were formed (Lei et al., 2021). Because of this, the *in vitro*-generated MI-spermatocytes only contained univalent chromosomes (Lei et al., 2020, 2021).

*In vivo*, when meiotic synapsis and recombination are ongoing, the aforementioned meiotic prophase checkpoints inhibit cell cycle progression to obtain an extended meiotic prophase (Lee & Amon, 2001; Roeder & Bailis, 2000). However, *in vitro*, we found the meiotic prophase stage to take a much shorter time (around 9 days) than is normal *in vivo* (14 days) (Lei et al., 2020; Oakberg, 1956). In addition, also other studies that used mouse SSCs as the starting point for *in vitro* spermatogenesis reported meiotic progression *in vitro* to proceed faster than that *in vivo* (Wang et al., 2016; Zhang et al., 2021). This shortened meiotic prophase may not provide sufficient time to complete meiotic synapsis and recombination, which may lead to premature exit from the pachytene stage and premature initiation of the MI stage before the timely activation of checkpoints can initiate the apoptosis of abnormal spermatocytes.

As presented in Table 1, while meiotic DSB formation and early DSB repair, as well as chromosome synapsis and XY-body formation, were investigated *in vitro* in several studies, the functioning of most meiotic checkpoints remains largely unclear in relation to *in vitro* gametogenesis. Considering that meiotic checkpoints prevent the generation of aneuploid sperm, an analysis of checkpoint activity should be performed when spermatogenesis is mimicked *in vitro*.

## 4. Prospects of *in vitro* spermatogenesis for research or clinical use

### 4.1 Evaluation of *in vitro*-derived germ cells

In the current review, we discuss the meiotic events that have been described in previous studies of *in vitro* spermatogenesis. As presented in Table 1, some studies have reported that complete meiosis can be achieved *in vitro*, although techniques involving only the detection of gene or protein expression, and not their function, were frequently used to identify meiotic cells. Previous spermatogenic cell type-specific transcriptome studies have revealed that spermatogonia express many meiosis-associated genes, such as *Spo11*, *Sycp3*, and *Dmc1*. However, their encoded proteins become functional during meiosis (Hermann et al., 2018; Jan et al., 2017). To prove that meiosis can be mimicked *in vitro* and provide convincing evidence beyond the available gene expression data for the generation of *in vitro*-derived gametes, Handel and colleagues (Handel et al., 2014) proposed rigorous criteria to evaluate *in vitro*-generated germ cells, including chromosome content and organization, synaptonemal complex formation, proper meiotic recombination, and viable euploid offspring. Although meiotic synapsis and XY-body formation could be achieved *in vitro*, we conclude that other meiotic events, such as complete meiotic DSB repair, *epi*genetic silencing of MSCI, and crossover formation, have not always been systematically described. Moreover, only one study has reported normal chiasmata frequency in MI spermatocytes *in vitro*. In addition, the generation of elongated spermatids could be achieved by *ex vivo* cultures of testis tissue fragments or after *in vitro*-generated germ cells were transplanted into recipient mouse testes. However, *in vitro* spermatogenesis by cell culture strategies only led to the formation of round spermatids, and thus, *in vitro* spermiogenesis remains unachieved by the available cell culture strategies.

One could argue that the generation of viable euploid offspring is sufficient to assess the quality of *in vitro*-generated sperm. This may be a valid argument for a model organism such as the mouse for which many embryos can be generated, in which the (epi)genetic stability and health of the offspring can be checked in hindsight. However, for the clinical use of *in vitro*-generated sperm, the quality of the gametes must be guaranteed. In many human studies (Easley IV et al., 2012; Eguizabal et al., 2011; Kee et al., 2009; Panula et al., 2011; Zhao et al., 2018), the evaluation of *in vitro*-generated spermatids was limited to the analysis of gene and protein expression or gross ploidy status. However, these types of analyses cannot verify gamete developmental potential (*i.e.*, fertilization capacity) or safety (*i.e.*, persistent DSBs or single aneuploidy). Therefore, to ensure safety, at least DSB repair, genetic imprinting, and ploidy status should be thoroughly analyzed by cell sorting and (epi)genetic sequencing. Moreover, as suggested by us in this review and Handel and colleagues (Handel et al., 2014), cytological analysis of the meiotic prophase, as well as the subsequent divisions, is needed to provide additional confirmation of proper DSB repair and chromosome segregation prior to spermatid formation. Importantly, we found that, despite incomplete synapsis, impaired recombination, and insufficient meiotic crossovers, round spermatid-like cells could still be formed *in vitro* (Lei et al., 2020). Apparently, the checkpoints that would eliminate such cells *in vivo* are not fully functional *in vitro*. This can be further investigated by challenging the existing culture systems, for instance by irradiation-induced DSB formation or genetic mutations that are known to activate these checkpoints *in vivo*.

## 4.2 Concluding remarks on possible future applications

*In vitro* culture systems have the potential to become a great tool to study mammalian spermatogenesis, and especially meiosis. Moreover, *in vitro* systems may reduce the need for animal experimentation, and they may be easier to manipulate and study. Research on human spermatogenesis would greatly benefit from a reliable *in vitro* culture system because (genetic) manipulation of the human testis or germ cells is often not possible or ethical. However, for the discovery of novel mechanisms involved in meiosis, *in vitro* systems may be problematic because not all meiotic processes occur *in vitro* as *in vivo*. For instance, we found that meiotic checkpoints may not be fully functional *in vitro* (Lei et al., 2020, 2021). The lack of knowledge on the *in vivo* situation may lead to incorrect conclusions, that is, to wrongly extrapolate mechanisms found *in vitro* automatically to the *in vivo* situation. For the clinical use of *in vitro*-generated gametes, this lack of knowledge can lead to serious safety issues, including the generation of aneuploid sperm with remaining DNA damage and (epi)genetic mutations. We suggest that all key meiotic events, both in the meiotic prophase and metaphase stages, including meiotic checkpoint activity, be investigated before *in vitro* spermatogenesis can be considered for clinical application.


# Acknowledgments

Q.L. is the recipient of a China Scholarship Council grant (201706300107).


# References


Bolcun-Filas, E., & Handel, M. A. (2018). Meiosis: The chromosomal foundation of reproduction. *Biology of Reproduction*, *99*(1), 112–126.

Bolcun-Filas, E., Rinaldi, V. D., White, M. E., & Schimenti, J. C. (2014). Reversal of female infertility by Chk2 ablation reveals the oocyte DNA damage checkpoint pathway. *Science*, *343*(6170), 533–536.

Carr, A. M. (1996). Checkpoints take the next step. *Science*, *271*(5247), 314–315.

Dann, C. T., Alvarado, A. L., Molyneux, L. A., Denard, B. S., Garbers, D. L., & Porteus, M. H. (2008). Spermatogonial stem cell self-renewal requires OCT4, a factor downregulated during retinoic acid-induced differentiation. *Stem Cells*, *26*(11), 2928–2937.

De Michele, F., Poels, J., Vermeulen, M., Ambroise, J., Gruson, D., Guiot, Y., et al. (2018). Haploid germ cells generated in organotypic culture of testicular tissue from prepubertal boys. *Frontiers in Physiology*, *9*, 1413.

Easley, C. A., IV, Phillips, B. T., McGuire, M. M., Barringer, J. M., Valli, H., Hermann, B. P., et al. (2012). Direct differentiation of human pluripotent stem cells into haploid spermatogenic cells. *Cell Reports*, *2*(3), 440–446.

Egozcue, J., Sarrate, Z., Codina-Pascual, M., Egozcue, S., Oliver-Bonet, M., Blanco, J., et al. (2005). Meiotic abnormalities in infertile males. *Cytogenetic and Genome Research*, *111*(3-4), 337–342.

Eguizabal, C., Montserrat, N., Vassena, R., Barragan, M., Garreta, E., Garcia-Quevedo, L., et al. (2011). Complete meiosis from human induced pluripotent stem cells. *Stem Cells*, *29*(8), 1186–1195.

Endo, T., Freinkman, E., de Rooij, D. G., & Page, D. C. (2017). Periodic production of retinoic acid by meiotic and somatic cells coordinates four transitions in mouse spermatogenesis. *Proceedings of the National Academy of Sciences*, *114*(47), E10132–E10141.

Feng, L.-X., Chen, Y., Dettin, L., Pera, R. A. R., Herr, J. C., Goldberg, E., et al. (2002). Generation and in vitro differentiation of a spermatogonial cell line. *Science*, *297*(5580), 392–395.

Geijsen, N., Horoschak, M., Kim, K., Gribnau, J., Eggan, K., & Daley, G. Q. (2004). Derivation of embryonic germ cells and male gametes from embryonic stem cells. *Nature*, *427*(6970), 148–154.

Gonsalves, J., Sun, F., Schlegel, P. N., Turek, P. J., Hopps, C. V., Greene, C., et al. (2004). Defective recombination in infertile men. *Human Molecular Genetics*, *13*(22), 2875–2883.

Gross, K. X., Hanson, B. M., & Hotaling, J. M. (2020). Round spermatid injection. *Urologic Clinics*, *47*(2), 175–183.

Handel, M. A., Eppig, J. J., & Schimenti, J. C. (2014). Applying "gold standards" to in-vitro-derived germ cells. *Cell*, *157*(6), 1257–1261.

Hassold, T., & Hunt, P. (2001). To err (meiotically) is human: The genesis of human aneuploidy. *Nature Reviews Genetics*, *2*(4), 280–291.

Hayashi, K., Ohta, H., Kurimoto, K., Aramaki, S., & Saitou, M. (2011). Reconstitution of the mouse germ cell specification pathway in culture by pluripotent stem cells. *Cell*, *146*(4), 519–532.

Hendriks, S., Dancet, E. A., van Pelt, A. M., Hamer, G., & Repping, S. (2015). Artificial gametes: A systematic review of biological progress towards clinical application. *Human Reproduction Update*, *21*(3), 285–296.

Hendriks, S., Dondorp, W., de Wert, G., Hamer, G., Repping, S., & Dancet, E. A. (2015). Potential consequences of clinical application of artificial gametes: A systematic review of stakeholder views. *Human Reproduction Update*, *21*(3), 297–309.

Hermann, B. P., Cheng, K., Singh, A., Roa-De La Cruz, L., Mutoji, K. N., Chen, I.-C., et al. (2018). The mammalian spermatogenesis single-cell transcriptome, from spermatogonial stem cells to spermatids. *Cell Reports*, *25*(6), 1650–1667. e1658.

Hulten, M., Eliasson, R., & Tillinger, K. (1970). Low chiasma count and other meiotic irregularities in two infertile 46, XY men with spermatogenic arrest. *Hereditas*, *65*(2), 285–290.

Ichijima, Y., Sin, H.-S., & Namekawa, S. H. (2012). Sex chromosome inactivation in germ cells: Emerging roles of DNA damage response pathways. *Cellular and Molecular Life Sciences*, *69*(15), 2559–2572.

Ishikura, Y., Ohta, H., Sato, T., Murase, Y., Yabuta, Y., Kojima, Y., et al. (2021). In vitro reconstitution of the whole male germ-cell development from mouse pluripotent stem cells. *Cell Stem Cell*, *28*(12), 2167–2179.e9.

Ishikura, Y., Yabuta, Y., Ohta, H., Hayashi, K., Nakamura, T., Okamoto, I., et al. (2016). In vitro derivation and propagation of spermatogonial stem cell activity from mouse pluripotent stem cells. *Cell Reports*, *17*(10), 2789–2804.

Jan, S. Z., Hamer, G., Repping, S., de Rooij, D. G., van Pelt, A. M., & Vormer, T. L. (2012). Molecular control of rodent spermatogenesis. *Biochimica et Biophysica Acta (BBA) - Molecular Basis of Disease*, *1822*(12), 1838–1850.

Jan, S. Z., Jongejan, A., Korver, C. M., van Daalen, S. K., van Pelt, A. M., Repping, S., et al. (2018). Distinct prophase arrest mechanisms in human male meiosis. *Development*, *145*(16), dev160614.

Jan, S. Z., Vormer, T. L., Jongejan, A., Röling, M. D., Silber, S. J., de Rooij, D. G., et al. (2017). Unraveling transcriptome dynamics in human spermatogenesis. *Development, 144*(20), 3659–3673.

Kar, F. M., & Hochwagen, A. (2021). Phospho-regulation of meiotic prophase. *Frontiers in Cell and Development Biology, 9*, 667073.

Kee, K., Angeles, V. T., Flores, M., Nguyen, H. N., & Pera, R. A. R. (2009). Human DAZL, DAZ and BOULE genes modulate primordial germ-cell and haploid gamete formation. *Nature, 462*(7270), 222–225.

Krausz, C., & Riera-Escamilla, A. (2018). Genetics of male infertility. *Nature Reviews Urology, 15*(6), 369–384.

Kumar, N., & Singh, A. K. (2015). Trends of male factor infertility, an important cause of infertility: A review of literature. *Journal of Human Reproductive Sciences, 8*(4), 191.

Laberge, R.-M., & Boissonneault, G. (2005). On the nature and origin of DNA strand breaks in elongating spermatids. *Biology of Reproduction, 73*(2), 289–296.

Lane, S., & Kauppi, L. (2019). Meiotic spindle assembly checkpoint and aneuploidy in males versus females. *Cellular and Molecular Life Sciences, 76*(6), 1135–1150.

Lee, B., & Amon, A. (2001). Meiosis: How to create a specialized cell cycle. *Current Opinion in Cell Biology, 13*(6), 770–777.

Lei, Q., & Hamer, G. (2021). The use of spermatogonial stem cells to correct a mutation causing meiotic arrest. *Asian Journal of Andrology, 0*.

Lei, Q., Lai, X., Eliveld, J., de Sousa Lopes, S. M. C., van Pelt, A. M., & Hamer, G. (2020). In vitro meiosis of male germline stem cells. *Stem Cell Reports, 15*(5), 1140–1153.

Lei, Q., Zhang, E., van Pelt, A. M., & Hamer, G. (2021). Meiotic chromosome synapsis and XY-body formation in vitro. *Frontiers in Endocrinology, 12*. https://doi.org/10.3389/fendo.2021.761249, 761249.

Li, N., Ma, W., Shen, Q., Zhang, M., Du, Z., Wu, C., et al. (2019). Reconstitution of male germline cell specification from mouse embryonic stem cells using defined factors in vitro. *Cell Death and Differentiation, 26*(10), 2115–2124.

MacQueen, A. J., & Hochwagen, A. (2011). Checkpoint mechanisms: the puppet masters of meiotic prophase. *Trends in Cell Biology, 21*(7), 393–400.

Marcet-Ortega, M., Pacheco, S., Martínez-Marchal, A., Castillo, H., Flores, E., Jasin, M., et al. (2017). p53 and TAp63 participate in the recombination-dependent pachytene arrest in mouse spermatocytes. *PLoS Genetics, 13*(6), e1006845.

Marcon, L., & Boissonneault, G. (2004). Transient DNA strand breaks during mouse and human spermiogenesis: New insights in stage specificity and link to chromatin remodeling. *Biology of Reproduction, 70*(4), 910–918.

Medrano, J. V., Vilanova-Pérez, T., Fornés-Ferrer, V., Navarro-Gomezlechon, A., Martínez-Triguero, M. L., García, S., et al. (2018). Influence of temperature, serum, and gonadotropin supplementation in short-and long-term organotypic culture of human immature testicular tissue. *Fertility and Sterility, 110*(6), 1045–1057. e1043.

Meistrich, M. L. (2013). Effects of chemotherapy and radiotherapy on spermatogenesis in humans. *Fertility and Sterility, 100*(5), 1180–1186.

Morimoto, H., Ogonuki, N., Kanatsu-Shinohara, M., Matoba, S., Ogura, A., & Shinohara, T. (2021). Spermatogonial stem cell transplantation into nonablated mouse recipient testes. *Stem Cell Reports, 16*, 1832–1844.

Mulder, C. L., Zheng, Y., Jan, S. Z., Struijk, R. B., Repping, S., Hamer, G., et al. (2016). Spermatogonial stem cell autotransplantation and germline genomic editing: A future cure for spermatogenic failure and prevention of transmission of genomic diseases. *Human Reproduction Update, 22*(5), 561–573.

Murry, C. E., & Keller, G. (2008). Differentiation of embryonic stem cells to clinically relevant populations: Lessons from embryonic development. *Cell, 132*(4), 661–680.

Musacchio, A. (2015). The molecular biology of spindle assembly checkpoint signaling dynamics. *Current Biology, 25*(20), R1002–R1018.

Nayernia, K., Nolte, J., Michelmann, H. W., Lee, J. H., Rathsack, K., Drusenheimer, N., et al. (2006). In vitro-differentiated embryonic stem cells give rise to male gametes that can generate offspring mice. *Developmental Cell*, *11*(1), 125–132.

Nolte, J., Michelmann, H. W., Wolf, M., Wulf, G., Nayernia, K., Meinhardt, A., et al. (2010). PSCDGs of mouse multipotent adult germline stem cells can enter and progress through meiosis to form haploid male germ cells in vitro. *Differentiation*, *80*(4-5), 184–194.

Oakberg, E. F. (1956). Duration of spermatogenesis in the mouse and timing of stages of the cycle of the seminiferous epithelium. *American Journal of Anatomy*, *99*(3), 507–516.

Odorisio, T., Rodriguez, T. A., Evans, E. P., Clarke, A. R., & Burgoyne, P. S. (1998). The meiotic checkpoint monitoring sypapsis eliminates spermatocytes via p53-independent apoptosis. *Nature Genetics*, *18*(3), 257–261.

Panula, S., Medrano, J. V., Kee, K., Bergström, R., Nguyen, H. N., Byers, B., et al. (2011). Human germ cell differentiation from fetal-and adult-derived induced pluripotent stem cells. *Human Molecular Genetics*, *20*(4), 752–762.

Perrard, M.-H., Sereni, N., Schluth-Bolard, C., Blondet, A., d'Estaing, S. G., Plotton, I., et al. (2016). Complete human and rat ex vivo spermatogenesis from fresh or frozen testicular tissue. *Biology of Reproduction*, *95*(4), 89. 81-10.

Petronczki, M., Siomos, M. F., & Nasmyth, K. (2003). Un menage a quatre: The molecular biology of chromosome segregation in meiosis. *Cell*, *112*(4), 423–440.

Portela, J. M., de Winter-Korver, C. M., van Daalen, S. K., Meißner, A., de Melker, A. A., Repping, S., et al. (2019). Assessment of fresh and cryopreserved testicular tissues from (pre) pubertal boys during organ culture as a strategy for in vitro spermatogenesis. *Human Reproduction*, *34*(12), 2443–2455.

Roeder, G. S., & Bailis, J. M. (2000). The pachytene checkpoint. *Trends in Genetics*, *16*(9), 395–403.

Royo, H., Polikiewicz, G., Mahadevaiah, S. K., Prosser, H., Mitchell, M., Bradley, A., et al. (2010). Evidence that meiotic sex chromosome inactivation is essential for male fertility. *Current Biology*, *20*(23), 2117–2123.

Sato, T., Katagiri, K., Gohbara, A., Inoue, K., Ogonuki, N., Ogura, A., et al. (2011). In vitro production of functional sperm in cultured neonatal mouse testes. *Nature*, *471*(7339), 504–507.

Sato, T., Katagiri, K., Kojima, K., Komeya, M., Yao, M., & Ogawa, T. (2015). In vitro spermatogenesis in explanted adult mouse testis tissues. *PLoS One*, *10*(6), e0130171.

Skakkebaek, N., Bryant, J. I., & Philip, J. (1973). Studies on meiotic chromosomes in infertile men and controls with normal karyotypes. *Reproduction*, *35*(1), 23–36.

Sou, I. F., Pryce, R. M., Tee, W.-W., & McClurg, U. L. (2021). Meiosis initiation: A story of two sexes in all creatures great and small. *Biochemical Journal*, *478*(20), 3791–3805.

Struijk, R. B., Mulder, C. L., Van Der Veen, F., Van Pelt, A. M., & Repping, S. (2013). Restoring fertility in sterile childhood cancer survivors by autotransplanting spermatogonial stem cells: Are we there yet? *BioMed Research International*, *2013*, 903142.

Subramanian, V. V., & Hochwagen, A. (2014). The meiotic checkpoint network: Step-by-step through meiotic prophase. *Cold Spring Harbor Perspectives in Biology*, *6*(10), a016675.

Sun, M., Yuan, Q., Niu, M., Wang, H., Wen, L., Yao, C., et al. (2018). Efficient generation of functional haploid spermatids from human germline stem cells by three-dimensional-induced system. *Cell Death and Differentiation*, *25*(4), 749–766.

Templado, C., Marina, S., & Egozcue, J. (1976). Three cases of low chiasma frequency associated with infertility in man. *Andrologia*, *8*(4), 285–289.

Tournaye, H., Krausz, C., & Oates, R. D. (2017). Novel concepts in the aetiology of male reproductive impairment. *The Lancet Diabetes & Endocrinology*, *5*(7), 544–553.

Turner, J. M. (2007). Meiotic sex chromosome inactivation. *Development*, *134*(10), 1823–1831.

Turner, J. M. (2015). Meiotic silencing in mammals. *Annual Review of Genetics*, *49*, 395–412.
Tüttelmann, F., Ruckert, C., & Röpke, A. (2018). Disorders of spermatogenesis. *Medizinische Genetik*, *30*(1), 12–20.
Uroz, L., Rajmil, O., & Templado, C. (2011). Meiotic chromosome abnormalities in fertile men: Are they increasing? *Fertility and Sterility*, *95*(1), 141–146.
Uroz, L., & Templado, C. (2012). Meiotic non-disjunction mechanisms in human fertile males. *Human Reproduction*, *27*(5), 1518–1524.
Van Der Steeg, J. W., Steures, P., Eijkemans, M. J., Habbema, J. D. F., Hompes, P. G., Kremer, J. A., et al. (2011). Role of semen analysis in subfertile couples. *Fertility and Sterility*, *95*(3), 1013–1019.
Wang, S., Wang, X., Ma, L., Lin, X., Zhang, D., Li, Z., et al. (2016). Retinoic acid is sufficient for the in vitro induction of mouse spermatocytes. *Stem Cell Reports*, *7*(1), 80–94.
Wang, Y.-H., Yan, M., Zhang, X., Liu, X.-Y., Ding, Y.-F., Lai, C.-P., et al. (2021). Rescue of male infertility through correcting a genetic mutation causing meiotic arrest in spermatogonial stem cells. *Asian Journal of Andrology*, *23*(6), 590–599.
Webster, A., & Schuh, M. (2017). Mechanisms of aneuploidy in human eggs. *Trends in Cell Biology*, *27*(1), 55–68.
Xu, H., Yang, M., Tian, R., Wang, Y., Liu, L., Zhu, Z., et al. (2020). Derivation and propagation of spermatogonial stem cells from human pluripotent cells. *Stem Cell Research & Therapy*, *11*(1), 1–15.
Yuan, Y., Li, L., Cheng, Q., Diao, F., Zeng, Q., Yang, X., et al. (2020). In vitro testicular organogenesis from human fetal gonads produces fertilization-competent spermatids. *Cell Research*, *30*(3), 244–255.
Zhang, X., Gunewardena, S., & Wang, N. (2021). Nutrient restriction synergizes with retinoic acid to induce mammalian meiotic initiation in vitro. *Nature Communications*, *12*(1), 1–15.
Zhao, Y., Ye, S., Liang, D., Wang, P., Fu, J., Ma, Q., et al. (2018). In vitro modeling of human germ cell development using pluripotent stem cells. *Stem Cell Reports*, *10*(2), 509–523.
Zhou, Q., Wang, M., Yuan, Y., Wang, X., Fu, R., Wan, H., et al. (2016). Complete meiosis from embryonic stem cell-derived germ cells in vitro. *Cell Stem Cell*, *18*(3), 330–340.

FSC
www.fsc.org
MIX
Papier | Fördert
gute Waldnutzung
FSC® C083411

Kolibri 360 GmbH
Ferdinand-Jühlke-Straße 7
99095 Erfurt, Deutschland
produktsicherheit@kolibri360.de